Minimally Processed Fruits and Vegetables

Fundamental Aspects and Applications

ASPEN FOOD ENGINEERING SERIES

Aspen Food Engineering Series

Jose M. Aguilera and David W. Stanley, *Microstructural Principles of Food Processing and Engineering*, second edition (1999)
Gustavo Barbosa-Cánovas and Humberto Vega-Mercado, *Dehydration of Foods* (1996)
Pedro Fito, Enrique Ortega-Rodríguez, and Gustavo Barbosa-Cánovas, *Food Engineering 2000* (1997)
P. J. Fryer, D. L. Pyle, and C. D. Rielly, *Chemical Engineering for the Food Industry* (1997)
S. D. Holdsworth, *Thermal Processing of Packaged Foods* (1997)
Michael Lewis and Neil Heppell, *Continuous Thermal Processing of Foods: Pasteurization and UHT Sterilization* (2000)
Rosana G. Moreira, M. Elena Castell-Perez, and Maria A. Barrufet, *Deep-Fat Frying: Fundamentals and Applications* (1999)
M. Anandha Rao, *Rheology of Fluid and Semisolid Foods: Principles and Applications* (1999)

Minimally Processed Fruits and Vegetables

Fundamental Aspects and Applications

Stella M. Alzamora
Research Professor
Facultad de Ciencias Exactas y Naturales
Universidad de Buenos Aires
Buenos Aires, Argentina

María S. Tapia, MS
Instituto de Ciencia y Technología de Alimentos
Facultad de Ciencias-Universidad Central de Venezuela
Caracas, Venezuela

Aurelio López-Malo, MSc
Professor
Departmento de Ingeniería Química y Alimentos
Universidad de las Américas-Puebla
Cholula, Mexico

AN ASPEN PUBLICATION
Aspen Publishers, Inc.
Gaithersburg, Maryland
2000

The author has made every effort to ensure the accuracy of the information herein. However, appropriate information sources should be consulted, especially for new or unfamiliar procedures. It is the responsibility of every practitioner to evaluate the appropriateness of a particular opinion in the context of actual clinical situations and with due considerations to new developments. The author, editors, and the publisher cannot be held responsible for any typographical or other errors found in this book.

Library of Congress Cataloging-in-Publication Data

Alzamora, Stella M.
Minimally processed fruits and vegetables: fundamental aspects and applications /
Stella M. Alzamora, María S. Tapia, Aurelio López-Malo.
p. cm.
Includes bibliographical references and index.
ISBN 0-8342-1672-8
1. Fruit—Storage. 2. Vegetables—Storage. I. Tapia, María S. II. López-Malo, Aurelio
III. Title
TP440 .A59 2000
634'.0468—dc21
00-038113

Orders: (800) 638-8437
Customer Service: (800) 234-1660

Editorial Services: Timothy Sniffin
Library of Congress Catalog Card Number: 00-038113
ISBN: 0-8342-1672-8

Printed in the United States of America

1 2 3 4 5

Contents

Contributors

Raija Ahvenainen, PhD
Chief Research Scientist, Minimal Processing and Packaging
VTT Biotechnology and Food Research
Espoo, Finland

Stella M. Alzamora, PhD
Research Professor
Departamento de Industrias
Facultad de Ciencias Exactas y Naturales
Universidad de Buenos Aires
Buenos Aires, Argentina

Efrén Parada Arias, PhD
Professor
Departamento de Graduados e Investigación en Alimentos
Escuela Nacional de Ciencias Biológicas (ENCB)
Instituto Politécnico Nacional (IPN)
Mexico D.F.
Mexico

Gustavo V. Barbosa-Cánovas, PhD
Professor of Food Engineering
Biological Systems Engineering Department
Washington State University
Pullman, Washington

Marjon H.J. Bennik, PhD
Research Scientist
Agrotechnological Research Institute (ATO)
Wageningen University Research Center
Wageningen, The Netherlands

Larry R. Beuchat, MS, PhD
Research Professor
Center for Food Safety and Quality Enhancement
Department of Food Science and Technology
University of Georgia
Griffin, Georgia

Guillermo Binstok, MS
Professor
Departamento de Industrias
Facultad de Ciencias Exactas y Naturales
Universidad de Buenos Aires
Ciudad Universitaria
Buenos Aires, Argentina

Carmen A. Campos, PhD
Professor
Departamento de Industrias
Facultad de Ciencias Exactas y Naturales
Universidad de Buenos Aires
Ciudad Universitaria
Buenos Aires, Argentina

María A. Castro, PhD
Professor
Laboratorio de Anatomia Vegetal
Departamento de Ciencias Biológicas
Facultad de Ciencias Exactas y Naturales
Universidad de Buenos Aires
Ciudad Universitaria
Buenos Aires, Argentina

Amparo Chiralt, PhD
Professor
Food Technology Department
Universidad Politécnica de Valencia
Valencia, Spain.

Rosa V. Díaz, MSc
Assistant Professor
Instituto de Ciencia y Tecnología de Alimentos
Universidad Central de Venezuela
Caracas, Venezuela

Lidia Dorantes-Alvarez, MSc, PhD
Professor in Food Technology
Departamento de Graduados en Alimentos
Escuela Nacional de Ciencias Biológicas
Mexico D.F.
Mexico

Pedro Fito, PhD
Professor
Food Technology Department
Universidad Politécnica de Valencia.
Valencia, Spain

Lía Noemí Gerschenson, PhD
Professor
Departamento de Industrias
Facultad de Ciencias Exactas y Naturales
Universidad de Buenos Aires
Ciudad Universitaria
Buenos Aires, Argentina

M. Marcela Góngora-Nieto, BS
PhD Candidate
Biological Systems Engineering Department
Washington State University
Pullman, Washington

Leon Gorris, PhD
Department of Microbiology
Unilever Research Vlaardingen
Vlaardingen, The Netherlands

Grahame W. Gould, MSc, PhD
Formerly Professor of Microbiology
Department of Food Science
University of Leeds
Leeds, United Kingdom
Unilever Research
Bedford, United Kingdom

Sandra Guerrero, PhD
Professor
Departamento de Industrias
Facultad de Ciencias Exactas y Naturales
Universidad de Buenos Aires
Ciudad Universitaria
Buenos Aires, Argentina

Mónica Haros
Professor
Departamento de Industrias
Facultad de Ciencias Exactas y Naturales
Universidad de Buenos Aires
Ciudad Universitaria
Buenos Aires, Argentina

Lothar Leistner, PhD
International Food Consultant
Former Director and Professor
Federal Centre of Meat Research
Kulmbach, Germany

Aurelio López-Malo, MSc
Professor
Departamento de Ingeniería Química y Alimentos
Universidad de las Américas-Puebla
Cholula, Mexico

Amaury Martínez, MSc, PhD
Titular Professor
Instituto de Ciencia y Tecnología de Alimentos
Universidad Central de Venezuela
Caracas, Venezuela

Andrea B. Nieto, MS
Professor
Departamento de Industrias
Facultad de Ciencias Exactas y Naturales
Universidad de Buenos Aires
Buenos Aires
Argentina

Enrique Palou, MSc, PhD
Professor
Departamento de Ingeniería Química y Alimentos
Universidad de las Américas-Puebla
Cholula, Mexico

Ana M. Rojas, PhD
Professor
Departamento de Industrias
Facultad de Ciencias Exactas y Naturales
Universidad de Buenos Aires
Ciudad Universitaria
Buenos Aires, Argentina

Clara O. Rovedo, PhD
Professor
Departamento de Industrias
Facultad de Ciencias Exactas y Naturales
Universidad de Buenos Aires
Ciudad Universitaria
Buenos Aires, Argentina

Daniela Salvatori, PhD
Professor
Departamento de Industrias
Facultad de Ciencias Exactas y Naturales
Universidad de Buenos Aires
Buenos Aires, Argentina

E.J. Smid, PhD
Department of Preservation Technology and Food Safety
Agrotechnological Research Institute (ATO-DLO)
Wageningen, The Netherlands

Barry G. Swanson, PhD
Professor of Food Science
Food Science and Human Nutrition Department
Washington State University
Pullman, Washington

Constantino Suárez, PhD
Professor
Departamento de Industrias
Facultad de Ciencias Exactas y Naturales
Universidad de Buenos Aires
Ciudad Universitaria
Buenos Aires, Argentina

María S. Tapia, MSc
Associate Professor
Instituto de Ciencia y Tecnología de Alimentos
Facultad de Ciencias
Universidad Central de Venezuela
Caracas, Venezuela

L.M.M. Tijskens, PhD
Senior Scientist
Agrotechnological Research Institute (ATO)
Wageningen, The Netherlands

Marcela P. Tolaba, PhD
Professor
Departamento de Industrias
Facultad de Ciencias Exactas y Naturales
Universidad de Buenos Aires
Ciudad Universitaria
Buenos Aires, Argentina

C. van Dijk, PhD
Head of Department
Agrotechnological Research Institute (ATO)
Wageningen, The Netherlands

J. Andrés Vasconcellos, MSc, PhD
Professor
Department of Food Science and Nutrition
Chapman University
Orange, California

Susana L. Vidales, PhD
Professor
Departamento de Tecnología
Universidad Nacional de Luján
Buenos Aires, Argentina

Pascual E. Viollaz
Professor
Departamento de Industrias
Facultad de Ciencias Exactas y Naturales
Universidad de Buenos Aires
Ciudad Universitaria
Buenos Aires, Argentina

Jorge Welti-Chanes, PhD
Professor
Departamento de Ingeniería Química y Alimentos
Universidad de las Américas-Puebla
Cholula, Mexico

Preface

This book intends to focus attention on selected advances of fundamental aspects related to the topic of minimal processing of fruits and vegetables. According to Professor Thomas Ohlsson's definition and for the purpose of this book, the term *minimal processing* will be used to include a wide range of technologies for a) preserving short shelf-life vegetable and fruit products while minimizing changes that would alter freshness characteristics and b) improving quality of long shelf-life products.

Although several reviews and books have been published, more basic and technical knowledge is needed on minimally processed products. Minimal processing should be seen as a multidisciplinary technology that requires inputs from academia, industry, government, and even consumers. Microbiologists, food engineers, biologists, refrigeration and packaging experts, government agencies, food processor associations, etc., need to work in a coordinated manner to produce sound information and a good understanding of the principles and practices that set the foundations of this widely scoped technology. This book attempts to focus on recent findings and some "hot" issues to encourage its integration and application to the broad field of minimal processing. With this approach, the book is divided into five parts.

Part I deals with microbiologic aspects. Contributions to this part will include a topic that might appear to be too traditional—hurdle technology. However, this topic, especially its particular application in the design of minimally processed foods, is important because severity of treatments in minimal processing should be avoided at any cost, leaving only the rational use of the concept of combinations of factors for assuring stability and safety. Induced tolerance of pathogenic microorganisms to stress can compromise the success of the minimal technologies. The effective use of sanitizers for removing microorganisms from the surfaces of whole and cut produce is an essential part for the success of minimal processing techniques. Contamination by food-borne pathogens and the microbial ecology that prevails in products and processes is, by all means, the main issue related to the microbiology of minimally processed fruits and vegetables. Finally, the integrated applications of traditional and modern microbiologic tools (such as predictive microbiology and risk assessment) for final safety and stability of minimal processing products are discussed.

Design of minimal processes must be accompanied by a profound and fundamental understanding of phenomena related to fruit and vegetable transformation due to processing. The high degree of structure of biologic tissues requires the consideration of what is happening not only at molecular scale but also at the macromolecular assemblies, the subcellular organelles, and the cell level. Part II, regarding physicochemical and structural aspects, provides advanced information on the relationships between structural modifications and quality and biochemical changes due to processing. The interactions between food matrices components and/or additives during processing and storage and their effects on stability are also covered.

Part III presents recent developments of novel and traditional preservation technologies for minimal processing and storage of fruit and vegetables, such as high hydrostatic pressure, pulsed electric fields, and the addition of natural antimicrobials of plant and microbial origin, along with a discussion of their advantages and disadvantages, potential commercial application, and enhancement or complementation of existing in-house processes. Novel and refined existing techniques of drying for developing minimal processes are explored. Vacuum osmotic dehydration, a new method of osmotic dehydration that takes advantage of the porous microstructure of vegetable tissues, is analyzed as a technique to reduce process time and improve the incorporation of additives. Recent trends for the improvement of quality and shelf life of living tissues are reviewed.

The results of the multinational projects for development of minimal processing technologies for food preservation (Project XI.3 of the CYTED Program) and development of preservation techniques for tropical fruits using vacuum impregnation (STD-3 Program of European Commission) are summarized and presented as new technologies for fruit preservation in Part 4. Finally, Part 5 examines the regulatory issues associated with minimally processed fruits and vegetables, particularly those referring to refrigerated products.

This book covers selected trends in minimal processing of fruit and vegetables and tries to summarize scientists' progress in the field that should be known and disseminated to government, industry, and the commercial community. It is designed to serve primarily as a reference book for those involved in the diverse aspects of the broad area of minimally processed fruits and vegetables and interested in its future potential. We hope that the book will be interesting to readers and will provide them with useful, enlightening, and up-to-date information.

The editors want to express their gratitude and acknowledge the support of the CYTED Program and Secretaría de Ciencia y Tecnología del Ministerio de Cultura y Educación de la República Argentina for funding Project XI.3 and the European Commission for funding Project TS3*-CT94–0333 (DG HSMU). Readers are presented with a good part of the scientific and technical results obtained by researchers in the context of both projects.

The editors also want to express their gratitude to Aspen Publishers for the trust deposited in this project and are indebted in particular to all contributors of the various chapters, whom we thank for their patience. We also wish to include a final word of appreciation to Leticia Scoccia for her professionalism and dedication in making possible the completion of this volume.

Stella M. Alzamora,
María. S. Tapia
Aurelio López-Malo

Overview

Stella M. Alzamora, Aurelio López-Malo, and María S. Tapia

INTRODUCTION

The increasing popularity of minimally processed fruit and vegetables has been attributed to the health benefits associated with fresh produce, combined with the ongoing consumer trend toward eating out and consuming ready-to-eat foods. The minimally processed fruit and vegetable industry was initially developed to supply hotels, restaurants, catering services, and other institutions. More recently, it was expanded to include food retailers for home consumption. In 1998, the sales volume of minimally processed fruits and vegetables in the United States has been estimated to be around $6 billion, and it is expected to increase to about $20 billion in the next 3–5 years.[1] A similarly increasing demand has been observed since the late 1980s in Europe, mainly in France and England.[2] Japan is a special country case; the demand for fresh food or freshlike products is continually increasing. Other countries that produce "exotic" foods have begun to consider the potential market of minimal processing, not only for export but also for internal consumption.[3,4]

Dr. Samuel O. Thier, president of the National Academy of Sciences' Institute of Medicine, stated in 1990 that safe and nutritious food is going to become progressively more important in the protection of health and the improvement of aging.[5] The increasing demand of these minimally processed products represents a challenge for researchers and processors to make them stable and safe. The increased time and distance between processing and consumption may contribute to higher risks of food-borne illness. Although chemical and physical hazards are of concern, the hazards specific to minimally processed and ready-to-eat fruit and vegetables reside mainly with microbial contaminants. Some of the microbial pathogens associated with fresh produce include *Listeria monocytogenes*, *Salmonella* species, *Shigella* species, enteropathogenic strains of *Escherichia coli*, hepatitis A virus, etc.[6,7] The possible sources of contamination in these products involve the incoming raw fruit and vegetables, the plant workers, and the processing environment. When fruit and vegetables are peeled, chopped, or shredded, the release of plant cellular fluids provides a nutritive medium for microbial growth and/or toxin production. The high moisture content of fresh plant tissues, the lack of lethal process to eliminate microbial pathogens, and the potential for temperature abuse during preparation, distribution, and handling further intensify the risk of food-borne illness. Continuous and increased research and development by academia, governments, and industry reflect the response to the increased consumption of minimally processed fruit and vegetables and the risks of food-borne illnesses associated with these products.

MINIMALLY PROCESSED FOODS

Consumer trends are changing, and high-quality foods with freshlike attributes are demanded.[8–10] Consequently, less extreme treatments and/or additives are being required. Within a wider and more modern concept of minimal processing, Gould[8,11] identified some food characteristics that must be attained in response to consumer demands, as presented in Figure O–1. These are less heat and chill damage, fresh appearance, and less acid, salt, sugar, and fat. To satisfy these demands, some changes or reductions in the traditionally used preservation techniques must be achieved (Figure O–1). From a microbiologic point of view, these possible changes or reductions have important and significant implications. Therefore, to satisfy market requirements, the safety and quality of foods will be based on substantial improvements in traditional preservation methods or on the use of emerging technologies.

Shank and Carson,[12] attempting to answer the question, What is safe food?, concluded that there is not a truly and complete answer. However, it was concluded that food is safer today than in the past and that scientific efforts will continue to make it safer in the future. Safe food may have different meanings to different people; consumers, for instance, regard a risk-free food as a safe one and associate increased risk with the increased use of added substances such as antimicrobials, synthetic additives, sodium, or fat, among others. Scientists, public health officers, and international organizations define a *safe food* as one that provides maximum nutrition and quality while revealing a minimal hazard to public health, and they assume any risk present to be minimal.[12]

DEFINITIONS

Those foods that have freshlike characteristics are known as *minimally processed foods*

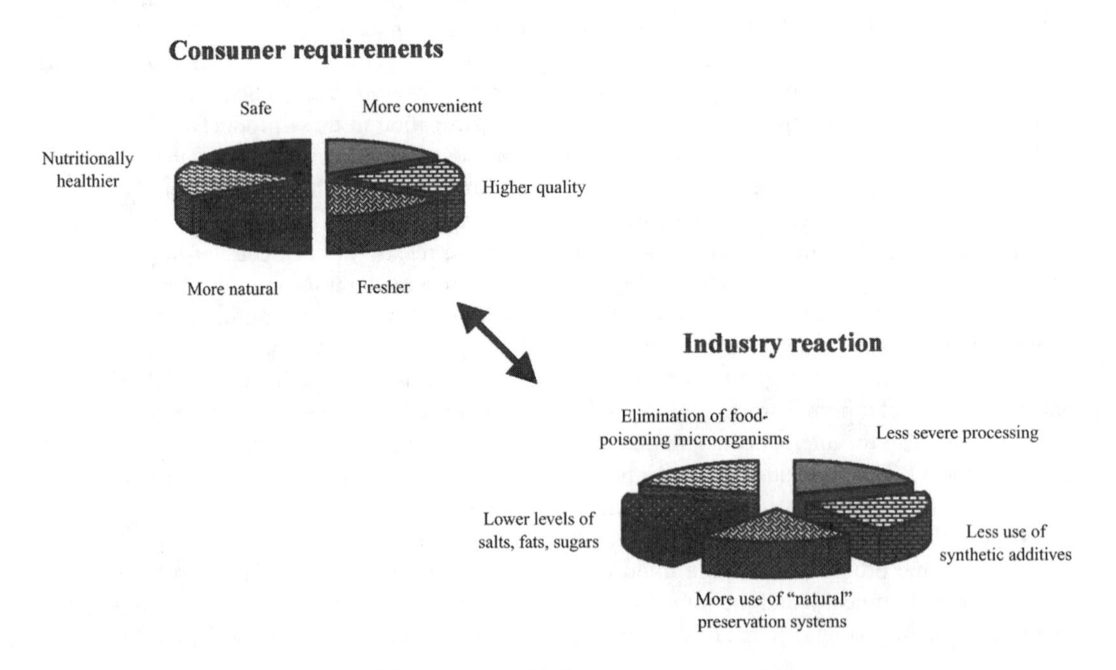

Figure O–1 Consumers' food qualifying factors and industry response to satisfy them.

(MPF) or *partially processed foods* (PPF) and satisfy, at least partially, the demand of fresh-like, high-quality foods. These types of products have been closely related to the changes in consumption patterns[13,14] and to certain needs of the catering industry.[15] Also, in many countries where there are no storage and transportation refrigerated facilities, some MPF or PPF may act as a mechanism to regulate fruit and vegetable production and their supply to the final transformation industries.[16–18] MPF and PPF include a series of products and processes that may be grouped in diverse food categories, such as minimally processed, using invisible processing, carefully processed, partially processed, and using high-moisture, shelf-stable processing. All of these terms, although they do not represent the same types of products, can be grouped as minimally processed, based on the hurdle technology approach,[13,14] giving a wider concept than those terms used by Rolle and Chism,[19] Shewfelt,[20] Huxsoll and Bolin,[21] Wiley,[22] and Ohlsson.[23]

The MPF definition has evolved as the minimal processing concepts have been better understood. Exhibit O–1 presents some definitions that have risen during recent years and that determine the trends in this area of food preservation. According to Rolle and Chism,[19] the manipulation and basic preparation of foods, as well as life permanence in the biologic tissue, are the elements that distinguish the minimal processing. Shewfelt[20] adds to the MPF definition the possible use of low-level irradiation and individual packaging as preservation factors that allow safety and high quality. Examples of MPF presented by Shewfelt[20] included vegetable stick snacks, packaged tossed salads, chilled peach halves, peeled and cored whole pineapple, shelled fresh legumes, microwaveable fresh vegetable trays, and gourmet chilled dinners, among others. The use of refrigeration appears as a fundamental preservation factor for this type of food. Huxsoll and Bolin[21] include in MPF those foods having tissues that can be not alive but whose freshness should be kept as an important objective of preservation. Another contribution of Huxsoll and Bolin[21] is that MPF are not necessarily products for direct consumption but can be considered as preserved foods that maintain their freshness characteristics and can be later transformed into processed products, using conventional techniques.

Exhibit O–1 Minimally Processed Food Definitions

Minimal processing includes all of the operations (washing, selecting, peeling, slicing, etc.) that must be carried out before blanching in a conventional processing line and that keep the food as a living tissue.[19]

Minimally processed foods include meat and fresh products, as well as any process that adds some value to the product, compared with conventional food preservation processes. Processes such as chopping, husking, coring, low-level irradiation, and individual packaging, form part of the minimal processing.[20]

Minimally processed fruits and vegetables are products that maintain their attributes and quality similar to those of fresh products. In some cases, a minimally processed product is a "raw" food, and the tissue cells are alive. However, these characteristics are not necessarily required if food freshness is kept.[21]

Minimally processed vegetables are products that contain living tissues or those that have slightly modified their freshness condition but keep their quality and character similar to those of fresh products.[22]

Minimal processing methods include those procedures that cause the least possible change in the food quality (keeping their freshness appearance) but at the same time provide the food with enough useful life to transport it from the production site to the consumer .[23]

A *minimal process* is "the least possible treatment" to achieve a purpose.[35]

Wiley[22] presents a very important alternative to the use of refrigeration alone—the modification and control of atmospheres in food packages as other preservation factors to be employed during the MPF elaboration. The reviews of Ahvenainen et al.,[15,24] Singh and Oliveira,[25] and Wiley[22] on minimal processing are useful as further reference on the subject. Ohlsson[23] adds, as an important element, the necessity that this processing should be enough and adequate to keep the product quality during the storage period between preservation and consumption.

The expansion of minimal processing concepts has been reflected in new, renewed, and improved products and processes formulated and designed to produce a greater diversity of MPF. Food processing technologies that can be used to minimize food damage while maintaining safety and convenience include thermal processes combined with airtight packaging and refrigeration, packaging in modified atmospheres combined with refrigeration, high-temperature/short-time processes, and postpackaging irradiation.[26] There is also a great interest in the application of new or emerging technologies to obtain MPF using nonthermal processes in the framework of the "hurdle" concept. It is clear that the present developments, as well as those being considered recently, are based on the combination of preservation factors, some of which use traditional preservation procedures such as refrigeration and others that apply nonthermal preservation factors such as high pressures and electric pulses.[27–29]

HIGH-MOISTURE FRUIT PRODUCTS AS AN EXAMPLE OF MINIMALLY PROCESSED FOODS USING THE HURDLE APPROACH

The original concept of minimal processing considered only food products that maintain their freshness by keeping alive the biologic tissues.[19,20] To date, it also considers those products that maintain the characteristics of fresh foods (or close to them) by inactivating the cellular metabolism in biologic tissues.[21–23] Therefore, high-moisture fruit and vegetable products (HMFP) preserved by hurdle technology can be classified as minimally processed, as are minimally processed refrigerated fruits (MPRF), for example.[4]

Product quality, the method of preservation, packaging procedure, and storage conditions are the main differences between MPRF, HMFP, and fresh fruits or those that have been preserved by cold, irradiation, dehydration, or thermal treatments. MPRF keep freshness characteristics using refrigeration as the main preservation factor and have an expected average shelf life of less than one month. In addition to the preparation and manipulation steps of MPRF, HMFP are foods in which other preservation factors such as blanching, reduction of water activity (a_w) and pH, and incorporation of antimicrobial agents and other additives are incorporated as refrigeration substitute.[3] HMFP represent an improved version of many intermediate-moisture (IMF) products and an alternative to the MPRF, because refrigeration does not need to be used as a factor to increase the stability and shelf life of the product. Tapia et al.[13,14] compared HMFP, MPRF, and IMF products developed from fruits and indicated that the a_w levels of HMFP are more similar to the ones of MPRF and very much higher than those of IMF. In HMFP, the incorporation of a_w depressing solutes should not affect the freshness of the fruit, and the lack of need of refrigeration as a preservation factor is set as an economic and technologic objective. Also, the use of other preservation factors (incorporation of additives and blanching) distinguishes HMFP from MPRF.[3]

MULTITARGET PRESERVATION OF FOODS AS MINIMAL PROCESSING TOOL

Table O–1 presents a summary of selected preservation factors that have been used in MPF, as well as those proposed for immediate or future application as minimal processing techniques. The preservation factors included in Table O–1 have been demonstrated to inhibit or inactivate microbial growth and delay other deteriorative reactions so that they can be used

Table O–1 Selected Preservation Factors Applied or Having Potential for Use in Minimal Processing of Fruits and Vegetables

Currently in Use	*Having Potential for Immediate Use*	*Having Potential for Future Use*
Superficial cleaning	Mild thermal treatment (blanching)	Irradiation
Treatment in water chlorine	Slight a_w depression (0.95–0.99)	Electric pulses
Air atmosphere packaging	pH controls (3.5–4.4)	Ultrasound
Modified-atmosphere packaging	Traditional antimicrobials (sorbates, benzoates)	Natural antimicrobials
Refrigeration	Other additives (calcium salts, ascorbic acid)	Light pulses
Vacuum cooking	High pressure	Magnetic fields
Edible film covering	Biocontrol with lactic acid bacteria	Microstructural control
	Active packaging	
Aseptic packaging	Treatment with ozone	

as tools for increasing shelf life. The great goals of safety, high quality, and convenience in the development of MPF can be reached only if the concepts of the combined use of preservation factors are understood and applied within the framework of multitarget preservation, hurdle technology, or combined factors technology.[13,14,29,30]

Homeostasis is the tendency to uniformity and stability within the microbial cell or spore. Food preservation is achieved by disturbing temporarily or permanently the homeostasis of microorganisms present in a food. Microbial homeostasis is an especially important event during food preservation; the microbial response to the selected factors or hurdles applied during processing determines whether the microorganisms initially present remain in the lag phase or die before their homeostasis is repaired.[31,32]

In "multitarget" preservation, the interference with microbial homeostasis takes place on different targets within the cell, due to the action of the combined use of different preservation factors, or "hurdles."[33,34] The rational selection of hurdles to be applied during minimal processing causes reduction of possible microbial risks without quality losses. Table O–2 presents several combinations of preservation factors evaluated on MPF or laboratory model systems. The appropriate combination of such factors leads to the inhibition or inactivation of pathogenic and spoilage microorganisms. The MPF listed have been divided into those requiring refrigeration or not as a preservation factor. Foods that have received a thermal or an equivalent treatment to inhibit or destroy pathogenic and spoilage flora, as well as to inhibit most of the food's biologic activity, are included in this last group. On the other hand, the use of refrigeration seems to be essential for the MPF when not thermally treated. Table O–2 also includes an example that combines chlorination and refrigeration with low-dose irradiation. Treatment with ionizing radiation has been suggested as an interesting factor to produce ready-to-eat fruits and vegetables with low levels of microorganisms. These products may be of value as food geared specially for those who are young, old, pregnant, or immunocompromised.

With the increasing popularity of ready-to-eat, minimally processed, fresh and processed fruits and vegetables that are preserved only by relatively mild techniques, new ecologic routes for microbial growth have emerged. To control microbial growth in these environments while keeping loss of quality at minimum, a hurdle

Table O–2 Selected Examples of Formulation of Minimal Processes, Based on the Combination of Preservation Factors

Preservation Factors	*Target Microorganisms and Test Medium*	*Reference*
A. With refrigeration		
MAP or packing under moderate vacuum (MVP)	*Listeria monocytogenes* (mungo bean sprouts, chicory endive)	Gorris[36]
MAP or MVP + bacteriocins	*L. monocytogenes* (model systems)	Gorris[36]
Chlorination + MAP + gamma irradiation	Aerobic microorganisms (carrots)	Hayenmaier & Baker[37]
Moisture absorbers + MAP	Aerobic microorganisms (mushrooms)	Roy et al.[38]
a_w reduction + pH + high pressure	Aerobic microorganisms, molds and yeasts (avocado purée)	López-Malo et al.[39]
Low pH + a_w reduction + sodium pyrophosphate	*Clostridium perfringens* (TPGY broth)	Juneja et al.[40]
B. Without refrigeration		
Ultrasound + heat + high pressure	*Bacillus subtilis, B. coagulans, Saccharomyces cerevisiae, Aeromonas hydrophila, B. stearothermophilus* (water)	Sala et al.[41]
Lysozyme or butylated hydroxyanizole + EDTA + pH + NaCl	Gram-positive and gram-negative microorganisms (BHI broth)	Razavi-Rohani & Griffiths[42]
Thermal sterilization + nisin	*B. stearothermophilus, C. thermosaccharolyticum* (low-acid canned vegetables)	Vas et al.[43]
Bacteriocins + high pressure or pulses electric field	Gram-negative bacteria (model systems)	Kalchayanand et al.[44]
Ultrasound + heat	*Zygosaccharomyces bailii* (orange juice) *B. subtilis, B. cereus, B. licheniformis* (water)	Hurst et al.[45]
a_w reduction + pH + sorbic acid + sodium bisulphite	Various molds, yeasts,, and aerobic microorganisms (fruits)	Alzamora et al.[17]
a_w reduction + pH + sorbic acid + high pressure	*Z. bailii* (model systems)	Palou et al.[46]

technology approach appears to be the preservation technology of choice to establish the combination of preservation factors that adequately ensure product safety and convenience. According to Alzamora et al.,[29] hurdle technology can be applied in the design of the preservation system of minimally processed foods in many ways:

- at various stages of the food distribution chain, during storage, processing, and/or packaging as a "backup" measure in existing minimally processed products with short shelf life to diminish microbial pathogenic risk and/or increase their shelf life[33,34] (i.e., use of natural antimicrobials or other stress factors in addition to refrigeration);
- as an important tool for improving quality of long shelf-life products without diminishing their microbial stability/safety (i.e., use of heat coadjuvants to reduce the severity of thermal treatments); or
- as a new preservation procedure deliberately designed to obtain novel minimally

processed fruits or vegetables (i.e., the new preservation methodologies based on the use of electric pulses and high pressures).

Within the framework of hurdle technology or a hurdle approach, a huge amount of scientific literature has been published in the last years, indicating the enormous popularity and potential application of the concept in the development of new or improved technologies to aid in minimally processiing foods. However, several aspects need to be clarified in relation to issues such as microbial response to the combined hurdles; adaptation of microorganisms to sublethal stresses; interaction between hurdles, food matrix, and microorganisms; and microbial response to hurdles in real foods, including the possible presence of competitive (synergic or antagonic) microorganisms. Also important are quality and sensory aspects of minimally processed fruits and vegetables. Quality optimization studies in terms of nutrient content, as well as sensory attribute retention, are needed.

FUTURE NEEDS

The design and application of processes based on the hurdle approach in food preservation have undergone a dramatic increase, and further research and development in the area are expected. The combination of hurdles required during processing, as well as those to prolong shelf life of fresh cut fruits and vegetables, needs to be further investigated to maintain safety and quality. In addition to making products stable and safe, the concept of hurdle technology may also contribute to maintain or improve the sensory attributes and the total quality of minimally processed fruits and vegetables to comply with consumers' expectations.

Microbial response to combined factors is a research area that still needs to be constantly addressed. Also, a predictive microbiologic approach is needed to select factors and levels with a statistical sense. For a better and more efficient use of hurdle technology in the minimally processed fruits and vegetables, further research is needed on the mechanism of action of traditional and emerging preservation factors on microorganisms, enzymes, and deteriorative reactions, which continues to be an important issue for the development of minimal processing systems. A better understanding regarding these areas will help to identify key factors and their combined effect on product safety, stability, and quality.

Emerging technologies, such as pulsed electric fields or high pressure in combination with other preservation factors, can be used to inactivate microorganisms as well as deteriorative enzyme systems in minimally processed fruit and vegetable products. Future work must be focused on the application of these technologies in the context of minimal processing, including studies on the interaction of emerging process variables with other preservation factors such as pH, ionic strength, and temperature, among others, and the effects on product quality and stability.

REFERENCES

1. Reyes VG. Improved preservation systems for minimally processed vegetables. *Food Austr.* 1996;48:87–90.
2. Güntensperger B. International Symposium on Minimal Processing of Foods, Conference Report. *Trends Food Sci Technol.* 1994;5:266–269.
3. Welti-Chanes J, Vergara-Balderas F, López-Malo A. Minimally processed foods: State of the art and future. In: Fito P, Ortega-Rodríguez E, Barbosa-Cánovas GV, eds. *Food Engineering 2000.* New York: Chapman & Hall; 1997:181.
4. Welti-Chanes J, Alzamora SM, López-Malo A, Tapia MS, et al. Role of water in the stability of minimally or partially processed foods. In: Roos YH, Leslie RB, Lillford PJ, eds. *Water Management in the Design and Distribution of Quality Foods.* Lancaster, PA: Technomic Publishing Co.; 1999:503–532.
5. Thier SO. The sciences of nutrition. *Food Technol.* 1990;44(8):26–34.
6. Breidt F, Fleming HP. Using lactic acid bacteria to im-

prove the safety of minimally processed fruits and vegetables. *Food Technol.* 1997;51(9):44–51.

7. Nguyen-the C, Carlin F. The microbiology of minimally processed fresh fruits and vegetables. *Crit Rev Food Sci Nutr.* 1994;34:371–401.
8. Gould GW. Overview. In: Gould GW, ed. *New Methods of Food Preservation.* London: Blackie Academic and Professional; 1995:XV–XIX.
9. Gould GW. The microbe as a high pressure target. In: Ledward DA, Johnston DE, Earnshaw RG, Hasting APM, eds. *High Pressure Processing of Foods.* Nottingham, UK: Nottingham University Press; 1995.
10. Gould GW. Industry perspectives on the use of natural antimicrobials and inhibitors for food applications. *J Food Prot.* 1996;Suppl:82–86.
11. Gould GW. Ecosystem approaches to food preservation. *J Appl Bacteriol.* 1992;73:58S–68S.
12. Shank FR, Carson KL. What is safe food? In: Finley JW, Robinson SF, Armstrong DJ, eds. *Food Safety Assessment.* Washington, DC: American Chemical Society, ACS Symposium Series 484; 1992:26–34.
13. Tapia MS, Alzamora SM, Welti-Chanes J. Combination of preservation factors applied to minimal processing of foods. *Crit Rev Food Sci Nutr.* 1996;36:629–659.
14. Tapia MS, Alzamora SM, Welti-Chanes J. Obtention of minimally processed high moisture fruit products by combined methods: Results of a multinational project. In: Barbosa-Cánovas GV, Fito P, Ortega E, eds. *Proceedings of the First Ibero-American Congress of Food Engineering.* New York: Chapman & Hall; 1996.
15. Ahvenainen, R. New approaches in improving shelf life of minimally processed fruit and vegetables. *Trends Food Sci Technol.* 1996;7:179–187.
16. Alzamora SM, Tapia MS, Argaiz A, Welti J. Application of combined methods technology in minimally processed fruits. *Food Res Intern.* 1993;26:125–130.
17. Alzamora SM, Cerrutti P, Guerrero S, López-Malo A. Minimally processed fruits by combined methods. In: Barbosa-Cánovas G, Welti-Chanes J, eds. *Food Preservation by Moisture Control. Fundamentals and Applications.* Isopow Practicum II. Lancaster, PA: Technomic Publishing Co.; 1995:463–492.
18. Argaiz A, López-Malo A, Welti-Chanes J. Considerations for the development and the stability of high moisture fruit products during storage. In: Barbosa-Cánovas G, Welti-Chanes J, eds. *Food Preservation by Moisture Control. Fundamentals and Applications.* Isopow Practicum II. Lancaster, PA: Technomic Publishing Co.; 1995:729–760.
19. Rolle RS, Chism GW. Physiological consequences of minimally processed fruits and vegetables. *J Food Qual.* 1987;10:187–193.
20. Shewfelt R. Quality of minimally processed fruits and vegetables. *J Food Qual.* 1987;10:143–156.
21. Huxsoll CC, Bolin HR. Processing and distribution alternatives for minimally processed fruits and vegetables. *Food Technol.* 1989;43(2):132–138.
22. Wiley RC. Introduction to minimally processed refrigerated fruits and vegetables. In: Wiley RC, ed. *Minimally Processed Refrigerated Fruits and Vegetables.* New York: Chapman & Hall; 1994.
23. Ohlsson T. Minimal processing-preservation methods of the future: An overview. *Trends Food Sci Technol.* 1994;5:341–344.
24. Ahvenainen R, Mattila-Sandholm T, Ohlsson T. Minimal processing of foods. *VTT Symposium Series No. 142.* Espoo, Finland: Technical Research Center of Finland (VTT); 1994.
25. Singh RP, Oliveira FAR, eds. *Minimal Processing of Foods and Process Optimization. An Interface.* Boca Raton, FL: CRC Press; 1994.
26. Ronk RJ, Carson KL, Thompson P. Processing, packaging and regulation of minimally processed fruits and vegetables. *Food Technol.* 1989;42(2):136–139.
27. Palou E, López-Malo A, Barbosa-Cánovas GV, Swanson BG. High pressure treatment in food preservation. In: Rahman MS, ed. *Handbook of Food Preservation.* New York: Marcel Dekker; 1999:533–576.
28. López-Malo A, Palou E, Barbosa-Cánovas GV, Swanson BG, et al. Minimally processed foods and high hydrostatic pressure. In: Lozano J, Añon C, Parada-Arias E, Barbosa-Cánovas GV, eds. *Current Trends in Food Engineering.* New York: Chapman & Hall; 1999: unpublished.
29. Alzamora SM, Tapia MS, Welti-Chanes J. New strategies for minimal processing of foods: The role of multitarget preservation. *Food Sci Technol Int.* 4;353–361:1998.
30. Leistner L, Gorris LGM. Food preservation by hurdle technology. *Trends Food Sci Technol.* 1995;6:41–46.
31. Gould GW. Interference with homeostasis-food. In: Whittenbury R, Gould GW, Banks JG, Board RG, eds. *Homeostatic Mechanisms in Microorganisms.* Bath: Bath University Press; 1988:220–228.
32. Gould GW. Homeostatic mechanisms during food preservation by combined methods. In: Barbosa-Cánovas GV, Welti-Chanes J, eds. *Food Preservation by Moisture Control: Fundamentals and Applications.* Lancaster, PA: Technomic Publishing Co.; 1995:397–410.
33. Leistner L. Use of hurdle technology in food processing: Recent advances. In: Barbosa-Cánovas GV, Welti-Chanes J, eds. *Food Preservation by Moisture Control. Fundamentals and Applications.* ISOPOW Practicum II. Lancaster, PA: Technomic Publishing Co.; 1995:377–396.
34. Leistner L. Principles and applications of hurdle technology, In: Gould GW, ed. *New Methods for Food Preservation.* London: Blackie Academic and Professional; 1995:1–21.

35. Manvell C. Minimal processing of food. *Food Sci Technol Today.* 1997;11:107.

36. Gorris LGM. Improvement of the safety and quality of refrigerated ready-to-eat foods using novel mild preservation techniques. In: Singh RP, Oliveira FAR, eds. *Minimal Processing of Foods and Process Optimization. An Interface.* Boca Raton, FL: CRC Press; 1994:57–72.

37. Hayenmaier RD, Baker RA. Microbial population of shredded carrot in modified atmosphere packaging as related to irradiation treatment. *J Food Sci.* 1998;63: 162–164.

38. Roy S, Anantheswaran RC, Beelman RB. Modified atmosphere and modified humidity packaging of fresh mushrooms. *J Food Sci.* 1996;61:391–397.

39. López-Malo A, Palou E, Barbosa-Cánovas GV, Welti-Chanes J, et al. Polyphenoloxidase activity and color changes during storage in high hydrostatic pressure treated avocado purée. *Food Res Int.* 1999;31:549–556.

40. Juneja VK, Marmer BS, Phillips JG, Palumbo SA. Interactive effects of temperature, initial pH, sodium chloride, and sodium pyrophosphate on the growth kinetics of *Clostridium perfringens*. *J Food Prot.* 1996; 59:963–968.

41. Sala FJ, Burgos J, Condón S, López P, et al. Effect of heat ultrasound on microorganisms and enzymes. In: Gould GW, ed. *New Methods of Food Preservation.* London: Blackie Academic and Professional; 1995:176–204.

42. Razavi-Rohani SM, Griffiths MW. The effect of lysozyme and butylated hydroxyanizole on spoilage and pathogenic bacteria associated with foods. *J Food Safety.* 1996;16:59–74.

43. Vas K, Kiss L, Kiss N. Use of nisin for shortening the heat treatment in the sterilization of green peas. *Z Lebensm Unters Forsch.* 1967;133:141–144.

44. Kalchayanand N, Sikes T, Dunne CP, Ray B. Hydrostatic pressure and electroporation have increased bacteriocidal efficiency in combination with bacteriocins. *Appl Environ Microbiol.* 1994;60:4174–4177.

45. Hurst RM, Betts GD, Earnshaw RG. The antimicrobial effect of power ultrasound. *Research and Development Report No. 4.* Chipping Campden: Gloucestershire; 1995.

46. Palou E, López-Malo A, Barbosa-Cánovas GV, Welti-Chanes J, et al. High hydrostatic pressure as a hurdle for *Zygosaccharomyces bailii* inactivation. *J Food Sci.* 1997;62:855–857.

Part I

Microbiological Aspects

CHAPTER 1

Hurdle Technology in the Design of Minimally Processed Foods

Lothar Leistner

INTRODUCTION

With increasing process severity, sensory quality of foods usually decreases. Therefore, the consumer prefers minimally processed foods, which should have freshlike characteristics and must be microbiologically safe and stable. An approach that is aiming to fulfill these somewhat conflicting goals is the application of combined preservative factors (called *hurdles*) in the preservation of foods. This is because if gentle hurdles of different preservative factors are combined, the sensory quality of a food is less reduced, and if these hurdles are combined intelligently, their effect on the microbial safety and stability of a food might even become synergistic. In the light of minimally processed foods, the hurdle effect, from which the hurdle technology has been derived, will be discussed, as will emerging concepts, such as the metabolic exhaustion of microorganisms and the multitarget preservation of foods, which are also instrumental for minimally processed foods. Furthermore, examples of minimally processed foods will be given in which different principles based on hurdle technology are decisive. A strategy for the design of hurdle technology foods will be outlined, and some future prospects for the application of hurdle technology in food preservation will be mentioned.

HURDLE APPROACH TO FOOD PRESERVATION

The microbial stability and the sensory quality of most foods are based on a combination of hurdles. This is true for traditional foods with inherent empiric hurdles, as well as for novel products for which the hurdles are intelligently selected and intentionally applied.[1,2]

Hurdles in Foods

The most important hurdles commonly used in food preservation are temperature (high or low), water activity (a_w), acidity (pH), redox potential (Eh), preservatives (e.g., nitrite, sorbate, sulfite), and competitive microorganisms (e.g., lactic acid bacteria). However, more than 60 potential hurdles for foods of animal or plant origin that improve the stability and/or quality of these products have been already described, and the list of possible hurdles for food preservation is by no means complete.[3] At present, physical, nonthermal processes (high hydrostatic pressure, mano-thermo-sonication, oscillating magnetic fields, pulsed electric fields, light pulses, etc.) especially receive considerable attention because, in combination with other conventional hurdles, they are of potential use for the microbial stabilization of freshlike food products, with little degradation of nutritional and sensory properties.[4] With these novel processes, it is often not a sterile product but only a reduction of the microbial load that is intended, and growth of the residual microorganisms is inhibited by additional, conventional hurdles. Another group of hurdles that is at present of special interest in industrialized as well as in developing countries is "natural preservatives" (spices and their extracts, hop extracts, lysozyme, chitosan, protamine, pectine hydrolysate, etc.). In most countries, these "green preservatives" are preferred because they are not synthetic chemicals. Moreover, in some African countries that are short of foreign currency, they are given preference because spices are available and cheaper than imported chemicals.

Some hurdles (e.g., Maillard reaction products) influence the safety as well as the quality of foods, because they have antimicrobial properties and at the same time improve the flavor of the products; this also applies to nitrite used in the curing of meat. The same hurdle could have a positive or a negative effect on foods, depending on its intensity. For instance, chilling to an unsuitably low temperature will be detrimental to some foods of plant origin ("chilling injury"), whereas moderate chilling is beneficial for the shelf life. Another example is the pH of fermented sausages, which should be low enough to inhibit pathogenic bacteria but not so low as to impair taste. If the intensity of a particular hurdle in a food is too small, it should be strengthened; on the other hand, if it is detrimental to the food quality, it should be lowered. By this adjustment, the hurdles in foods can be kept in the optimal range, considering safety as well as quality of the product and, thus, the total quality of a food.[1,5]

Hurdle Effect

For each stable and safe food, a certain set of hurdles is inherent, which differs in quality and intensity, depending on the particular product; however, in any case, the hurdles must keep the usual ("normal") population of microorganisms in this food under control. The microorganisms present ("at the start") in a food should not be able to overcome ("leap over") the hurdles present during storage of the product; otherwise, the food will spoil or even cause food poisoning. This situation is illustrated by the so-called hurdle effect, first introduced in 1978,[6] which is of fundamental importance for the preservation of intermediate-moisture[7] as well as high-moisture foods.[8] Leistner and co-workers acknowledged that the hurdle effect illustrates only the well-known fact that complex interactions of temperature, water activity, acidity, redox potential, preservatives, etc., are significant for the microbial stability and safety of most foods.

In previous publications,[1,6,9] eight examples illustrating the hurdle effect in foods have been presented, which will not be repeated here. However, it should be mentioned that the hurdle effect is important for the ultraclean or aseptic packaging of foods, because if there are only a few microorganisms present at the start, then a few or low hurdles are sufficient for the stability of the product. The same proves true if the initial microbial load of a food (e.g., in high-moisture fruits or on meat carcass) is substantially reduced (e.g., by the application of steam), because after such decontamination procedures, fewer microorganisms are present and are more easily inhibited. The number and intensity of the hurdles needed for microbial stability are also lower if the microorganisms present are sublethally injured, because they then lack "vitality" and, thus, are easier to inhibit. On the other hand, a food rich in nutrients and vitamins will foster the growth of microorganisms (this is called the *booster* or *trampoline effect*), and, thus, the hurdles in such a product must be enhanced; otherwise, they will be overcome. The latter also happens: If, due to bad hygienic conditions, too many undesirable microorganisms are initially present and the usual hurdles inherent to a product may be unable to prevent spoilage or food poisoning. In fermented foods (salami, cheese, pickled vegetables, etc.), a sequence of hurdles is active, which is important in different stages of the ripening process, finally leading to a microbiologically stable and safe finished product. Further research in this respect is promising, which should clear up and optimize the sequence of hurdles in various fermented foods.

Hurdle Technology

The better understanding of the impact and interaction of different preservative factors (hurdles) in foods is the basis for improvements in food preservation because, if the hurdles in a food are known, the microbial stability and safety of this food might be optimized by changing the intensity or quality of these hurdles. Therefore, from an understanding of the hurdle effect, the hurdle technology has been derived,[10] which means that hurdles are deliberately and intentionally combined to improve the microbial stability and safety of foods. However, the hurdle technology approach is applicable to a wider concept of food preservation than just microbial safety.[11] The sensory quality is also determined by a number of positive and negative hurdles. By an intelligent combination of hurdles, the microbial stability and safety, as well as the sensory, nutritive, and economic properties of a food, are secured. For example, for the economy of a food item, the amount of water in a product that is compatible with the microbial stability of this food is important. Thus, hurdle technology aims to improve the total quality of foods by application of an intelligent mix of hurdles.

Over the years, the insight into the hurdle effect has been broadened and the application of hurdle technology extended. In industrialized countries, the hurdle technology approach is currently of particular interest for minimally processed foods that are mildly heated or fermented and for underpinning the microbial stability and safety of foods coming from future lines, such as healthful foods with less fat and/or salt[12] or advanced hurdle-technology foods that require less packaging.[13,14] For refrigerated foods, chill temperatures are the major and sometimes the only hurdle. However, if exposed to temperature abuse during distribution, this hurdle breaks down, and spoilage or even food poisoning could occur. Therefore, additional hurdles are incorporated as safeguards for chilled foods, using an approach called *invisible technology*.[3]

In developing countries, the intentional application of hurdle technology for foods that remain stable, safe, and tasty even if stored without refrigeration is of paramount importance and has made impressive strides, especially in Latin America, with the development of novel high-moisture fruit products. However, much interest in intentional hurdle technology is also emerging for meat products in China, as well as for dairy products in India. There is a general trend in developing countries to move gradually away from intermediate-moisture foods because they are often too salty or too sweet and have a less appealing texture and appearance than do high-moisture foods. This goal is achieved by the application of intentional hurdle technology. The progress made in the application of intentional hurdle technology in developing countries of Latin America, China, India, and Africa has recently been reviewed.[15]

Deliberate and intelligent application of hurdle technology is advancing worldwide; even various expressions are used for the same concept in different languages (e.g., *Hürden-Technologie* in German, *hurdle technology* in English, *technologie des barrières* in French, *barjernaja technologija* in Russian, *technologija prepreka* in Serbian, *tecnologia degli ostacoli* in Italian, *tecnología de obstáculos* in Spanish, *shogai gijutsu* in Japanese, and *zanglangishu* in Chinese). Many groups of researchers are working at present in different countries on the application of hurdle technology to a variety of foods, and a treasure of results that foster advanced food preservation is anticipated.

BASIC ASPECTS

Food preservation implies putting microorganisms in a hostile environment to inhibit their growth, shorten their survival, or cause their death. The feasible responses of microorganisms to such a hostile environment determine whether they may grow or die. In view of these responses, more research is needed; however, recent advances have been made by considering the homeostasis, metabolic exhaustion, and stress reactions of microorganisms in relation to hurdle technology, as well as by introducing the concept of multitarget preservation for a gentle but most effective preservation of hurdle-technology foods.[1,16] The physiologic response of microorganisms to food preservation is also the key issue for improvements of the safety and stability of minimally processed foods.

Homeostasis

Homeostasis is the tendency to uniformity and stability in the internal status (internal environment) of organisms. For instance, the maintenance of a defined pH in narrow limits is a prerequisite and feature of all living cells; this applies to higher organisms, as well as to microorganisms.[17] Much is already known about the homeostasis in higher organisms at the molecular, subcellular, cellular, and systematic levels in the fields of molecular biology, biochemistry, physiology, pharmacology, and medicine.[17] This knowledge should now be transferred to microorganisms important for the poisoning and spoilage of foods. In food preservation, the homeostasis of microorganisms is a key phenomenon that deserves much attention, because if the homeostasis of these organisms is disturbed by preservative factors (hurdles) in foods, they will not multiply, i.e., they remain in the lag phase or even die before their homeostasis is reestablished ("repaired"). Thus, food preservation is achieved by disturbing the homeostasis of microorganisms in a food temporarily or permanently, and gentle food preservation means using an intelligent mix of hurdles that secures the safety and stability, as well as the quality, of foods.[1] Gould was the first to draw attention to the interference by the food with the homeostasis of the microorganisms present in this food,[18,19] and, in the following chapter of this book, the interrelationship of homeostasis of microorganisms and preservative factors in foods will be discussed competently and comprehensively.

Metabolic Exhaustion

Another phenomenon of certainly practical importance is the metabolic exhaustion of microorganisms, which could lead to an "autosterilization" of foods. This was first observed in experiments with mildly heated (95°C core temperature) liver sausage adjusted to different water activities by the addition of salt and fat; the products were then inoculated with *Clostridium sporogenes* and stored at 37°C. Clostridial spores that survived the heat treatment vanished in the product during storage if the products were stable.[20] Later, this behavior of *Clostridium* and *Bacillus* spores was regularly observed during storage of shelf-stable meat products if the products were stored at room temperature.[21] The most likely explanation is that bacterial spores that survive the heat treatment are able to germinate in these foods under less favorable conditions than those under which vegetative bacteria are able to multiply.[9] Therefore, during storage of these products, some viable spores germinate, but the germinated spores or the vegetative cells derived from these spores die. Thus, the spore counts in stable hurdle technology foods actually decrease during storage, especially in unrefrigerated foods. Also, during studies in our laboratory with Chinese dried meat products, we observed the same behavior of microorganisms.[22] If these meats were contaminated after processing with staphylococci, salmonellae, or yeasts, the counts of these microorganisms on stable products decreased quite fast during unrefrigerated storage, especially on meats with a water activity close to the threshold for microbial growth. Again, the same phenomenon was observed by Latin American researchers[23–26] in their studies with high-moisture fruit products (HMFP). These studies showed that the counts of a variety of bacteria, yeasts, and molds that survived the mild heat treatment decreased quickly in the products during unrefrigerated storage because the hurdles applied (pH, a_w, sorbate, sulfite) did not allow growth.

A general explanation for this behavior might be that vegetative microorganisms that cannot grow will die, and they die more quickly if the stability is close to the threshold for growth, storage temperature is elevated, antimicrobial substances are present, and the organisms are sublethally injured.[1] Apparently, microorganisms in stable hurdle-technology food strain every possible repair mechanism for their homeostasis to overcome the hostile environment. By doing this, they completely use up their energy and die if they become metabolically exhausted. This leads eventually to an autosterilization of such foods.[16] Thus, due to autosterilization, the hurdle-technology foods, which are microbiologically stable, become even safer during storage, especially at ambient temperatures. For example, salmonellae that survive the ripening process in fermented sausages will vanish more quickly if the products are stored at ambient temperature, and they will survive longer and possibly cause food-borne illness if the products are stored under refrigeration.[1] It is also well known that salmonellae survive in mayonnaise at chill temperatures much better than at ambient temperature. Unilever laboratories in Vlaardingen have confirmed metabolic exhaustion in water-in-oil emulsions (resembling margarine) inoculated with *Listeria innocua*. In these products, *Listeria* vanished faster at ambient (25°C) than at chill (7°C) temperature, at pH 4.3 > pH 6.0, in fine emulsions more quickly than in coarse emulsions, and under anaerobic conditions more quickly than under aerobic conditions. From these experiments, it was concluded that metabolic exhaustion is accelerated if more hurdles are present, and this might be caused by increasing energy demands to maintain internal homeostasis under stress conditions.[27] Therefore, it could be concluded that refrigeration is not always beneficial for the microbial safety and stability of foods. However, this is true only if the hurdles present in a food inhibit the growth of microorganisms also without refrigeration; if this is not true, refrigeration is essential. Certainly, the survival of microorganisms in stable hurdle-technology foods is much shorter without refrigeration.

Stress Reactions

Some bacteria become more resistant or even more virulent under stress, because they generate stress shock proteins. The synthesis of protective stress shock proteins is induced by heat, pH, a_w, ethanol, oxidative compounds, etc., as well as by starvation. Stress reactions might have a nonspecific effect because, due to particular stress, microorganisms become also more tolerant to other stresses, i.e., they acquire a "cross-tolerance." This induced tolerance of bacteria toward stress will be discussed in detail in the following chapter of this book. Here, only the relationship of stress reactions to hurdle technology will be mentioned. The responses of microorganisms under stress might hamper food preservation and could turn out to be problematic for the application of hurdle technology. On the other hand, the activation of genes for the synthesis of stress shock proteins, which helps organisms to cope with stress situations, should be more difficult if different stresses are received at the same time. Simultaneous exposure to different stresses will require energy-consuming synthesis of several or at least much more protective stress shock proteins, which, in turn, may cause the microorganisms to become metabolically exhausted.[28] Therefore, multitarget preservation of foods could be the way to avoid synthesis of stress shock proteins, which otherwise could jeopardize the microbial stability and safety of hurdle-technology foods.[16] Further research on stress shock proteins and the different mechanisms that govern their formation seems warranted in relation to hurdle-technology foods.

Multitarget Preservation

The concept of multitarget preservation of foods has been introduced recently by the author of this chapter.[1,16] Multitarget preservation of foods should be the ultimate goal for gentle but most effective preservation of foods.[16] It has been suspected for some time that different hurdles in a food might not have just an additive effect on microbial stability, but might act synergistically.[6] A synergistic effect could be achieved if the hurdles in a food hit different targets (e.g., cell membrane, DNA, enzyme systems, pH, a_w, Eh) within the microbial cells at the same time and, thus, disturb the homeostasis of the microorganisms present in several respects. If so, the repair of homeostasis, as well as the activation of stress shock proteins, becomes more difficult.[1] Therefore, employing different hurdles simultaneously in the preservation of a particular food should achieve optimal microbial stability. In practical terms, this could mean that it is more effective to use different preservatives in small amounts, rather than one preservative in larger amounts, because different preservatives might act synergistically.[5]

It is anticipated that the targets in microorganisms of different preservative factors (hurdles) for foods will be elucidated and that hurdles could then be grouped in classes, according to their targets. A mild and effective preservation of foods, i.e., a synergistic effect of hurdles, is likely if the preservation measures are based on intelligent selection and a combination of hurdles taken from different target classes.[1] This approach seems valid not only for traditional food-preservation procedures, but for modern processes (e.g., food irradiation, ultra-high pressure, pulsed technologies, etc.) as well.

Food microbiologists could learn in this respect from pharmacologists, because the mechanisms of action of biocides have been studied extensively in the medical field. At least 12 classes of biocides are already known that have different targets, and often more than one, within the microbial cell. Often, the cell membrane is the primary target (it becomes leaky and unzips the organism), but biocides also impair the synthesis of enzymes, proteins, and DNA.[29] Multidrug attack has proven successful in the medical field to fight bacterial infections (e.g., tuberculosis), as well as viral infections (e.g., AIDS), and, thus, a multitarget attack of microorganisms should also be a promising approach in food microbiology.[16]

EXAMPLES FOR MINIMALLY PROCESSED FOODS

Consumers prefer minimally processed foods because they have appealing, freshlike characteristics and, thus, a superior sensory quality. However, at the same time, these foods must be microbiologically safe and stable. These somewhat conflicting goals are achievable by the application of advanced hurdle technology. Therefore, this concept is now used in the production of a variety of minimally processed foods, and, depending on the type of food, different hurdles are employed and combined. Different approaches are successfully used for minimally processed meats, fruits, or vegetables; therefore, these three examples will be chosen and discussed.

Minimally Processed Meats

Popular meat products that are convenient and suitable as snack foods are shelf-stable products (SSP). These are heat-processed foods based on hurdle technology and are stable at ambient temperature.[10] They offer the following advantages: the mild heat treatment (70–110°C) improves the sensory and nutritional properties of the food, and the lack of refrigeration simplifies distribution and saves energy during storage. SSP are heated in sealed containers (casings or pouches), which avoid recontamination after processing. However, because of the mild heat treatment, these foods still contain viable spores of bacilli and clostridia, which are inhibited by the adjustment of a_w, pH, and Eh and, in the case of autoclaved sausages (F-SSP), by sublethal injury of the spores. At present, four different types of SSP are distinguished—F-SSP, a_w-SSP, pH-SSP, and Combi-SSP—depending on their primary hurdles, although additional hurdles will foster the safety and stability of these products.[1,9,21]

In F-SSP,[9,10,21] the sublethal damage of bacterial spores is the primary hurdle, which is achieved by a mild heat treatment. Such sausages with an adjusted water activity (bologna-type sausage $a_w < 0.97$; liver and blood sausage $a_w < 0.96$) filled in polyvinylidenechloride (PVDC) casings impermeable to water vapor and oxygen are heated in counterpressure autoclaves to $F_o > 0.4$. They are storable unrefrigerated for several weeks and have caused no problems with regard to food poisoning or spoilage because guidelines for their processing have been established.[30] F-SSP is due to metabolic exhaustion of the microorganisms present, even autosterilized during storage, as has been mentioned in the section on basic aspects.

The stability of a_w-SSP[9,10,21] is primarily caused by a reduction of a_w below 0.95, and guidelines for the processing of such products have been suggested.[31] Examples of traditional a_w-SSP meats are Italian Mortadella and German Brühdauerwurst. A large variety of such meat products is today on the market, most being snack items ("finger foods"). The shelf life of a_w-SSP at ambient temperature is even better than that of fermented sausages because, in a_w-SSP due to the heat treatment (internal temperature > 75°C), lipases are inactivated and, thus, these products are less prone to become rancid.

In pH-SSP,[9,10,31] increased acidity is the primary hurdle. This is the principle applied in Gelderse Rookworst, to which glucono-delta-lactone is added.[21] Other traditional meat products of the pH-SSP type are brawns (jelly sausages), which are adjusted to an appropriate pH by the addition of acetic acid. Such products are composed of a brine (pH < 4.8) made of water, gelatin, salt, sugar, agar-agar (2%), acetic acid, spices, and a solid phase made of bologna-type sausage in cubes with an a_w of < 0.98. Both components are mixed (two parts brine to three parts meat), filled in casings, and heated to an internal temperature of > 72°C but not higher than 80°C; otherwise the sausage will not solidify after chilling. If the product is in equilibrium, it should have a final pH < 5.2 and is then storable for several days at ambient temperature. Outside the meat industry, pH-SSP are common as heat-pasteurized fruit and vegetable preserves with pH < 4.5, which are bacteriologically stable and safe, despite a mild heat treatment. In such products, vegetative microorganisms are inactivated by heat, and the multiplication of surviving bacilli and clostridia is inhibited by the low pH.

In Combi-SSP,[9,21,32] a combination of rather equal hurdles is applied. Our experimental work suggests that even small enhancements of individual hurdles have a distinct effect on the microbial stability of a product. For instance, for the stability and safety of a food, it is of significance whether the F_o is 0.3 or 0.4, the a_w is 0.975 or 0.970, the pH is 6.5 or 6.3, and the Eh value is somewhat higher or lower. Every small improvement or reinforcement of a hurdle brings some weight to the balance, and the sum of these weights determines whether a food is microbiologically unstable, uncertain, or stable. In other words, many small steps in the direction of stability can swing the balance from an unstable to a stable state.[9,21] We followed this procedure in our product design of bologna-type sausages as Combi-SSP. Different types of emulsion-type sausages (wieners, bockwurst, fleischwurst, fleischkäse, etc.) have been developed that proved microbiologically stable and safe for at least one week at 30°C. The initial spore load of the sausage mix must be low, and, therefore, spice extracts instead of natural spices are used and nitrite (100 ppm) with curing salt must be added. These products are heated to a core temperature of > 72°C, and the sausages are adjusted to an a_w and pH of < 0.965 and < 5.7, respectively. They are repasteurized after vacuum-packaging for 45–60 minutes (depending on the diameter of the product) at 82–85°C.[33] Combi-SSP require strict rules for food design and control, and, thus, Hazard Analysis Critical Control Points (HACCP) plans for these products have been established.[21] The Combi-SSP concept is applicable not only to meat products, but to other foods as well. For instance, an Italian pasta product (tortellini) has been stabilized using as hurdles an a_w reduction and a mild heat treatment, as well as modified atmosphere or ethanol vapor during storage, combined with moderate chilling temperatures.[34] Another example is Paneer, a dairy product of India, which was also developed as Combi-SSP.[21,35] In both cases, the thesis work of young scientists was groundbreaking.

In conclusion, it might be stated that, by the application of advanced hurdle technology, a new line of minimally processed meat products was introduced with superior sensory, freshlike properties that, nevertheless, are microbiologically safe and stable, even without refrigeration. These products are heat-pasteurized in sealed containers, and the surviving bacterial spore-formers are inhibited by carefully selected hurdles (a_w, pH, Eh, and nitrite), which are applied in combination. Because the bacterial spores are able to germinate at less favorable conditions than those at which vegetative bacilli and clostridia are able to multiply, in F-SSP, a_w-SSP, pH-SSP, and Combi-SSP, the number of spores tends to decrease during storage of the products, due to metabolic exhaustion of microorganisms, and, thus, the microbial stability of the products even increases during storage without refrigeration. For several years, these SSP meats have been available in large quantities and variety in Germany, and they have caused no problems in relation to spoilage or food poisoning, if the established guidelines[21] for their production are observed.

Minimally Processed Fruits

During the last decade, minimally processed, high-moisture, ambient stable fruit products (HMFP; $a_w > 0.93$) have been developed in seven Latin American countries, under the leadership of Argentina, Mexico, and Venezuela. This novel process has already been applied to peach halves, pineapple slices, mango slices and purée, papaya slices, chicozapote slices, purée of banana, plum, passion fruit, and tamarind, as well as to whole figs, strawberries, and pomalaca.[24,25] The new technologies were based on the combination of a mild heat treatment (blanching for 1–3 minutes with saturated steam), slight reduction in a_w (to 0.98–0.93 by addition of glucose or sucrose), lowering of pH (to 4.1–3.0 by addition of citric or phosphoric acid), and the addition of antimicrobials (1000 ppm potassium sorbate or sodium benzoate plus 150 ppm sodium sulfite or sodium bisulfite) to the syrup of the products. During storage of HMFP, the sorbate and, in particular, the sulfite levels decreased, and the a_w fell (i.e., the a_w hurdle increased), due to the hydrolysis of sucrose.[25]

Thus, combined methods technology (hurdle technology) was applied in these novel processes.[23–25,36,37] The minimal processes proved inexpensive, energy-efficient, simple to carry out (little capital investment), and satisfactory for preserving fruits in situ. The resulting freshlike products were still scored high by a consumer panel after 3 months of storage at 35°C for taste, flavor, color, and, especially, texture, which is often problematic for canned fruits. According to Latin American researchers,[24,25,36,38,39] combined methods allow the storage of fruits, without losses between seasonal harvest peaks, for direct domestic consumption or for further processing to confectionery, bakery goods, and dairy products or for preserves, jams, and jellies. Fruit pieces can also be utilized as ingredients in salads, pizzas, yogurt, and fruit drink formulations.[25,38] Moreover, these novel HMFP will open new possibilities for export markets.

The novel high-moisture fruit products stabilized by hurdle technology proved shelf-stable and safe for at least 3–8 months of storage at 25–35°C. Due to the blanching process, the initial microbial counts were substantially reduced, and during the storage of the stabilized HMFP, the number of surviving bacteria, yeasts, and molds decreased further, often below the detection limit.[23–26,36–39] Banana purée challenged with yeasts, molds, clostridia, and bacilli (known to spoil fruits) and stored at ambient temperatures for 120 days remained stable if proper hurdles were applied (mild heat treatment, adjustment of a_w to 0.97 and pH to 3.4, addition of 100 ppm potassium sorbate, 400 ppm sodium bisulfite, and 250 ppm ascorbic acid). The inoculated microorganisms declined and often vanished below the detection limit.[37] Because HMFP during storage at ambient temperature become apparently sterile, pathogenic and toxigenic microorganisms are not likely to be a hazard for these foods.[15]

Alzamora and co-workers expressed the opinion that HMFP technologies, as developed in Latin America, will attract much attention in many developing countries, because they are easy to implement and will improve considerably the quality of stored fruits.[24] They even believe that the usefulness of combined methods preservation (hurdle technology) for HMFP may give rise to an explosion of research on minimally processed fruits and the application of this innovative process by the food industry.[25] The advances in Latin America in fruit preservation are impressive and recently have been confirmed by Indian researchers, who concluded that "hurdle technology is seen as a promising technique for the preservation of fresh fruits and vegetables."[40] However, the preservation of HMFP must certainly be based on guidelines for good manufacturing practices (GMP) to be successful in the long term under industrial or even artisan conditions.[41] For instance, the reuse of syrup may become a risk for buildup of spoilage flora (e.g., *Zygosaccharomyces bailii*, which could become sorbate-resistant), and, therefore, the reuse of syrup in HMFP processes should occur only after pasteurization.

In conclusion, it might be stated that the application of advanced hurdle technology in Latin America has created a new line of minimally processed, freshlike fruits that are microbiologically safe and stable at ambient temperature. The hurdles that proved suitable for this group of food products are mild heat treatment (blanching), slight reduction of a_w and pH, and the moderate addition of preservatives (sorbate or benzoate and sulfite or bisulfite). The blanching (partial decontamination) of the fruits is very important for the microbial stability. However, even vegetative microorganisms might survive this mild heat treatment, but their number is reduced, and, thus, fewer and lower hurdles are essential. The number of surviving bacteria, yeasts, and molds is decreasing fast during ambient storage of the products because they are not able to multiply in the stable hurdle-technology products. The surviving bacteria, yeasts, and molds decline during storage, even below the detection limit, and this is probably due to metabolic exhaustion of the microorganisms. Also, the added sulfite and sorbate deplete during storage of the fruits, and this is beneficial for the consumer but diminishes the inhibition of the microorganisms in the products. Therefore, a recontamination of the hurdle-technology fruits during storage should be avoided by suitable measures (closed jars or bulk tanks with a lid). It is anticipated that this innovative HMFP process will gain ground worldwide, after suitable guidelines for a safe and efficient processing have been established.

Minimally Processed Vegetables

As a third example, minimally processed vegetables will be briefly discussed, because they represent a category of hurdle-technology foods that might harbor viable microorganisms, due to the fact that they are not or only mildly heated and subsequently stored under refrigeration. Two major groups of minimally processed vegetables should be distinguished: raw vegetables (washed, trimmed, sliced) and cooked *sous vide* preparations (vegetable- and potato-based dishes). Raw vegetables harbor a variety of bacteria, yeast, and mold species; most of them could cause spoilage. Of safety concern in this group of minimally processed vegetables are psychrotrophic pathogens, such as *L. monocytogenes*, *Aeromonas hydrophila,* and *Yersinia enterocolitica*, which might proliferate in this produce if refrigeration is not adequate. In cooked *sous vide* preparations, the mild heat treatment eliminates vegetative microorganisms, but sporeformers may survive. For the safety of this group of minimally processed vegetables, nonproteolytic *C. botulinum* is of particular concern, because it might grow well in refrigerated products in which the competitive flora has been inactivated by heat. Hurdle technology is applied in both groups of minimally processed vegetables to secure microbial stability and safety, but the types of hurdles differ somewhat. For both groups, refrigeration is the most important hurdle, and the additional hurdle in raw vegetables of modified-atmosphere packaging (MAP) is important, whereas in cooked *sous vide* preparations, the heat treatment and vacuum packaging are important additional hurdles. However, the limitations of the hurdles in both groups of minimally processed vegetables must be recognized and will be discussed in more detail.

Raw Vegetables

Modified-atmosphere packaging and subsequent storage at refrigeration temperature (5–7 days at 1–8°C) have been developed over the past decade as an adequate process to prolong high-quality shelf life of raw vegetables. The MAP system employed should be carefully tailored to the physiologic and microbiologic characteristics of the different produce to achieve high-quality products that are microbiologically stable and safe. The occurrence and behavior of microorganisms in minimally processed raw vegetables have been extensively studied,[42–44] and it was revealed that *Enterobacteriaceae* and *Pseudomonas* species constitute the major populations of raw vegetables before and after controlled-atmosphere (MAP) storage at 8°C. It was found that the modified-atmosphere conditions that were favorable for product quality retard growth of spoilage microorganisms during low-temperature storage. Growth of *L. monocytogenes* was inhibited, depending on the initial number of the pathogen, the type of produce, and the size of the competitive spoilage flora. Reducing the initial microbial load by decontamination of the raw material minimized microbial spoilage and might improve the safety of the product. However, *L. monocytogenes* grew better on disinfected produce than on nondisinfected or water-rinsed produce, indicating the practical importance of avoiding recontamination with pathogens after disinfection. It was concluded that, on raw, minimally processed vegetables stored at 8°C under typical modified-atmosphere conditions (1–5% O_2, 5–10% CO_2, made up to 100% with N_2), *L. monocytogenes* might grow,[42] and, thus, if refrigeration is not optimal, additional hurdles are advisable. For instance, lactic acid bacteria can be added that produce low amounts of acids, which do not much diminish sensory properties but synthesize sufficient quantities of antilisterial bacteriocins. Another option is natural preservatives, derived from herbs and spices, which inhibit food poisoning and spoilage bacteria, yeasts, and molds. Also, edible coatings prepared from carbohydrates, proteins, or fats, applied directly to the surface of the product, could be employed because they act as a physical barrier to food contamination and inhibit microbial growth if antimicrobial compounds are added to these coatings.[45]

Cooked Sous Vide *Preparations*

Minimally processed foods with a longer shelf life (up to 42 days at 1–8°C) are the so-called refrigerated processed foods of extended durability (REPFEDs), and these could be various food items, including *sous vide* vegetable and potato dishes.[46] REPFEDs are heated to 65–95°C, which should eliminate cells of vegetative bacteria but not bacterial spores. Surviving spore-formers that germinate during storage of the products will be able to proliferate without competition from the bacteria previously present. REPFEDs are mostly packed under vacuum or in anaerobic atmosphere; thus, growth of aerobic spore-formers is restricted, whereas clostridia are favored. Therefore, REPFEDs favor microorganisms that produce heat-resistant spores and grow in the absence of oxygen at refrigeration temperatures. Of most concern for REPFEDs is nonproteolytic *C. botulinum*, which can multiply and form toxin at temperatures as low as 3.0°C.[47,48] Because optimal refrigeration is not always guaranteed under practical conditions of food distribution and in the home, additional hurdles, such as organic acids, bacteriocins, natural preservatives derived from herbs and spices, or ultra-high pressure treatment, are recommended for REPFEDs.[46] With REPFEDs, heating is frequently slow because of large portion sizes and, thus, the synthesis of heat shock proteins might be induced that increase the heat resistance of microorganisms. Furthermore, it has been demonstrated that vegetable juice can aid the recovery of heat-damaged spores. Moreover, psychrotrophic *Bacillus* species might promote the growth and toxin production of nonproteolytic *C. botulinum.*[46] Therefore, the microbial stability and safety of REPFEDs demand strictly controlled process conditions. A number of relevant national and international guidelines and codes of practice have been drawn up with respect to the safe production of REPFEDs; most are targeted at preventing growth and toxin production by nonproteolytic *C. botulinum.*[46] In effect, for most REPFEDs, the principal factors controlling microbiologic safety and quality are likely to be the heat process, the storage temperature, and the maximum shelf life.[45]

In conclusion, it might be stated that, by application of additional hurdles (MAP, vacuum packaging, and possible application of bacteriocins, natural preservatives, edible coatings, ultra-high pressure, etc.), minimally processed vegetables (raw vegetables or cooked *sous vide* preparations) that are mildly heated and subsequently stored under refrigeration are microbiologically stable and safe only under strictly controlled processing and storage conditions. In this respect, minimally processed vegetables are similar to minimally processed meats and fruits, as discussed above. However, minimally processed vegetables are more risky with respect to safety, compared with these other product groups because they are not shelf-stable hurdle-technology foods and, thus, have to be stored under strict refrigeration. This has the consequence that pathogenic bacteria survive longer in minimally processed vegetables than in minimally processed meats or fruits, because metabolic exhaustion (autosterilization) of the pathogens present barely takes place during storage of these vegetables. It would be an ambitious goal to design minimally processed vegetables that are shelf stable without refrigeration, due to the application of proper hurdles, because storage and distribution would then be facilitated, and, at the same time, the safety of the products could be improved due to metabolic exhaustion of pathogenic microorganisms present. As they exist today, minimally processed vegetables are microbiologically more risky than are minimally processed meats or fruits.

DESIGN OF MINIMALLY PROCESSED FOODS

The application of hurdle technology is useful for the optimization of traditional products, as well as in the development of novel products. There are similarities to the concepts of predictive microbiology and hazard analysis critical control point (HACCP). The three concepts have related but different goals: hurdle technology is primarily used in food design, predictive microbiology for process refinement, and HACCP for process control. In product development, these three concepts should be combined.

User Guide to Food Design

We have suggested a 10-step procedure that encompasses hurdle technology, predictive microbiology, and HACCP[49] for the optimization of traditional foods or the design of new hurdle-technology foods. This approach proved suitable when solving real product development tasks in the food industry.[5,21] Predictive microbiology is a promising concept that involves computer-based and quantitative predictions of microbial growth, survival, and death in foods and, thus, should be an integral part of advanced food design. However, the predictive models constructed so far are applicable only to pathogenic bacteria, and models to predict the behavior of the spoilage flora are just emerging. Furthermore, the available predictive models handle only up to four different factors (hurdles) simultaneously. There are numerous hurdles to be considered that are important for the stability, safety, and quality of a particular food, and more than 60 different hurdles have been already described. It is unlikely that all or even a majority of these hurdles could be covered by predictive modeling. Thus, predictive microbiology cannot be a quantitative approach to the totality of hurdle technology. However, it does allow quite reliable predictions of the fate of microorganisms in food systems, while considering few but the most important hurdles. Because several hurdles are not taken into account, the predicted results are, fortunately, often on the safe side, i.e., the limits indicated for growth of pathogens in foods by the models available are generally more prudent than are the limits in real foods. Nevertheless, predictive microbiology will be an important tool for advanced food design, because it can narrow considerably the range over which challenge tests with relevant microorganisms need to be performed. Although predictive microbiology will never render challenge testing obsolete, it may greatly reduce both the time for and the costs of product development.[1,5,21] After the food has been properly designed, its manufacturing process must be effectively controlled, for which purpose the application of HACCP might be suitable. However, in a strict sense, the HACCP concept controls only the hazards of foods and not their stability or quality. Even in commercial practice, safety and quality issues will often overlap if HACCP is applied.[21] Because microbial safety and stability as well as sensory quality (i.e., total quality) for hurdle-technology foods are essential, the HACCP concept might be too narrow for this purpose if it relates only to biologic, chemical, and physical hazards. Therefore, the HACCP concept should be broadened to cover the microbial safety (food poisoning) and stability (spoilage) of foods, as well as their sensory quality. If this is not acceptable, the production process should be controlled by GMP, and rules or guidelines for the production of each food item must be defined.[31] For hurdle-technology foods of developing countries where many small producers prevail, GMP guidelines are often more acceptable because the application of HACCP poses practical difficulties.

Multidisciplinary Food Design

In food design, different groups of researchers, including microbiologists and technologists, must work together. The microbiologist should determine which types and intensities of hurdles are needed for the necessary safety and stability of a particular food product, and the technologist should determine which ingredients or processes are proper for establishing these hurdles in a food, taking into account the legal, technologic, sensory, and nutritive limitations. Because the engineering, economic, and marketing aspects must also be considered, food design is indeed a multidisciplinary endeavor.[1,21]

CONCLUSIONS AND FUTURE PROSPECTS

For the design of minimally processed foods, the application of intelligent hurdle technology is the most promising option. Therefore, hurdle technology is today frequently applied for minimally processed foods in industrialized countries, as well as in developing countries. However, the types and the combinations of hurdles that are suitable for different categories of minimally processed foods are product-specific and have to be carefully selected. It is useful to study the hurdle combinations that are effective for different product groups because they might become a source of innovation for other categories of food. In general, minimally processed foods are more vulnerable than are traditional foods, which are often overprocessed. Thus, in the design of minimally processed foods, skillful and intelligent hurdle technology must be employed. In the design of advanced hurdle-technology foods, a combination of hurdle technology with predictive microbiology and HACCP (or GMP) has proved useful.

Future prospects are that the physiology of microorganisms is taken more into account in the design of hurdle-technology foods, and this relates, in particular, to the homeostasis and metabolic exhaustion of microorganisms, but also to their stress reactions. Metabolic exhaustion of microorganisms is already instrumental for the microbial stability and safety of minimally processed meats and fruits that are storable without refrigeration and could also be beneficial for minimally processed vegetables if their process would be designed accordingly. The ultimate goal is the multitarget preservation of foods, in which gentle hurdles might have a synergistic effect. After the targets of different preservative factors within the microbial cells have been elucidated, and this should definitely become a major research priority, the future preservation of foods could progress far beyond the state of the art of hurdle technology as applied today. The food industry answers the demands of the consumer by providing convenient foods that are freshlike, have superior sensory properties, and, indeed, are minimally processed foods. This is, at present, the best option. However, even as the sensory quality of foods becomes increasingly important, microbial safety and stability should still have the highest priority.

REFERENCES

1. Leistner L. Principles and applications of hurdle technology, In: Gould GW, ed. *New Methods for Food Preservation.* London: Blackie Academic and Professional; 1995:1–21.
2. Leistner L, Gorris LGM. Food preservation by hurdle technology. *Trends Food Sci Technol.* 1995; 6:41–46.
3. Leistner L. Combined methods for food preservation, In: Shafiur Rahman M, ed. *Food Preservation Handbook.* New York: Marcel Dekker; 1999:457–485.
4. Barbosa-Cánovas GV, Pothakamury UR, Palou E, Swanson BG. *Nonthermal Preservation of Foods.* New York: Marcel Dekker; 1998.
5. Leistner L. Further developments in the utilization of hurdle technology for food preservation. *J Food Engr.* 1994;22:421–432.
6. Leistner L. Hurdle effect and energy saving. In: Downey WK, ed. *Food Quality and Nutrition.* London: Applied Science Publishers; 1978:553–557.
7. Leistner L, Rödel W. The stability of intermediate moisture foods with respect to micro-organisms. In: Davies R, Birch GG, Parker KJ, eds. *Intermediate Moisture Foods.* London: Applied Science Publishers; 1976:120–137.
8. Leistner L, Rödel W, Krispien K. Microbiology of meat and meat products in high- and intermediate-moisture ranges. In: Rockland LB, Stewart GF, eds. *Water Activity: Influences on Food Quality.* New York: Academic Press; 1981:855–916.
9. Leistner L. Food preservation by combined methods. *Food Res Int.* 1992;25:151–158.
10. Leistner L. Hurdle technology applied to meat products of the shelf stable product and intermediate moisture food types. In: Simatos D, Multon JL, eds. *Properties of Water in Foods in Relation to Quality and Stability.* Dordrecht: Martinus Nijhoff Publishers; 1985:309–329.
11. Stanley DW. Biological membrane deterioration and associated quality losses in food tissues. *Crit Rev Food Sci Nutr.* 1991;30:487–553.
12. Leistner L. Microbial stability and safety of healthy meat, poultry and fish products. In: Pearson AM, Dutson TR, eds. *Production and Processing of Healthy Meat, Poultry and Fish Products.* London: Blackie Academic and Professional; 1997:347–360.
13. Ono K. *Packaging Design and Innovation.* Material for a third country training programme in the field of food packaging, conducted at Singapore; 1994.
14. Ono K. *Personal communication.* Tokyo: Snow Brand Co.; 1996.
15. Leistner L. Use of combined preservative factors in foods of developing countries. In: Lund BM, Baird-Parker AC, Gould GW, eds. *The Microbiological Safety and Quality of Food.* Vol. 1. Gaithersburg, MD: Aspen Publishers; 2000:294–314.
16. Leistner L. Emerging concepts for food safety. *Proceedings 41st ICoMST,* San Antonio, TX: 1995:321–322.
17. Häussinger D, ed. *pH Homeostasis: Mechanisms and Control.* London: Academic Press; 1988.
18. Gould GW. Interference with homeostasis—food. In: Whittenbury R, Gould GW, Banks JG, Board RG, eds. *Homeostatic Mechanisms in Micro-organisms.* Bath: Bath University Press; 1988:220–228.
19. Gould GW. Homeostatic mechanisms during food preservation by combined methods. In: Barbosa-Cánovas GV, Welti-Chanes J, eds. *Food Preservation by Moisture Control: Fundamentals and Applications.* Lancaster, PA: Technomic Publishing Co.; 1995:397–410.
20. Leistner L, Karan-Djurdjić S. Beeinflussung der Stabilität von Fleischkonserven durch Steuerung der Wasseraktivität. *Fleischwirtschaft.* 1970;50:1547–1549.
21. Leistner L. *Food Design by Hurdle Technology and HACCP.* Kulmbach, Germany: Adalbert-Raps-Foundation; 1994.

22. Shin H-K. *Energiesparende Konservierungsmethoden für Fleischerzeugnisse, abgeleitet von traditionellen Intermediate Moisture Foods*. Stuttgart-Hohenheim, Germany: Universität Hohenheim; 1984. Ph.D. thesis.
23. Sajur S. *Preconservación de Duraznos por Métodos Combinados*. Argentina: Universidad Nacional de Mar del Plata; 1985. M.S. thesis.
24. Alzamora SM, Tapia MS, Argaiz A, Welti J. Application of combined methods technology in minimally processed fruits. *Food Res Internat.* 1993;26:125–130.
25. Alzamora SM, Cerrutti P, Guerrero S, López-Malo A. Minimally processed fruits by combined methods. In: Barbosa-Cánovas GV, Welti-Chanes J, eds. *Food Preservation by Moisture Control: Fundamentals and Applications*. Lancaster, PA: Technomic Publishing Co., 1995:463–492.
26. Tapia de Daza MS, Argaiz A, López-Malo A, Díaz, RV. Microbial stability assessment in high and intermediate moisture foods: Special emphasis on fruit preservation. In: Barbosa-Cánovas GV, Welti-Chanes J, eds. *Food Preservation by Moisture Control: Fundamentals and Applications*. Lancaster, PA: Technomic Publishing Co.; 1995:575–601.
27. ter Steeg PF. *Personal communication*. Vlaardingen: Unilever Research; 1995.
28. Leistner L. Food protection by hurdle technology. *Bull Jpn Soc Res Food Prot.* 1996; 2:2–27.
29. Denyer SP, Hugo WB, eds. *Mechanisms of Action of Chemical Biocides: Their Study and Exploitation*. Oxford, England: Blackwell Scientific Publications; 1991.
30. Hechelmann H, Leistner L. Mikrobiologische Stabilität autoklavierter Darmware. *Mitteilungsblatt Bundesanstalt Fleischforschung*, Kulmbach 1984; Nr. 84:5894–5899.
31. Leistner L. Shelf-stable products and intermediate moisture foods based on meat, In: Rockland LB, Beuchat LR, eds. *Water Activity: Theory and Applications to Food*. New York: Marcel Dekker; 1987:295–327.
32. Leistner L, Hechelmann H. Food preservation by hurdle-technology. In: *Proceedings of Food Preservation 2000*, Vol. II. Natick, MA: U.S. Army Natick, Research, Development and Engineering Center; 1993:511–520.
33. Hechelmann H, Kasprowiak R, Reil S, Bergmann A, Leistner L. *Stabile Fleischerzeugnisse mit Frischprodukt-Charakter für die Truppe*, BMVg FBWM 91–11, Dokumentations- und Fachinformationszentrum der Bundeswehr 1991; Bonn, Germany.
34. Giavedoni P. *Azioni Combinate nella Stabilizzazione degli Alimenti*. Udine, Italy: Università degli Studi di Udine; 1994. Ph.D. thesis.
35. Rao KJ. *Application of Hurdle Technology in the Development of Long Life Paneer-Based Convenience Food*. Karnal, India: National Dairy Research Institute; 1993. Ph.D. thesis.
36. López-Malo A, Palou E, Welti J, Corte P, Argaiz A. Shelf-stable high moisture papaya minimally processed by combined methods. *Food Res Int.* 1994; 27:545–553.
37. Guerrero S, Alzamora SM, Gerschenson LN. Development of a shelf-stable banana purée by combined factors: Microbial stability. *J Food Prot.* 1994;57:902–907.
38. Argaiz A, López-Malo A, Welti-Chanes J. Considerations for the development and the stability of high moisture fruit products during storage. In: Barbosa-Cánovas GV, Welti-Chanes J, eds. *Food Preservation by Moisture Control: Fundamentals and Applications*. Lancaster, PA: Technomic Publishing Co., 1995:729–760.
39. Tapia de Daza MS, Alzamora SM, Welti-Chanes J. Combination of preservation factors applied to minimal processing of foods. *Crit Rev Food Sci Nutr.* 1996; 36:629–659.
40. Rastogi NK, Sandhi JS, Viswanath P, Saroja S. Application of hurdle/combined method technology in minimally processed long-term non-refrigerated preservation of banana and coconut. Mysore, India: *Abstracts for ICFoST'95;* 1995:109.
41. Leistner L. Use of hurdle technology in food processing: recent advances. In: Barbosa-Cánovas GV, Welti-Chanes J, eds. *Food Preservation by Moisture Control: Fundamentals and Applications*. Lancaster, PA: Technomic Publishing Co.; 1995:377–396.
42. Bennik MHJ, Smid EJ, Rombouts FM, Gorris LGM. Growth of psychrotrophic foodborne pathogens in a solid surface model system under the influence of carbon dioxide and oxygen. *Food Microbiol.* 1995;12:509–519.
43. Bennik MHJ, Peppelenbos HW, Nguyen-the C, Carlin F, Smid EJ, Gorris LGM. Microbiology of minimally processed, modified-atmosphere packaged chicory endive. *Postharvest Biol Technol.* 1996; 9:209–221.
44. Bennik MHJ, Vorstman W, Smid EJ, Gorris LGM. The influence of oxygen and carbon dioxide on the growth of prevalent *Enterobacteriaceae* and *Pseudomonas* species isolated from fresh and controlled-atmosphere-stored vegetables. *Food Microbiol.* 1998;15:459–469.
45. Gorris LGM. Hurdle technology, a concept for safe, minimal processing of foods. In: Robinson R, Batt C, Patel P, eds. *Encyclopedia of Food Microbiology*. London: Academic Press; 2000. In press.
46. Gorris LGM, Peck MW. Microbiological safety considerations when using hurdle technology with refrigerated processed foods of extended durability. In: Ghazala S, ed. *Sous Vide and Cook-Chill Processing for the Food Industry*. Gaithersburg, MD: Aspen Publishers; 1998:206–233.
47. Carlin F, Peck MW. Growth of, and toxin production by, non-proteolytic *Clostridium botulinum* in cooked vegetables at refrigeration temperatures. *Appl Environ Microbiol.* 1996;62:3069–3072.

48. Peck MW. *Clostridium botulinum* and safety of refrigerated processed foods of extended durability. *Trends Food Sci Technol.* 1997;8:186–192.
49. Leistner L. User guide to food design. In: Leistner L, Gorris LGM, eds. *Food Preservation by Combined Processes*, FLAIR Final Report, EUR 15776 EN. Brussels: European Commission; 1994:25–28.

Chapter 2

Induced Tolerance of Microorganisms to Stress Factors

Grahame W. Gould

INTRODUCTION

The most widely used of the food preservation techniques act by inhibiting the growth of microorganisms, rather than by inactivating them. Inhibitory techniques include reduction in temperature (chilling, freezing); reduction in water activity (a_w) (curing, conserving, drying); reduction in pH (addition of inorganic and organic acids, lactic and acetic fermentations); modified-atmosphere packaging (vacuum, nitrogen, carbon dioxide, oxygen); and addition of preservatives. In contrast, few techniques act primarily by inactivating microorganisms in foods. Heat is by far the most employed inactivation technique (thermization, pasteurization, sterilization). There is increasing but still far less use being made of alternatives, such as ionizing radiation and the new and "emerging" technologies, such as high hydrostatic pressure, high-voltage electric discharge, and high-intensity light pulses.

Minimal processing technologies for fruits and vegetables rely to a large extent on the application of inhibitory techniques. These offer the possibility of use in combinations in such a way as to achieve a reduction in the severity of the overall preservation process, thus helping to meet consumers' requirements for high-quality foods that are convenient to store and use. However, there is a potential problem that must be taken into account in the development of mild techniques. This is the reaction of microorganisms to the inhibitory procedures. Some mild preservation techniques can encourage microorganisms to undergo "stress reactions."

Stress reactions have evolved to become widespread in plants, animals, and microorganisms. Particularly effective stress reactions have been selected during microbial evolution, probably because, being very small, microorganisms are normally in intimate contact with their environment. They must, therefore, react quickly and effectively to the sudden changes that often occur if they are to compete well and survive. Stress reactions occur in bacteria, yeasts, and molds; in microorganisms responsible for food spoilage; as well as in pathogens that cause food poisoning. The stress reactions result generally in increases in resistance of the microorganisms to the particular stress that was applied, but also often to other seemingly unrelated stresses as well. Furthermore, other important changes may result, even including increases in the pathogenicity of food poisoning bacteria.

Knochel and Gould[1] pointed out that the adoption of milder preservation techniques could be expected to lead to more stress responses in food-borne microorganisms and, therefore, in more problems in the control of food spoilage and food poisoning, although there are alternative views. For instance, Archer[2] suggested that the opposite might be true, i.e., that milder processing might be expected to lead to less expression of stress responses than those that occur with severe processes. However, whatever the outcome, in the development of minimal processes, it is important to know about the stress reactions that can occur, in particular, so that such potentially damaging increases in tolerance to preservation factors and in pathogenicity can be avoided.

MAJOR STRESS FACTORS IN MINIMAL PROCESSING

The major inhibitory stresses to which microorganisms may react in the minimal processing of fruits and vegetables are increased H^+ concentration (lowered pH); the presence of weak organic acid preservatives, such as sorbate and benzoate; the presence of inorganic acid preservatives, especially sulfite; and reduction of a_w, especially by high concentrations of sugars. The major inactivating stress employed is heat at substerilizing levels.

Microorganisms react homeostatically to all of these stresses. That is, when their environment is perturbed by the stress factor, they usually react in ways that maintain some key element of their physiology constant, i.e., unperturbed. Microorganisms undergo many important homeostatic reactions—many more than are relevant to minimally processed fruits and vegetables. The wide range is illustrated in Table 2–1, where the nature of the reaction is summarized against the particular stress. It can be seen that most stress reactions are active processes. That is, in addition to the genetic changes that underly many of them, they include some level of altered metabolism, and this often involves the expenditure of energy, e.g., to transport protons across the cell membrane, to maintain high cytoplasmic concentrations of "osmoregulatory," or "compatible" solutes, etc. Some stress reactions are, in a sense, built into cells and are, therefore, passive, with no requirement for the expenditure of energy to operate. Gerhardt[4] used the term *refractory homeostasis* to describe, for example, the built-in mechanisms that operate to maintain the resistance of bacterial spores to heat (Table 2–1). To these homeostatic responses, one can logically add "population homeostasis." This refers to the maintenance of stable associations of microorganisms that often function to exclude potential competitors. These associations are often composed of aggregates of microorganisms showing some degree of symbiosis. Examples include the microbial communities in dental plaque on teeth, some lactic and yeast-lactic acid bacterial fermentations, and, of particular relevance to the preservation of minimally processed fruits and vegetables, the formation and persistence of biofilms in food processing plants and on food contact surfaces in processing areas.

REACTIONS TO STRESS

General Stress Response Mechanisms

There is a general response mechanism underlying many of the apparently specific responses of microorganisms to the stresses imposed on them in foods. It has become clear recently that this underlying or "global response" includes many of the elements of microorganisms' reactions to starvation. Starvation usually occurs during the stationary phase of growth in culture, so the reaction is regarded to be part of a complex "stationary-phase response." This is known now to be mediated predominantly by the stationary-phase regulator RpoS (previously known as KatF), a sigma factor (σ^s; a subunit of RNA polymerase that determines promoter specificity) that regulates the expression of many important stationary-phase stress resistance genes.[5,6] Rees et al.[7] listed more than 20 genes known to be regulated by RpoS.

The central role of RpoS helps to explain the cross-resistances to different stresses that have commonly been found to occur in reaction to a single stress. Most of the stresses relevant to minimally processed fruits and vegetables are included. For instance, the global response mechanism underlies the reactions of microbial cells to heat: the "heat shock" response and the synthesis of "heat shock proteins."[8] It occurs in response to oxidative stress, e.g., in *Escherichia coli* (high oxygen levels; presence of hydrogen

Table 2–1 Major Stresses and Homeostatic Reactions of Relevance to Preserved Foods

Stress Factor	*Homeostatic Response*
Active homeostasis	
Low levels of nutrients	Nutrient scavenging; oligotrophy; "stationary-phase response"; generation of "viable nonculturable" forms
Lowered pH	Extrusion of protons across the cell membrane; maintenance of cytoplasmic pH; maintenance of transmembrane pH gradient
Presence of weak organic acid preservatives (e.g., sorbate, benzoate)	As above, and sometimes extrusion of the organic acid
Lowered water activity	Osmoregulation; accumulation of "compatible solutes"; avoidance of water loss; maintenance of membrane turgor
Lowered temperature for growth	"Cold shock" response; changes in membrane lipids to maintain satisfactory fluidity
Raised temperature for growth	"Heat shock" response; membrane lipid changes
Raised levels of oxygen	Enzymic protection (catalase, peroxidase, superoxide dismutase) from H_2O_2 and oxygen-derived free radicals
Presence of biocides and some preservatives	Phenotypic adaptation; reduction in cell wall/membrane permeability
Ultraviolet radiation	Excision of thymine dimers and repair of DNA
Ionizing radiation (β, γ, X)	Repair of single-strand breaks in DNA
Passive homeostasis	
High temperature	Spore structure; inbuilt mechanisms maintaining low water content in the spore protoplast
High hydrostatic pressure	Uncertain; possibly low spore water content
High voltage electric discharge	Low electrical conductivity of the spore protoplast
Ultrasonication	Structural strength of cell or spore wall
Population homeostasis	
Competition from other microorganisms	Formation of interacting communities; aggregates of cells showing some degree of symbiosis; biofilms

peroxide[9]), and in response to anaerobiosis, e.g., in yeast.[10] Osmoregulating cells growing at low a_w share many of the core reactions of the global response mechanism,[11] as do cells adapted to grow at low pH, e.g., streptococci[12] and eukaryotic cells affected by respiratory uncouplers.[10] Even the reaction of microorganisms to exposure to ethanol triggers part of the response, e.g., in *Candida utilis,*[13] leading to the otherwise surprising finding that treatment of some microorganisms (e.g., *Listeria monocytogenes*) with sublethal concentrations of ethanol protected them from what otherwise would have been lethal doses of other preservation processes, significantly including reduced pH and reduced a_w (raised levels of salt[14]). Furthermore, the connection of the stress response to virulence in pathogenic bacteria became clear when it was found that RpoS also regulates virulence determinants, e.g., in salmonellae.[15]

RpoS is normally unstable during the rapid growth of unstressed cells, being broken down by proteolysis. However, this breakdown ceases as cells enter the stationary phase of growth, so that its concentration rises and its regulatory activity increases.[16,17]

It has been suggested that, under conditions of cell crowding, e.g., in biofilms, the cell-to-cell signaling that occurs at high concentrations, possibly mediated by homoserine lactones and other "messenger" metabolites,[18] rather than by starvation,[19] may be involved in enhancing RpoS activity and, therefore, also in enhancing the resistance of biofilm microorganisms, e.g., to heat[20,21] and other stresses. Such reactions are of obvious potential importance in the hygienic operation of food handling areas where mild preservation procedures are applied and microorganisms with enhanced tolerance must be avoided.

Thus, the involvement of RpoS and related processes in a wide range of stress responses has changed the perception of those responses from what appeared to be distinct and unrelated phenomena to an interrelated group of stress survival mechanisms with strong connections to survival of starvation and to survival in the stationary phase, an overall "safety net."[11]

Low pH

In many foods, and certainly in minimally preserved fruits and vegetables, the cytoplasmic pH of microorganisms will normally be one or two units higher than that of their environment.[22,23] The ability of microorganisms to grow at low pH values depends on their ability to prevent protons from crossing the cell membrane and entering the cytoplasm and, to a large degree, to their ability to export any protons that do gain access, thus maintaining a satisfactorily high intracellular pH. If net proton influx cannot be prevented, cytoplasmic pH will fall, leading to cessation of growth and death of the cell.

However, many bacteria, yeasts, and molds can react to acid conditions, so as to enhance their survival. The stress response following exposure to acid conditions results in adaptation such that the efficacy of proton export increases. This acid tolerance response then results in microorganisms that have been exposed to even mildly acidic conditions becoming able to grow or survive at low pH values that would otherwise have been lethal.[24] This adaptation is important with respect to the potential for growth of spoilage microorganisms in acid-preserved foods and is particularly important with respect to the survival of pathogenic bacteria. For example, acid-adapted *Salmonella* species survived much longer than did unadapted cells in cheese,[25] and similar acid adaptation has been shown to operate in a number of other food poisoning bacteria, e.g., *E. coli,*[26,27], *Salmonella,*[28] *L. monocytogenes,*[29] *Aeromonas hydrophila.*[30] Outbreaks of food poisoning have resulted from survival of salmonellae in unpasteurized apple juice[31] and in orange juice,[32,33] as well as from *E. coli* O157 surviving in unpasteurized apple juice and cider.[34–36] Furthermore, the ability to survive at low pH is of particular importance in the pathogenicity of some food poisoning microorganisms because the low pH values in the stomach and in phagocytosing cells are important animal defense mechanisms. Thus, well-acid-adapted microorganisms may be expected to have increased virulence.

As mentioned earlier, the stress response cross-reactions that often occur are important in acid adaptation. Acid-adapted cells may become tolerant to a range of other environmental stresses, e.g., in *S. typhimurium.*[37] Acid-shocked *L. monocytogenes* became more acid tolerant, as expected, but also exhibited a rise in heat resistance.[38] Conversely, the heat shock response that follows mild heating can result in cells becoming more acid tolerant.[39] An impressive example was given by Wang and Doyle,[40] who exposed seven strains of *E. coli* O157:H7 to mild heat (48°C; 10 minutes) at neutral pH prior to incubation at pH 2.5 and 37°C. The heated cells were between 10 and 100 times more acid tolerant than were controls that had not been heated. The rise in acid tolerance was as great as that induced by exposure of cells to acid at pH 5. The authors pointed out that such adaptation could clearly lead to the unexpected survival of this pathogen in foods and ingredients that receive a mild heat treatment, then rely on pH reduction for their preservation, e.g., fruit-based drinks, mayonnaises and dressings, and minimally processed fruits and vegetables, unless the heat treatments they are given are sufficiently severe.

Weak Organic Acids

The ability of yeasts and molds to adapt to high concentrations of the weak acid preservatives, sorbate and benzoate, is of major importance in the preservation of minimally processed fruits and vegetables, particularly with respect to spoilage yeasts. As indicated earlier, pH homeostasis in yeast is maintained predominantly due to the activity of membrane-bound H^+ATPase, which catalyzes the export of protons from the cell, coupled to the hydrolysis of ATP. The importance of the reaction is illustrated by the fact that activity of the H^+ATPase is essential for growth[41] and can account for up to 60% of cellular ATP consumption.[42] Yeast cells normally maintain their internal pH values between about 5 and 6.5, even if the external pH is greatly reduced.[43,44] The weak organic acid preservatives interfere with pH homeostasis.

The interference derives to a large extent from the lipid solubility of the acids, in particular, in their protonated, undissociated forms, which can readily cross the cell membrane and enter the cytoplasm. The undissociated forms are favored at low pH (the pKa of sorbic acid is 4.67, and that of benzoic acid is 4.19) and, therefore, predominate at the pH values of minimally preserved fruits and vegetables. One element of their modes of action, having equilibrated across the cell membrane, is to dissociate within the cytoplasm, which has a higher pH than does the external environment, so delivering protons and preservative anions to the interior of the cell. Numerous studies have indicated that, as a consequence of this, the preservative acids bring about acidification of the cytoplasm and that this, perhaps along with specific effects of the rising concentrations of free anion, leads to inhibition of growth.[45–47] However, more recently, careful measurements of the internal pH of yeast cells in the presence of sorbic acid demonstrated little correlation with reduction in growth rate.[48] Rather, inhibition correlated well with an increase in the intracellular ADP/ATP ratio, due to the consumption of ATP. This was attributed to a stress reaction in which increased export of protons by the membrane H^+ATPase occurred, thus ensuring that the internal pH did not fall. The effects on the yeast cells include large extensions of the lag phase,[49] slower growth, lower yield,[50] and, eventually, complete inhibition, resulting mainly from increasing energy demand on the cell to maintain its cytoplasmic pH, such that insufficient ATP is available for normal growth. In addition to this energy-demanding stress reaction, there is now evidence that active extrusion of weak acids from the cell, first proposed by Warth,[51,52] may also occur.[53] In support of this, Piper et al.[54] recently identified a membrane pump protein that is induced by sorbic acid in *Saccharomyces cerevisiae*. Furthermore, Stratford and Anslow[50] presented evidence that low-molecular-weight (MW) organic acids, including sorbate, also act more directly by partitioning into cell membranes and interfering with their functionality. That the ionic form of sorbate contributed to the antimicrobial activity of this preservative was illustrated in modeling studies by Eklund.[55]

Overall, these responses to the stress of low pH, coupled with the presence of sorbate or benzoate, can allow some contaminating microorganisms, particularly spoilage yeasts, such as *S. cerevisiae* and *Zygosaccharomyces bailii*, to acquire greatly enhanced tolerance to the preservatives. For example, increases in the resistance of adapted yeasts to benzoate exceeding 100% are common, i.e., to levels in excess of 1,200 ppm at pH 3.5.[56] *Zygosaccharomyces bailii* can grow in soft drinks containing 500 mg l^{-1} of sorbic acid.[57] Chipley[58] reported strains of *Z. rouxii* and *Z. bailii* with minimum inhibitory concentrations (MICs) at pH 4.8 of 1,000 and 4,500 mg l^{-1}, respectively. Prior exposure of *Z. bailii* to preservative residues in the factory can easily result in it resisting concentrations 4–5 times the permitted levels.[59]

In addition to yeasts and molds, some bacteria that are able to grow at low pH are also relatively resistant to weak organic acid preservatives and can adapt to become even more resistant. For example, strains of *Gluconobacter oxydans* isolated from spoiled fruit had an MIC of 1,000 ppm for sorbic acid at pH 3.8 and 500 ppm at pH 3.3. Growth at pH 3.8 in the presence of sorbic acid at 400 ppm brought about a rapid rise in resistance.[60] Sorbate-resistant strains from grapes were hardly inhibited in apple juice at pH 3.4 by sorbic acid at 1,000 ppm.[61]

The levels of resistance of well-adapted yeasts, in particular, are such that legally allowed concentrations of weak acid preservatives are quite unable to prevent spoilage by them. This is of critical importance in the minimal preservation of fruits and vegetables. An especially high degree of hygiene in processing and packaging areas is vital. Otherwise, buildup of phenotypically superresistant strains must be expected.

Sulfite

Analogous to the weak organic acid preservatives, the undissociated form of sulfite, present as hydrated sulfur dioxide ($SO_2 \cdot H_2O$),[62] readily permeates the cell membrane.[63] In addition, there are additional uptake systems. For instance, in *S. cerevisiae,* there is evidence for a high-affinity system for the active uptake of sulfite, which saturates at low concentrations. This activity could be due to the low specificity of other anion uptake systems, such as those for sulfate or phosphate. There are carrier-mediated, energy-dependent uptake processes for sulfite in *Candida utilis,*[64] *Penicillium chrysogenum,*[65] and other *Penicillium* and *Aspergillus* species.[66] Although of no significance in the preservation of minimally processed fruits and vegetables, it has been shown that *Desulfovibrio desulfuricans* possesses a specific sulfite transport system.[67]

The reaction leading to ionization of SO_2 has a pK_a value of 1.86, whereas the value for the reaction leading to production of the bisulfite anion (HSO_3^-) is 7.18.[68] Consequently, at the pH value of minimally preserved fruits and vegetables, the predominant species present will be bisulfite, generally above 90% of the total present, with hydrated sulfur dioxide, generally at less than 10%.

As with the organic acid preservatives, when molecular SO_2 reaches the cell cytoplasm, it will re-equilibrate at the higher pH to generate protons, bisulfite, and some sulfite. In this way, SO_2 is removed, and it is the bisulfite and sulfite that are accumulated intracellularly and exert their antimicrobial effects, at least until the intracellular pH can no longer be maintained and acidification of the cytoplasm occurs. The intracellular reactions undergone by sulfite, which is a very reactive compound, are numerous, interfering with many metabolic processes.[62,69–71] It is unlikely that there is just one specific reaction that is predominant in its effects on diverse microorganisms.

Microbial reactions to sulfur dioxide stress include two distinct strategies: first, inactivation of sulfite before it can reach target molecules within the cell; second, reduction in the rate at which it reaches those targets.[62] As a consequence, the inherent sensitivities of different microorganisms to sulfite vary greatly. Yeasts, such as *Kloeckera opiculata* and *Hansenula anomala*, commonly involved in the spontaneous fermentation of grape musts[72] and apple juice,[73] are much more sensitive than are strains of *S. cerevisiae,*[74] which is generally regarded to be a sulfite-tolerant yeast. As with the organic acid preservatives, *Z. bailii*[75] and *Saccharomycodes ludwigii*[76] are even more tolerant.

A major reaction limiting the toxic effects of sulfite for microorganisms is removal of the anion. In particular, sulfite may be detoxified by reaction with binding compounds, such as acetaldehyde, to overcome inhibition during the lag phase and during subsequent growth.[76] Sulfite-induced production and excretion of acetaldehyde have been known for nearly a century. Intracellular formation of an α-hydroxysulfonate from acetaldehyde prevents conversion of the acetaldehyde into ethanol and, hence, the regeneration of NAD^+. Instead, the regeneration process involves reduction of glyceraldehyde 3-phosphate to glycerol 3-phosphate, which is then dephosphorylated to generate glycerol. Different strains vary in their sulfite-binding ability.[77] The ability to excrete acetaldehyde was shown to correlate with sulfite resistance of strains of *S. cerevisiae, S. ludwigii,*[76] and *Z. bailii.*[78] Sensitive strains are inhibited by about 0.2–1 mg l^{-1}, whereas some of the more resistant strains are able to grow in the presence of about 20 mg l^{-1}. These three yeast strains also differ in their capacity to accumulate sulfite from the medium around them, although the reason for these differences remains uncertain.[76,78]

Low Water Activity

The response of microorganisms to lowered a_w is essentially a response to osmotic stress and is, therefore, often referred to as *osmoregulation* or *osmoadaptation*. The reaction is widespread in animals, plants, and microorganisms but is most developed in microorganisms and, particularly, in the most osmo-tolerant of the yeasts and molds. The key reaction of cells to reduced a_w is the accumulation of low-MW solutes in their cytoplasm at concentrations sufficient to just exceed the osmolality of the surrounding medium. In this way, the cells regain or avoid loss of water by osmosis and maintain the turgor in the cell membrane that is essential for its proper functioning.[11] Because of their central role, research has concentrated on the compatible solutes. These (reviewed by Galinski,[79] Galinski and Truper,[80] Booth et al.,[81] and Gutierrez et al.[82]) share a number of properties. They are small organic molecules, highly soluble and accumulable to high concentrations (i.e., molar and above); they are usually neutral or zwitterionic molecules; the cell membrane exhibits controlled permeability to them, allowing their concentration in the cytoplasmic pool to be determined by the external osmotic pressure; they have little effect on the activity of cytoplasmic enzymes and may protect them from denaturation by salts[83]; they are usually end-product metabolites, rather than intermediaries in biosynthetic pathways, and, therefore, tend to be stable, once formed. They include a wide range of chemical types, including proline, betaine (N,N,N-trimethylglycine), ectoine, small peptides, trehalose, glycerol, sucrose, mannitol, and arabitol, with the amino acids and amino acid derivatives more common in the bacteria and the polyols more common in the yeasts and molds.

Compatible solutes may be synthesized within the cell or transported from the environment, if present, e.g., in a particular food. For example, in *E. coli*, trehalose is synthesized, whereas betaine can be transported from the environment or derived by the oxidation of choline that has been transported from the environment.[84,85] Key elements of the osmoregulatory systems operating in most of the food poisoning bacteria and in some food spoilage microorganisms have been elucidated, e.g., in enteric bacteria,[11,79,82] *L. monocytogenes,*[86–89] *Yersinia enterocolitica,*[90]; *Pseudomonas aeruginosa,*[91] *Staphylococcus aureus,*[92,93] *Bacillus subtilis,*[94–96] *Lactobacillus plantarum,*[97] *Lactococcus lactis,*[98] and *Brevibacterium linens.*[99] For all of these organisms, most foods contain some level of the relevant compatible solutes or suitable precursors that can be metabolized to generate them (Figure 2–1).

Much of the genetic basis of osmoregulation has been elucidated. As with many other stresses (discussed earlier), in some microorganisms, it includes stabilization of the stationary-phase sigma factor (σ^S, RpoS) that controls the expression of many genes important in stress survival.[100] Also, as with many other stress reactions, cross-adaptation to other stresses accompanies the adaptation of some microorganisms to low a_w environments, e.g., in *E. coli.*[101,102]

The reduction in a_w resulting from the addition of sugars to foods such as minimally preserved fruits and vegetables means that microorganisms surviving processing in such foods or contaminating the food after processing will be osmoregulating at some level or other. However, if osmoregulation diverts resources away from normal cell biosynthetic processes, then it is clear that the stressed cells may be more vulnerable to other stresses in combination preservation systems, particularly when the other

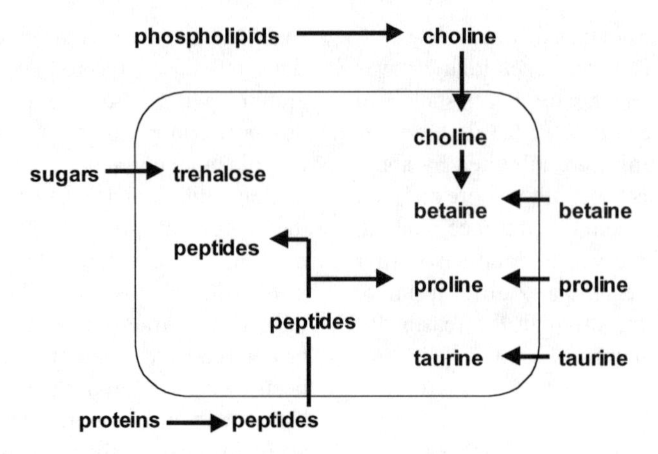

Figure 2–1 Potential food sources of compatible solutes.

stresses also divert away from the synthesis of new cell material.

Heat

The heat shock response is the most studied of the microbial stress reactions. Analogous to RpoS, there is a heat shock sigma factor, RpoH, that is regulated by proteolysis in response to growth phase and stress.[103] RpoH is required for normal growth, e.g., of *E. coli,* but, as the temperature is increased, proteolysis is repressed, and there is a consequent increase in expression of the system. A sudden temperature rise causes transient amplification of the system, prior to it settling back into a new steady state. Among the many changes occurring in heat shocked cells, increased synthesis of chaperone proteins plays a major role. These influence the kinetics of protein folding, so as to minimize denaturation and to assist in the refolding of proteins that may have partly unfolded under the influence of the heat stress.

The practical implications of the heat shock response derive from the fact that it can lead to increases in the resistance of microorganisms to heat and other stresses. Microorganisms exposed to temperatures near to their maxima for growth may exhibit very large increases in heat resistance. Even relatively short exposures of about 30 minutes to 2 hours or so may result in a fewfold up to more than 100-fold increases in heat resistance.[40,104,105] Furthermore, the shapes of survivor curves may change dramatically following heat shock. For example, *S. typhimurium* heated at 57°C without a prior heat shock exhibited exponential inactivation kinetics over at least a 5 log range. However, following preincubation at 48°C for 30 minutes, the inactivation kinetics exhibited a long shoulder before exponential inactivation commenced.[106] The overall result was that, whereas heating unshocked cells at 57°C for 10 minutes achieved at least a 10^6-fold kill, heating shocked cells for the same length of time reduced numbers by less than 10-fold (Figure 2–2).

The importance of the heat shock phenomenon in food processing depends, of course, on the rates of heating that are applied. Generally, if heating rates are rapid, microorganisms will be inactivated before the opportunity for adaptation occurs. However, some commercial processes utilize slow cooking, and, under these circumstances, heat adaptation may be significant because the greatest increases in resistance occur when the temperature rises very slowly, e.g., at a rate of less than about 0.7°C min^{-1} for *S. typhimurium*[104] and *L. monocytogenes.*[107]

Of particular relevance to the minimal preservation of fruits and vegetables is the fact that

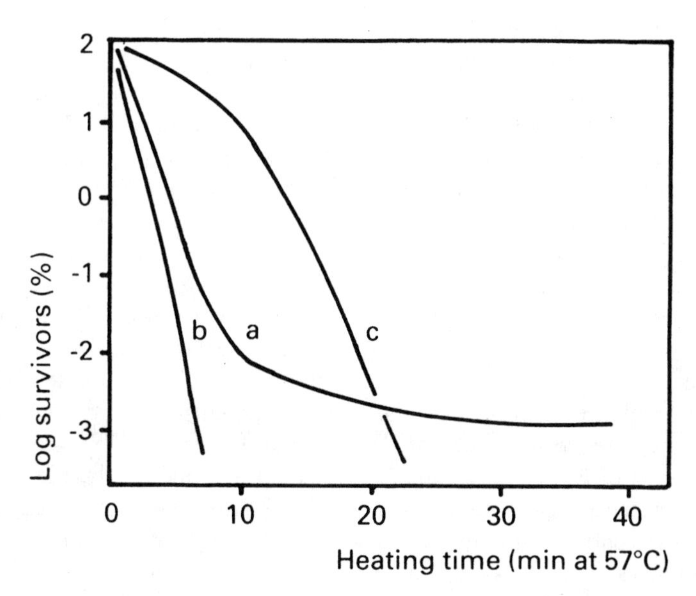

Figure 2–2 Heat-inactivation curves of *Salmonella typhimurium* at 57°C. Curve (a) is grown in trypticase soy broth plus yeast extract (TSYB), washed, and heated in ground beef. Curves (b) and (c) are grown in TSYB and heated in the same medium without prior heat shock (curve b) following incubation at 48°C for 30 minutes (curve c).

heat-shocked microorganisms may acquire increased tolerance to acid. This occurs with spoilage microorganisms and also with pathogens, e.g., with *E. coli* O157:H7.[40] Heating should always be sufficiently rapid, and of a high enough temperature, to minimize the opportunity for this response to occur.

VIRULENCE OF STRESSED MICROORGANISMS

With respect to the control of food poisoning, one aspect of the stress response of microorganisms is of special relevance. This is the enhancement of virulence that can be part of a stress response. For example, it has been shown that RpoS may be a key component of the regulation of the expression of some virulence genes. Some of these are highly expressed in *Salmonella* during entry into the stationary phase of growth,[108–110] so that, bearing in mind the relationship of RpoS and the stationary phase response to many stress responses, it is not surprising that virulence genes are also induced by stress, e.g., in *Salmonella.*[108–112] Osmoregulation genes are involved in the expression of virulence in *S. typhimurium*[113] and in *Shigella flexneri.*[114] The *RpoS* gene is necessary for the expression of heat-stable toxin in stationary phase cells of *Y. enterocolitica.*[115] Stationary-phase cells of *E. coli* synthesize surface fibers ("curli") that help the bacteria to bind to the surfaces of eukaryotic cells. Expression of the *curli* gene is prevented in *RpoS* mutants.[116] The induction of virulence genes at the onset of the stationary phase is also common in gram-positive bacteria, e.g., in *S. aureus*[117] and in *L. monocytogene.*[118]

These examples again emphasize the importance of avoiding the survival of stress-responding microorganisms in mildly preserved foods, e.g., by careful process design and by operating with a high degree of hygiene.

CONCLUSION

A wide range of more or less rapidly acting homeostatic and stress response mechanisms has developed in microorganisms during evolution. These act first to keep key physiologic systems operating and relatively unperturbed, even when the environment around the cell is greatly perturbed. Second, if the environmental stress is severe, they react further in ways that increase resistance to the stress. Unfortunately, many of the milder preservation techniques that have been developed include stresses that elicit these responses. It is, therefore, important that this is taken into account in the design of new preservation procedures.

As a first line of defense against microbial stress responses, processes can be designed, through the application of HACCP techniques, to avoid them; for instance, by avoiding slow heating stages and by ensuring no preincubation of ingredients, premixes, etc., at low pH or at low a_w for times long enough for the development of a stress response. Second, for many processes, a danger is that microorganisms will undergo stress reactions within food processing areas where food debris, ingredients, preservatives, etc., are allowed to contaminate stores, floors, chill rooms, food contact surfaces, etc. Adapted organisms may find their way from these foci into finished products and cause problems in preservation or in safety. Strict hygienic operation is necessary to avoid such buildup of adapted microorganisms in food processing environments. Third, homeostatic and stress responses often require the expenditure of energy by the stressed cells. Restriction of the availability of energy is then a sensible target to pursue. This probably forms the basis of many of the successful, empirically derived, mild combination preservation procedures, exemplified by the "hurdle technology" and "multitarget" preservation approaches of Leistner.[119] For example, if a food can be preserved by lowering the pH, then it is sensible also to include a weak acid preservative, if possible, that will amplify the effect of the protons or to allow a milder, higher pH to be employed. It is sensible also to reduce the a_w as much as is organoleptically acceptable, so that the energy-demanding proton export is made more difficult by the additional requirement of the cells to osmoregulate. Then, if the food can be enclosed in oxygen-free vacuum or modified-atmosphere packaging, facultative anaerobes will be further energy-restricted at a time when the various stress and homeostatic reactions are demanding more energy if growth is to proceed.[120] Fourth, if the processes applied are sufficiently severe to inactivate the microorganisms of concern or to deliver stresses of such magnitude that they cannot be overcome, any problem arising from stress responses will be removed. However, severe processing is not generally compatible with milder preservation, so that the alternative strategies to avoid the problems of the stress response in the minimal processing of fruits and vegetables will, of course, generally be preferred.

REFERENCES

1. Knochel S, Gould GW. Preservation microbiology and safety: Quo vadis? *Trends Food Sci Technol.* 1995;7:91–95.
2. Archer DL. Preservation microbiology and safety: Evidence that stress enhances virulence and triggers adaptive mutations. *Trends Food Sci Technol.* 1996;7:91–95.
3. Gould GW. Methods for preservation and extension of shelf life. *Int J Food Microbiol.* 1996;33:51–64.
4. Gerhardt P. The refractory homeostasis of bacterial spores. In: Whittenbury R, Gould GW, Banks JG, Board RG, eds. *Homeostatic Mechanisms in Microorganisms.* Bath: Bath University Press; 1988:41–49.
5. Hengge-Aronis R. Survival of hunger and stress: the role of *rpoS* in early stationary phase gene regulation. *Cell.* 1993;72:165–168.
6. Loewen PC, Hengge-Aronis R. The role of the sigma factor σ^S (KatF) in bacterial global regulation. *Ann Rev Microbiol.* 1994;48:53–80.
7. Rees CED, Dodd CER, Gibson PT, Booth IR, Stewart GSAB. The significance of bacteria in stationary

phase to food microbiology. *Int J Food Microbiol.* 1995;28:263–275.

8. Schlessinger MJ, Ashburner M, Tissieres A, eds. *Heat Shock: From Bacteria to Man.* New York: Cold Spring Harbor Laboratory; 1992.
9. Demple B, Halbrook J. Inducible repair of oxidative DNA damage in *Escherichia coli. Nature.* 1983;304: 466–468.
10. Ananthan J, Goldberg AL, Voellmy R. Abnormal proteins serve as eukaryotic stress signals and trigger the activation of heat shock genes. *Science.* 1986;232: 522–524.
11. Booth IR. Bacterial responses to osmotic stress: diverse mechanisms to achieve a common goal. In: Reid DS, ed. *Properties of Water in Foods ISOPOW 6.* London: Blackie Academic and Professional; 1998:456–485.
12. Kobayashi H, Suzuki T, Kinoshita N, Unemoto T. Streptococcal cytoplasmic pH is regulated by changes in amount and activity of proton-translocating ATP-ase. *J Biol Chem.* 1986;261:627–630.
13. Zeuthen ML, Dabrowa N, Aniebo CM, Howard DH. Ethanol tolerance and the induction of stress proteins in *Candida albicans. J Gen Microbiol.* 1988;134:1375–1384.
14. Lou Y, Yousef AE. Adaptation to sublethal environmental stresses protect *Listeria monocytogenes* against lethal preservation factors. *Appl Environ Microbiol.* 1997;63:1252–1255.
15. Heiskanen P, Taira S, Rhen M. Role of RpoS in the regulation of *Salmonella* plasmid virulence (*spv*) genes. *FEMS Microbiol Lett.* 1994;123:125–130.
16. Muffler A, Fischer D, Altuvia S. The response regulator RssB controls the σ^S subunit of RNA polymerase. *EMBO J.* 1996;15:1333–1339.
17. Zgurskaya HI, Keyhan M, Matin A. The σ^S level in starving *Escherichia coli* cells increases solely as a result of its increased stability, despite decreased synthesis. *Mol Microbiol.* 1997;24:643–651.
18. Huisman G, Kolter R. Sensing starvation: A homoserine lactone-dependent signalling pathway in *Escherichia coli. Science.* 1994;265:537–539.
19. Aldsworth TG, Dodd CER, Stewart GSAB. Induction of RpoS in *Salmonella typhimurium* by nutrient-poor and depleted media is slower than that achieved by a competitive microflora. *Lett Appl Microbiol.* 1999;28:255–257.
20. Aldsworth TG, Sharman RL, Dodd CER, Stewart GSAB. A competitive microflora increases the resistance of *Salmonella typhimurium* to inimical processes: Evidence for a suicide response. *Appl Environ Microbiol.* 1998;64:1323–1327.
21. Duffy GA, Ellison A, Anderson W, Cole MB, Stewart GSAB. Use of bioluminescence to model the thermal inactivation of *Salmonella typhimurium* in the presence of a competitive microflora. *Appl Environ Microbiol.* 1995;61:3463–3465.
22. Booth IR. Regulation of cytoplasmic pH in bacteria. *Microbiol Rev.* 1995;49:359–378.
23. Booth IR, Kroll RG. The preservation of food by low pH. In: Gould GW, ed. *Mechanisms of Action of Food Preservation Procedures.* Amsterdam: Elsevier Science; 1989:119–160.
24. Hill C, O'Driscoll B, Booth IR. Acid adaptation and food poisoning microorganisms. *Int J Food Microbiol.* 1995;28:245–254.
25. Leyer GJ, Johnson EA. Acid adaptation promotes survival of *Salmonella* spp. in cheese. *Appl Environ Microbiol.* 1992;58:2075–2080.
26. Goodson M, Rowbury RJ. Habituation to normally lethal acidity by prior growth of *Escherichia coli* at a sublethal pH value. *Lett Appl Microbiol.* 1989;8:77–79.
27. Miller LG, Kaspar W. *Escherichia coli* O157:H7. Acid tolerance and survival in apple cider. *J Food Prot.* 1994;57:460–464.
28. Foster JW. Beyond pH homeostasis: the acid tolerance response of salmonellae. *ASM News.* 1993;58:266–270.
29. Kroll RG, Patchett RA. Induced acid tolerance in *Listeria monocytogenes. Lett Appl Microbiol.* 1992;14: 224–227.
30. Karem KL, Foster JW, Bej AK. Adaptive acid tolerance response (ATR) in *Aeromonas hydrophila. Microbiol.* 1994;140:1731–1736.
31. Centers for Disease Control. *Salmonella typhimurium* outbreak traced to apple cider. *MMWR Morb Mortal Wkly Rep.* 1975;24:87–88.
32. Cook KA, Dobbs TE, Hlady WG. Outbreak of *Salmonella* serotype Hartford infections associated with unpasteurized orange juice. *JAMA.* 1997;280:1504–1509.
33. Parish ME. Public health and unpasteurized fruit juices. *Crit Rev Microbiol.* 1997;23:109–119.
34. Besser RE, Lett SM, Weber JT. An outbreak of diarrhea and hemolytic uremic syndrome from *Escherichia coli* O157:H7 in fresh-pressed apple cider. *JAMA.* 1993;269:2217–2220.
35. Centers for Disease Control and Prevention. Outbreak of *Escherichia coli* O157:H7 infections associated with drinking unpasteurized commercial apple juice—British Columbia, California, Colorado and Washington, October 1996. *MMWR Morb Mortal Wkly Rep.* 1996;45:975.
36. Centers for Disease Control and Prevention. Outbreaks of *Escherichia coli* O157:H7 infection and cryptosporidiosis associated with drinking unpasteurized apple cider—Connecticut and New York, October 1996. *MMWR Morb Mortal Wkly Rep.* 1997;46:4–8.
37. Leyer GJ, Johnson EA. Acid adaptation induces cross-protection against environmental stresses in

Salmonella typhimurium. Appl Environ Microbiol. 1993;59:1842–1847.

38. Farber JM, Pagotto F. The effect of acid shock on the heat resistance of *Listeria monocytogenes. Lett Appl Microbiol. 1992;*15:197–201.
39. Rowbury RJ. An assessment of the environmental factors influencing acid tolerance and sensitivity in *Escherichia coli, Salmonella* spp. and other enterobacteria. *Lett Appl Microbiol.* 1995;20:333–337.
40. Wang G, Doyle MP. Heat shock response enhances acid tolerance of *Escherichia coli* O157:H7. *Lett Appl Microbiol.* 1998;26:31–34.
41. Serrano R, Keilland-Brandt MC, Fink GR. Yeast plasma membrane ATPase is essential for growth and has homology with (Na^+ & K^+) and Ca^{2+} ATPases. *Nature.* 1986;319:689–693.
42. Serrano R. Transport across yeast vacuolar and plasma membranes. In: Strathern JN, Jones EW, Broach JR, eds. *The Molecular Biology of the Yeast Saccharomyces. Genome Dynamics, Protein Synthesis, and Energetics.* Cold Spring Harbor, NY: Cold Spring Harbor Laboratory Press; 1991:523–585.
43. Cimprich P, Slavik J, Kotyk A. Distribution of individual cytoplasmic pH values in a population of the yeast *Saccharomyces cerevisiae. FEMS Microbiol Lett.* 1995;130:245–252.
44. Imai T, Ohno T. Measurement of yeast intracellular pH by image processing and the change it undergoes during growth phases. *J Biotechnol.* 1995;38:165–172.
45. Salmond CV, Kroll RG, Booth IR. The effect of food preservatives on pH homeostasis in *Escherichia coli. J Gen Microbiol.* 1984;130:2845–2850.
46. Eklund T. The effect of sorbic acid and esters of p-hydroxybenzoic acid on the proton-motive force in *Escherichia coli* membrane vesicles. *J Gen Microbiol.* 1985;131:73–76.
47. Cole MB, Keenan MHJ. Effects of weak acids and external pH on the intracellular pH of *Zygosaccharomyces bailii,* and its implications in weak acid resistance. *Yeast.* 1987;3:23–32.
48. Bracey D, Holyoak CD, Coote PJ. Comparison of the effect of sorbic acid and amphotericin B on *Saccharomyces cerevisiae*: Is growth inhibition dependent on reduced intracellular pH? *J Appl Microbiol.* 1998;85:1056–1066.
49. Lambert RJ. Stratford M. Weak-acid preservatives: modelling microbial inhibition and response. *J Appl Microbiol.* 1999;86:157–167.
50. Stratford M, Anslow PA. Evidence that sorbic acid does not inhibit yeast as a classic "weak acid preservative." *Lett Appl Microbiol.* 1998;27:203–206.
51. Warth AD. Mechanism of resistance of *Saccharomyces bailii* to benzoic, sorbic and other weak acids used as food preservatives. *J Appl Bacteriol.* 1977;43:215–230.
52. Warth AD. Transport of benzoic and propionic acids by *Zygosaccharomyces bailii. J Gen Microbiology.* 1989;135:1383–1390.
53. Henriques M, Quintas C, Louveiro-Dias M. Extrusion of benzoic acid in *Saccharomyces cerevisiae* by an energy-dependent mechanism. *Microbiology.* 1997;143: 1877–1883.
54. Piper PW, Ortiz-Calderon C, Holyoak C, Coote PJ, Cole MB. Hsp 30, the integral plasma membrane heat shock protein of *Saccharomyces cerevisiae*, is a stress-inducible regulator of plasma membrane H^+ATPase. *Cell Stress Chap.* 1997;2:12–24.
55. Eklund T. Organic acids and esters. In: Gould GW, ed. *Mechanisms of Action of Food Preservation Procedures.* Amsterdam: Elsevier Science; 1989:161–200.
56. Warth AD. Effect of benzoic acid on growth yield of yeasts differing in their resistance to preservatives. *Appl Environ Microbiol.* 1988;54:2091–2095.
57. Neves L, Pampulha ME, Loureiro-Dias MC. Resistance of food yeasts to sorbic acid. *Lett Appl Microbiol.* 1994;19:8–11.
58. Chipley JR. Sodium benzoate and benzoic acid. In: Davidson PM, Branen AL, eds. *Antimicrobials in Foods.* New York: Marcel Dekker; 1993:11–48.
59. Stratford M, Hoffman PD, Cole MB. Fruit juices, fruit drinks, and soft drinks. In: Lund BM, Baird-Parker AC, Gould GW, eds. *The Microbiological Safety and Quality of Food.* Gaithersburg, MD: Aspen Publishers; 2000: 836–869.
60. Eyles MJ, Warth AD. The response of *Gluconobacter oxydans* to sorbic and benzoic acids. *Int J Food Microbiol.* 1989;8:335–342.
61. Splittstoesser DF, Churney JJ. The incidence of sorbic acid resistant gluconobacters and yeasts on grapes grown in New York State. *Am J Enol Vitic.* 1994;43:290–293.
62. Rose AH, Pilkington BJ. Sulphite. In: Gould GW, ed. *Mechanisms of Action of Food Preservation Procedures.* Amsterdam: Elsevier Science; 1989:201–223.
63. Stratford M, Rose AH. Transport of sulphur dioxide by *Saccharomyces cerevisiae. J Gen Microbiol.* 1986; 132:1–6.
64. Garcia M, Benitez J, Delgado J, Kotyk A. Isolation of sulphate transport defective mutants of *Candida utilis*: Further evidence for a common transport system for sulphate, sulphite and thiosulphate. *Folia Microbiol.* 1983;28:1–5.
65. Yamamoto LA, Segel JH. The inorganic sulphate transport system of *Penicillium chrysogenum. Arch Biochem Biophys.* 1966;114:523–531.
66. Tweedie JW, Segel IW. Specificity of transport processes for sulfur, selenium and molybdenum ions by filamentous fungi. *Biochim Biophys Acta.* 1970;196:95–106.

67. Cypionka H. Uptake of sulfate, sulfite and thiosulfate by proton-anion symport in *Desulfovibrio desulfuricans*. *Arch Microbiol.* 1987;148:144–149.
68. Wedzicha BLC. *Chemistry of Sulphur Dioxide in Foods*. Amsterdam: Elsevier Science; 1984.
69. Gould GW, Russell NJ. Sulphite. In: Russell NJ, Gould GW, eds. *Food Preservatives*. Glasgow: Blackie Academic and Professional; 1991:72–88.
70. Ough CS. Sulfur dioxide and sulfites. In: Davidson MP, Branen AL, eds. *Antimicrobials in Foods*. New York: Marcel Dekker; 1993:137–190.
71. Gould GW. The use of other chemical preservatives: sulfite and nitrite. In: Lund BM, Baird-Parker AC, Gould GW, eds. *The Microbiological Safety and Quality of Foods*. Gaithersburg, MD: Aspen Publishers; 2000: 200–213.
72. Kunkee RE, Goswell RW. Table wines. In: Rose AH, ed. *Economic Microbiology,* Vol. 1. London: Academic Press: 1977:315–386.
73. Beech FW, Carr JG. Cider and perry. In: Rose AH, ed. *Economic Microbiology*. London: Academic Press; 1977:139–313.
74. Warth AD. Resistance of yeast species to benzoic and sorbic acids and to sulfur dioxide. *J Food Protect.* 1985;48:564–569.
75. Thomas DS, Davenport RR. *Zygosaccharomyces bailii*: A profile of characteristics and spoilage activities. *Food Microbiol.* 1985;2:157–169.
76. Stratford M, Morgan P, Rose AH. Sulphur dioxide resistance in *Saccharomyces cerevisiae* and *Saccharomycodes ludwigii*. *J Gen Microbiol.* 1987;133:2173–2179.
77. Rankine BC, Pocock KF. Influence of yeast strain on binding of sulphur dioxide in wines and on its formation during fermentation. *J Sci Food Agric.* 1969;20:104–109.
78. Pilkington BJ, Rose AH. Reactions of *Saccharomyces cerevisiae* and *Zygosaccharomyces bailii* to sulphite. *J Gen Microbiol.* 1988;134:2823–2830.
79. Galinski EA. Compatible solutes of halophilic eubacteria—molecular principles, water-solute interaction, stress protection. *Experient.* 1993;49:487–496.
80. Galinski EA, Truper HG. Microbial behaviour in salt-stressed ecosystems. *FEMS Microbiol Rev.* 1994;15: 95–108.
81. Booth IR, Pourkomalian BP, McLaggan D, Koo SP. Mechanism controlling compatible solute accumulation: A consideration of the genetics and physiology of bacterial osmoregulation. In: Fito P, Mulet A, McKennan B, eds. *Water in Foods*. London: Elsevier Science;1994:381–388.
82. Gutierrez C, Abee T, Booth IR. Physiology of the osmotic stress response in microorganisms. *Int J Food Microbiol.* 1995;28:233–244.
83. Brown AD. Microbial water stress. *Bacteriol Rev.* 1976;40:803–846.
84. Andersen PA, Kaasen I, Styrvold O, Boulnois G, Strom AR. Molecular cloning, physical mapping and expression of *bet* genes governing the osmo-regulatory choline-glycine-betaine pathway of *Escherichia coli*. *J Gen Microbiol.* 1988;134:1737–1746.
85. Giaever HM, Styrvold OB, Kaasen I, Strom AR. Biochemical and genetic characterization of osmoregulatory trehalose synthesis in *Escherichia coli*. *J Bacteriol.* 1988;170:2841–2849.
86. Patchett RA, Kelly AF, Kroll RG. Effect of sodium chloride on the intracellular pools of *Listeria monocytogenes*. *Appl Environ Microbiol.* 1992;58:3959–3963.
87. Beumer RR, te Giffel MC, Cox LJ, Rombouts FM, Abee T. Effect of exogenous proline, betaine, and carnitine on growth of *Listeria monocytogenes* in a minimal medium. *Appl Environ Microbiol.* 1994;60:1359–1363.
88. Amezega M-R, Davidson I, McLaggan D, Verheul A, Abee T, Booth IR. The role of peptide metabolism in the growth of *Listeria monocytogenes* at high osmolarity. *Microbiology.* 1995;141:41–49.
89. Verheul A, Rombouts FM, Beumer RR, Abee T. A novel, ATP-dependent L-carnitine transporter in *Listeria monocytogenes* Scott A is involved in osmoregulation. *J Bacteriol.* 1995;177:3205–3212.
90. Park S, Smith LT, Smith GM. Role of glycine betaine and related osmolytes in osmotic stress adaptation in *Yersinia enterocolitica* ATCC 9610. *Appl Environ Microbiol.* 1995;61:4378–4381.
91. Dsouzaalt MR, Smith LT, Smith GM. Roles of N-acetylglutaminylglutamine amide and glycine betaine in adaptation of *Pseudomonas aeruginosa* to osmotic stress. *Appl Environ Microbiol.* 1993;59:473–478.
92. Armstrong-Buisseret L, Cole MB, Stewart GSAB. A homologue to the *Escherichia coli* alkyl hydroperoxide reductase AhpC is induced by osmotic upshock in *Staphylococcus aureus*. *Microbiology.* 1995;141:1655–1661.
93. Pourkomalian B, Booth IR. Glycine betaine transport by *Staphylococcus aureus*: Evidence for feedback regulation of the activity of two transport systems. *Microbiology.* 1994;140:3131–3138.
94. Boch J, Kempf B, Bremer E. Osmoregulation in *Bacillus subtilis*: synthesis of the osmoprotectant glycine betaine from exogenously provided choline. *J Bacteriol.* 1994;176:5364–5371.
95. Kempf B, Bremer E. OpuA, an osmotically regulated binding protein-dependent transport system for the osmoprotectant glycine betaine in *Bacillus subtilis*. *J Biol Chem.* 1995;270:16701–16713.
96. Deuerling E, Paeslack B, Schumann W. The *ftsH* gene of *Bacillus subtilis* is transiently induced after osmotic and temperature upshift. *J Bacteriol.* 1995;177:4105–4112.

97. Glaasker E, Konings WN, Poolman B. Osmotic regulation of intracellular solute pools in *Lactobacillus plantarum*. *J Bacteriol*. 1996;178:575–582.

98. Molenaar D, Hagting A, Alkena H, Dreissen AJM, Konings WN. Characteristics and osmoregulatory roles of uptake systems for proline and glycine betaine in *Lactococcus lactis*. *J Bacteriol*. 1993;175:5438–5444.

99. Bernard T, Jebbar M, Rassouli Y, Himdi-Kabbab S, Hammelin J, Blanco C. Ectoine accumulation and osmotic regulation in *Brevibacterium linens*. *J Gen Microbiol*. 1993;139:129–136.

100. Hengge-Aronis R. Back to log phase: Sigma (σ) as a global regulator in the osmotic control of gene expression in *Escherichia coli*. *Mol Microbiol*. 1996;21:887–893.

101. Jenkins DE, Chaisson SA, Matin A. Starvation-induced cross-protection against osmotic challenge in *Escherichia coli*. *J Bacteriol*. 1990;172:2779–2781.

102. Makin G, Lapidot A. Induction of tetrahydropyrimidine derivatives in *Streptomyces* strains and their effect on *Escherichia coli* in response to osmotic stress and heat stress. *J Bacteriol*. 1996;178:385–395.

103. Tomayasu T, Gamer J, Bukau B, Kanemori M, Mori H, Rutman AJ. *Escherichia coli* FtsH is a membrane-bound, ATP-dependent protease which degrades the heat-shock transcription factor σ^{32}. *EMBO J*. 1995;14:2551–2560.

104. Mackey BM, Derrick CM. Changes in the heat resistance of *Salmonella typhimurium* during heating at rising temperatures. *Lett Appl Microbiol*. 1987; 4:13–16.

105. Farber JM, Brown BE. Effect of prior heat shock on heat resistance of *Listeria monocytogenes* in meat. *Appl Environ Microbiol*. 1990;56:1584–1587.

106. Mackey BM, Derrick CM. Elevation of the heat resistance of *Salmonella typhimurium* by sublethal heat shock. *J Appl Bacteriol*. 1986;61:389–393.

107. Stephens PJ, Cole MB, Jones MV. Effect of heating rate on the thermal inactivation of *Listeria monocytogenes*. *J Appl Bacteriol*. 1994;77:702–708.

108. Fang FC, Kraume M, Roudier C. Growth regulation of a *Salmonella* plasmid gene essential for virulence. *J Bacteriol*. 1991;173:6783–6789.

109. Krause M, Fang FC, Guiney DG. Regulation of plasmid virulence gene expression in *Salmonella dublin* involves an unusual operon structure. *J Bacteriol*. 1992;174:4482–4489.

110. Norel F, Robbe-Saule V, Popoff MY, Coynault C. The putative sigma factor KatF (RpoS) is required for the transcription of the *Salmonella typhimurium* virulence gene *spvB* in *Escherichia coli*. *FEMS Microbiol Lett*. 1992;99:271–276.

111. Spink JM, Pullinger GD, Wood MW, Lax AJ. Regulation of *spvR*, the positive regulatory gene of *Salmonella* plasmid virulence genes. *FEMS Microbiol Lett*. 1994;116:113–121.

112. Valone SE, Chikami GK, Miller VL. Stress induction of the virulence proteins (SpvA, SpvB and SpvC) from native plasmid pSD12 of *Salmonella dublin*. *Infect Immunol*. 1993;61:705–713.

113. Foster JW, Spector MP. How salmonellae survive against the odds. *Ann Rev Microbiol*. 1995;49: 145–174.

114. Mekalanos JJ. Environmental signals controlling expression of virulence determinants in bacteria. *J Bacteriol*. 1992;174:1–7.

115. Iriate M, Stanier I, Cornelis GR. The rpoS gene from *Yersinia enterocolitica* and its influence on expression of virulence factors. *Infect Immunol*. 1995;63:1840–1847.

116. Hengge-Aronis R, Lang R, Henneberg N, Fischer D. Osmotic regulation of *rpoS*-dependent genes in *Escherichia coli*. *J Bacteriol*. 1993;175:259–265.

117. Sheehan BJ, Foster TJ, Dorman CJ. Osmotic and growth phase dependent regulation of the *eta* gene of *Staphylococcus aureus*: A role for DNA supercoiling. *Mol Gen Genet*. 1992;232:49–57.

118. Sokolovic Z, Geobel W. Synthesis of listeriolysin in *Listeria monocytogenes* under heat shock conditions. *Infect Immunol*. 1989;57:295–298.

119. Leistner L. Combined methods for food preservation. In: Rahman S, ed. *Handbook of Food Preservation*. New York: Marcel Dekker; 1999:457–485.

120. Gould GW. Homeostatic mechanisms during food preservation by combined methods. In: Barbosa-Canovas GV, Welti-Chanes J, eds. *Food Preservation by Moisture Control: Fundamentals and Applications*. Lancaster, PA: Technomic; 1995:397–410.

Chapter 3

Microbial Ecology of Spoilage and Pathogenic Flora Associated to Fruits and Vegetables

Amaury Martínez, Rosa V. Díaz, and María S. Tapia

INTRODUCTION

Virtually all fruits and vegetables in their natural state are susceptible to spoilage by microorganisms at a rate that depends on various intrinsic and extrinsic factors. Salunkhe[1] estimates that one-fourth of all produce harvested is not consumed before spoilage. Spoilage of fresh fruits and vegetables usually occurs during storage, transport, and while waiting to be processed. The preservation of plant materials is now achieved by drying, salting, fermentation, freezing, refrigeration, canning, and irradiation. In certain instances, two or more processes may be combined. However, consumers are increasingly demanding convenient, ready-to-use and ready-to-eat fruits and vegetables with freshlike quality and containing only natural ingredients.[2]

In general, the pH of plant tissues is in the range of 5–7, which is very favorable for the growth of numerous microbial species. Microflora in fruits is different from that of vegetables. Many fruits also possess natural defense mechanisms, such as thick skin, or antimicrobial substances, such as essential oils.[3] Fruits, in contrast to vegetables, have a good record from a public health standpoint. Fruits contain organic acids in sufficient quantities for getting a pH value of 4.6 or lower. However, certain fruits have a higher pH (watermelons, cucumbers, bananas). The low pH and type of acid itself are the major influences that select for the predominant microflora of fruits.[3]

CONTAMINATION AND SPOILAGE

As soon as fruits and vegetables are gathered into boxes, lugs, baskets, or trucks during harvesting, they are subject to cross-contamination with spoilage organisms from other fruits and vegetables and from containers. Mechanical damage may increase susceptibility to decay and growth of microorganisms. Some operations, such as washing, can reduce the microbial load; however, they may also help to distribute spoilage microorganisms and moisten surfaces enough to permit growth of microorganisms during holding periods.[4]

Authors gratefully acknowledge support of Consejo Nacional de Investigaciones Científicas y Tecnológicas de Venezuela (CONICIT), Project S1–2722 "Incidencia, comportamiento y patogenicidad de *Listeria monocytogenes* y *Aeromonas hydrophila* en productos de origen vegetal"; Consejo de Desarrollo Científico y Humanístico de la Universidad Central de Venezuela, "Incidencia y Sobrevivencia de *E. coli* O157:H7 en productos de origen animal y vegetal"; CYTED Program Project XI-3 "Development of minimal processing techniques for food preservation"; and European Commission Project TS3*-CT94–0333(DG HSMU) "Development of preservation techniques for tropical fruits using vacuum impregnation techniques."

The means by which microbes penetrate the tissues have not been clearly established. External surfaces of produce in the field will bear the heaviest microbial load, but inner tissues will, on occasion, contain certain bacteria that, by unknown mechanisms, have gained entry. Plant tissues themselves, even having been invaded, represent a well-defined ecosystem, and different preservation methods should be used to guarantee the stability and safety of the food. Additionally, decontamination procedures to reduce the initial microbial load of raw material and methods to avoid postprocess contamination change the original microflora of the ecosystem, affecting the kind of microbial association that can be established after that.[5] Microbial association is specific for each type of food that has been invaded, and it is affected by intrinsic and extrinsic factors that play an important ecologic role in the establishment of the saprophytic microflora and in the colonization of food by pathogens.

Most of the organisms present in fresh vegetables are saprophytes such as coryniforms, lactic acid bacteria, spore-formers, coliforms, micrococci, and pseudomonads derived from the soil, air and water. Pseudomonads and the group of *Klebsiella-Enterobacter-Serratia* from the Enterobacteriaceae are the most frequent. Fungi, including *Aureobasidium*, *Fusarium,* and *Alternaria,* are often present but in relatively lower numbers than are bacteria. Because of the acidity of raw fruits, the primary spoilage organisms are fungi, predominantly molds.

Psychrothrophs are able to grow in vegetable products; some of them are *Erwinia carotovora, Pseudomonas fluorescens*, *P. aeruginosa, P. luteola, Bacillus* species, *Cytophaga johnsonae*, *Xanthomonas campestri*, and *Vibrio fluvialis.*[6–8] The pectinolitic species cause soft tissue. Approximately 10–20% of isolates among the mesophilic bacteria of shredded lettuce were found to be pectinolytic.[9] A high proportion of pseudomonads (20–60%) were also found to be pectinolytic in many samples of shredded carrots and shredded chicory salads.[10] Pectinolytic isolates were usually identified as *P. fluorescens, P. paucimobilis, P. viridiflora, P. luteola, Xanthomonas malthopila*, *Flavobacterium* species, *Cytophaga* species, *V. fluvialis.*[10–13] Some pectinolytic molds and yeasts have been isolated from shredded carrots.[12] Despite the fact that *Enterobacter* species and some lactic acid bacteria have not been associated with diarrhea and are considered unpathogenic, some of them have been associated with human disease causing infections of nosocomial origin.

PATHOGENIC FLORA

Demand for fresh, minimally processed vegetables has led to an increase in the quantity and variety of products available to the consumer. The ready-to-use vegetables retain much of their indigenous microflora after minimal processing. Pathogens may form part of this microflora, posing a potential safety problem.[14]

Listeria monocytogenes

The potential for contamination of fruits and vegetables is high because of the large variety of conditions to which produce is exposed during growth, harvest, processing, and distribution. These considerations actually acquire great significance through the new processing techniques that offer attributes of convenience and freshlikeness in response to changes in consumption patterns and increased demand of fresh and minimally processed fruits and vegetables. Thus, reliance on low-temperature storage and on improved packaging materials techniques has increased.

Listeria monocytogenes is widely distributed on plant vegetation.[15] Plants and plant parts used as vegetable salads play a key role in disseminating the pathogen from natural habits to the human food supply. Confirmed outbreaks of human listeriosis in North America in recent years have incriminated foods from both plant and animal origin as causative vehicles. Schlech et al.[16] identified coleslaw as the probable vehicle of transmission of *L. monocytogenes* that caused 7 adult and 34 perinatal cases of infection in Canada in 1981. Ho et al.[17] suggested that consumption of raw celery, tomatoes, and lettuce might be linked to listeriosis in hospitalized, immunosuppressed patients. *Listeria monocytogenes* septicemia associated with consumption of salted mushrooms[18] and alfalfa tablets[19] has also been reported. The information concerning the incidence of listeriae in raw vegetables marketed in Europe and also in Latin America is of relatively recent origin, and the knowledge is restricted to a few reports.

Table 3–1[20–38] shows a list of the incidence of *L. monocytogenes* and *Listeria* species, and Table 3–2[22,23,39–52] presents the growth and survival of *L. monocytogenes* in fresh and minimally processed fruits and vegetables, both with emphasis on data from Venezuela.

Of particular concern is that the organism is psychrotrophic, capable of growing at refrigeration temperatures.[53] It is also facultative anaerobic, capable of survival/growth under the low O_2 concentrations within modified-atmosphere packages.[42] *Listeria monocytogenes* has been isolated from intact vegetables, such as asparagus,[20] cabbage, cucumbers, lettuce and lettuce juice, potatoes, radishes, mushrooms,[25,27] tomatoes,[27] bean sprouts and leafy vegetables,[29] raw celery,[32] and other unprocessed vegetables.[26,31]

Sizmur and Walker[22] evaluated the incidence of *L. monocytogenes* on salads and found that two raw salad types contained *L. monocytogenes*. One salad consisted of cabbage, celery, raisins, onions, and carrots, and the other consisted of lettuce, cucumbers, radishes, fennel, watercress, and leeks. The population increased twofold when salads were held at 4°C for 4 days. It is clear from these data that there is a risk for health, because *L. monocytogenes* has also been isolated from a wide range of other ready-to-eat vegetables.[23,24,28–32] Because the primary vegetable in coleslaw was and still is the only vegetable that has been directly linked to an actual outbreak of listeriosis in humans, it is not surprising that growth and survival of *L. monocytogenes* on cabbage have been investigated. Beuchat et al.[39] determined the behavior of *L. monocytogenes* on inoculated samples of shredded, raw, and autoclaved cabbage and found that *L. monocytogenes* presented similar behavior patterns with a reduction of 3 log CFU/g during 42 days of refrigerated storage.

The ability of *L. monocytogenes* to survive and grow on other vegetables has also been demonstrated. *Listeria monocytogenes* can survive and grow at refrigeration temperatures on many raw or processed vegetables, such as ready-to-eat fresh salad vegetables, including cabbage, celery, raisins, onions, carrot salad, lettuce, cucumber, radish, fennel, watercress, leek salad,[22] asparagus, broccoli, cauliflower,[36] lettuce, lettuce juice, minimally processed lettuce,[23,41,42] butterhead lettuce salad, broad-leaved and curly-leaved endive,[43] minimally processed fresh endive,[44–46] freshly peeled Hamlin orange,[47] and vacuum-packaged prepeeled potatoes.[48]

Listeria monocytogenes supported elevated salt concentrations over a wide pH range.[54] The antibacterial effect of a natural component has also been demonstrated. Beuchat and Brackett[55] studied the survival and growth of *L. monocytogenes* on raw and cooked carrots and found an anti-*Listeria* effect of carrots, which was eliminated when carrots were cooked. In others studies,[56–59] it has been observed that carrot juice has a lethal effect on *L. monocytogenes*. Inhibitory effects of raw tomatoes[60] and processed lamb lettuce have been proven. A variety of naturally occurring plant components also has antilisterial effect; for example, diacetyl, benzaldehyde, pyruvic aldehyde and piperonal,[61] cinnamic acid[62] and allicin, or diallyl thiosulphinic acid.[63]

Information regarding incidence and fate of *L. monocytogenes* in fresh and processed fruits and vegetables in Latin America is limited. Raybaudi[34] did not find *L. monocytogenes* on cabbage and celery harvested in Venezuela but found a relatively high incidence of *Listeria* species. *Listeria grayi* was the main *Listeria* species identified. The fate of *L. monocytogenes* (4 log CFU/g) on celery and cabbage showed that, during the first 7 days of storage under refrigeration, the counts of *L. monocytogenes* remained at the same levels, and after 26 days

Table 3–1 Incidence of *Listeria monocytogenes* and/or *Listeria* spp. in Fresh and Minimally Processed Fruits and Vegetables

Product	*Frequency (%)*	*Country*	*References*
Asparagus	0	Canada	20
Lettuce, celery, tomatoes, radish, mushroom, cabbage	0		21
Ready-to-eat fresh salad vegetables, including cabbage, celery, raisins, onions, and carrot salad; lettuce, cucumber, radish, fennel, watercress, and leek salad	6.7	London, UK	22
Lettuce salad	4.06	US	23
Freshly cut vegetables	44	Netherlands	24
Cabbage	2.2		
Cucumbers and lettuce juice	10.9		
Lettuce and lettuce juice	1.1		
Potatoes	25.8		
Radishes	30.3		
Mushrooms	12	US	25
Vegetables	12.2	Taiwan	26
Tomato	13.3		
Cucumber	6.7	Pakistan	27
Salads vegetables and prepared salads	10.9	Ireland	28
Bean sprouts	85		
Sliced cucumbers	80		
Leafy vegetables	22.7	Malaysia	29
Coleslaw/vegetable salad	2	Singapore	30
Ready-to-eat packaged salad and other fresh vegetables	0	London, UK	31
Ready-to-use vegetables: salad mix, coleslaw mix, green peppers, broccoli florets, and chopped lettuce	2,8		
Unprocessed raw celery	1,4	Canada	32
Cucumber	11.7		
Bell pepper	5		
Apple	60		
Grape	20		
Prune	25		
Tomato	25	Venezuela*	33
Cabbage	60		
Celery	47	Venezuela*	34
Minimally processed fruits	12.5	Venezuela*	35
Minimally processed salads with and w/o dressing	9.4	Venezuela†	36
American salad	37.5		
European salad	16.6	Venezuela*	37
Escarole	29	Venezuela*	38

**L. grayi; L. welshimerii; L. murrayi; L. ivanovii.*
†L. monocytogenes, L. grayi; L. welshimerii; L. murrayi; L. ivanovi.

Table 3–2 Growth and Survival of *Listeria monocytogenes* in Fresh and Minimally Processed Fruits and Vegetables

Product	*Reference*
Cabbage and cabbage juice	39
Ready-to-eat fresh salad vegetables, including cabbage, celery, raisins, onions, and carrot salad; lettuce, cucumber, radish, fennel, watercress, and leek salad	22
Asparagus, broccoli, and cauliflower	40
Lettuce, lettuce juice, and minimally processed lettuce	23, 41, 42
Green salads; Butterhead lettuce salad, broad-leaved endive, and curly-leaved endive	43
Minimally processed fresh endive	44, 45, 46
Freshly peeled Hamlin orange	47
Vacuum-packaged prepeeled potatoes	48
Cabbage, celery	49
American salad, European salad	50
Minimally processed salads	51
Minimally processed fruits	52

of storage, a reduction of the microorganism count was observed. Diaz et al.[49] determined that celery and cabbage did not have an inhibitory effect per se upon *L. monocytogenes.*[49]

The incidence of *Listeria* species in minimally processed salads with or without dressing was studied by Brión.[36] The pH of samples ranged from 4.47 to 5.12, and *Listeria* species were found in levels of 10^4 CFU/g. Salads with dressing had lower levels. A significant relationship between pH and *Listeria* species count was found. Three of the samples were positive for *L. monocytogenes*. Other *Listeria* species isolated were *L. welshimeri* and *L. ivanovii*. The results show the ability of *L. monocytogenes* to survive at low pH. Donnelly[64] reported that *Listeria* can grow at pH between 4.3 and 9.6. The survival of *L. monocytogenes* in both types of salads also was studied[51] by challenge tests at 5°C for 8 days. In salads with dressing, the levels of *L. monocytogenes* (10^4 CFU/g) did not change during storage. The same fate was observed in minimally processed salads without dressing. A significant decrease of pH during storage was observed in both cases (Figure 3–1). These results suggest that minimally processed salads could be a potential risk for public health.

The incidence *L. monocytogenes* in American style and European style salads has been evaluated by Díaz et al.[37] The American style salads were made of lettuce, watercress, carrots, and cabbage, whereas European style salads were prepared with different varieties of lettuce, watercress, and escarole. *Listeria grayi* and *L. welshimeri* were the specimens most frequently isolated. American style salad was challenged with *L. monocytogenes* and stored at 5°C for 15 days. Results showed that the organism was able to grow in this type of salad.[50] A reduction of 1 log CFU/g was observed at day 15. During that period, *Pseudomonas* species increased 3 log CFU/g, but a significant correlation between both microorganisms was not observed.

The incidence of *L. monocytogenes* on minimally processed fruit salads prepared with papaya, apple, watermelon, grape, guava, and pineapple was also determined. The main species of *Listeria* isolated were *L. innocua* and *L. grayi.*[35] A reduction of 0.5 log CFU/g of *L. monocytogenes* at the sixth day of refrigerated storage (5°C) was reported by Mejía and Díaz[52] when studying the fate of the organism in these products. However, the initial inoculated level of *L. monocytogenes* was reached at the eighth day of storage. No significant correlation was observed between *L. monocytogenes* and pH and natural microflora variations (Figure 3–2).

Data from literature indicate that *L. monocytogenes* and *Listeria* species have been isolated from products such as raw fish, smoked fish, vegetables, and fruits minimally processed that were supposed to undergo listericidal processes. This suggests postprocessing recontamination. In recent years, evidence has helped to raise interest in evaluating sanitizers traditionally used in the food industry. Chlorine has been widely used to treat drinking water, as well as a sanitizer of processing environment and equipment.

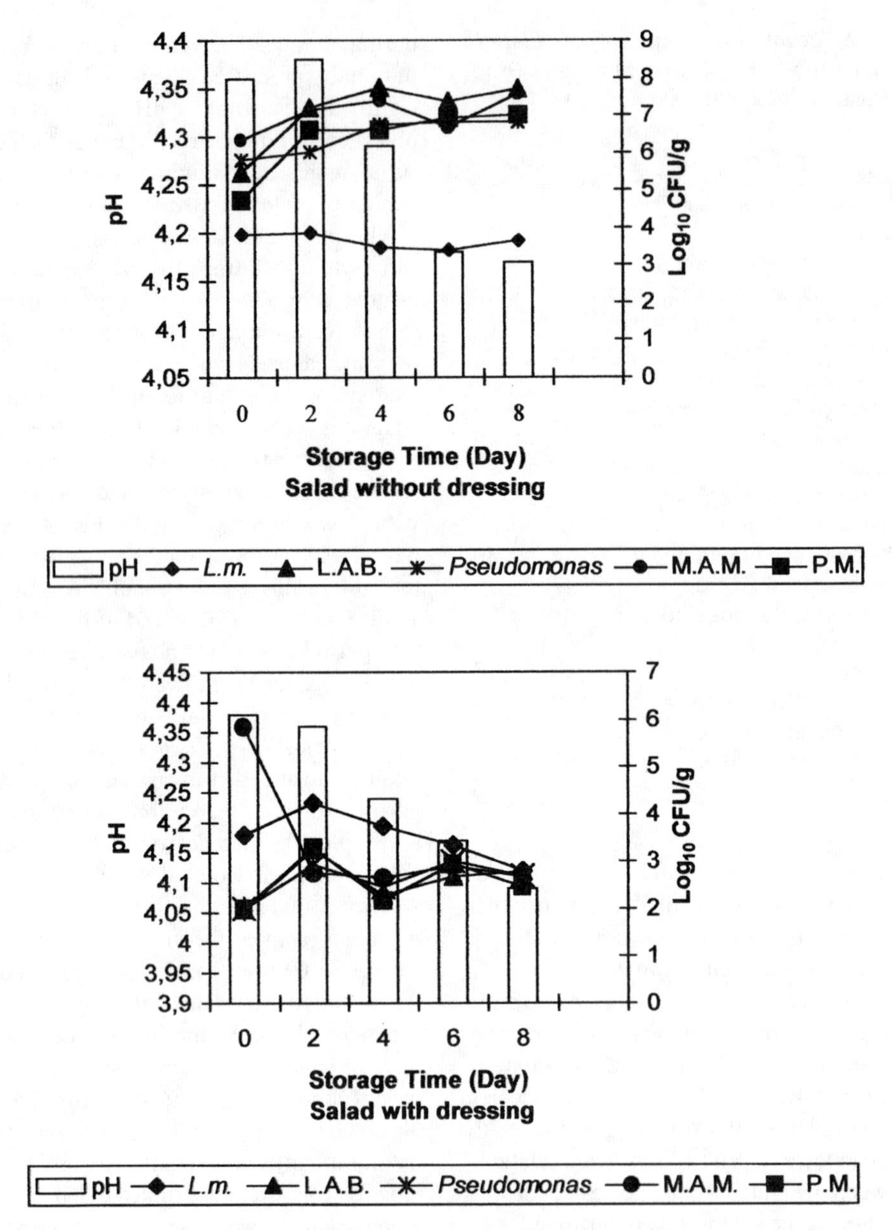

Figure 3–1 Behavior of *Listeria monocytogenes (L.m.)*, lactic acid bacteria (L.A.B.), *Pseudomonas*, mesophilic aerobic microorganisms (M.A.M.), psychrotrophic microorganisms (P.M.), and pH in ready-to-eat minimally processed salads held at 5°C.

The effect in vitro of sodium hypochlorite on *L. monocytogenes* has been studied by Brackett,[65] who evaluated the effect of chlorine on brussel sprouts contaminated with *L. monocytogenes*, reducing the counts approximately 100-fold, 10-fold more than those treated with water. Reduction of 1.3–1.7 log CFU/g and 0.9–1.2 log CFU/g on shredded lettuce and cabbage treated

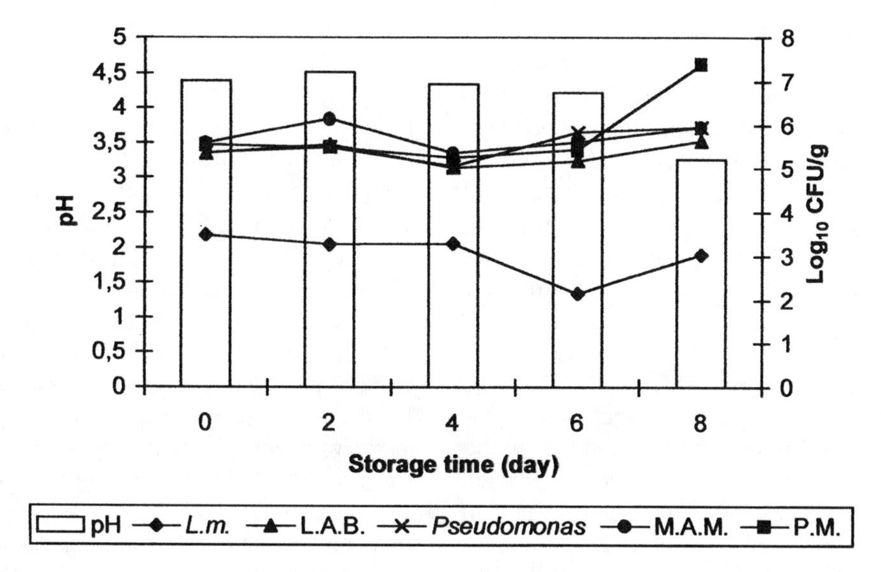

Figure 3–2 Behavior of *Listeria monocytogenes (L.m.)*, lactic acid bacteria (L.A.B.), *Pseudomonas*, mesophilic aerobic microorganisms (M.A.M.), psychrotrophic microorganisms (P.M.), and pH in minimally processed fruit salads held at 5°C.

with 200 ppm of chlorine for 10 minutes was reported by Zhang and Farber.[66] Beuchat and Brackett[41,60] found that disinfection of salad leaves and tomato slices with chlorine prior to inoculation with *L. monocytogenes* did not affect its growth. The elimination of *L. monocytogenes* from the surface of vegetables by chlorine is limited and unpredictable.[67] *Listeria monocytogenes* inoculated on endive leaves disinfected with hydrogen peroxide grew better than on water-rinsed produce.[45]

Like chlorine, iodine compounds are widely used as sanitizers for food processing equipment and surfaces. However, the iodophors are more effective as vibriocidal agents than are the same concentrations of chlorine. Trisodium phosphate (TSP) has been approved by U.S. Food and Drug Administration to reduce populations of *Salmonella* and other microorganisms in poultry.[68] The use of TSP to remove *L. monocytogenes* in vegetables is less promising because it has been observed that *L. monocytogenes* is resistant to TSP.[69] The effects of lactic and acetic acid, either alone or in combination with chlorine, on survival of *L. monocytogenes* on shredded lettuce were studied by Zhang and Farber.[66] Lactic acid (0.75% or 1%), in combination with 100 ppm chlorine, was more effective in reducing levels of *L. monocytogenes* than were either lactic acid or chlorine alone. Beuchat[70] points out that sanitizers under investigation for their efficacy in killing pathogens on produce include chlorine dioxide, organic acids, trisodium phosphate, and ozone. The use of ozone to inactivate microorganisms on lettuce shows that efficient ozone delivery to microorganisms on lettuce requires a combination of ozone bubbling and high-speed stir.[71] To date, none of these chemicals has been shown to eliminate pathogenic bacteria completely from produce without adversely affecting sensory quality.[70]

Escherichia coli O157:H7

Escherichia coli O157:H7 has emerged as a highly significant food-borne pathogen. The principal reservoir of *E. coli* O157:H7 is believed to be the bovine gastrointestinal tract.[70]

Thus, contamination of associated food products with feces is a significant risk factor, particularly if untreated contaminated water is consumed directly or is used to wash uncooked foods.[42]

Abdul-Raouf et al.[72] studied the effect of the storage temperature, use of modified atmosphere, and storage time on the survival and/or death of *E. coli* O157:H7 in vegetable salads prepared with cucumbers, shredded carrots, and lettuce. *E. coli* O157:H7 decreased rapidly and remained only a short time in salads stored at 5°C, whereas salads stored at 12 and 21°C showed an increase in the population of the pathogen. The authors pointed out the inhibitory effect of carrot juice on *L. monocytogenes,* but this effect was not significant on *E. coli* O157:H7, and they suggested that 6-metoxy-mellein, naturally present on carrots, inhibits the growth of some bacteria and molds and, therefore, can have an inhibitory effect on *E. coli* O157:H7. The atmospheric composition did not affect the survival of *E. coli* O157:H7 in the conditions of their experiment. The rapid growth of psychrophiles and mesophilic microorganisms did not have an important effect on the growth rate of the pathogen.

In 1993, Zhao et al.[73] showed that unpasteurized apple cider was an appropriate substrate for growing *E. coli* O157:H7. The microorganism was able to survive from 10 to 31 days at 8°C. The use of 0.1% potassium sorbate or 0.1% sodium benzoate alone had a minimal effect on *E. coli* O157:H7. The combination of 0.1% of sodium benzoate and 0.1% of potassium sorbate significantly inhibited the growth of *E. coli* O157:H7. The survival and growth of *E. coli* O157:H7 in cantaloupe and watermelon have been studied,[74] and *E. coli* O157:H7 was observed in both types of fruit, reaching levels of 6.8 $\log_{10}$ CFU/g in melon and 8.51 $\log_{10}$ CFU/g in watermelon stored at 25°C, but the number of viable cells did not suffer changes at 5°C until 34 hours. The microorganism was able to survive in the peel of both fruits when they were stored at 25° or 5°C.[74]

Diaz and Hotchkiss[75] compared the microbiologic spoilage with survival of *E. coli* O157:H7 in shredded lettuce stored in modified atmosphere and observed that growth rates of the aerobic plate count and *E. coli* O157:H7 were higher in air at 22°C. Temperature and level of CO_2 had no significant effect on the growth of *E. coli* O157:H7. The extended shelf life provided by the modified atmosphere allowed *E. coli* O157:H7 to grow to higher numbers, compared with air-held shredded lettuce. Similar results have been reported by other authors.[76,77]

Challenge tests in freshly peeled Hamlim[47] orange inoculated with selected pathogenic bacteria (*Salmonella* species, *E. coli* O157:H7, *L. monocytogenes,* and *Staphylococcus aureus*) to study the survival and growth of these microorganisms showed that *E. coli* O157:H7 counts remained constant throughout storage at refrigeration temperatures (4 and 8°C). Growth was observed with all tested pathogens only when stored at abuse temperature (24°C).

Itoh et al.[78] emphasized the importance of using seeds free from *E. coli* O157:H7 in the production of radish sprouts because, according to their results, they found *E. coli* O157:H7 not only in the outer surface but also in the inner tissues and stomata of cotyledons of radish sprouts grown from seed experimentally contaminated with the pathogen.

The tolerance of *E. coli* O157:H7 to acid pH has been widely studied.[79,80] The pH can play an important role in the survival of *E. coli* O157:H7 in vegetables. Weagant et al.,[80] working with salads dressed with mayonnaise and acidified with different acids, acetic, citric, and lactic acid, and stored at 5, 21 or 30°C for 72 hours, found that the storage temperature played an important role in the effectiveness of the acids: At low temperature (5°C), all acids had the same effectiveness. At 5°C, the population of *E. coli* O157:H7 was significantly reduced during the first 4 hours, and the mortality of the microorganisms was higher when the pH of the media decreased.

The tolerance of some strains of *E. coli* O157:H7 has been studied in yogurt,[81] apple cider,[73] and mayonnaise.[80,82,83] Abdul-Raouf et al.[84] reported the acid tolerance and survival characteristics of *E. coli* O157:H7 in ground and roasted beef. Cutter and Siragusa[85] evaluated the effects of organic acids to control the presence of *E. coli* O157:H7 in meat, revealing that the type of acid is not an important factor in the efficiency of the treatment and that the concentration, type of tissue, and microbial strains are elements that influence the reduction of the microbial population.[85] Levels of 5% of lactic, acetic, or citric acid were more effective to reduce *E. coli* O157:H7 and *P. fluorescens*. Levels of citric, acetic, and lactic acid lower than 1.5% were not effective in the reduction of *E. coli* O157:H7 in meat, according to Brackett et al.[86] Recently, it has been shown that *E. coli* O157:H7 was able to survive more than 35 days in salads with mayonnaise-based dressing stored at 5°C.[80] The survival and acid tolerance of *E. coli* O157:H7 in acid foods such as mayonnaise and cider can be due to a preexistent mechanism inside the cell responsible for the resistance to the toxic effects of the acids. Also, it has been shown that *E. coli* O157:H7 produces mucoid colonies with layers of exopolysaccharides, which could also explain the tolerance to low acidity.[87] The survival and growth characteristics of acid-adapted, acid-shocked, and control cells of *E. coli* O157:H7 inoculated into tryptic soy broth acidified with acetic acid and lactic acid in apple cider and orange juice indicate that the pathogen shows an extraordinary tolerance to the low pH of apple cider and orange juice held at 5 or 25°C for up to 42 days.[88] Deng et al.[89] pointed out that tolerance of acid-adapted and control cells on subsequent exposure to low pH is influenced by the type of acidulant. They suggested that when performing acid challenge studies to determine survival and growth characteristics of *E. coli* O157:H7 in foods, consideration should be given to the type of acid to which the cells have been previously exposed, the procedure used to achieve acidic environments, and possible differences in response among strains.

The incidence of *E. coli* O157:H7 in some fresh vegetables purchased in local markets in Venezuela was studied by León.[90] The vegetables studied were lettuce, watercress, celery, parsley, cabbage, carrot, radish, and coriander, and different culture media were evaluated. The microorganism was not detected in the 37 samples analyzed. Carrot was challenged with the pathogen and stored for 15 days at 5°C.[91] Inoculated carrots exhibited an increase in the population of *E. coli* O157:H7 during the first two days of storage, maintaining stable levels for two more days, and slowly declining during the rest of the storage period, even if the levels were higher than the original inoculum (Figure 3–3). Variations in the mesophilic aerobic microorganism, psychrotrophic microorganism levels, and pH did not affect the growth of *E. coli* O157:H7. This suggests that, even if carrots contain bactericidal substances, it is an adequate substrate for *E. coli* O157:H7. This is supported also by results of Abdul-Raouf et al.,[72] who did not observe a significant effect of carrots on *E. coli* O157:H7. However, naturally occurring plant components, such as diacetyl, benzaldehyde, pyruvic aldehyde, and piperonal,[61] and α-methoxyphenol and o-ethylphenol[92] have also been demonstrated to be effective inhibitors of *E. coli* O157:H7 or of production of verotoxin.

Aeromonas hydrophila

Aeromonas hydrophila has characteristics of concern in vegetables. It is a psychrotrophic and facultative anaerobe.[42,64] *Aeromonas hydrophila* is considered to be an ubiquitous microorganism, and it has been isolated from many sources. The best known sources are treated and untreated water, and animals associated with water, such as fish and shellfish.[93] *Aeromonas* strains are susceptible to disinfectants, including chlorine, although recovery of *Aeromonas* from chlorinated water supplies has been reported. This could be the result of recontamination, unavailability of chlorine by organic matter, the presence of a high initial number of cells, or possibly the persistence of the organism in a viable but not cultivable state following treatment.[94,95] However, there have been surprisingly few studies on inactivation. The response of *Aeromonas*

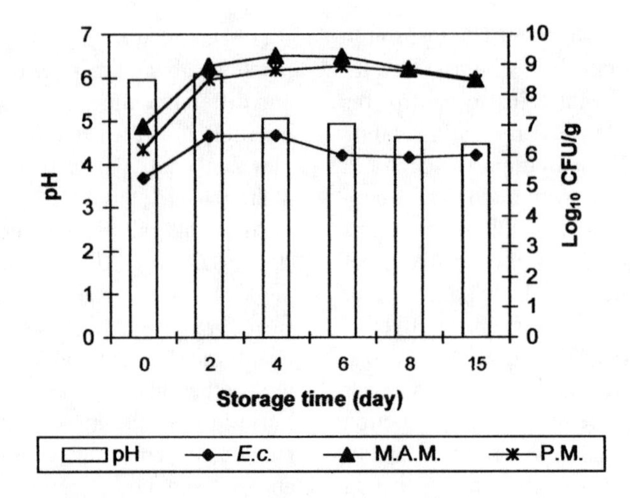

Figure 3–3 Behavior of *Escherichia coli* O157:H7 (*E.c.*), mesophilic aerobic microorganisms (M.A.M.), psychrotrophic microorganisms (P.M.), and pH in shredded carrot held at 5°C.

species to sodium hypochlorite, iodoform, and 2-chlorophenol revealed the sensitivity of these organisms to disinfectants. Treatments with 5 ppm hypochlorite inactivated *Aeromonas* cells in 1 minute at 25°C.[42,93] Recently, Massa et al.[96] studied the behavior of various strains of *A. hydrophila* after exposure to a range of chlorine concentrations equivalent to that found in tap water in most distribution systems. These authors found that certain strains were resistant to the usual chlorine concentration used for purified drinking water (0.1–0.3 mg/L) passing into public water supplies.

Aeromonas hydrophila was identified as the possible cause of an outbreak of gastroenteritis due to the consumption of raw oysters.[97] However, to date, the implications of *A. hydrophila* in a food-borne related outbreak have not been completely demonstrated. More recently, Kirov[95] reported that *Aeromonas* species were implicated in an outbreak in Switzerland due to the consumption of a dish with shrimp, smoked sausages, liver pâté, and cooked ham. In Venezuela, there do not exist reports of gastroenteritis outbreaks by *A. hydrophila,* possibly due to the fact that the microorganism is not evaluated in feces of the patients.[98]

One of the most common problems when studying the incidence of *Aeromonas* species in vegetable products is that the majority of *Aeromonas* species isolated are not enteropathogenic, such as *A. caviae*. The pathogenic species are mainly *A. veronii* sobria and *A. hydrophila* serotypes 0:11 and 0:34, which are in very low numbers.[99] Table 3–3 shows the occurrence of *Aeromonas* species on fresh and minimally processed fruits and vegetables, with emphasis on data from Venezuela.

Callister and Agger[100] evaluated the incidence of *Aeromonas* species in vegetables such as parsley, spinach, celery, alfalfa sprouts, lettuce, broccoli, cauliflower, escarole, and endive. Levels of *Aeromonas* species ranged from 1.00×10^2 to 2.30×10^4 CFU/g at the moment of purchase. Levels of *Aeromonas* reached more than 10^5 CFU/g after 14 days of storage at 5°C. *Aeromonas hydrophila* has been found in fresh asparagus, broccoli, and cauliflower,[101] and in commercial vegetable salads.[102] *Aeromonas* species were also recovered from commercial mixed vegetable salads.[103]

Incidence of *Aeromonas* species in vegetable products was studied in Venezuela by Díaz et al.[104] The products analyzed were lettuce, water-

Table 3–3 Occurrence of *Aeromonas* spp. on Fresh and Minimally Processed Fruits and Vegetables

Product	*Total Count Log CFU/g (Time of Purchase)*	*Recovery after Enrichment (%)* or on Day 7 of Storage under Refrigeration (Log CFU/g)*	*Reference*
Alfalfa sprouts	4.36	3.89	
Celery	3.55	2.74	
Red-leaf lettuce	3.20	2.96	
Green-leaf lettuce	3.86	ND	
Spinach	3.70	3.07	
Escarole	ND	3.99	
Broccoli	ND	2.74	100
Parsley	ND	5.30	
Endive	ND	ND	
Romaine	2.00	3.21	
Boston lettuce	ND	ND	
Kale	ND	2.30	
Asparagus	ND	5.0	
Broccoli	2.00	5,0	101
Cauliflower	2.00	ND	
Red and green chicory	3.79–5.52		
Carrot	4.40–6.07		102
Mixed salad	3.87–5.74	NR	
Commercial mixed salads	NR	NR	103
Watercress	4.79–6.40	6.61–7.83	
Escarole	2.65–4.87	2.60–5.30	
Lettuce	ND–2.54	ND–2.85	104
Parsley	2.90–4.77	2.60–5.30	
Cabbage	ND–2.40	ND	
Pepper	ND	ND[a]	
Tomato	ND	ND[a]	
Cucumber	ND	7.4[a]	105
Apple	ND	ND[a]	
Plum	ND	ND[a]	
Grape	ND	3.7[a]	
Celery	2.30 B 6.35	93.3[a]	34
Cabbage	4.94 B 6.92	33.3[a]	
Salads with dressing	NDB>6.48	100[a]	36
European style salad	ND–4.30	100[a]	
American style salad	ND–4.17	100[a]	
Sprout beans	5.67	100[a]	
Watercress	5.26	100[a]	
Alfalfa sprouts	ND	100[a]	
"Radichio"	ND	100[a]	
Minimally processed fruit salads	ND B3.17	100[a]	107
Purple-leaf cabbage	ND–4.40		
White-leaf cabbage	ND–3.20		
Carrot	ND–4.00	90[a]	108
Escarole	ND–3.30		
Lettuce	ND–3.00		

*ND, $<1.00 \times 10^2$; NR, Not reported.

cress, cabbage, escarole, and parsley. *Aeromonas* species were detected in all products analyzed at least by one of the two methods evaluated (direct count on starch ampicillin agar [SAA] and most probable number [MPN] in trypticase soy broth with subsequent isolation on SAA). The highest incidence of *Aeromonas* organisms was detected in watercress, followed by parsley, escarole, lettuce, and cabbage. At purchase, the levels of *Aeromonas* species ranged from $< 1.00 \times 10^2$ CFU/g and < 3.00 MPN/g to 2.50×10^6 CFU/g and 4.60×10^6 MPN/g. After 7 days of storage at 3°C, the levels of *Aeromonas* spp. increased 10–100 times, with 6.75×10^7 CFU/g and 4.60×10^7 MPN/g as the lowest levels detected. The main species identified were *A. hydrophila* (20.5%), *A. caviae* (43.6%), and *A. sobria* (2.6%). In this study, *Klebsiella oxytoca* (5.1%) was also isolated, showing very similar characteristics to *Aeromonas* species on SAA plates.

Ortiz[105] analyzed the incidence of *Aeromonas* at purchase (day 0) and after 24 hours of enrichment in trypticase soy broth (TSB) in grapes, apples, plums, bell peppers, tomatoes, and cucumbers, obtaining a 3.7% of recovery of *Aeromonas* in grapes and 7.4% in cucumbers. An incidence of 93.3% and 33.3% of *Aeromonas* in celery and cabbage at purchase has been reported.[34] Levels of *Aeromonas* ranged from 2.00×10^2 to 3.40×10^6 CFU/g in celery and 8.90×10^4 to 8.30×10^6 CFU/g in cabbage. The main species isolated were identified as *A. caviae* (80%), *A. hydrophila* (10%), and 10% could not be assigned to any of the three *Aeromonas* species.

Minimally processed vegetables have also been studied for the incidence of *Aeromonas*. Brión[36] evaluated salads expended in the metropolitan area of Caracas (Venezuela). In all salads analyzed using an enrichment step, *Aeromonas* species were detected. The counts ranged from $< 1.00 \times 10^2$ to $> 3.00 \times 10^6$ CFU/g. The highest incidence was detected in salads without dressing, possibly because the pH was higher than in salads with dressing. In this study, 82.6% of isolates were *A. hydrophila*, 8.7% *A. caviae,* and 8.7% could not be identified.

Minimally processed ready-to-eat salads (European style salads and American style salads) were analyzed by Rodríguez[106] for incidence of *Aeromonas*. Rodríguez[106] also analyzed samples of "radichio," watercress, bean sprouts, and alfalfa sprouts. *Aeromonas* species were detected in all samples analyzed using an enrichment step. Bean sprouts had the highest levels of *Aeromonas* (5.67 log CFU/g). A relationship between incidence of *Aeromonas* and pH of samples was not found. In this study, 29.9% of isolates were identified as *A. hydrophila*, 41.1% as *A. caviae,* and 10.5% as *A. sobria*. Low levels (> 2.00–3.18 log CFU/g)) of *Aeromonas* species in minimally processed fruit salads have been reported,[107] using an enrichment step in their detection. However, 62.5% of the strains isolated were identified as *A. hydrophila,* using either one multitest API 20E or traditional biochemical tests. The results suggest that fruit salads are a potential source of *A. hydrophila* and a casual agent of disease.

Incidence of *Aeromonas* in vegetable products was evaluated by Méndez.[108] The products analyzed were purple-leaf-cabbage, white-leaf-cabbage, carrot, escarole, and lettuce. In 45% of the salads analyzed, presumptive *Aeromonas* species were detected, with levels ranging from nondetected (ND) to 2.5×10^4 CFU/g. However, using an enrichment step for 24 hours, the qualitative recovery of *Aeromonas* species was 90%. Biochemical traditional tests used to identify the presumptive *Aeromonas* species revealed that 65% of the isolates were identified as *A. hydrophila*, 8.69% as *A. caviae,* 8.69% as *A. salmonicida,* and 17.4% could not be included in the genus. Using a multitest system (API 20E), 65.2% of the isolates were *A. hydrophila,* and the remaining 34.75% were dubious.

The potential virulence of *Aeromonas* strains isolated from vegetable samples (lettuce, watercress, and escarole), their adhesive and invasive abilities, and in vivo and in vitro production of thermolabile enterotoxins suggest that *Aeromonas* species isolated from vegetables may represent a risk to consumers' health.[109] Several studies have also demonstrated that *A. hydro-*

phila can grow and survive in vegetable products. Table 3–4 shows the growth and survival of *A. hydrophila* in fruits and vegetables, with special emphasis on data from Venezuela.

Berrang et al.[101] studied the effect of controlled-atmosphere storage (CAS) on the growth and survival of *A. hydrophila* in fresh asparagus, broccoli, and cauliflower. In all cases, the storage under controlled atmosphere extended the shelf life of the vegetable. However, the process did not affect the population of *A. hydrophila*. This behavior was also observed in celery and cabbage stored under refrigeration.[49] Ready-to-eat salads made with shredded carrots and red and green achicoria and stored for one week at 5°C were analyzed by Marchetti et al.,[102] who found unusually high levels of *A. hydrophila* in the salads at the moment of purchase. However, at the third day of storage, the carrot samples did not show the presence of *A. hydrophila*, suggesting that high levels of lactic acid bacteria, together with a decrease of pH, could be the cause of such as effect. Levels of 10^6 cells/g of *A. hydrophila* in chicory salads were detected at end of storage, but the microbial growth of yeast and lactic acid bacteria was low. The authors suggested that intrinsic characteristics of the products are very important to support the deterioration by *Aeromonas* species. One of these intrinsic factors is the possible interaction with other microorganisms that can play a synergistic or antagonistic effect. Marchetti et al.,[102] using clarified carrot juice and inoculating it with *A. hydrophila* and *P. fluorescens,* observed a decrease in the *A. hydrophila* population, whereas levels of *P. fluorescens* increased during the experiment. García-Gimeno et al.[103] evaluated the behavior of *A. hydrophila* in commercial mixed salads made with lettuce, red cabbage, and carrot and stored under modified atmosphere at 4 and 15°C. They detected levels of 10^8 CFU/g of *A. hydrophila* the first 24 hours in the salads stored at 15°C, and a reduction in the population was observed on further storage. Apparently, the low pH values, along with the high levels of CO_2, were responsible for this reduction. The low values of pH could be due to high levels of lactic acid bacteria, together with a high concentration of CO_2. The authors pointed out that ingredients can play an important role in the composition of the initial flora, also influenced by the evolution of the gases inside the package as consequence of the different respiratory rate and pH. The modified atmosphere extended the shelf life of the product but permitted the growth and survival of potentially dangerous pathogens.

Salads (white cabbage and carrots) with mayonnaise, vinegar, and sugar and salads without dressing were inoculated with 10^3–10^4 CFU/g of *A. hydrophila* and stored at 5°C, showing significant differences ($p < 0.05$) between both types of salads.[110] An increase of 3 log of *A. hydrophila* at 8 days of storage at 5°C was observed in salads without dressing (Figure 3–4). The population of mesophilic aerobic microorganisms increased in both types of salads, and levels of *A. hydrophila* were not affected by psychrotrophic microorganisms, lactic acid bacteria, and *Pseudomonas*. However, in salads with dressing, the levels of *A. hydrophila* remained constant during 2 days at 5°C, then the population of the pathogen decreased 3 log at the eighth day of storage. Significant changes were not observed in the pH values during storage. Moreno[111] did not observe an inhibitory effect

Table 3–4 Growth and Survival of *Aeromonas hydrophila* in Fresh and Minimally Processed Fruits and Vegetables

Product	*Reference*
Asparagus, broccoli, cauliflower under controlled atmosphere	101
Red and green chicory, carrot, mixed salad	102
Commercial mixed salads: lettuce, red cabbage, and carrot	103
Celery, cabbage	49
Minimally processed salads with and w/o dressing	110
Minimally processed fruits	114
American salad, European salad	108

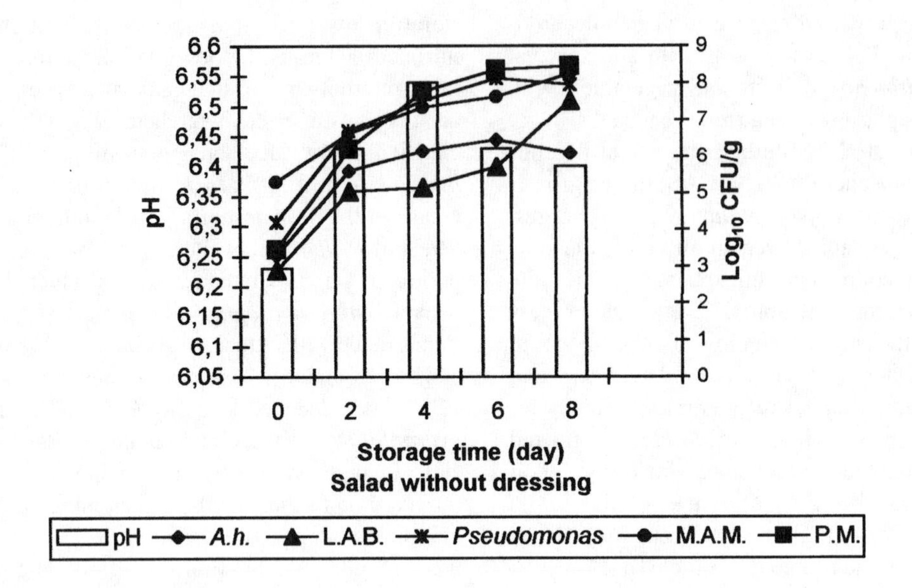

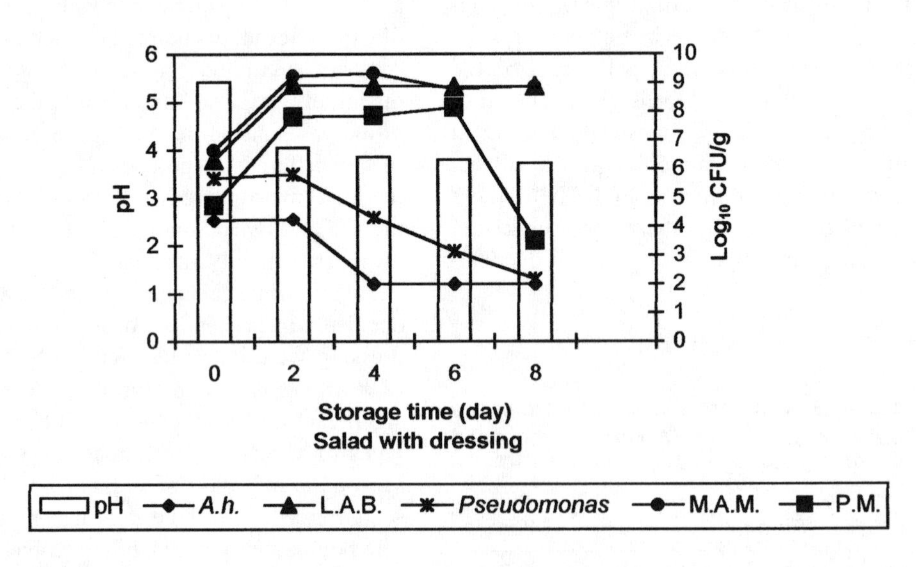

Figure 3–4 Behavior of *Aeromonas hydrophila (A.h.)*, lactic acid bacteria (L.A.B.), *Pseudomonas,* mesophilic aerobic microorganisms (M.A.M.), psychrotrophic microorganisms (P.M.), and pH in ready-to-eat minimally processed salads held at 5°C.

of carrots on *A. hydrophila*, a fact that was supported by data of Kabara in 1987 (cited by Babic et al.[112]), which found that carrot extract is more effective against gram-positive than against gram-negative bacteria, such as *A. hydrophila*. A high reduction of *A. hydrophila* growing in TSB with different levels of cabbage juice was observed by Raybaudi.[34] However, levels of 10% and 50% of cabbage juice significantly reduced the population of *A. hydrophila* at 24 and 7 hours of incubation, respectively.[111] However, naturally occurring plant components, such as diacetyl, benzaldehyde, pyruvic aldehyde, piperonal,[61] and basil methyl chavicol,[113] are bactericidal to *A. hydrophila.*

Challenge studies[114] inoculating *A. hydrophila* (10^3 CFU/g) in minimally processed fruit salads showed that *A. hydrophila* was able to grow at 5°C during the first 6 days, then the pathogen count decreased after 8 days of storage (Figure 3–5). Levels of mesophilic aerobic microorganisms, psychrotrophic microorganisms, lactic acid bacteria, and *Pseudomonas* increased on the order of 1 to 5 log. A significant correlation between these microorganisms and *A. hydrophila* was not observed. These results suggest that the pathogen can grow and survive under refrigeration on fruit salads when it is still in very low numbers at the moment of storage.[114] The fate of *A. hydrophila* under refrigeration at 5°C during 15 days in minimally processed salads showed that the growth and survival of the microorganism were progressive, and the population was significantly higher at the end of the storage period. Correlation between the growth of *A. hydrophila* and psychrotrophic microorganisms and molds was observed during storage but not between *A. hydrophila* and lactic acid bacteria. The pH showed a significant correlation with levels of *A. hydrophila.*[108]

CONCLUSION

Human diseases associated with the consumption of raw fruits and vegetables used to occur more frequently in less industrialized countries, where raw fruits and vegetables, as well as foods of animal origin, have always been known as vehicles for transmission of infectious microorganisms. The number of confirmed cases of illness associated with raw fruits and vegetables in industrialized countries had been traditionally relatively low. During the past decade, this situation has changed in developed countries, and this risk has increased markedly with the advent and popularity of minimally processed fruits and vegetables. Agronomic practices, handling, processing, distribution, and marketing have ex-

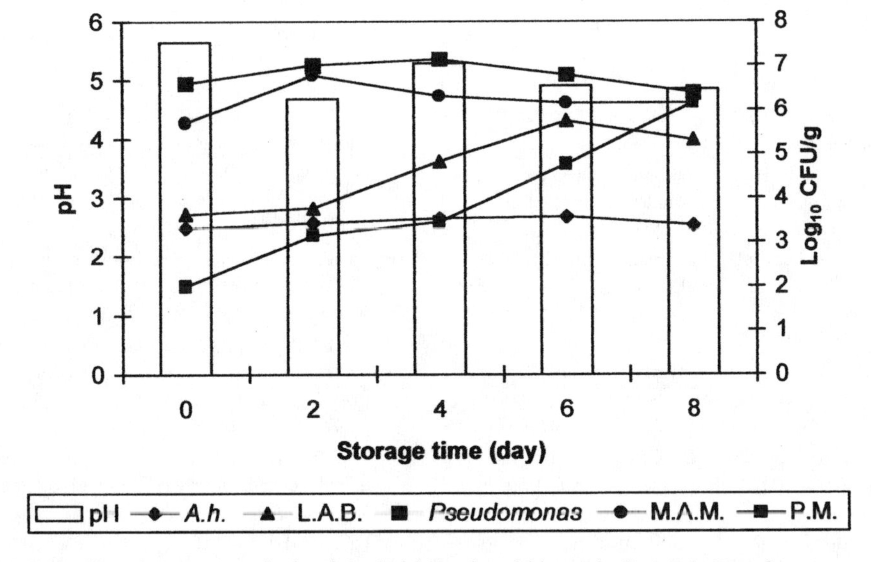

Figure 3–5 Behavior of *Aeromonas hydrophila (A.h.)*, lactic acid bacteria (L.A.B.), *Pseudomonas*, mesophilic aerobic microorganisms (M.A.M.), psychrotrophic microorganisms (P.M.), and pH in minimally processed fruit salads held at 5°C.

erted an important effect on risks associated with microbiologic hazards of raw and minimally processed fruits and vegetables.

Incidence of pathogenic organisms such as *L. monocytogenes*, *A. hydrophila,* and *E. coli* O157:H7 can be low in products of plant origin and in minimally processed vegetables and fruits (even if the latter cannot be considered good sources of these organisms). However, all of these products can be considered potential substrates that can support growth with the corresponding risks for public health, especially knowing the psychrotrophic nature of the organisms that allows their competitive growth at refrigeration temperatures. These products are kept at refrigeration temperatures that can serve as enrichment conditions for the organisms. A huge amount of information, mostly generated by traditional challenge testing, has demonstrated that these pathogens can survive on minimally processed fruits and vegetables, even if they may not proliferate during the time of normal refrigerated storage—probably due to low pH values along with the presence of possible inhibitory compounds. The levels of the initial inoculum are maintained through the period of storage. In other cases, competitive growth occurs speedily. The development of an international epidemiologic surveillance system for better understanding the role of raw and minimally processed fruits and vegetables as vehicles for diseases is very important.

REFERENCES

1. Shalunkhe DK. Developments in technology of storage and handling of fresh fruits and vegetables. *Crit Rev Food Technol.* April 1974; 15–54.
2. Lund DB. Food processing; from art to engineering. *Food Technol.* 1989;43:242–247.
3. Goepfert JM. Vegetables, fruits, nuts, and their products. In: Silliker JH, Elliot RP, Baird-Parker AC, et al., eds. *Microbial Ecology of Foods: Food Commodities*, Vol. II. New York: Academic Press, Inc.; 1980:606–642.
4. Frazier WC, Westhoff DC, eds. *Food Microbiology.* New York: McGraw-Hill, Inc.; 1978:194–217.
5. Gould GW. Ecosystem approaches to food preservation. *J Appl Bacteriol.* 1992;73:585–685.
6. King AD, Bolin HR. Physiological and microbiological storage stability of minimally processed fruits and vegetables. *Food Technol.* 1989;43:132–135.
7. Barriga MI, Tracay G, Willemont C, Simard RE. Microbial changes in shredded iceberg lettuce stored under controlled atmosphere. *J Food Sci.* 1991;56: 1586–1588,1599.
8. Liao CH, Wells JM. Diversity of pectolytic, fluorescent pseudomonas causing soft rots of fresh vegetables at produce market. *Phytopathology.* 1987;77:673–677.
9. Magnusson JA, King AD, Torok T. Micoflora of partially processed lettuce. *Appl Environ Microbiol.* 1990;56:3851–3854.
10. Nguyen-the C, Prunier JP. Involvement of pseudomonas in the deterioration of "ready-to-use" salads. *Int J Food Sci Technol.*1989;24:47–58.
11. Babic L, Hilbert G, Nguyen-the C, Guiraud J. The yeast flora of store ready-to-use carrots and their role in spoilage. *Int J Food Sci Technol.*1992;27:473.
12. Carlin F, Nguyen-the C, Cudennet P, Reich M. Microbiological spoilage of fresh ready-to-use grated carrots. *Sci Alim.*1989;9:371–384.
13. Denis C, Picoche B. Microbiologíe des légumes frais prédécoupés. *Ind Agr Alim.*1986;103:547.
14. Francis GA., Thomas C, O'Beirne D. The microbiological safety of minimally processed vegetables. *Int J Food Sci Technol.*1999;34:1–22.
15. Beuchat LR, Berrang ME, Brackett RE. Presence and public health implications of *Listeria monocytogenes* on vegetables. In: Miller AJ, Smith JL, Somkuti GA, eds. *Foodborne Listeriosis.* Amsterdam: Elsevier; 1990:175–181.
16. Schlech WF, Lavigne PM, Bortolussi RA, Allen AC, et al. Epidemic listeriosis—evidence for transmission by food. *N Engl J Med.* 1983;308:203–206.
17. Ho JL, Shands KN, Friedland C, Eckind P et al. An outbreak of type 4b *Listeria monocytogenes* infection involving patients from eight Boston hospitals. *Arch Intern Med.* 1986;146:520–524.
18. Juntilla J, Brander M. *Listeria monocytogenes* septicemia associated with consumption of salted mushrooms. *Scand J Infect Dis.* 1989;21:339–342.
19. Farber JM, Carter AO, Varughese PV, Ashton FE, et al. Listeriosis traced to the consumption of alfalfa tablets and soft cheese. *N Engl J Med.* 1990;332, 338.

20. Langlois BE, Bastin S, Akers K, O'Leary I. Microbiological quality of foods produced by an enhanced cook-chill system in a hospital. *J Food Prot.* 1987;60:655–661.
21. Petran R, Zottola EA, Gravani RB. Incidence of *Listeria monocytogenes* in market samples of fresh and frozen vegetables. *J Food Sci.* 1988;53:1238–1240.
22. Sizmur K, Walker CW. *Listeria* in prepacked salads. *Lancet.* 1988;1:1167.
23. Steinbruegge EG, Maxcy RB, Liewen MB. Fate of *Listeria monocytogenes* on ready to serve lettuce. *J Food Prot.* 1988;51:596–599.
24. Beekers HJ, in't Veld PH, Soentoro PSJ, Delfgou-van Ash EHH. The occurrence of *Listeria* in food. *Proceedings of a Symposium on Foodborne Listeriosis.* September 7, Weisbaden, Germany, Behr. Hamburg. 1989;85–97.
25. Heisick JE, Wagner DE, Nierman ML, Peeler JT. *Listeria* spp. found on fresh market produce. *Appl Environ Microbiol.* 1989;55:1925–1927.
26. Wong HC, Chao WL, Lee SJ. Incidence and characterization of *Listeria monocytogenes* in foods available in Taiwan. *Appl Environ Microbiol.* 1990;56:3101–3104.
27. Vahidy R. Isolation of *Listeria monocytogenes* from fresh fruits and vegetables (abstracts). *Hort Sci.* 1992;27: 628.
28. Harvey J, Gilmour A. Occurrence and characteristics of *Listeria* in foods produced in Northern Ireland. *Int J Food Microbiol.* 1993;19:193–205.
29. Arumgaswamy RKG, Rusul Rahamat Ali G, Nadzriah Bte Abd Hamid S. Prevalence of *Listeria monocytogenes* in foods in Malaysia. *Int J Food Microbiol.* 1994;23:117–121.
30. Ng DLK, Seah HL. Isolation and identification of *Listeria monocytogenes* from a range of foods of Singapore. *Food Control.* 1995;6:171–173.
31. Tortorello ML, Reineke KF, Stewart DS. Comparison of antibody-direct epifluorescent filter technique with the most probable number procedure for rapid enumeration of *Listeria* in fresh vegetables. *J AOAC Int.* 1997;80:1208–1214.
32. Odumuru JA, Mitchell SJ, Alves DM, Lynch JA, et al. Assessment of the microbiological quality of ready-to-use vegetables for health-care food services. *J Food Prot.* 1997;954–960.
33. Ortiz R. *Incidencia de* Aeromonas *spp. y* Listeria *spp. en Frutas y Vegetales*. Caracas, Venezuela: Universidad Central de Venezuela; 1993. Thesis.
34. Raybaudi R. *El Célery* (Apium graveolen *L.) y el Repollo (*Brassica oleracea *L. var. Capitata L.) Como Sustratos para el Desarrollo de* Listeria monocytogenes *y* Aeromonas hydrophila. Caracas, Venezuela: Universidad Central de Venezuela; 1994. Thesis.
35. Mejía L. *Incidencia y Sobrevivencia de* Listeria monocytogenes *en Ensaladas de Frutas con Mínimo Procesamiento*. Caracas, Venezuela: Universidad Central de Venezuela; 1997. Thesis.
36. Brión D. *Incidencia de* Listeria monocytogenes *y* Aeromonas hydrophila *en Productos Vegetales Mínimamente Procesados*. Caracas, Venezuela: Universidad Central de Venezuela; 1994. Thesis.
37. Díaz RV, Guevara LM, Rojas B. Incidencia de *Listeria* sp. en ensaladas de vegetales con mínimo procesamiento empacadas en bolsas de polipropileno. Presented at V Congresso Latino-Americano de Microbiologia e Higiene de Alimentos. COMBHAL. VI Simpósio Brasileiro de Microbiología de Alimentos; November 22–26, 1998; Águas de Lindóia, SP, Brasil.
38. Marconi A. *Incidencia y Comportamiento de* Listeria monocytogenes *en Escarola Lisa* (Cichorium endivia). Caracas, Venezuela: Universidad Central de Venezuela; 1999. Thesis.
39. Beuchat LR, Brackett RE, Hao DY, Conner DE. Growth and thermal inactivation of *Listeria monocytogenes* in cabbage and cabbage juice. *Can J Microbiol.* 1986;32:791–795.
40. Berrang ME, Brackett RE, Beuchat LR. Growth of *Listeria monocytogenes* on fresh vegetables stored under controlled atmosphere. *J Food Prot.* 1989;52:702–705.
41. Beuchat LR, Brackett RE. Survival and growth of *Listeria monocytogenes* on lettuce as influenced by shredding, chlorine treatment, modified atmosphere packaging and temperature. *J Food Sci.* 1990;55:755–758, 870.
42. Francis GA, Beirne DO. Effect of gas atmosphere, antimicrobial dip and temperature on the fate of *Listeria innocua* and *Listeria monocytogenes* on minimally processed lettuce. *Int J Food Sci Technol.* 1997;32:141–151.
43. Carlin F, Nguyen-the C. Fate of *Listeria monocytogenes* on four types of minimally processed green salads. *Lett Appl Microbiol.* 1994;18:222–226.
44. Carlin F, Nguyen-the C, Abreu da Silva A. Factors affecting the growth of *Listeria monocytogenes* on minimally processed fresh endive. *J Appl Bacteriol.* 1996;78:636–648.
45. Carlin F, Nguyen-the C, Morris CE. Influence of background microflora on *Listeria monocytogenes* on minimally processed fresh broad-leaved endive (*Cichorium endivia* var. *latifolia*). *J Food Prot.* 1996;59:698–703.
46. Vankerschaver K, Willcox F, Smout C, Hendrickx M, et al. The influence of temperature and gas mixture on the growth of the intrinsic microorganisms on cut endive: Predictive versus actual growth. *Food Microbiol.* 1996;13:427–440.

47. Pao S, Brown GE, Schneider K. Challenge studies with selected pathogenic bacteria on freshly peeled Hamlin orange. *J Food Sci.* 1998;63:359–362.
48. Juneja VK, Martin ST, Sapers GM. Control of *Listeria monocytogenes* in vacuum-packaged pre-peeled potatoes. *J Food Sci.* 1998;63:911–914.
49. Díaz RV, Raybaudi R, Martínez AJ. Survival and growth of *Aeromonas hydrophila* and *Listeria monocytogenes* on raw cabbage and celery. Presented at annual meeting of the International Association of Milk, Food, Sanitation (IAMFES). Seattle: June 30–July 03, 1996.
50. Guevara L. *Ensaladas con Mínimo Procesamiento Como Sustrato para el Crecimiento y Sobrevivencia de* Listeria monocytogenes. Caracas, Venezuela: Universidad Central de Venezuela; 1997. Thesis.
51. Guevara L, Brión D, Díaz R, Alvarado R. Incidencia y sobrevivencia de *Listeria monocytogenes* en ensaladas aderezadas y sin aderezar con mínimo procesamiento. Presented at V Congresso Latino-Americano de Microbiologia e Higiene de Alimentos. COMBHAL. VI Simpósio Brasileiro de Microbiología de Alimentos; November 22–26, 1998; Águas de Lindóia, SP, Brasil.
52. Mejía L, Díaz RV. Crecimiento y sobrevivencia de *Listeria monocytogenes* en ensaladas de frutas con mínimo procesamiento. Presented at V Congresso Latino-Americano de Microbiologia e Higiene de Alimentos. COMBHAL. VI Simpósio Brasileiro de Microbiología de Alimentos; November 22–26, 1998; Águas de Lindóia, SP, Brasil.
53. Palumbo AS. Is refrigeration enough to restrain foodborne pathogens? *J Food Prot.* 1986;49:1003–1009.
54. Conner DE, Brackett RE, Beuchat LR. Effect of temperature, sodium chloride, and pH on growth of *Listeria monocytogenes* in cabbage juice. *Appl Environ Microbiol.* 1986;52:59–63.
55. Beuchat LR, Brackett RE. Inhibitory effects of raw carrots on *Listeria monocytogenes. Appl Environ Microbiol.* 1990;56:1734–1742.
56. Nguyen-the C, Lund BM. The lethal effect of carrot on *Listeria* species. *J Appl Bacteriol.* 1990;479–488.
57. Nguyen-the C, Lund BM. An investigation of the antibacterial effect of carrot on *Listeria monocytogenes*. *J Appl Bacteriol.* 1992;73:23–30.
58. Beuchat LR, Doyle MP. Survival and growth of *Listeria monocytogenes* in foods treated or supplemented with carrot juice. *Food Microbiol.* 1995;12:73–80.
59. Finn MJ, Upton ME. Survival of pathogens on modified-atmosphere-packaged shredded carrot and cabbage. *J Food Prot.* 1997;60:1347–1350.
60. Beuchat LR, Brackett RE. Behavior of *Listeria monocytogenes* inoculate into raw tomatoes and processed tomato products. *Appl Environ Microbiol.* 1991;57:1367–1371.
61. Bowles BL, Juneja VK. Inhibition of foodborne bacterial pathogens by naturally occurring food additives. *J Food Safety.* 1998;18:101–112.
62. Kouassi Y, Shelef LA. Inhibition of *Listeria monocytogenes* by cinnamic acid: possible interaction of the acid with cysteinyl residues. *J Food Safety.* 1998;18:231–242.
63. Kumar M, Berwal JS. Sensitivity of food pathogens to garlic (*Allium sativum*). *J Appl Microbiol.* 1998;84: 213–215.
64. Donnelly CW. *Listeria monocytogenes*. In: Hui YD, Gorham JR, Murrel KD, Cliver DO, eds. *Foodborne Disease Handbook*, Vol. 1. New York: Marcel Dekker; 1994:215–252.
65. Brackett RE. Antimicrobial effect of chlorine on *Listeria monocytogenes. J Food Prot.* 1987;50:999–1003.
66. Zhang S, Farber JM. Effects of various disinfectants against *Listeria monocytogenes* on fresh-cut vegetables. *Food Microbiol.* 1996;22:97–100.
67. Nguyen-the C, Carlin F. The microbiology of minimally processed fresh fruits and vegetables. *Crit Rev Food Sci Nutr.* 1994;34:371–401.
68. Giese, J. *Salmonella* reduction process receives approval. *Food Technol.* 1993;47:110.
69. Somers EB, Schoeni JL, Wong ACL. Effect of trisodium phosphate on biofilm and planktonic cells of *Campylobacter jejuni*, *Escherichia coli* O157:H7, *Listeria monocytogenes* in cold-smoked fishery products and processing plants. *Int J Food Microbiol.* 1994;22:269–276.
70. Beuchat LR. Use of sanitizers in raw fruit and vegetable processing. In: Alzamora SM, Tapia MS, López-Malo A, eds. *Design of Minimal Processing Technologies for Fruit and Vegetables*. Gaithersburg, MD: Aspen Publishers. See Chapter 4.
71. Kim J, Yousef AE, Chism GW. Use of ozone to inactivate microorganism on lettuce. *J Food Safety.* 1999;19:17–34.
72. Abdul-Raouf UM, Beuchat LR, Ammar MS. Survival and growth of *Escherichia coli* O157:H7 on salad vegetables. *Appl Environ Microbiol.* 1993;59:1999–2006.
73. Zhao T, Doyle MP, Besser RE. Fate of enterohemorrhagic *Escherichia coli* O157:H7 in apple cider with and without preservatives. *Appl Environ Microbiol.* 1993;59:2526–2530.
74. Del Rosario BA., Beuchat LR. Survival and growth of enterohemorrhagic *Escherichia coli* O157:H7 in cantaloupe and watermelon. *J Food Prot.* 1995;58:105–107.
75. Diaz C, Hotchkiss JH. Comparative growth shredded iceberg lettuce stored under modified atmospheres. *J Sci Food Agric.* 1996;70:433–438.
76. Barriga MI, Trachy G, Willemot C, Simard RE. Microbial changes in shredded iceberg lettuce stored under

controlled atmospheres. *J Food Sci.* 1991;56:1586–1588, 1599.

77. Hao YY, Brackett RE. Influence of modified atmosphere on growth of vegetable spoilage bacteria in media. *J Food Prot.* 1993;56:223–228.
78. Itoh Y, Sigita-Konishi Y, Kasuka F, Iwaki M, et al. Enterohemorrhagic *Escherichia coli* O157:H7 present in radish sprouts. *Appl Environ Microbiol.* 1998;64: 1532–1535.
79. Conner DE, Kotrola JS. Growth and survival of *Escherichia coli* O157:H7, under acid conditions. *Appl Environ Microbiol.* 1995;61:392–385.
80. Weagant SD, Bryant JL, Bark DH. Survival of *Escherichia coli* O157:H7 in mayonnaise and mayonnaise-based sauces at room and refrigerated temperatures. *J Food Prot.* 1994;57:629–631.
81. Massa S, Altieri C, Quaranta V, DePace R. Survival of *Escherichia coli* O157:H7 in yogurt during preparation and storage at 4°C. *Lett Appl Microbiol.* 1997; 24:347–350.
82. Zhao T, Doyle MP. Rate of enterohemorrhagic *Escherichia coli* O157:H7 in commercial mayonnaise. *J Food Prot.* 1993;57:780–783.
83. Hathcox AL, Beuchat LR, Doyle MP. Death of enterohemorrhagic *Escherichia coli* O157:H7 in real mayonnaise and reduced-caloric mayonnaise dressings as influenced by initial population and storage time. *Appl Environ Microbiol.* 1995;61:4172–4177.
84. Abdul-Raouf UM, Beuchat LR, Ammar MS. Survival and growth of *Escherichia coli* O157:H7 in ground, roasted beef as affected by pH, acidulants and temperature. *Appl Environ Microbiol.* 1993;59:2364–2368.
85. Cutter CN, Siragusa GR. Efficacy of organic acids against *Escherichia coli* O157:H7 attached to beef carcass tissue using a pilot scale model carcass washer. *J Food Prot.* 1994;57:97–103.
86. Brackett RE, Hao YY, Doyle MP. Ineffectiveness of hot acid sprays to decontaminate *Escherichia coli* O157:H7 on beef. *J Food Prot.* 1994. 57:188–203.
87. Erickson JP, Stamer JW, Hayes M, Mckenna DN, et al. An assessment of *Escherichia coli* O157:H7 contamination risks in commercial mayonnaise from pasteurized eggs and environmental sources and behavior in low-pH dressings. *J Food Prot.* 1995;58:1059–1064.
88. Ryu JH, Beuchat LR. Influence of acid tolerance responses on survival, growth, and thermal cross-protection of *Escherichia coli* O157:H7 in acidified media and fruit juices. *Int J Food Microbiol.* 1998;45:185–193.
89. Deng Y, Ryu JH, Beuchat LR. Tolerance of acid-adapted and non-adapted *Escherichia coli* O157:H7 cells to reduced pH as affected by type of acidulant. *J Appl Microbiol.* 1999;86:203–210.
90. León MJ. *Incidencia de* Escherichia coli *O157:H7 Enterohemorrágica en Productos Vegetales*. Caracas, Venezuela: Universidad Central de Venezuela; 1997. Thesis.
91. Martínez A, León MJ, Díaz RV. Incidencia y sobrevivencvia de *E. coli* O157:H7 Enterohemorrágica en productos vegetales. Presented at V Congresso Latino-Americano de Microbiologia e Higiene de Alimentos. COMBHAL. VI Simpósio Brasileiro de Microbiología de Alimentos; November 22–26, 1998; Águas de Lindóia, SP, Brasil.
92. Sakagami Y, Ichise R, Kajimura K, Yokoyama H. Inhibitory effect of creosolte and its main components on production of verotoxin of enterohaemorrhagic *Escherichia coli* O157:H7. *Lett Appl Microbiol.* 1999;28:118–120.
93. International Commission on Microbiological Specifications for Foods. *Aeromonas*. In: Roberts TA, Baird-Parker AC, Tompkin RB, eds. *Microorganisms in Food. 5 Microbiological Specifications of Food Pathogens.* London: Blackie Academic and Professional; 1996:5–19.
94. Abeyta C, Palumbo SA, Stelma GN. *Aeromomas hydrophila* group. In: Hui YD, Gorham JR, Murrel KD, Cliver DO, eds. *Foodborne Disease Handbook*, Vol. 1. New York: Marcel Dekker; 1994:215–252.
95. Kirov SM. *Aeromonas* and *Plesiomonas* species. In: Doyle MP, Beuchat LR, Montville TJ, eds. *Food Microbiology. Fundamentals and Frontiers.* Washington, DC: ASM Press; 1997:265–287.
96. Massa S, Armuzzi R, Tosques M, Canganella F, et al. Susceptibility of chlorine of *Aeromonas hydrophila* strains. *J Appl Bacteriol.* 1999;86:169–173.
97. Abeyta C, Kaysner CA, Wekell MM, Sullivan JJ, et al. Recovery of *Aeromonas hydrophila* from oysters implicated in an outbreak of illness. *J Food Prot.* 1986;49:643–646.
98. Díaz RV. *Enumeración y Caracterización de* Aeromonas *sp. en Productos de Origen Animal y Vegetal.* Caracas, Venezuela: Universidad Central de Venezuela; 1997. Thesis.
99. Kirov SM, Hudson JA, Hayward LJ, Mott SJ. Distribution of *Aeromonas hydrophila* hybridization groups and their virulence properties in Australian clinical and environmental strains. *Lett Appl Microbiol.* 1994:18:71–73.
100. Callister S, Agger W. Enumeration and characterization of *Aeromonas hydrophila* and *Aeromonas caviae* isolated from grocery store produce. *Appl Environ Microbiol.* 1987;53:249–253.
101. Berrang ME, Brackett RE, Beuchat LR. Growth of *Aeromonas hydrophila* on fresh vegetables stored

under a controlled atmosphere. *Appl Environ Microbiol.* 1989b;55:2167–2171.

102. Marchetti R, Casadei MA, Guerzoni ME. Microbial population dynamics in ready-to-use vegetable salads. *Ital J Food Sci.*1992;2:97–108.

103. García-Gimeno RM, Sanchez-Pozo MD, Amaro-López MA, Zurera-Cosano G. Behavior of *Aeromonas hydrophila* in vegetable salads stored under modified atmosphere at 4 and 15ºC. *Food Microbiol.* 1996;13:369–374.

104. Díaz R, Martínez A, Tapia M, Tablante A. Enumeración y caracterización de *Aeromonas* sp. en productos de origen animal y vegetal. Presented at II Congreso Latino-Americano de Microbiología de Alimentos. November 5–18, 1989; Caracas. Venezuela.

105. Ortiz R. *Incidencia* de Aeromonas *sp. y* Listeria *sp. en Frutas y Vegetales.* Caracas, Venezuela: Universidad Central de Venezuela; 1997. Thesis.

106. Rodríguez C. *Incidencia de* Aeromonas sp. *en productos vegetales con mínimo procesamiento.* Caracas, Venezuela: Universidad Central de Venezuela; 1996. Thesis.

107. Revette M, Díaz R. Incidencia de *Aeromonas hydrophila* en ensaladas de frutas con mínimo procesamiento. Presented at II Congreso Venezolano de Ciencia y Tecnología de Alimentos. Dr. Asher Ludin. April 24–28, 1999; Caracas, Venezuela.

108. Méndez J. *Ensaladas con mínimo procesamiento como sustrato para el crecimiento y sobrevivencia de* Aeromonas hydrophila. Caracas, Venezuela: Universidad Central de Venezuela; 1999. Thesis.

109. Pedroso DMM, Iaria ST, Cerqueira Campos ML, Heidtmann S, et al. Virulence factors in motile *Aeromonas spp.* isolated from vegetables. *Revista de Microbiología* 1997;28:49–54.

110. Díaz R, Moreno I, Brión D. Incidencia y sobrevivencia de *Aeromonas hydrophila* en ensaladas de vegetales aderezadas y sin aderezar con mínimo procesamiento. Presented at V Congresso Latino-Americano de Microbiologia e Higiene de Alimentos. COMBHAL. VI Simpósio Brasileiro de Microbiología de Alimentos; November 22–26, 1998; Águas de Lindóia, SP, Brasil.

111. Moreno I. *Comportamiento de* Aeromonas hydrophila *en Ensaladas de Vegetales con Mínimo Procesamiento Almacenadas a 5°C.* Caracas, Venezuela: Universidad Central de Venezuela; 1997. Thesis.

112. Babic I, Nguyen-the C, Amiot MJ, Aubert S. Antimicrobial activity of shredded carrot extracts on food-borne bacteria and yeast. *J Appl Bacteriol.*1994;76:135–141.

113. Wan J, Wilcock A, Coventry MJ. The effect of essential oils of basil on growth of *Aeromonas hydrophila* and *Pseudomonas fluorescens*. *J Appl Microbiol.* 1998;84:152–158.

114. Revette M. *Incidencia y Sobrevivencia de* Aeromonas hydrophila en *Ensaladas de Frutas con Mínimo Procesamiento.* Caracas, Venezuela: Universidad Central de Venezuela; 1998. Thesis.

Chapter 4

Use of Sanitizers in Raw Fruit and Vegetable Processing

Larry R. Beuchat

INTRODUCTION

Losses in the fruit and vegetable industry resulting from microbiologic deterioration occur in the field and at every step of handling, including transient time from the field to the packinghouse or processor, holding at the point of packing or processing, shipping and holding after processing, and in the consumer's home. The natural presence of numerous genera of spoilage bacteria, molds, and yeasts, as well as occasional pathogenic bacteria, such as *Listeria monocytogenes*, *Salmonella*, *Escherichia coli* O157:H7, *Clostridium botulinum,* and others, on fresh fruits and vegetables has been recognized for many years.[1] Fresh produce as a vehicle of viruses and parasites capable of causing human illness has been recognized more recently. Progressive fruit and vegetable handling and processing operations are developing hazard analysis critical control point (HACCP) programs that will, in combination with good manufacturing practices (GMPs), reduce the risk of illness associated with pathogens that may be present on their products. The effective use of sanitizers for containers and equipment and for removing microorganisms from the surfaces of whole and cut produce is an essential part of these programs.

Sanitizers (chemicals) that may be used in the United States to wash or to assist in lye peeling of fruits and vegetables are listed in the U.S. Code of Federal Regulations (CFR), Title 21, Ch. 1, Section 173.315 (U.S. Food and Drug Administration [FDA], Department of Health and Human Services, 4/1/96 edition, p. 127–128). These chemicals may be safely used to wash or to assist in lye peeling of fruits and vegetables, in accordance with stipulated conditions.

Indirect food additives (adjuvants, production aids, and sanitizers) used to control the growth of microorganisms on food processing equipment and utensils and on other food contact articles are specified in CFR, Title 21, Ch. 1, Section 178.1010, 4/1/96 edition, p. 327–334. When sanitizers are used in the processing of fruits and vegetables, their declaration on the label is not required (CFR, Title 21, Section 101.100 (a) (3)). They are considered to be processing aids, being present in an insignificant amount and having no functional or technical effect on the product.

In the United States, chemical sanitizers are also regulated by the U.S. Environmental Protection Agency (EPA), which recognizes them as pesticides that must meet the requirements of the Federal Insecticide, Fungicide and Rodenticide Act (FIFRA). Sanitizers expected to come in contact with fruits and vegetables are regulated by the FDA in accordance with the Federal Food, Drug and Cosmetic Act (FFDCA) as outlined in CFR, Title 21. As noted by Barmore,[2] there is no equally effective chlorine substitute for use in the washing of fruits and vegetables that is permitted by the FDA. There are numerous alternatives to chlorine for sanitizing equipment (Regulation 21, CFR 178.1010), which can be used in a total sanitation program, but none has as broad a spectrum of activity as does chlorine. Chemical sanitizers currently used in the food industry are reviewed in detail by Cords and Dychdala.[3] The balance of this chapter will summarize characteristics of chlorine and other chemicals, some of which may not be permitted for use in some countries, as sanitizers in fruit and vegetable handling and processing.

CHLORINE

Chlorine has been widely used for many years to treat drinking water and waste water, as well as to sanitize food processing equipment and surfaces in processing environments. It is routinely used as a sanitizer in wash, spray, and flume waters used in the fresh fruit and vegetable industry. There are three major groups of chlorine compounds (liquid chlorine, hypochlorites, and chlorine dioxide) that exhibit various degrees of antimicrobial activity.

Liquid Chlorine and Hypochlorites

Chlorine is an effective sanitizer for surfaces that may come in contact with fruits and vegetables during harvesting and handling, as well as for processing equipment. It is commonly used at 200 ppm (available chlorine) and at a pH below 8.0, with a contact time of 1–2 minutes. When liquid chlorine (Cl_2), sodium hypochlorite, and calcium hypochlorite are added to water, they undergo the following reactions:

$$Cl_2 + H_2O \rightarrow HOCl + H^+ + Cl^-$$
$$NaOCl + H_2O \rightarrow NaOH + HOCl$$
$$Ca(OCl)_2 + 2H_2O \rightarrow Ca(OH)_2 + 2\ HOCl$$
$$HOCl \leftrightarrow H^+ + OCl^-$$

Microbicidal activity depends on the amount of free available chlorine (as hypochlorous acid, HOCl) in the water that comes in contact with microbial cells. The dissociation of HOCl depends on pH, and the equilibrium between HOCl and OCl^- is maintained, even when HOCl is constantly consumed upon contact with organic matter. As the pH of the solution is reduced, the equilibrium is in favor of HOCl (Table 4–1). However, because processing equipment is often susceptible to corrosion at low pH, a pH of 6.0–7.5 is more appropriate for effective sanitizing activity without damaging equipment surfaces. Deadly chlorine gas (Cl_2) is formed at pH below 4. At a given pH, equilibrium is in favor of HOCl as the temperature is decreased. This is because chlorine vaporizes as water temperature increases. Chlorine rapidly loses activity upon contact with organic matter or exposure to air, light, or metals. A concern among workers is that prolonged exposure to chlorine vapors in processing environments can cause irritation to the skin and mucous membranes.

Because maximum chlorine solubility is achieved in water at about 4°C (39°F), a cooler water temperature is more effective in sanitizing processing equipment. The same phenomenon would hold true for sanitizing the surfaces of fruits and vegetables. However, the temperature of the chlorinated water should ideally be at least 10°C higher than fruits or vegetables to achieve a positive temperature differential, thereby minimizing uptake of wash water through stem tissues and open areas in the skin or leaves, whether they be natural, e.g., lenticels and stomata, or due to mechanical assault. Elimination of uptake of wash water that may contain microorgan-

Table 4–1 Percentage of Chlorine as Hypochlorous Acid (HOCl), as Affected by pH and Temperature

	HOCl (%)	
pH	*0°C (32°F)*	*20°C (68°F)*
4.0	100	100
5.0	100	99.7
6.0	98.2	96.8
7.0	83.2	75.2
8.0	32.2	23.2
8.5	13.7	8.8
9.0	4.5	2.9
10.0	0.5	0.3

isms, including those that may cause human illnesses, should be considered as a critical control point in the handling and processing of fruits and vegetables.

Numerous theories have been advanced concerning the mode of action through which HOCl exerts its lethal effects on microorganisms. Lethality has been attributed to chlorine combining with cell membrane proteins to form N-chloro compounds, which interfere with glucose oxidation or oxidation of sulfhydryl groups. Chlorine impairs membrane permeability[4] and transport of extracellular nutrients.[5] Spore permeability is altered by chlorine, resulting in release of Ca^{++}, dipicolinic acid, RNA, and DNA.[6]

Chlorinated water is widely used to sanitize whole fruits and vegetables, as well as freshly cut produce. Possible uses in packinghouses and during washing, cooling, and transport for the purpose of controlling postharvest diseases of whole produce have been reviewed by Eckert and Ogawa.[7] The effects of chlorine concentration on aerobic microorganisms and fecal coliforms on leafy salad greens were studied by Mazollier.[8] Total counts were markedly reduced with increased concentrations of chlorine up 50 ppm, but further increase in concentrations up to 200 ppm did not have an additional substantial effect. A standard procedure for washing lettuce leaves in tap water was reported to reduce populations (ca. 10^7/g) of microflora by 92%.[9] Inclusion of 100 ppm available free chlorine (pH ca. 9) reduced the count by 97.8%. Adjusting the pH from 9 to 4.5–5.0 with inorganic and organic acids resulted in a 1.5- to 4.0-fold increase in microbicidal effect. Increasing the washing time in hypochlorite solution from 5 to 30 minutes did not decrease microbial numbers further; whereas extended washing in tap water produced a reduction comparable to that seen with hypochlorite. Addition of 100 ppm of a surfactant (Tween 80) to hypochlorite washing solution enhanced lethality but adversely affected sensory qualities of lettuce. Somers[10] reported that wash water with about 5 ppm chlorine reduced microbial populations on several fruits and vegetables by less than 1 $\log_{10}$, from initial populations of 10^4–10^6 CFU/g. In-plant treatment had no effect on flavor after canning of fruits.

Aside from the effects of pH and temperature on the effectiveness of chlorine in killing microorganisms naturally occurring on fruits and vegetables, the type of produce and diversity of microorganisms that they contain can also greatly influence efficacy. For example, Garg et al.[11] observed that dipping lettuce in water containing 300 ppm chlorine reduced total microbial counts by about 3 $\log_{10}$ CFU/g on lettuce but had no effect on carrots or red cabbage. Treatment of whole and shredded lettuce leaves in water containing 200–250 ppm chlorine reduced populations of aerobic microorganisms by 1–2 $\log_{10}$ CFU/g and psychrotropic microorganisms and yeasts and molds by 0.5–1 $\log_{10}$ CFU/g. Several studies have been done to determine the effectiveness of chlorine in killing bacterial pathogens inoculated onto produce. Dipping brussels sprouts into a 200 ppm chlorine solution for 10 seconds decreased the number of viable *Listeria monocytogenes* cells[12] ($\log_{10}$ 6.1 CFU/g) by about 2 $\log_{10}$. However, dipping inoculated sprouts in sterile water reduced the number of viable cells by about 1 $\log_{10}$ CFU/g. The maximum $\log_{10}$ reduction of *L. monocytogenes* on shredded lettuce and cabbage treated with chlorine (200 ppm for 10 minutes) was reported to be 1.3–1.7 $\log_{10}$ CFU/g and 0.9–1.2 $\log_{10}$ CFU/g, respectively.[13] Initial populations ranged from $\log_{10}$ 5.4 to 5.7 CFU/g. Reductions were greater when treatment was at 22°C than at 4°C. Treatment was more effective against *L. monocytogenes* on lettuce than on cabbage at both temperatures, indicating that antimicrobial activity is influenced by the nature of the vegetable being treated. Numbers decreased only marginally with increased exposure time from 1 to 10 minutes, which agrees with observations made by others[12,14] that the action of chlorine against *L. monocytogenes* occurs primarily during the first 30 seconds of exposure. Nguyen-the and Carlin[15] concluded that the elimination of *L. monocytogenes* from the surface of vegetables by chlorine is limited and unpredictable.

The efficacy of chlorine treatment on inactivation of *Salmonella montevideo* on mature green tomatoes has been studied. Populations on the surface and in the stem core tissue were significantly reduced by dipping tomatoes 2 minutes in a solution containing 60 or 110 ppm chlorine, respectively; however, treatment in a solution containing 320 ppm chlorine did not result in complete inactivation[16] (Table 4–2). The ineffectiveness of 100 ppm chlorine against *S. montevideo* inoculated into cracks in the skin of mature green tomatoes was demonstrated by Wei et al.[17]

Bartz and Showalter[18] showed that warm (26–40°C) tomatoes immersed for 10 minutes or longer in cool (20–22°C) suspensions of bacteria resulted in infiltration by water and these bacteria. Infiltration was associated with a negative temperature difference between the water and the tomato, i.e., the water temperature was less than the tomato temperature. When the differential was shifted to a positive relationship (water temperature higher than tomato temperature), the extent of infiltration was reduced. A significantly higher number of *S. montevideo* cells have been shown to be taken up by the core tissue when tomatoes at 25°C are dipped in suspensions at 10°C, compared with the number of cells taken up by tomatoes dipped in suspensions at 25 or 37°C.[16] Thus, the level of free chlorine reaching viable cells of *S. montevideo* that have infiltrated the core tissues is reduced to the point that lethality is substantially diminished. The uptake of pathogens and spoilage microorganisms from water during washing of other fruits and vegetables has not been investigated. However, infiltration of microbial cells due to a negative temperature differential between the water and the fruit or vegetable would appear to be quite possible.

Treatment of alfalfa seeds inoculated with *Salmonella stanley* (10^2–10^3 CFU/g) in 100 ppm chlorine solution for 10 minutes has been reported to cause a significant reduction in population, and treatment in 290 ppm chlorine solution resulted in a significant reduction, compared with treatment with 100 ppm chlorine.[19] However, initial free chlorine concentrations up to 1000 ppm failed to result in further significant reductions. Treatment of seeds containing 10^1–10^2 CFU/g of *S. stanley* for 5 minutes in a solution containing 2,040 ppm chlorine reduced the population to <1 CFU/g. The FDA has since recommended that alfalfa seeds destined for sprout production be treated with 2,000 ppm chlorine for 30 minutes.[20] The sensory quality of sprouts produced from seeds receiving this treatment is not adversely affected. In another study,[21] alfalfa sprouts inoculated with a five-serovar (*S. agona, S. enteritidis, S. hartford, S. poona, S. montevideo*) cocktail of *Salmonella* were dipped in chlorine solutions (200, 500, or 2,000 ppm) for 2 minutes. The pathogen was reduced by about 3.4 $\log_{10}$ CFU/g after treatment with 500 ppm chlorine and to an undetectable level (< 1 CFU/g) after treatment with 2,000 ppm chlorine. Chlorine treatment (2,000 ppm) of cantaloupe cubes inoculated with the same *Salmonella* serovars resulted in less than 1 $\log_{10}$ reduction in viable cells. The very high level of

Table 4–2 Effectiveness of Chlorine on Inactivating *Salmonella montevideo* on the Surface and in the Core Tissues of Mature Green Tomatoes*

	Population†	
Free Cl^- Concentration (ppm)	*Surface ($\log_{10}$ CFU/cm^2)*	*Core ($\log_{10}$ CFU/g)*
0	4.81 a	5.97 a
60	4.17 b	5.74 a
110	3.59 c	5.28 b
210	3.58 c	5.21 b
320	3.45 c	4.92 b

*Tomatoes (25°C) were dipped in a suspension (10°C) of *S. montevideo* to achieve surface inoculation and uptake of cells in the core tissue before drying, then dipped in chlorine solution (37°C) for 2 min.

†Mean values in the same column that are not followed by the same letter are significantly different ($\alpha = 0.05$).

organic matter in the juice released from cut cantaloupe tissue apparently neutralizes the chlorine before its lethality can be manifested.

Failure to maintain adequate chlorine in wash water may lead to apparent increases in microbial populations on produce. In a study designed to determine microbiologic changes in fresh market tomatoes during packing operations, Senter et al.[22] observed that total plate counts and populations of Enterobacteriaceae were higher, compared to controls, on tomatoes washed in water containing an average of 114 ppm (range 90–140) chlorine; decreases were noted when tomatoes were treated in water containing 226 ppm chlorine (range 120–280). Recontamination of tomatoes occurred in the waxing operation, as evidenced by increased total plate counts and mold populations.

In addition to the neutralizing effect of fruit and vegetable tissue components on chlorine, thereby rendering it inactive against microorganisms in general, as noted by Nguyen-the and Carlin,[15] the inaccessibility of chlorine to microbial cells in creases, crevices, pockets, and natural openings in the skin undoubtedly contributes to its overall lack of effectiveness. The hydrophobic nature of the waxy cuticle protects surface contaminants from exposure to chlorine, and, undoubtedly, other chemicals used as produce sanitizers that do not penetrate or dissolve these waxes. Surface-active agents, such as detergents and ethanol, lessen the hydrophobicity of fruit and vegetable skins, as well as the surfaces of edible leaves, stems, and flowers, and tend to cause deterioration of sensory qualities.[9,13] Sanitizers that contain a solvent that would remove the waxy cuticle layer and, with it, surface contaminants, without adversely affecting sensory characteristics would appear to hold greater potential in reducing microbial populations on the surface of raw produce. Such sanitizers may be limited for use on produce that is to be further processed into juice or cut products or on whole fruits, vegetables, or plant parts that are destined for immediate consumption, e.g., in a food service setting or the home.

The formation of trihalomethanes (THM) and haloacetic acid (HAA) when chlorine and other halogenated sanitizers react with some types of organic matter has prompted potential regulatory action by the EPA. The current draft of this action, aimed at drinking water, is the Disinfectant-Disinfection Byproduct (D-DBP) Rule and would not apply to food processing operations unless potable water is produced by the processor. If the EPA begins to examine discharge water from fruit and vegetable processing operations, possible regulatory action could result.

Chlorine Dioxide

Chlorine dioxide (ClO_2) has received recent attention as a sanitizer from the food processing industry, largely because its efficacy is less affected by pH and organic matter, it is less corrosive to stainless steel, and it does not react with ammonia to form chloramines, as do liquid chlorine and hypochlorites. A disadvantage of ClO_2 is that it is unstable and can be explosive when concentrated. Chlorine dioxide decomposes at temperatures greater than 30ºC (86ºF) when exposed to light. Potentially hazardous THMs are not formed as reaction products of ClO_2, but reaction with bromide ion can occur to form bromate, which may be carcinogenic. Its potential for use as a sanitizer has become more attractive in recent years, due to the development of technologies permitting shipment to areas of use, rather than requiring on-site generation. Compared with information available on the effectiveness of chlorine as a sanitizer, however, much less is known about the efficacy of ClO_2.

The oxidizing power of ClO_2 is reported to be about 2.5 times that of chlorine,[23] and its activity is not affected by pH. Its mechanism of action involves disruption of cell protein synthesis and membrane permeability control. Chlorine dioxide is more effective than chlorine in killing *Escherichia coli* in sewage effluent (pH 8.5).

The FDA permits the use of ClO_2 for sanitizing equipment at a maximum of 200 ppm. It can be used for washing whole fresh fruits and vegetables and shelled beans and peas with intact cuticles at a concentration not to exceed 5 ppm. For peeled potatoes, the maximum permitted wash concentration is 1 ppm. The use of ClO_2 to sanitize other freshly cut fruits and vegetables is not permitted by the FDA.

Lillard[24] reported that 5 ppm ClO_2 is equivalent to 20 ppm chlorine in maintaining low levels of bacteria in chilled water used in poultry processing. A ClO_2 concentration one-seventh that of chlorine was needed to attain the same bacterial level. Treatment of broiler carcasses with ClO_2 has been reported to extend shelf life[25] and to reduce the incidence of *Salmonella* organisms.[26] More recently, Cutter and Dorsa[27] evaluated ClO_2 spray washes for reducing fecal contamination of raw beef. Populations of aerobic microorganisms on carcass tissue sprayed (10 seconds, 520 kPa, 16°C) with ClO_2 at concentrations up to 20 ppm were reduced by no more than 0.93 $\log_{10}$ CFU/cm^2 and were not significantly different from those on water-treated beef. Longer spray times further reduced populations but treatments with ClO_2 were no more effective than water for reducing fecal contamination. Other researchers have investigated the use of ClO_2 as a sanitizer of fresh fruits and vegetables. Control of postharvest fungal pathogens on pears[28] and protozoa in water[29] has been studied. In vitro tests with conidia and sporangiospores of several fungal pathogens of apples and other fruits demonstrated > 99% mortality resulting from a 1-minute treatment in water containing 3 or 5 ppm ClO_2.[30] Longer exposure times were necessary to achieve similar mortalities by treatment with 1 ppm. Of the fungi tested, *Botrytis cinerea* and *Penicillium expansum* were least sensitive to ClO_2. Treatment of belts and pads in a commercial apple and pear packinghouse with 14–18 ppm ClO_2 in a foam formulation resulted in significantly lower numbers of fungi. It was concluded that ClO_2 has desirable properties as a sanitizing agent for postharvest decay management when residues of postharvest fungicides are not desired or allowed.

The efficacy of ClO_2 in preventing buildup of microorganisms in water used to handle cucumbers and on the microorganisms present in fresh cucumbers has been studied.[31] At 2.5 ppm, ClO_2 was effective in killing microorganisms in wash water but, at concentrations up to 105 ppm, failed to reduce the population of microorganisms present in or on fresh cucumbers. It was concluded that many microorganisms were so intimately associated with the cucumber fruit that they were unaffected by chlorine and ClO_2. Reina et al.[32] observed a similar phenomenon. They evaluated the efficacy of ClO_2 in controlling microorganisms in recycled water in a spray-type hydrocooler used to treat pickling cucumbers. Residual ClO_2 at 1.3 ppm was found to control (2–6 $\log_{10}$ CFU/mL reduction) the number of microorganisms in the water optimally. At 0.95 ppm ClO_2, the population was relatively static, whereas at 2.8 and 5.1 ppm, the odor became excessive. Microbial populations on and in cucumbers were not greatly influenced by ClO_2, even at 5.1 ppm. It was concluded that the use of ClO_2 in water used to cool cucumbers seems to be an effective means of controlling microbial buildup but it has little effect on viability of microorganisms on cucumbers.

The effectiveness of ClO_2 in killing *L. monocytogenes* inoculated onto the surface of shredded lettuce and cabbage leaves has been studied.[13] A 10-minute exposure of lettuce to 5 ppm ClO_2 caused a maximum reduction of 1.1 and 0.8 logs in numbers of *L. monocytogenes* at 4 and 22°C, respectively, as compared with a tap water control (0 ppm ClO_2) (Table 4–3). Generally, the reduction was greater at 22°C than at 4°C, at each level of ClO_2 tested. Similar results were obtained with cabbage. Thus, the maximum reduction in populations of *L. monocytogenes* on shredded lettuce and cabbage treated with ClO_2 at target concentrations up to 5 ppm was only a little more than 1 log, which tends to confirm observations on the lack of effectiveness of ClO_2 on populations of microorganisms on and in cucumbers.[31,32]

BROMINE

Bromine has been used alone and in combination with chlorine compounds in water treatment programs. *Pseudomonas aeruginosa* appears to be more resistant than are some other gram-negative and gram-positive bacteria. Free bromine (200 ppm) does not kill *P. aeruginosa* within 15 minutes at 24°C but does kill *E. coli, Salmonella typhosa,* and *Staphylococcus aureus.*[33] Chlorine (NaOCl) is more lethal against *B. cereus* spores than is dibromodimethyl hydrantoin[34] but is equally effective against *Streptococcus faecalis.*[35] Others[36–38] have observed that addition of bromine to solutions containing chlorine compounds increases antimicrobial activity, the effectiveness sometimes being synergistic.

IODINE

Iodine compounds, like chlorine, are widely used as sanitizers for food processing equipment and surfaces. Under most conditions, free elemental iodine and hypoiodous acid are believed to be the active antimicrobial agents.[39] The major iodine compounds are ethanol-iodine solutions, aqueous iodine solutions, and iodophors, which are combinations of elemental iodine and nonionic surfactants, such as nonylphenol ethoxylates,[40] or a carrier, such as polyvinylpyroliodone.[41] Iodophors are approved by the FDA as no-rinse food-contact sanitizer for use at concentrations up to 25 ppm titratable iodine. Iodophors have greater solubility in water and are less volatile and irritating to the skin, compared with ethanolic or aqueous solutions of iodine[42] and have a broad spectrum of activity, including yeasts and molds. Iodophors are less corrosive than chlorine at low temperatures but vaporize at temperatures above about 50°C (120°F), where they can be highly corrosive. The efficacy of iodophors is reduced at low temperatures.

Table 4–3 Effect of ClO_2 Treatment on Survival of *Listeria monocytogenes* on Shredded Lettuce at 4 and 22°C

Initial Target Total Chlorine Concentration (ppm)	*Exposure Time (min)*	*L. monocytogenes (log_{10} CFU/g) Treatment Temperature**		*pH of Solutions*		*Total/Free Chlorine (ppm)†*	
		4°C	*22°C*	*4°C*	*22°C*	*4°C*	*22°C*
0	1	5.1 a	5.3 a	8.35	7.53	0.5/50.5	0.5/05
	2	5.1 a	5.1 a				
	5	5.0 a	5.2 a				
	10	5.1 a	5.0 a				
2	1	4.7 b	4.9 b	7.85	7.29	4.78/1.38	4.50/1.50
	2	4.8 b	4.8 b				
	5	4.8 a	4.8 b				
	10	4.5 ab	4.5 ab				
5	1	4.5 c	4.7 c	7.39	6.82	10/2.80	11.13/3.50
	2	4.8 b	4.7 b				
	5	4.7 a	4.6 b				
	10	4.0 b	4.2 bb				

*Values of the same exposure time (within each temperature) not followed by the same letter are significantly different ($\alpha = 0.05$).

†Initial total/free chlorine at 4° and 22°C.

Iodophors are most effective at pH 2–5 but can also be active at slightly alkaline pH, depending on other conditions. At concentrations of 6–13 ppm available iodine (pH 6.6–7.0), the time to reduce populations of vegetative cells of gram-negative and gram-positive bacteria by 90% ranges from 3 to 15 seconds.[43–45] Bacterial spores are very resistant to iodine, as compared with vegetative cells. Spores of *B. cereus*, *B. subtilis,* and *Clostridium botulinum* (type A) have D values 10- and 1,000-fold greater than do vegetative cells of bacteria treated with 10–100 ppm iodophor.[39]

Although iodophors are not affected by organic matter, they may stain equipment (particularly polypropylene belts) and react with starch to form a blue-purple color. For the latter reason, the use of iodophors for direct-contact sanitation of many fruits and vegetables has limited potential. Iodophors have, however, been promoted as a sanitizer of seafoods. Gray and Hsu[43] demonstrated that iodophors were superior to chlorine in killing *Vibrio parahaemolyticus* at 0°C. In a simulated oyster blowing process, iodophors (25 ppm) proved to be more effective as a vibriocidal agent than was the same concentration of chlorine. Odor and color of oysters exposed to iodophor solution were not affected.

TRISODIUM PHOSPHATE

Trisodium phosphate (TSP) has been approved by the FDA for use in reducing populations of *Salmonella* and other microorganisms on poultry.[46] This alkali is known to be effective in killing *Salmonella* on poultry[47] and red meats,[48] when applied to carcasses in the form of a chill or rinse water during processing. Zhuang and Beuchat[49] investigated the effectiveness of TSP in wash water in killing *S. montevideo* on the surface and in core tissue of inoculated mature green tomatoes. Complete inactivation of *Salmonella* (5.18 $\log_{10}$ CFU/cm^2) on the tomato surface was achieved by dipping tomatoes in 15% TSP solution for 15 seconds. Significant reductions were obtained by dipping tomatoes in a 1% solution for 15 seconds. Populations (5.58 $\log_{10}$ CFU/g) were significantly reduced in core tissue of tomatoes dipped in 4–15% TSP. However, even at 15%, only about a 2 $\log_{10}$ reduction was achieved. Upon ripening, the hue and chroma of tomatoes, indices of color and brightness, respectively, were unaffected by TSP treatment. It was concluded that the use of TSP as a sanitizer for removal of *Salmonella* from the surface of mature green tomatoes has good potential.

The use of TSP to remove *L. monocytogenes* from shredded lettuce is less promising. Zhang and Farber[13] reported that treatment of lettuce with 2% TSP had almost no effect on reducing the population of *L. monocytogenes*. Solutions containing more than 10% TSP damaged the sensory quality of lettuce. Other investigators have observed that *L. monocytogenes* is resistant to TSP.[50] *Escherichia coli* O157:H7, on the contrary, was sensitive to 1% TSP, 10^6 CFU/mL or 10^5 CFU/cm^2 of biofilm being killed within 30 seconds at room temperature or 10°C. *Campylobacter jejuni* was only slightly more resistant than *E. coli* O157:H7.

QUATERNARY AMMONIUM COMPOUNDS

Quaternary ammonium compounds (quats) are cationic surfactants used largely to sanitize floors, walls, drains, and equipment and other food-contact surfaces in processing plants. No water rinse is required unless the solution exceeds 200 ppm active ingredient. Quats are noncorrosive to metals and stable at high temperature. They are more effective against yeasts and molds and gram-positive microorganisms, such as *L. monocytogenes,* than is chlorine and less effective in killing gram-negative bacteria, such as coliforms, *Salmonella*, pathogenic *E. coli*, and *Pseudomonas* and *Erwinia* species, the latter two being among the major spoilage bacteria of raw vegetables. Because of their surfactant activity, quats have good penetrating ability and appear to form a residual antimicrobial film when applied to most hard surfaces. Quats are relatively stable in the presence of organic matter, and properly diluted solutions are odorless and colorless. Their effectiveness is greatest in a pH range of 6–10, thus limiting their potential as sanitizers in acid environments. Quats are not compatible with soaps and anionic detergents. Because most cleaners are anionic, surfaces must be thoroughly rinsed between cleaning and sanitizing.

ACIDS

Organic acids naturally present in fruits and vegetables or accumulated as a result of fermentation are relied on to retard the growth of some microorganisms and prevent the growth of others. Food-borne bacteria capable of causing human illness cannot grow at pH less than about 4.0, so the acidic pH of the edible portions of most fruits (Table 4–4) precludes their involvement as substrates for proliferation of human pathogens. The pH of many vegetables and a few fruits, e.g., melons, however, is in a range at which pathogens can grow.

Some organic acids naturally found in or applied to fruits and vegetables behave primarily as fungistats, whereas others are more effective at inhibiting bacterial growth. Acetic, citric, succinic, malic, tartaric, benzoic, and sorbic acids are the major organic acids naturally occurring in many fruits and vegetables. The mode of action of these acids is attributed to direct pH reduction, depression of internal pH of microbial cells by ionization of the undissociated acid molecule, or disruption of substrate transport by alteration of cell membrane permeability. In addition to inhibiting substrate transport, organic acids may also inhibit NADH oxidation, thus eliminating supplies of reducing agents to electron transport systems. Because the undissociated portion of the acid molecule is primarily responsible for antimicrobial activity, effectiveness at a given pH depends largely on the dissociation constant(s) (pK_a) of the acid (Table 4–5). Because the pK_a of most organic acids is between pH 3 and 5, surface application would be most effective on fruits. However, treatment of vegetables with an organic acid wash followed by washing with potable water to achieve removal of the acid may also be a means of partial disinfection.

The use of washes and sprays containing organic acids, particularly lactic acid, has been successful in decontaminating beef, lamb, pork, and poultry carcasses. Application of organic acid washes to the surface of fruits and vegetables for the purpose of reducing populations of viable microorganisms also has potential. Effective procedures as simple as applying lemon juice, which contains citric acid as the major acid, to cut fruits have been shown to kill or retard the growth of pathogens. Escartin et al.[51] reported that application of lemon juice to the surface of papaya (pH 5.69) and jicama (pH 5.97) cubes inoculated with *Salmonella typhi* reduced populations, compared with the control, but growth resumed after several hours. Survival of *Campylobacter jejuni* inoculated onto watermelon (pH 5.5) and papaya (pH 5.6) cubes, as affected by treatment with lemon juice, was investigated by Castillo and Escartin.[52] The percentage of survivors 6 hours after treatment ranged from 7.7% to 61.8% in fruits not treated with lemon juice acid and from 0% to 14.3% in fruits with lemon juice added. Treatment appeared to be more effective in killing *Campylobacter* on papaya than on watermelon.

Use of citric acid to reduce microbial populations on salad vegetables has also been studied. Shapiro and Holder[53] observed that treatment with up to 1,500 ppm (0.15%) citric acid did not affect bacterial growth during a subsequent 4-day storage period at 10°C. Treatment with 1,500 ppm tartaric acid reduced total counts by 10-fold. Treatment of cut lettuce, endive, carrots, celery, radishes, and green onions with 2,000 ppm sorbate or 10,000 ppm ascorbic acid, alone and in combination, resulted in less than 1 log difference in populations of aerobic microorganisms after 10 days of storage in 4.4°C (40°F).[54]

Table 4–4 pH of Vegetables and Fruits

Vegetables

Product	*pH*	*Reference*
Asparagus	5.7–6.1, 5.0–6.1, 5.4–5.8	72, 73, 78
Bean (string)	4.6, 5.0–6.0	73, 78
Bean (Lima)	6.5, 5.4–6.5, 6.2	72, 73, 79
Beet	4.9–5.8	72
Beet (sugar)	4.2–4.4	78
Broccoli	6.5, 5.2–6.0	73, 78
Brussels sprout	6.3, 6.3–6.6	72, 78
Cabbage (green)	5.4–6.0, 5.2–6.3, 5.8, 5.2–5.4	72, 73, 78, 79
Carrot	4.9–5.2, 6.0, 4.9–6.3, 5.2–5.8	72, 73, 78, 79
Cauliflower	6.0–6.7	72
Celery	5.7–6.0, 5.6	72, 78, 79
Corn (sweet)	7.3, 5.9–6.5, 7.1	72, 78, 79
Eggplant	4.5	78
Lettuce	6.0, 6.0–6.4	72, 78
Mushroom	6.0–6.5	73
Olive	3.6–3.8	78
Onion (red)	5.3–5.8, 5.0	72, 78, 79
Parsley	5.7–6.0, 5.8	78, 79
Parsnip	5.3	78
Peas	5.8–6.5	73
Pepper (green cherry)	5.6–7.0	75
Pepper (red cherry)	5.3–5.8	75
Pepper (several types)	5.0–5.6	79
Pimento	4.3–5.2	72
Potato tuber	5.6–6.2, 6.0, 5.4–6.0	73, 74, 79
Pumpkin	4.8–5.2, 4.8–5.5	73, 78
Spinach	5.5–6.0, 5.1–6.8, 4.8–5.8	72, 73, 78
Squash	5.0–5.4, 5.0–5.4	73, 78
Sweet potato	5.3–5.6, 5.3–5.6	72, 73
Tomato whole	4.2–4.3, 3.7–4.9	72, 78
Tomato (under ripe)	3.2–4.5	81
Tomato (ripe)	3.4–4.7, 4.0–4.4	
Tomato (over ripe)	4.0–4.8	81
Turnip	5.2–5.5, 5.2–5.6	72, 73, 78

Fruits

Product	*pH*	*Reference*
Apple	2.9–3.3, 3.1–3.9	78, 80
Apricot	3.5–4.0, 3.3–4.4	78, 80
Banana	4.5–4.7, 4.5–5.2	78,80
Blackberry	3.0–4.2	80
Blueberry	3.2–3.4	80
Cherry	3.2–4.7, 3.2–4.0	78, 80
Cranberry	2.5–2.7	80
Fig	4.6, 4.8-5.0	78, 80
Grape	3.4–4.5, 3.0–4.0, 3.9	76, 78, 80
Grapefruit	2.9–3.6	80
Jicama	6.0	51
Lemon	2.2–2.6	80
Lime	2.3–2.4	80
Mango	3.8–4.7	80
Melon		
Cantaloupe	6.2–6.5, 6.3, 7.0	72, 76, 77
Honeydew	6.3–6.7	72
Watermelon	5.2–5.6, 5.5, 5.6	52, 77, 78
Orange	3.6–4.3, 3.3–4.0, 3.5	76, 78, 80
Papaya	5.6, 5.7	51, 52
Peach	3.3–4.2	80
Pear	3.4–4.7, 3.7–4.6	72, 80
Pineapple	3.4–3.7	80
Plum	2.8–4.6, 3.2–4.0	78, 80
Raspberry	2.9–3.5	80
Rhubarb	3.1–3.4, 2.9–3.3	73, 78
Strawberry	3.0–3.9, 4.2	76, 80

Table 4–5 Undissociation of Acids Affected by pH

Acid	pK_a	Percentage Undissociated Acid at pH:						
		2.5	3.5	4.5	5.0	5.5	6.0	7.0
Acetic	4.74	99	95	63	35	14	5.2	0.55
Citric	3.13	81	30	4.1	1.3	0.4	0.13	0.01
Formic	3.75	95	64	15	5.3	1.7	0.56	0.06
Lactic	2.74	64	15	1.7	0.5	0.2	0.06	0.01
Malic	3.40	89	44	7.4	2.5	0.8	0.25	0.03
Tartaric	2.98	75	23	2.9	0.9	0.3	0.10	0.01
Benzoic	4.19	98	83	33	13	4.7	1.5	0.15
Propionic	4.87	100	96	70	43	19	6.9	0.74
Sorbic	4.76	99	95	65	37	15	5.4	0.57
Sulfurous acid	1.81	17	2	0.2	0.06	0.02	0.01	0.00

Removal of pathogenic bacteria from leafy salad greens by treating with other organic acids has been studied. Reduction in counts of *Yersinia enterocolitica* inoculated onto parsley leaves from 10^7 CFU/g to < 1 CFU/g by washing in a solution of 2% acetic acid or 40% vinegar for 15 minutes was achieved by Karapinar and Gonul.[55] No viable aerobic bacteria were recovered after a 30-minute dip in 5% acetic acid, whereas a vinegar dip led to a 3–6 $\log_{10}$ decrease in the number of aerobic bacteria, depending on vinegar concentration and holding time. Sensory quality of treated parsley was not described.

The effects of lactic and acetic acid, either alone or in combination with chlorine, on survival of *L. monocytogenes* inoculated onto shredded lettuce were studied by Zhang and Farber.[13] Compared with lettuce washed in tap water, only 1% lactic acid and combinations of 0.5% or 1% lactic acid and 100 ppm chlorine reduced numbers of *L. monocytogenes*. Lactic acid (0.75% or 1%), in combination with 100 ppm chlorine, was more effective in reducing levels of *L. monocytogenes* than were either lactic acid or chlorine alone. Results obtained with acetic acid were similar to those of lactic acid. Centrifuging lettuce immediately after removal from acid solutions was necessary to reduce deterioration of sensory quality, i.e., loss of texture.

Acid anionic sanitizers are regulated by FDA as no-rinse food-contact surface sanitizers. They contain a mixture of anionic surfactants and acid, thus having antimicrobial activity and preventing mineral buildup simultaneously. Acid anionic sanitizers are not corrosive to stainless steel unless water is high in chlorides. They are effective only at low pH (2–3), rapidly losing activity above pH 3. They are more effective against gram-positive bacteria than against gram-negative bacteria and have low activity against yeasts and molds. Acid anionic sanitizers have limited use in most clean-in-place systems because of their high foaming characteristics.

Carboxylic acid sanitizers, also known as fatty acid sanitizers, are regulated by FDA as no-rinse food-contact surface sanitizers. They consist of free fatty acids, sulfonated fatty acids, other organic acids, and, often, a mineral acid, phosphoric being preferred. Like acid anionic sanitizers, fatty acid sanitizers also function as an acid rinse. They are stable in the presence of organic matter and at high temperature, and have a broad spectrum of activity against gram-negative and gram-positive bacteria but have limited activity against molds. Activity is greatly decreased above pH 4 and by cationic surfactants. Fatty acid sanitizers perform poorly at temperatures below 10°C (50°F) and tend to damage plastics and rubber material, particularly at high temperatures.

Treatment of ready-to-use salads with 90 ppm peracetic acid has been shown to reduce total counts and fecal coliforms by nearly 100-fold, similar to reductions with 100 ppm chlorine.[56] Reduced growth of microflora during subsequent storage was attributed to residual effects of acetic acid released by degradation of peracetic acid.

Peroxyacetic acid is regulated by FDA as a no-rinse food-contact surface sanitizer. Although the acidity of peroxyacetic acid is lower than that of carboxylic acid sanitizers, it has antimicrobial activity on unclean surfaces and acid rinse at the same time. Peroxyacetic acid compounds have low foam characteristics and leave no residues. Being relatively tolerant of organic matter, they are exceptionally effective against biofilms. Peroxyacetic acid sanitizers have a broad spectrum of bactericidal activity over a broad pH range, up to pH 7.5, but vary in effectiveness against yeasts and molds. On the down side, peroxyacetic acid sanitizers lose effectiveness in the presence of some metals. They are corrosive to brass, copper, mild steel, and galvanized steel, especially at high temperatures.

HYDROGEN PEROXIDE

Hydrogen peroxide (H_2O_2) can have a lethal or inhibitory effect on microorganisms, depending on the pH, temperature, and other environmental factors. Toxicity of H_2O_2 is due to its capacity as an intermediate in oxygen reduction to generate more reactive oxygen species, such as the hydroxyl radical (HO^-), which is a powerful antioxidant that can initiate oxidation and cause damage to nucleic acids, proteins, and lipids.[57] The antibacterial activity of H_2O_2 was recently reviewed by Juven and Pierson.[58]

The efficacy of H_2O_2 as a bactericide in poultry chiller was investigated by Lillard and Thomson.[59] At 6,600 ppm H_2O_2 or higher, aerobic microorganisms were reduced by 95–99%; *E. coli* counts were reduced by 97–99% with 5,300 ppm H_2O_2. Because the reaction of H_2O_2 with catalase from blood resulted in a bleached and bloated carcass, it was concluded that H_2O_2 treatment would be commercially undesirable for fresh or frozen retail sales but may not be objectionable when used for deboned meats. Spray washing with H_2O_2 (5%) has also been reported to reduce populations of aerobic microorganisms on beef tissue.[60]

The same problems associated with catalase reactions in animal products may be minimal in fruits and vegetables, at least when application of H_2O_2 is for the purpose of sanitizing the surface of whole produce. Sapers[61] has studied the efficacy of H_2O_2 in improving the microbiologic quality and extending the shelf life of minimally processed fruit and vegetable products. Hydrogen peroxide vapor treatments were highly effective in reducing microbial numbers on whole cantaloupes, table grapes, prunes, raisins, walnuts, and pistachios. However, similar treatment induced browning in mushrooms exposed to levels necessary to delay spoilage by *Pseudomonas tolaasii*. Exposure to H_2O_2 vapor caused bleaching of anthocyanins in strawberries and raspberries. Dipping freshly cut green bell pepper, cucumber, zucchini, cantaloupe, and honeydew melon in H_2O_2 solution had no adverse effect on appearance, flavor, or texture but induced severe browning of shredded lettuce. Dip treatment significantly reduced the population of *Pseudomonas* on these products but had no measurable effect on yeasts and molds.

Reductions in populations of *Salmonella* on alfalfa sprouts are similar by dipping sprouts in 200 and 500 ppm chlorine or 2% and 5% H_2O_2, respectively, for 2 minutes.[21] Slightly more than 2 $\log_{10}$ CFU/g reduction was observed after treatment with 200 ppm chlorine or 2% H_2O_2. The effectiveness of the same concentrations of chlorine and H_2O_2 in eliminating *Salmonella* from cantaloupe cubes was much less, however, with reductions being less than 1 $\log_{10}$ CFU/g. Alfalfa sprouts and, to a lesser extent, cantaloupe cubes took on a mildly bleached appearance after treatment with 5% H_2O_2. Results from studies on fruits and vegetables indicate that H_2O_2 has potential for use as a sanitizer. Additional work to determine its effectiveness in killing pathogens on a wide range of produce is warranted.

OZONE

Treatment of drinking water with ozone for the purpose of killing microorganisms has been practiced for nearly a century. Ozone is permitted by the FDA for treatment of drinking water (21 CFR 184.1563) and for recycled water in poultry plants at a level not to exceed 0.1 ppm. Use in poultry operations requires approval by the U.S. Department of Agriculture (USDA). *Salmonella typhimurium, Y. enterocolitica, S. aureus,* and *L. monocytogenes* are among the pathogens sensitive to treatment in ozonated (20 ppm) water.[62] Enteric viruses[63] and oocysts of parasites, such as *Cryptosporidium parvum,*[64] are also sensitive to ozone. Greater than 90% inactivation of *C. parvum* was achieved by treatment with 1 ppm ozone for 5 minutes.[65] *C. parvum* oocysts are about 30 times more resistant to ozone and 14 times more resistant to ClO_2 than *Giardia* cysts exposed to these disinfectants under the same conditions.

The use of ozone to decontaminate various types of foods has been investigated. Preservation of fish,[66] reduction of aflatoxin in peanuts and cottonseed meals,[67] and reduction of microbial populations on poultry,[68] bacon, beef, butter, cheese, eggs, mushrooms, potatoes, and fruits[69,70] using gaseous ozone have been studied. High relative humidity or aqueous conditions generally favor microbicidal activity.

The lethal effect of ozone is a consequence of its strong oxidizing power. For this reason, physiologic injury of produce, for example, in bananas, can result from exposure to concentrations as low as 1.5 ppm without damage to sensory qualities.[71] After 8 days, black spots appear on the skin of bananas if the ozone concentration is maintained at 25–30 ppm. Extension of shelf life of oranges, strawberries, raspberries, grapes, apples, and pears can be achieved by treatment with ozonated water. In addition to the antimicrobial effects of ozone, oxidation of ethylene also occurs, thus retarding metabolic processes associated with ripening. Effective concentrations range from 2 to 3 ppm for berry fruits to 40 ppm for oranges.

Because of its instability, ozone must be generated at the usage site. Users should also be aware that, because of the strong oxidizing power of ozone, metal and other types of surfaces with which it comes in contact are subject to corrosion or other deterioration processes. The rate of deterioration depends on the concentration of ozone. Nevertheless, use of ozonated wash and flume waters in fruit and vegetable processing operations could provide a method to control buildup of microbial numbers, particularly in recycled water, and deserves further attention as a sanitizer for the produce industry.

REFERENCES

1. Beuchat LR. Pathogenic microorganisms associated with fresh produce. *J Food Prot.* 1996;59:204–206.
2. Barmore CR. *Chlorine—Are There Alternatives?* Arlington, VA: Cutting Edge. International Fresh-Cut Produce Association; Spring 1995 issue:4–5.
3. Cords BR, Dychdala GR. Sanitizers: Halogens, surface-active agents and peroxides. In: Davidson PM, Branen AL, eds. *Antimicrobials in Foods,* 2nd ed. New York: Marcel Dekker; 1993:469–537.
4. Freiberg L. Further quantitative studies on the reaction of chlorine with bacteria in water disinfection. *Acta Pathol Microbiol Scand.* 1957;40:67–80.
5. Camper AK, McFecters GA. Chlorine injury and enumeration of waterborne coliform bacteria. *Appl Environ Microbiol.* 1979;37:633–641.
6. Kulikovsky A, Pankratz HS, Sadoff HL. Ultrastructural and chemical changes in spores of *Bacillus cereus* after action of disinfectants. *J Appl Bacteriol.* 1975;38:39–46.
7. Eckert JW, Ogawa JM. The chemical control of postharvest diseases: Deciduous fruits, berries, vegetables and root/tuber crops. *Ann Rev Phytopathol.* 1988;26:433.
8. Mazollier J. IVè gamme. Lavage-desinfection des salades. *Infros-Ctifl.* 1988;41:19.

9. Adams MR, Hartley AD, Cox LJ. Factors affecting the efficiency of washing procedures used in the production of prepared salads. *Food Microbiol.* 1989;6:69–77.
10. Somers EB. Studies on in-plant chlorination. *Food Technol.* 1963;5(1):46–51.
11. Garg N, Churey JJ, Splittstoesser DF. Effect of processing conditions on the microflora of fresh-cut vegetables. *J Food Prot.* 1990;53:701–703.
12. Brackett RE. Antimicrobial effect of chlorine on *Listeria monocytogenes. J Food Prot.* 1987;50:999–1003.
13. Zhang S, Farber JM. The effects of various disinfectants against *Listeria monocytogenes* on fresh-cut vegetables. *Food Microbiol.* 1996;13:311–321.
14. Beuchat LR, Brackett RE. Survival and growth of *Listeria monocytogenes* on lettuce as influenced by shredding, chlorine treatment, modified atmosphere packaging and temperature. *J Food Sci.* 1990;55:755–758, 870.
15. Nguyen-the C, Carlin F. The microbiology of minimally processed fresh fruits and vegetables. *Crit Rev Food Sci Nutr.* 1994;34:371–401.
16. Zhuang R-Y, Beuchat LR, Angulo FJ. Fate of *Salmonella montevideo* on and in raw tomatoes as affected by temperature and treatment with chlorine. *Appl Environ Microbiol.* 1995;61:2127–2131.
17. Wei CI, Huang TS, Kim JM, Lin WF, Tamplin ML, Bartz JA. Growth and survival of *Salmonella montevideo* on tomatoes and disinfection with chlorinated water. *J Food Prot.* 1995;58:829–836.
18. Bartz JA, Showalter RK. Infiltration of tomatoes by bacteria in aqueous suspension. *Phytopathology.* 1981;71:515–518.
19. Jaquette CB, Beuchat LR, Mahon BE. Efficacy of chlorine and heat treatment in killing *Salmonella stanley* inoculated onto alfalfa seeds and growth and survival of the pathogen during sprouting and storage. *Appl Environ Microbiol.* 1996;62:2212–2215.
20. Department of Health and Human Services. *Memorandum: Microbiological safety of alfalfa sprouts.* From J. Madden to S. Altekrause, March 1, 1996: personal correspondence, 2 pp.
21. Beuchat LR, Ryu J-H. Produce handling and processing practices. *Emerg Infect Dis.* 1997;3:459–465.
22. Senter SD, Cox NA, Bailey JS, Forbus WR. Microbiological changes in fresh market tomatoes during packing operations. *J Food Sci.* 1985;50:254–255.
23. Benarde MA, Snow WB, Olivieri OP, Davidson B. Kinetics and mechanism of bacterial disinfection by chlorine dioxide. *Appl Microbiol.* 1967;15:257–265.
24. Lillard HS. Levels of chlorine and chloride dioxide of equivalent bacterial effect on poultry processing water. *J Water Sci.* 1979;44:1594–1597.
25. Thiessen GP, Usborne WR, Orr HL. The efficacy of chlorine dioxide in controlling *Salmonella* contamination and its effect on product quality of chicken broiler carcasses. *Poultry Sci.* 1984;63:647–653.
26. Villarreal ME, Baker RC, Regenstein JM. The incidence of *Salmonella* on poultry carcasses following the use of slow release chlorine dioxide (Alcide). *J Food Prot.* 1990;53:465–467.
27. Cutter CN, Dorsa WJ. Chlorine spray washes for reducing fecal contamination on beef. *J Food Prot.* 1995;58:1294–1296.
28. Spotts RA, Peters BB. Chlorine and chlorine dioxide for control of d'Anjou pear decay. *Plant Dis.* 1980;64: 1095–1097.
29. Chen YSR, Sproul OJ, Rubin AJ. Inactivation of *Naegleria gruberi* cysts by chlorine dioxide. *Water Res.* 1985;19:783–790.
30. Roberts RG, Reymond ST. Chlorine dioxide for reduction of postharvest pathogen inoculum during handling of tree fruits. *Appl Environ Microbiol.* 1994;60: 2864–2868.
31. Costilow RN, Uebersax MA, Ward PJ. Use of chlorine dioxide for controlling microorganisms during handling and storage of fresh cucumbers. *J Food Sci.* 1984;49:396–401.
32. Reina LD, Fleming HP, Humphries EG. Microbiological control of cucumber hydrocooling water with chlorine dioxide. *J Food Prot.* 1995;58:541–546.
33. Gershenfeld L, Witlin B. Evaluation of the antibacterial efficiency of dilute solutions of free halogens. *J Am Pharm Assoc Sci Ed.* 1949;38:411–414.
34. Cousins CM, Allan CD. Sporicidal properties of some halogens. *J Appl Bacteriol.* 1967;30:168–174.
35. Ortenzio LF, Stuart LS. A standard test for efficacy of germicides and acceptability of residual disinfecting activity in swimming pool water. *J Assoc Off Agric Chem.* 1964;47:540–547.
36. Farkas-Himsley H. Killing of chlorine-resistant bacteria with chlorine-bromine solutions. *Appl Microbiol.* 1964;12:1–6.
37. Kristofferson T. Mode of action of hypochlorite sanitizers with and without sodium bromide. *J Dairy Sci.* 1958;41:942–949.
38. Shere L, Kelley MJ, Richardson JH. Effect of bromide hypochlorite bactericides on microorganisms. *Appl Microbiol.* 1962;10:538–541.
39. Odlaug TE. Antimicrobial activity of halogens. *J Food Prot.* 1981;44:608–613.
40. Bartlett PG, Schmidt W. Surface-iodine complexes as germicides. *Appl Microbiol.* 1957;5:355–359.
41. Lacey RW. Antibacterial activity of providone iodine towards non-sporing bacteria. *J Appl Bacteriol.* 1979;46: 443–449.

42. Lawrence CA, Carpenter CM, Naylor-Foote AWC. Iodophors as disinfectants. *J Am Pharm Assoc.* 1957;46: 500–505.
43. Gray RJ, Hsu D. Effectiveness of iodophor in the destruction of *Vibrio parahaemolyticus*. *J Food Sci.* 1979;44:1097–1100.
44. Hays H, Elliker PR, Sandine WR. Microbial destruction by low concentrations of hypochlorite and iodophor germicides in alkaline and acidified water. *Appl Microbiol.* 1967;15:575–581.
45. Mosley EB, Elliker PR, Hays H. Destruction of food spoilage indicator and pathogenic organisms by various germicides in solution and on a stainless steel surface. *J Milk Food Technol.* 1976;39:830–836.
46. Giese J. *Salmonella* reduction process receives approval. *Food Technol.* 1993;47(1):110.
47. Lillard HS. Effect of trisodium phosphate on salmonellae attached to chicken skin. *J Food Prot.* 1994;57:465–469.
48. Dickson JS, Nettles-Cutter CG, Siragusa GR. Antimicrobial effects of trisodium phosphate against bacteria attached to beef tissue. *J Food Prot.* 1994;57:952–955.
49. Zhuang R-Y, Beuchat LR. Effectiveness of trisodium phosphate for killing *Salmonella montevideo* on tomatoes. *Lett Appl Microbiol.* 1996;22:97–100.
50. Somers EB, Schoeni JL, Wong ACL. Effect of trisodium phosphate on biofilm and planktonic cells of *Campylobacter jejuni, Escherichia coli* O157:H7, *Listeria monocytogenes* and *Salmonella typhimurium*. *Int J Food Microbiol.* 1994;22:269–276.
51. Escartin EF, Castillo Ayala A, Lozano JS. Survival and growth of *Salmonella* and *Shigella* on sliced fresh fruit. *J Food Prot.* 1989;52:471–472.
52. Castillo A, Escartin EV. Survival of *Campylobacter jejuni* on sliced watermelon and papaya. *J Food Prot.* 1994;57:166–168.
53. Shapiro JE, Holder IA. Effect of antibiotic and chemical dips on the microflora of packaged salad mix. *Appl Microbiol.* 1960;8:341.
54. Priepke PE, Wei LS, Nelson AI. Refrigerated storage of prepackaged salad vegetables. *J Food Sci.* 1976;41: 379–385.
55. Karapinar M, Gonul SA. Removal of *Yersinia enterocolitica* from fresh parsley by washing with acetic acid or vinegar. *Int J Food Microbiol.* 1992;16:261–264.
56. Masson RB. Recherche de Nouveax disinfectants pour les produits de 4ème gamme. *Proc. Congress Produits de 4ème Gamme et de 5ème Gamme*. Brussels, Belgium: C.E.R.I.A.; 1990:101.
57. Halliwell B, Gutteridge JMC. Biologically relevant metal ion-dependent hydroxyl radical generation: An update. *FEBS Lett.* 1992;307:108–112.
58. Juven BJ, Pierson MD. Antibacterial effects of hydrogen peroxide and methods for its detection and quantitation. *J Food Prot.* 1996;59:1233–1241.
59. Lillard HS, Thomson JE. Efficacy of hydrogen peroxide as a bactericide in poultry chiller water. *J Food Sci.* 1983;48:125–126.
60. Gorman BM, Sofos JN, Morgan JB, Schmidt GR, Smith GC. Evaluation of hand-trimming, various sanitizing agents and hot water spray-washing as decontamination interventions for beef brisket adipose tissue. *J Food Prot.* 1995;58:899–907.
61. Saper GM. Hydrogen peroxide as an alternative to chlorine. Abstract 59–4, *1996 IFT Ann. Mtg: Book of Abstracts*. Chicago: Institute of Food Technology; 1996:140.
62. Restaino L, Frampton EW, Hemphill JB, Palnikar P. Efficacy of ozonated water against various food-related microorganisms. *Appl Environ Microbiol.* 1995;61: 3471–3475.
63. Finch GR, Fairbairn N. Comparative inactivation of poliovirus type 3 and MS2 coliphage in demand-free phosphate buffer by using ozone. *Appl Environ Microbiol.* 1991;57:3121–3126.
64. Korich DG, Mead JR, Madore MS, Sinclair NA, Sterling CR. Effects of ozone, chlorine dioxide, chlorine, and nonochloramine on *Cryptosporidium parvum* oocyst viability. *Appl Environ Microbiol.* 1990;56:1423–1428.
65. Peeters JE, Mazas EA, Masschelein WJ, de Maturana IVM, Debacker E. Effect of disinfection of drinking water with ozone or chlorine dioxide on survival of *Cryptosporidium parvum* oocysts. *Appl Environ Microbiol.* 1989;55:1519–1522.
66. Haraguchi T, Slimidu U, Aiso K. Preserving effect of ozone on fish. *Bull Jpn Soc Sci Fish.* 1969;35:915–920.
67. Dwankanath CT, Rayner ET, Mann GE, Dollar FG. Reduction of aflatoxin levels in cottonseed and peanut meals by ozonation. *J Am Oil Chem Soc.* 1968;45:93–95.
68. Sheldon BW, Brown AL. Efficacy of ozone as disinfectant for poultry carcasses and chill water. *J Food Sci.* 1986;51:305–309.
69. Gammon R, Kerelak K. Gaseous sterilization of foods. *Am Inst Chem Engr Symp Ser.* 1973;69:91.
70. Kaess G, Weidemann, JF. Ozone treatment of chilled beef. *J Food Technol.* 1968;3:325–333.
71. Horvath M, Billitzky L, Huttner J. *Ozone*. Amsterdam: Elsevier Science; 1985.
72. Banwart GJ. *Basic Food Microbiology*, 2nd ed. New York: Van Nostrand Reinhold; 1989: 117.
73. Brackett RE. Vegetables and related products. In: Beuchat LR, ed. *Food and Beverage Mycology*, 2nd ed. New York: Van Nostrand Reinhold;1987:129–154.
74. Burton WG. *The Potato,* 2nd ed. Holland: H. Veenman and N. V. Zonen; 1966:162.

75. Daeschel MA, Fleming HP, Pharr DM. Acidification of brined cherry peppers. *J Food Sci.* 1990;55:186–192.

76. Deak T, Beuchat LR. *Handbook of Foodborne Spoilage Yeasts.* Boca Raton, FL: CRC Press; 1996:62.

77. del Rosario BA, Beuchat LR. Survival of enterohemorrhagic *Escherichia coli* O157:H7 in cantaloupe and watermelon. *J Food Prot.* 1995;58:105–107.

78. Jay JM. *Modern Food Microbiology*, 3rd ed. New York: Van Nostrand Reinhold; 1986:36.

79. Sapers GM, Phillips JG, DiVito AM. Correlation between pH and composition of foods comprising mixtures of tomatoes and low-acid ingredients. *J Food Sci.* 1984;49:233–235, 238.

80. Splittstoesser DF. Fruits and fruit products. In: Beuchat LR, ed. *Food and Beverage Mycology,* 2nd ed. New York: Van Nostrand Reinhold; 1987:101–128.

81. Wolf ID, Schwartau CM, Thompson DR, Zottola EA, Davis DW. The pH of 107 varieties of Minnesota-grown tomatoes. *J Food Sci.* 1979;44:1008–1010.

CHAPTER 5

Tools for Safety Control: HACCP, Risk Assessment, Predictive Microbiology, and Challenge Tests

María S. Tapia, Amaury Martínez, and Rosa V. Díaz

INTRODUCTION

Media have recently paid much attention to emerging microbiologic problems in foods of plant origin. The potential for microbiologic contamination of fruits and vegetables is high because of the wide variety of conditions to which produce is exposed during growth, harvest, processing, and distribution. These considerations acquire great significance in the current scenario of the new processing techniques that offer attributes of convenience and fresh-like quality in response to the changes in consumption patterns and to an increased demand for fresh and minimally processed fruits and vegetables. As a consequence, reliance on low-temperature storage and on improved packaging materials and packaging techniques has increased. Even if produce had not been considered as a major vector for food-borne disease, technologies that extend shelf-life by decreasing the rate of product deterioration might increase the risks associated with pathogenic microorganisms—especially of psychrotropic nature—by allowing sufficient time for their growth while retarding the development of competitive spoilage organisms.[1–3] Additionally, processing steps that modify the food microenvironment open new possibilities to support pathogens that, for ecologic reasons, would have never been present naturally in produce. Food-borne disease outbreaks traceable to produce have been reportedly due to *Salmonella* and *Shigella* species, *Listeria monocytogenes, Clostridium botulinum, Aeromonas hydrophila,* and *Campylobacter jejuni.*[4,5]

Raw and minimally processed fruits and vegetables cannot be excluded from the application of any of the modern tools that prevail in today's food industry for ensuring safety of produce and products. Traditionally, quality control of food products relied on inspection, which may fail to detect contaminated batches, and on end-

Authors gratefully acknowledge support of the European Commission Project TS3*-CT94–0333(DG HSMU) "Development of preservation techniques for tropical fruits using vacuum impregnation techniques"; CYTED Program Project XI-3 "Development of minimal processing techniques for food preservation"; Consejo Nacional de Investigaciones Científicas y Tecnológicas de Venezuela (CONICIT), Project S1–2722 "Incidencia, comportamiento y patogenicidad de *Listeria monocytogenes* y *Aeromonas hydrophila* en productos de origen vegetal"; and Consejo de Desarrollo Científico y Humanístico de la Universidad Central de Venezuela, Project "Incidencia y Sobrevivencia de *Escherichia coli* O157:H7 en productos de origen animal y vegetal."

product testing, which is expensive and time-consuming. The reliance on microbial testing and certification to maintain microbiologic quality is not practical and is often of limited value. This situation was improved by the introduction of good manufacturing practices (GMPs), which are preventive measures that have evolved from general principles of hygiene, based on practical experience over a long of period time.[6,7] This rather subjective, qualitative, inspectional approach, relying on personal opinions and expertise, is not enough for an objective risk assessment. This is particularly true in actual times when there is a markedly increased desire and need for quantitative data on the microbial risks associated with different classes of foods. These data would help regulatory authorities in reliable decision making for control programs of foodborne microbial hazards. Microbial hazards are almost always present, and "acceptable" levels should be defined. Recognition of this increased regulatory responsibility has been a major factor in the endorsement of the Hazard Analysis Critical Control Points (HACCP) system of food control,[8] which is one very important aspect of GMP.

The HACCP is a systematic approach to the identification, assessment, and control of hazards in a food operation by identifying problems before they occur and establishing measures for their control in those stages critical for safety.[9] However, even the HACCP is often used qualitatively and subjectively. A quantitative approach to HACCP should provide a better way to set proper criteria for critical process steps (critical control points [CCPs]), to execute control measures, and to optimize process. The quantitative approach can be created by the implementation of quantitative risk analysis (QRA) in existing HACCP systems.[6,10–12] Thus, terms and concepts originating from quantitative risk analysis are now being introduced, although there has been little experience in applying these terms in practice, and available information is insufficient at present for QRA to be applied in controlling food safety.[13]

Within HACCP plans, risk assessment involves two basic components: the identification and the assessment of hazards. The former may consist of a literature review of likely pathogens, epidemiologic data, surveys of the microbial composition of raw materials, etc. Once identification has been performed, it is necessary to determine which hazards can be present in raw materials and at the point of consumption. Bearing in mind that all steps, from production through consumption, will affect the food microflora, an assessment should be made of the impact of intrinsic, extrinsic, and other preservation factors used on the growth and survival of identified hazards[14] and, more importantly, which associated risk is or is not acceptable. Although a hazard may be present, the risk of illness related to that hazard may not necessarily be great.[15] Therefore, the new task of the food industry is to keep the level of risk to a minimum that is practically and technologically feasible.[15,16] The interaction between the diverse factors that affect microbial response in a food and the probability that a given microorganism will grow, survive, or die under these conditions can be studied by use of microbial challenge testing (MCT), storage tests (STs), and predictive microbiology (PM) tools. All of these instruments are very useful but are continuously discussed in terms of their limited validity for the conditions under which the experiments are conducted (MCT and ST) or on their extremely theoretical and predictive nature. These modern and traditional food microbiology concepts and their application to raw and minimally processed fruits and vegetables will be discussed in this chapter.

HACCP AND RISK ASSESSMENT

The conceptual approach to the production of microbiologically safe food and the glossary of terms presented in the fine work of Notermans et al.[13] are amply used in this section and recommended to the reader.

Good manufacturing processes and process control based on HACCP are important tools used for the food industry in achieving microbiologically safe food. The food industry applies GMPs and HACCP for controlling the food production process. Identification of CCPs may lead to a quantifiable reduction that a hazard will occur. Regulatory agencies, however, now face the task of inspection of control systems as the introduction of the HACCP is becoming frequent and integrated into legislation.

More directly related to the health of the consumers is the risk analysis. Regulatory agencies are primarily responsible for setting realistic microbiologic specifications by application of risk assessment, which is one step of risk analysis that also includes management and communication of risk.

A hazard is always potential, rather than actual, and may be associated with any agent in food or even a property of a food that may have an adverse effect on human health, whereas risk is a statistical concept directly related to a hazard. If the risk were zero (it will never be zero), no hazards will arise (hazards are always possible, and small risks must be accepted). Thus, for safety purposes, the use of GMPs will reduce the risks, but by applying the HACCP system, the risks should be reduced in a quantifiable manner. HACCP, in terms of risk analysis, should focus on those operations (practices, procedures, etc.) that can be managed, so that a desired level of safety (with an acceptable risk) can be attained. HACCP has a direct influence on the safety of the product, and the benefits should be quantifiable. Growth of microorganisms and recontamination of the product should no longer be considered as "hazards" but rather as "risk-increasing events." Thus, processing conditions should result in an acceptably safe product for which the risk of occurrence of adverse effects can be calculated in advance.[13]

Notermans et al.[17] state that, although the HACCP concept meets the needs of both food producers and legislative authorities, its use does not necessarily result in a food free from pathogens. Buchanan[14] indicates that, even if the HACCP approach may be simple conceptually, there are a number of sophisticated concepts that are neither well defined nor widely considered. The relationship between HACCP and microbiologic criteria illustrates this weakness well, whereas the use of risk assessment would establish this interaction on a quantitative basis, enhancing the refinement and evolution of the traditionally qualitative HACCP approach into a quantitative one.

Risk assessment is an analytical tool used to help define priorities for establishing public policies and has been used for managing several types of risks, including radiation control, chemical diseases, contamination of the environment and foods, water quality, and cancer prevention, such as in breast cancer. Application to microbiologic food safety is a recent focus, especially to specific food safety issues, such as the hazard of *L. monocytogenes* in milk, *E. coli* O157:H7 in ground beef, and *Salmonella* in egg.[18] Risk analysis is a means of making consistent, objective, and reliable assessment of risks. There is then a reliance on numerical expression of the risk implicit in its definition because it involves quantifying the probability of occurrence of an adverse health effect. This is the reason for the term *quantitative risk analysis* because it is based on quantitative data and models. QRA consists of six activities:

1. hazard identification
2. exposure assessment
3. dose-response assessment
4. risk characterization
5. risk management
6. risk communication.

Steps 1–4 are termed *risk assessment*. For hazard identification, data from consumer complaints, results of epidemiologic studies, microbiologic data, predictive models, etc., can be used. It is qualitatively acknowledged in this step that, for instance, *C. botulinum* may cause botulism, if present, and the toxin is formed in modified-atmosphere-packaged cabbage; *Shigella sonnei* may be present in shredded lettuce; and *Salmonella chester* may contaminate the surface of melons and cause shigellosis and salmonellosis, respectively.

Exposure assessment is determining the extent of human exposure before or after application of regulatory or voluntary controls.[18] What is the distribution of *Salmonella* species in fruits such as melons, and how is it affected by the quality of the irrigation water, the growing zone, the season, etc.? For exposure assessment to pathogenic microorganisms, data can be obtained from product surveillance, storage testing, challenge tests, and mathematical models that predict the likely number of microorganisms present in a food at the time of consumption. In practice, there has been little experience on exposure assessment for food-borne pathogens.[13]

Before regulatory agencies can conduct risk assessments for food-borne pathogens, dose-response models must be developed to predict infection at low doses.[19] Dose-response assessment defines the relationship between the magnitude of exposure and the probability of occurrence of the spectrum of possible health effects. What is the likelihood of becoming ill if 10 *Salmonella* cells are consumed, and how severe will the illness be?[18] In dose-response assessment, the quantitative estimation of risk at the time of consumption is made, based on the information obtained from dose-response relationships determined in human volunteer and animal model studies, as well as from epidemiologic analysis of food-borne diseases. This is one of the weakest components of risk analysis.[13] Because of ethical considerations, it is unlikely that adequate human dose-response data will become available for highly infectious agents, and animal models must be carefully reviewed for applicability to humans because of the inherent variability host/microorganism interaction.[19]

In the absence of actual data, results obtained from well-described outbreaks of food-borne disease can be used. To conduct a risk assessment for food-borne pathogens, a suitable dose-response model is necessary. Recently, Holcomb et al.[19] conducted a study to compare several models for their ability to fit microbial dose-response data. Comparison was accomplished by fitting the models to a range of dose-response data from the published literature for food-borne microbial pathogens. The Weibull-Gamma model, proposed by Farber et al.[20] to estimate infectious doses for *L. monocytogenes*, was the only model capable of describing the observed dose-response data for the range of food-borne microbial pathogens examined: *Shigella dysenteriae, Shigella flexneri, Salmonella typhosa,* and *Campylobacter jejuni.*

The risk assessment culminates then with risk characterization, which is intended to integrate the steps described above into a quantitative estimate (probability) of the adverse effects likely to occur in a given population and may identify additional economic and social impacts of human risk.[18]

This approach, however, is not yet well developed in relation to the microbiologic safety of food and has had little impact on food-borne pathogens. The reason for this is that, even if the goal is to be quantitative in the assessment of risk, this is often hindered by the lack of available data on microbiologic food safety risks.[13,18] This is particularly true for fruit and vegetable products.

As stated by Beuchat,[21] a quantitative microbiologic risk assessment of human infections and intoxications that can be linked to the consumption of contaminated raw fruits, vegetables, and plant materials should be undertaken. The development of a highly efficient, international epidemiologic surveillance system for better understanding the role of raw fruits and vegetables as vehicles for diseases is critical. Information generated by such a system would be valuable in establishing more meaningful practices and guidelines for preventing contamination and for decontamination.

As for HACCP, it has been extensively discussed elsewhere, and its seven steps, as set out by the Codex Alimentarius Commission[22] in 1991, have evolved and become part of the common terminology of food industry and government:

1. Conduction of a hazard analysis;
2. Determination of CCPs in the process;

3. Specification of criteria (establishment of target levels—critical limits—and tolerances for preventive measures associated with each identified CCP);
4. Implementation of CCP monitoring systems;
5. Corrective action (establish corrective actions to be taken when monitoring indicates that a particular CCP is not under control);
6. Verification (establish procedures for verification that HACCP system is working correctly); and
7. Documentation (establish documentation concerning all procedures).

The first step, or hazard analysis, is to identify potential hazards associated with food production at all stages to the point of consumption. It consists, in turn, of hazard identification, assessment of the likelihood of occurrence of hazards, and identification of preventive measures for their control. In relation to the HACCP concept, a *hazard* had been identified as "any aspect of the food production chain that is unacceptable because it is a potential cause of food safety problems."[23] More recently, the World Health Organization (WHO)[24] defined the *HACCP-hazard* as "a biological, chemical or physical agent with the potential to cause an adverse health effect when present at an unacceptable level." This definition complies with the actual HACCP approach of becoming "a quantitative" system.

Stier,[25] in the important reference produced by Pearson and Corlett in 1992 on principles and applications of HACCP, presents an HACCP model developed for shredded lettuce packaged in gas-permeable bags, based on an actual food operation used by California lettuce packers. (Figure 5–1). By the time of the publication, the model was still under consideration. In the model, there are CCPs for chemical hazards (pesticides, chemicals, fertilizers), and physical hazards (metals and foreign objects), but the majority are of microbial nature. Lettuce is a product extremely sensitive to abuse conditions, so controls to ensure safety will also help maintain quality. CCPs are identified at the coring operation—trimming and sorting and chopping—to control potentially harmful organisms and physical hazards. The keys to the former are proper equipment design and cleaning, which includes rinsing during operation. There are also other CCPs related to maintaining chlorine levels in water systems, which are monitored continuously and adjusted as needed. Basket loading and centrifuge operations are timed and considered as CCPs, and if the established times are exceeded, the product should be discarded. The system also includes a general sanitation CCP, which may be considered a GMP and not a CCP, because it includes education of cleaning crews, compliance with cleaning protocols, maintenance staff, etc. The operation cannot be started until established cleaning and sanitizing protocols have been completely reviewed by management. There are also two CCPs related to packaging and coding. Final CCPs deal with checking of package integrity and with maintenance and monitoring of temperatures of refrigerated vans and at retail level, with possible use of time–temperature indicators on the package. As can be seen, many of these CCPs have a strong qualitative component.

To date, even if HACCP is not yet mandatory for freshly cut produce, the industry has been conducting significant efforts in the areas of HACCP to ensure the safety and wholesomeness of minimally processed fruits and vegetables. The International Fresh-cut Produce Association has suggested HACCP guidelines for fresh produce. However, apart from internal standards of companies and with the exception that the product can contain no viable pathogens, there are no microbial criteria for such products. The model described by Stier[25] is a generic model for many fresh fruits and vegetables. Most large produce companies are using some form of HACCP, but most have been developed by their own staffs. Most companies have adopted only two CCPs: a) chlorine concentration in wash and flume waters and b) metal detection after bagging. Other important control points (temperature control, film type, etc.) are covered under

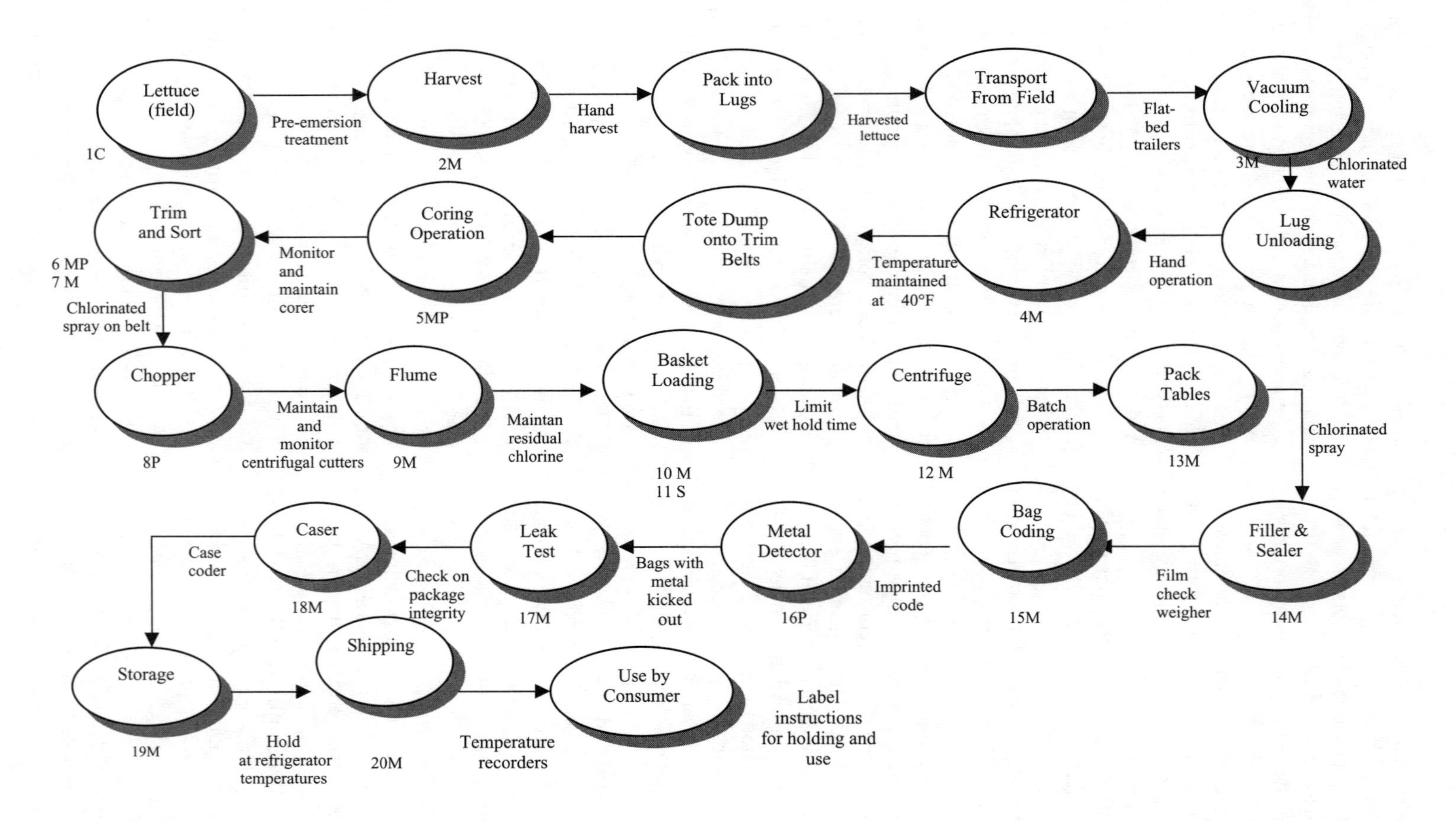

Figure 5–1 HACCP model proposed for production of minimally processed (shredded) lettuce. Critical control points: M = Microbiological, C = Chemical, P = Physical, and S = Sanitation.

standard operating procedures (SOPs).[26] The U.S. Food and Drug Administration (FDA) has published a guidance document for producers that is intended to help reduce microbiologic hazards common to the growing, harvesting, washing, sorting, packing, and transporting of fruits and vegetables. It was prepared in consultation with the U.S. Department of Agriculture (USDA) in response to a Presidential directive on food safety issued in October 1997. Once the produce has gone through the steps outlined in the guide, it may be minimally processed, canned, frozen, or dried.[27]

Notermans et al.[17] propose an approach for setting criteria at CCPs for bacterial hazards associated with food-borne disease and address such methods as tools to calculate or predict the numbers of organisms expected to be present in final food products, leading to decision making on the basis of levels of acceptability. Thus, human exposure to potentially hazardous pathogens can be evaluated to provide a quantitative risk assessment. In relation to quantitative risk analysis, CCPs can be considered as operations (e.g., steps, processes), where risks can be reduced through control of procedures, practices, etc. CCPs will be meaningful if they can be managed in such a way that a risk is reduced and the reduction can be quantified. In this context, the authors present an approach to establish the identity of quantitative CCPs. A list of operations and variables that exert a quantitative control on potential hazards should be made so that parameters important for a particular product and its production process are defined and will lead to posterior decision on whether or not they can be utilized to reduce or stabilize a potential hazard. In the case that it is utilized, the next question will be whether the effect is nullified by a subsequent process or procedure. If this is not the case, it is necessary to decide whether the operation controls the hazard in a quantifiable manner. Control does not necessarily mean a decrease in the number of hazardous organisms. In some cases, stabilization of numbers in prevention of growth may be sufficient.[17]

Table 5–1 presents the list of operations, variables, etc., that control potential hazards quantitatively, as proposed by Notermans et al.,[17] and presents a same type of list adapted to some minimally processed fruit and vegetable operations and variables. It also shows some examples of quantitative effects observed by application of some such operations or variables that control potential hazards quantitatively in minimal processing.[28–36]

Conventional evaluation of sanitizing treatments for plant foods is based on colony count data, which compare the level of a target microorganism on the plant tissue before and after the sanitizing treatment. Generally, counts are obtained after separating the target microorganism from the tissue, usually by stomaching. This result provides information on the treatment effect and serves well for calculation of quantitative effects but provides little information on how or why some treatments are more effective than others. An interesting work conducted by Seo and Frank[37] uses confocal scanning laser microscopy to observe the attachment and the location of artificially inoculated *E. coli* O157:H7 (ca.10^7–10^8 CFU/mL) over and within lettuce leaves. Viability was also evaluated by differential staining of living and dead cells in response to chlorine treatment as a technique that can be used to evaluate the conventional colony enumeration method and to quantitatively confirm the preferential attachment to the cut edges and the effectiveness of the decontamination treatments.

Disinfectants are effective in reducing numbers of disease-causing microorganisms on raw fruits and vegetables, but their efficacy depends on the types of fruits or vegetables and characteristics of their surfaces, the methods and procedures used for disinfection, and the types of pathogen. *Listeria monocytogenes* is generally more resistant to disinfectants than is *Salmonella,* pathogenic *Escherichia coli*, or *Shigella*, but little is known about the efficacy of disinfectants in killing parasites and viruses on fruits and vegetables. Vigorously washing fruits and vegetables with potable water reduces the

Table 5–1 Operations or Variables that Control Potential Hazards Quantitatively in Minimal Processing of Fruits and Vegetables

List of Operations, Variables, etc., that Control Potential Hazards Quantitatively		*Examples of Quantitative Effects Observed by Application of Some Operations or Variables that Control or Affect Potential Hazards Quantitatively in Minimal Processing of Fruits and Vegetables*
As presented by Notermans et al.[17]	Adapted to some minimally processed fruit and vegetable operations and variables	
Presence of potentially hazardous organisms in raw materials	Presence of potentially hazardous organisms in raw materials	*Aeromonas* spp. on fresh vegetables on the day of purchase vs. 7 days on refrigerated storage (3°C)[28] –Lettuce 1.00×10^2–3.50×10^2 vs. < 1.00×10^2–7×10 –Watercress 6.10^4–2.5×10^6 vs. 4.10^6–6.75×10^7 –Cabbage 1.10^2–2.5×10^2 vs. < 1×10^2 –Chicory 4.5×10^2–7.45×10^4 vs. 4×10^2–2×10^5 –Parsley 8×10^2–5.85×10^4 vs. 2×10^5–1×10^6 –Celery 2×10^2–3.4×10^6 vs. < 1×10^2–1.04×10^4
Processing operations –heating (pasteurization), etc. –irradiation –bactofugation –high-pressure treatment –washing	Processing operations –coring –trimming and sorting –chopping –washing or dipping in chlorinated water antimicrobial solutions	–*L. monocytogenes* on brussels sprouts[29] after dipping in 200 ppm chlorine: 2 log decrease vs. 1 log decrease after dipping in water –*L. monocytogenes* on freshly cut vegetables[30] after dipping in 200 ppm chlorine, 10 min.: 4°C:1.3 log decrease/22°C:1.7 log decrease (lettuce); 4°C:0.9 log decrease/ 22°C:1.2 log decrease (cabbage) –*L. innocua* (as model for *L. monocytogenes*) on shredded lettuce[31] after dipping in 100 ppm chlorine, 5 min.: 1 log decrease at time of treatment –*L. innocua* (as model for *L. monocytogenes*) on shredded lettuce[31] through 14 days of storage undip vs. dip in 100 ppm chlorine/1% citric acid: 3°C:1–1.5 log decrease/ 8°C; no decrease observed (no significant change in numbers) vs. significant increase at 8°C in dipped lettuce –*Salmonella* on alfalfa sprouts[32] after dipping on 500 and 2,000 ppm chlorine, 2 min.: 2 log decrease and reduction to undetectable levels, respectively

List of Operations, Variables, etc., that Control Potential Hazards Quantitatively		*Examples of Quantitative Effects Observed by Application of Some Operations or Variables that Control or Affect Potential Hazards Quantitatively in Minimal Processing of Fruits and Vegetables*
Product formulation (intrinsic factors) –drying (a_w) –acidification (pH) –preservatives (nitrite, bacteriocins) –starter cultures	Product formulation (intrinsic factors) –acidification (pH)	–*Aeromonas hydrophila*[33] on minimally processed green vegetable salads through 8 days of refrigerated storage at 3°C. MP green vegetable salads seasoned with acidified dressing: 2.23 log decrease vs. 2.77 log increase in nonseasoned green vegetable salads –pH < 4.0 is considered as a growth-controlling factor in foods of *Y. enterocolitica*[34] –pH can be a noncontrolling growth factor, due to evidence of acid tolerance of pathogens such as *E. coli* O157:H7[35,36]
Extrinsic factors –storage temperature –controlled atmosphere –smoking –storage time	Extrinsic factors –storage temperature –modified atmosphere (MAP) –N_2 flushing	–*L. monocytogenes* on minimally processed shredded lettuce through 14 days of storage[31] –Temperature 3°C: 1–1.5 log decrease/8°C; no decrease observed (no significant change in numbers). –Temperature and N_2 flushing: 3°C: 1 log decrease/8°C 1.5–2.5 log increase. –Temperature and MAP:3°C: 1 log decrease/8°C slight or no increase (no significant change in numbers).*

*Note: Storage temperature is the single most important factor affecting the growth of microorganisms in MP&V. Adequate refrigeration temperatures will limit pathogen growth to those which are psychrotrophic. Temperature abuse may reduce lag and generation time of such organisms.

number of microorganisms by 10- to 100-fold and is often as effective as treatment with 200 ppm chlorine. Treatment with chlorine dioxide, trisodium phosphate, organic acids, ozone, and irradiation has potential for removing pathogenic microorganisms from raw fruits and vegetables. Avoidance of contamination at all points of the food chain, from primary production to the consumer, is preferred over the application of disinfectant after contamination occurs.[21]

Risk assessment concerns the overall product safety and is applied to analysis of the food product, as presented to the consumer (analysis at end point), whereas HACCP enhances overall product safety by assuring day-to-day process control and may be applied at any point in the processing/handling chain.[18] Confusion arises often between risk assessment and HACCP in the point of hazard identification being the first activity in both QRA and HACCP—recognition of microorganisms of concern. As mentioned above, Notermans et al.[6,17] and Buchanan[14] have suggested the use of the four steps of risk assessment for specifying the microbiologic criteria of HACCP systems (in hazard analysis and in the setting of critical limits), i.e., to use QRA as a part of HACCP. Foegeding[18] broadens this approach by considering both risk assessment and HACCP as part of the risk analysis, with HACCP representing one management strategy. On the other side, the Codex Alimentarius Commission has focused its attention on risk analysis in the elaboration of standards and guidelines for the international trade in food for all classes of food-borne hazards.[8]

HAZARD IDENTIFICATION

The importance of hazard identification as the first common step of HACCP and QRA is recognized elsewhere. It is considered a step of both risk assessment and HACCP, where a lot of comprehensive knowledge exists. Good general information exists about food-borne microorganisms that may cause disease and the severity of the disease, even if there are still obscure areas in emerging pathogens, as well as on viral pathogens.[18]

Notermans et al.[6] present an approach to identify potentially hazardous bacteria in a given situation, based on a list of all bacteria known to cause food-borne disease. This list is used for determining whether the microorganisms are likely to be present in the raw materials used, and those organisms that had never been found were deleted. Of the remaining organisms, it must be established whether they are destroyed by processing (if so, they will be removed from the list) and whether recontamination is likely to occur with a pathogen (if so, it must be included in the list). The next question is whether the listed organisms have ever caused a food-borne disease involving either an identical or related product. If this is not the case, the organism is to be deleted from the list. All remaining organisms are discriminated into infectious or toxinogenic. All infections are regarded as potentially hazardous, and only those toxinogenic pathogens capable of growing are considered to be potentially hazardous. A more precise evaluation of the hazards is to be made during the identification of the CCPs, setting control criteria at each one of them, and during the step of verification.

van Gerwen et al.[12] implemented a hazard identification procedure as a computer program, based on the general approach of Notermans et al.,[6] described above, differing by its stepwise identification of important hazards and its interactive character. It is the first step of a procedure for quantitative risk assessment to be completely developed as a computer-aided system. It will be described in the following section.

A major concern with minimally processed vegetables is the survival and growth of several pathogenic organisms that can be traced to raw produce, plant workers, and processing environment. *Listeria monocytogenes* has been isolated or can survive and grow on many raw or processed vegetables, such as shredded iceberg lettuce, at refrigeration temperature.[38–41]

Odumer et al.[42] assessed the microbiologic quality of ready-to-use (RTU) vegetables for health care food service, including chopped lettuce, salad mix, carrot sticks, cauliflower florets, and sliced green peppers, before and after processing and after 7 days of storage in hospital coolers. The authors also determined the effects of storage temperature and time on the microbial population of RTU vegetables. Microbial profiles were obtained 24 hours after processing and on days 4, 7, and 11 after storage at 4°C and 10°C to simulate temperature abuse. *Listeria monocytogenes* was isolated from 6 of the 8 vegetable types tested, representing 2.8% of samples stored at 4°C and 13 of 129 samples at 10°C. Recommendations regarding processing, distribution, and storage of such products are presented by the authors. Temperature abuse resulted in a significantly higher incidence, as well as increased *L. monocytogenes* in RTU vegetables. Temperature is a CCP for the maintenance of the quality and safety of these products. Under optimum storage conditions, these results do not suggest an increased risk of *L. monocytogenes* isolation from RTU vegetables, but the need for continued surveillance for this pathogen and strict monitoring of CCPs during processing and handling is stressed.

PREDICTIVE MICROBIOLOGY

Historically, food microbiology had been an active area for mathematical modeling (i.e., the calculation of thermal resistance and process time). Today, the use of mathematical models to describe the growth, survival, and inactivation responses of food-borne microorganisms under specific environmental conditions has given rise to one of the most rapidly advancing subspecialties in food microbiology—predictive microbiology. Microorganisms are cultured under a variety of intrinsic factors, such as water activity (a_w), pH, etc., and extrinsic factors, such as temperature and gaseous atmosphere; their response is measured, and the resulting data are fitted to a mathematical equation. A large number of factors affect the microbe, but in most foods, only a few exert most of the control. Modeling does not usually reveal unexpected microbial behavior but quantifies the effect from the interaction between two or more factors and allows interpolation of combinations of factors not explicitly tested. The systematic quantification and understanding of these factors in model systems and prototype products make possible the generation of effective models that can estimate microbial behavior in a range of products. These models can subsequently provide the industry with an important tool for making objective initial assessment to establish priorities in relation to both product design and evaluation.[43]

Bacterial numbers can change at all stages during food production and processing, depending on the way that a food is handled. Predictive microbiology can then be used to assess the effects of processing, product composition, storage conditions, etc., on the final contamination rate of a product at the time of consumption.[44] A number of predictive mathematical models havebeen developed for use in food microbiology. Several useful reviews describe the development in this field and explain the various models developed.[43,45] The adoption of the new techniques in predictive microbiology by the food industry will ultimately be dependent on the development of user-friendly software applications for personal computers that make it easy for nonresearch personnel to employ the mathematical models. Two modeling programs currently exist. One is the Pathogen Modeling Program,[46] developed by USDA's Eastern Regional Research Center at the Microbial Safety Food Research Unit (Wyndmoor, PA), currently in its latest version of October 19, 1998. This software automates response–surface models for the effects of storage temperature, initial pH, NaCl content (a_w), sodium nitrite concentration, and oxygen availability on most of the common bacterial pathogens. The other available software is the Food MicroModel,[47] developed by the United Kingdom's Ministry of Agriculture, Fisheries and Food, and available from Leatherhead Research Laboratory (Surrey, England), with predictive equations for growth, survival, and death of pathogens. It utilizes the context of an extensive expert system/database for food microbiology.

Current and newly obtained knowledge, facts, and expert opinion, along with mathematical models, can be used to build computerized systems related to food safety and quality issues. *Expert systems* are computer programs that attempt to emulate the performance of human experts within a special domain of expertise. Even if they are no longer a topic of research in artificial intelligence per se, their application in food microbiology continues to be pertinent and useful, especially because the systematic approach to problem solving is being used in this field, making knowledge more organized and structured. This computer modeling technique formalizes the thinking process of experts in a field such as food microbiology, coupling this with objective tools such as mathematical modeling.

The structure most commonly used to represent expert knowledge in software is described as a *knowledge base*, which normally stores the expertise in the form of rules and/or objects. Other components of expert systems are an interface and an inference engine with an interpreter and scheduler that applies the inference rules in an appropriate order. Rules, or conditional "if/then" statements, are common in scientific fields, and expert systems have been successfully applied in areas where knowledge is well established and organized, such as medical diagnosis, chemical analysis, and oil prospecting. Such rules also work successfully in expert systems developed for tax regimes, pension plans, and computer configuration.[48,49]

Several corporations are currently developing impressive expert systems for assessing microbiologic relations within food products. An interesting example is presented by Adair and Briggs[49] of Unilever Research (Colworth House, England), describing a prototype system that assesses the microbiologic safety of chilled, ready-to-eat meals. The system contains knowledge for these meals and concentrates on the design, rather than on the diagnosis or faults, of the products.

This system is currently being applied in many food products manufactured by Unilever.[50]

In 1992, Zwietering et al.[51] published an excellent work on a decision support system that combines quantitative and qualitative information to predict possible spoilage reactions with an estimate of their kinetics, based on models. The system consists of a database with characteristics of foods, based on their physical properties, a database of microorganisms, with growth limits for the same physical variables. A combined model developed makes kinetic estimations, based on data in the database. Addition of qualitative reasoning is made in the form of knowledge rules concerning products and microorganisms to diminish the number of organisms in the list or to improve the value of the prediction. The system combines all of this information. Because it is impossible to collect quantitative data for all possible deterioration reactions on different products, a prediction is made on the basis of the data and knowledge collected in the system. The program can help to determine possible spoilage organisms and to estimate the change in growth rates of organisms when the physical properties are changed.

A second paper on a computerized decision support system was published by Wijtzes et al.[52] that simulates composition, production, and distribution of foods. The computer system uses both mathematical models and expert knowledge of food technologists for the determination of the microbial numbers in foods, with quality and safety of foods under production or distribution prone for prediction. This system uses databases of the system developed by Zwietering et al.[51] but calculates the effects of processing and distribution, and is able to determine shelf life of foods.

van Gerwen et al.[12] developed a stepwise identification procedure for food-borne microbial hazards, implemented as a computer program that performs systematically the first step of QRA. It consists of a food database, a pathogen database, and a knowledge database. The food database reported by Zwietering et al.[51] is extended, with information on presence (of groups) of microorganisms and food-borne outbreaks in the past; the organism database is changed into one containing only pathogens. Relevant hazards are identified by several levels of detail: rough (the most obvious hazards, having caused food-borne outbreaks via the specified product), detailed (pathogens that have been reported present in the ingredients of the product), and comprehensive (all pathogens are identified as hazardous) hazard identification. The knowledge database introduces pathogen-related knowledge rules whose sources are literature knowledge and expert knowledge. Knowledge rules were developed from the literature, then experts in the field of food microbiology were asked for opinions on these rules, and the rules were changed and reworded accordingly. Clearly defined, explicit rules that can be criticized and refined are used to reduce impractically long lists of pathogens and are of three types:

1. rules concerning presence or absence and survival or inactivation of pathogens, allowing selection of those organisms capable of doing so in the end product under normal and hygienic circumstances;
2. general rules on pathogen characteristics that allow for selection of pathogens likely to cause problems in the practice; and
3. rules concerning growth opportunities and toxin production to select for pathogens capable of doing so in the product.

It is recommended that risk assessments begin with selection of pathogens with these procedures. The computer program starts with a selection of a product and product characteristics, then constructs a process flow sheet. The user must then choose the level of detail. A list of pathogens that, according to the user and the information from the databases, are hazardous is produced. This list can be modified by adding or removing pathogens and by applying the knowledge rules.

van Gerwen[53] has recently applied the hazard identification procedure to vegetables and fruits. When applied to vegetables, a list of 10 organisms was obtained as a result of a rough identification. The application of a detailed level increased the list to 24 pathogens, and when the three types of knowledge rules were applied, the list of hazards varied significantly. If rules of type 1 are not included, for example, because it is known that there will be recontamination of the product after heat treatment, the combination of other two types of rules (2 and 3) resulted in the identification of the following organisms as hazards in vegetables: *B. cereus, E. coli, L. monocytogenes, Y. enterocolitica*, and *Salmonella*, *Shigella*, and *Staphylococcus* species. The results of the hazard identification procedure depend on the data available in the databases, which contain information on outbreaks or presence of microorganisms reported in the literature in the products of concern and codified as part of the food database.

There is a good amount of information on cabbage. Table 5–2 shows the results of van Gerwen[53] after applying the rough identification procedure to cabbage. *Listeria monocytogenes* was found to be the only identified hazard. When the detailed hazard identification was run, the number of hazards increased to make up a list of 27 pathogens, to which the three types of rules were applied. The search in the food database was made under green, red, white, and savoy cabbage.

Particularly for fruit, the limitations of the hazard identification procedure were obvious because, apparently, there are very little data in the food database for fruits or for fruits and nuts. The databases are not complete, and they should be updated regularly to improve hazard identification in the future. However, the databases contain much information useful in performing reliable hazard identification for several products.[53]

MICROBIAL CHALLENGE TESTS AND STORAGE TESTS

General information on spoilage and pathogenic organisms can be found in the literature and in databanks, and can be extrapolated to a large range of products through predictions that, if not exact, can constitute a good approximation of the effects of product composition, processing, storage time, and temperature, etc., on numbers of organisms. In the case of pathogenic organisms, model predictions should not be used for "fine-tuning" of processes because of their inherent inaccuracy. Storage tests and MCT then become necessary.[9]

Storage tests are performed on finished products in situations where the organisms of interest are present in sufficient number and provide information that is pertinent only to that particular product. Storage tests can be carried out under normal and/or abuse conditions, taking precautions on the drawing of conclusions from the results obtained. Performing challenge tests aiming to study the fate of pertinent microorganisms has been one of the traditional ways to assess product stability. Microbial challenge testing has become an established technique in the food industry to simulate what can happen to a product during processing, distribution, and subsequent handling, following inoculation with one or more relevant microorganisms and further holding under controlled conditions. Depending on the type of product and potential risk microorganisms, challenge studies will have different degrees of complexity. An essential reference on challenge testing is that of Notermans et al.[54] Both ST and MCT lack an important predictive component in assessing the changes in product formulation, processing, or packaging.

Predictive modeling, ST, and MCT constitute methods to generate information on the effects of different factors controlling safety of food products and can also be used for setting quantitative criteria.

Carlin and Nguyen-the[55] performed CT with *L. monocytogenes* on four types of minimally processed green salads widely consumed in Europe and found that the pathogen was able to survive and grow on broad-leaved endive, butterhead lettuce, and, to a lesser extent, on curly-leaved endive, whereas lamb's lettuce did not seem to be a good substrate for the organism. From CT, it can be assumed that the risk of intoxication by consuming contaminated salads increases with storage time for broad-leaved endive, butterhead lettuce, and, to a lesser extent, curly-leaved endive, but not for lamb's lettuce.

Marconi[56] challenged curly-leaved endive with *L. monocytogenes* (5.27 log CFU/g), stored it for 9 days at 3°C, and found that the organism survived with no significant decrease of the initial population. No correlation was found with development of lactic acid bacteria, mesophilic aerobic, psychrophilic aerobic, or pH changes during the period examined (Figure 5–2).

Carlin et al.[57] challenged minimally processed broad-leaved endive with *L. monocytogenes* and studied the effects of temperature (3, 6, 10, and 20°C), characteristics of the leaves, and characteristics of the concentration and strain of the inoculum on the fate of the organism. It can be calculated from this work that an initial population of 10 CFU/g^{-1} *L. monocytogenes* on leaves of endive could increase to 5×10^3 CFU/g^{-1} or 5×10^4 CFU/g^{-1} after storage of leaves at 10°C for 4 days without extensive spoilage. Setting a limit of 100 CFU/g^{-1} at the manufacturing level, as has been proposed for minimally processed fresh vegetables,[50] could result in products containing up to 5×10^5 CFU/g^{-1} after a 4-day storage period at 10°C. A limit of 100/g^{-1} *L. monocytogenes* in minimally processed fresh vegetables at the level of consumption is recommended in France. According to results of this work, this limit can be achieved only by using a strictly controlled chain of refrigeration at processing, storage, and retail, with temperatures never exceeding 3–4°C.[58] In Germany, official recommendations for microbiologic criteria have been published by the Federal Health Office, which include a quantitative test for *L. monocytogenes*. If the organism is present but numbers are $< 10^2$

Table 5–2 Application to Cabbage of the Computerized Stepwise Identification Procedure for Food-Borne Microbial Hazards Developed by van Gerwen et al.[12] Includes results after application of the three types of knowledge rules† to the list of hazards resulting from the detailed hazard identification.

		Knowledge Rules			
Rough Hazard Identification	*Detailed Hazard Identification*	*Type 1*	*Type 2*	*Type 3*	*Types 1&2&3*
Listeria monocytogenes	*Acinobacter* spp.		X		
	Alcaligenes spp.		X		
	Bacillus spp.	X	X		
	Bacillus anthracis	X		X	
	Bacillus cereus	X	X	X	X
	Chromobacterium spp.		X		
	Chromobacterium violaceum				
	Clostridium spp.	X	X		
	Clostridium botulinum type A	X	X		
	Clostridium perfringens	X	X		
	Corynebacterium spp.		X		
	Enterococcus spp.		X		
	Escherichia coli		X	X	
	Flavobacterium spp.		X		
	Klebsiella spp.				
	Listeria monocytogenes		X	X	
	Pasteurella multocida		X		
	Plesiomonas shigelloides				
	Pseudomonas spp.		X		
	Pseudomonas aeruginosa			X	
	Salmonella spp.		X	X	
	Serratia spp.		X		
	Shigella spp.		X	X	
	Staphylococcus spp.		X	X	
	Streptococcus spp.		X		
	Vibrio cholerae		X		
	Yersinia enterocolitica		X	X	

*Green, red, white, and Savoy cabbage are part of the food database under specific codes. The detailed hazard identification also searched for pathogens at other codes: cabbage, leafs, flowers, roots, vegetables, and ingredients.

†Type 1: knowledge rules concerning presence or absence and survival or inactivation of pathogens that allow selection for those organisms capable of doing so in the end product, under normal and hygienic circumstances.

Type 2: general knowledge rules on pathogen characteristics that allow selection for pathogens likely to cause problems in the practice.

Type 3: knowledge rules concerning growth opportunities and toxin production that select for pathogens capable of doing so in the product.

g^{-1}, no further action is taken. If numbers are $>10^2$ g^{-1}, further investigations are made. The current target in the United Kingdom and the United States is absence of *L. monocytogenes* in 25 g of ready-to-eat-foods.[59]

Survival of *Staphylococcus aureus*, *Bacillus cereus*, *L. innocua*, *Salmonella typhimurium,* and *C. perfringens*, inoculated in separated experiments of MCT performed on modified-atmosphere shredded carrot and cabbage, was

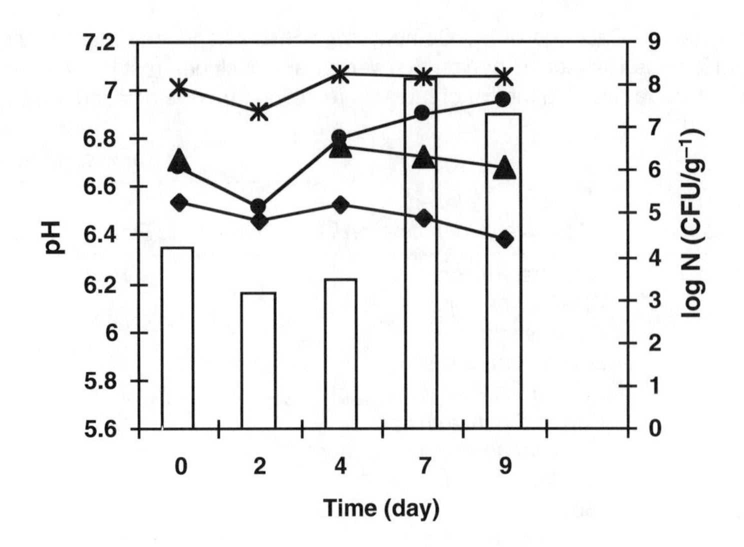

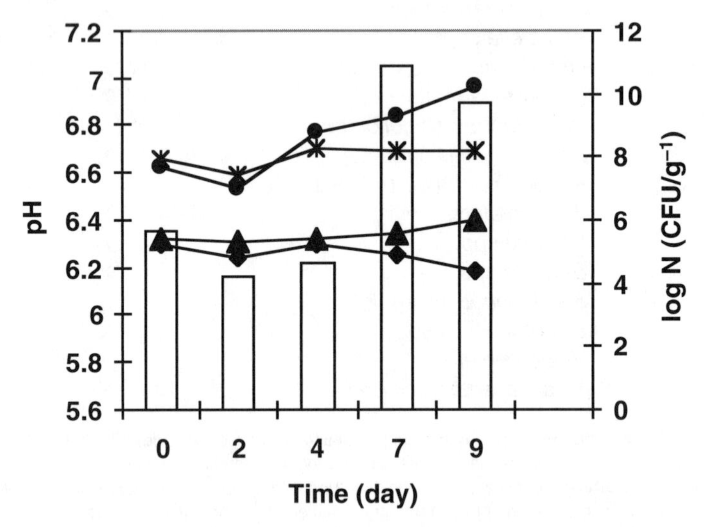

Figure 5–2 Growth of *Listeria monocytogenes* (◆), lactic acid bacteria (▲), mesophilic aerobics (✻); psychrotrophic aerobics (●), and pH (*bars*) evolution in two batches of minimally processed curly-leaved endive (*Cichorium endivia* L.) stored at 4°C.

studied by Finn and Upton.[60] In all cases, significant decreases were obtained, due probably to the high numbers of lactic acid bacteria that might have exerted an antimicrobial effect. The antilisterial activity of carrot is well documented.[61,62]

The ability of *C. botulinum* types A, B, and E spores to grow and produce botulinal toxin has been investigated through the performance of CT on five vegetables (lettuce, cabbage, broccoli, carrots, and green beans) packaged under vacuum or in air, using polyethylene bags of different transmission rates for O_2 and CO_2.[63] Seven proteolytic and three nonproteolytic strains were inoculated and incubated at 4, 12, or 21°C. No botulinal toxin was detected in any cabbage, carrot, or green bean samples, or in any inoculated control samples, although it was detected in some of the grossly spoiled samples maintained at abuse temperatures. Thus, consumer risk does not appear to be significantly increased due to the lack of toxin production before gross organoleptic spoilage had occurred.

CONCLUSION

Minimally processed fruit and vegetable products occupy an important niche in the modern food industry. They are not absent of neither hazards nor risks. An attempt was made to revise and place into a coordinated frame traditional and emerging approaches in the search for safety of such products. The food safety challenge that represents minimal processing and packaging, tailored to retain quality and microbiologic safety, demands maximum and continuous investigation efforts by scientists as well as regulators who are primarily responsible for setting realistic microbiologic specifications. Only through sustained research with cooperation and education of all parties involved can the microbial safety of minimally processed fruits and vegetables be obtained. Therefore, it is of paramount importance that food processors become familiar with the different tools that allow them to handle the elements of processing, packaging, and distribution. Those tools should be optimized to deliver safe minimally processed fruit and vegetable products for which the risk of occurrence of adverse effects could be decreased. Contamination should be prevented to minimize risks through the entire process, from field to store and to the homes of the consumers. There is a need to understand the importance of how pathogens can compromise the safety of final products during the different steps of the mild preservation technologies and by possible abuses that may have hazardous results. Predictive modeling, ST, and MCT allow generation of information for such purposes and for helping to set quantitative criteria for critical process steps to execute control measures and to optimize process in a quantitative approach to HACCP. Fruit and vegetable growers and handlers should be informed of the risks associated with pathogenic microorganisms related to raw fruits and vegetables to control microbiologic hazards that may be influenced by current and changing practices in aquaculture, agronomy, processing, marketing, and preparation. A quantitative microbiologic risk assessment of human infections and intoxications that can be linked to the consumption of contaminated raw fruits, vegetables, and plant materials, as well as minimally processed fruits and vegetables, should be undertaken.

REFERENCES

1. Tapia de Daza MS, Díaz de Tablante RV. Consideraciones ecológicas y de inocuidad alimentaria en productos de origen vegetal. *Arch Latin Nutr*. 1994;44:232–241.
2. Gould GW. Ecosystem approaches to food preservation. *J Appl Bacteriol Symp Suppl.* 1992;73:58S–68S.
3. Wolf I. Critical issues in food safety, 1991–2000. *Food Technol*. 1992;46:4–70.
4. Lund B. Ecosystems in vegetables foods. *J Appl Bacteriol Symp Suppl.* 1992;73:115S–126S.
5. Hotchkiss J, Banco M. Influence of new packaging technologies on the growth of microorganisms in produce. *J Food Prot.* 1992;55:815–820.
6. Notermans S, Gallhoff G, Zwietering MH, Mead GC. The HACCP concept: Identification of potentially hazardous micro-organisms. *Food Microbiol.* 1994;11: 203–214.
7. Notermans S, Jouve JL. Quantitative risk analysis and HACCP: Some remarks. *Food Microbiol.* 1995;12: 425–429.
8. Hathaway SC, Cook RL. A regulatory perspective on the potential uses of microbial risk assessment in international trade. *Int J Food Microbiol.* 1997;36:127 133.
9. Notermans S, Gallhoff G, Zwietering MH, Mead GC. The HACCP concept: Specification of criteria using quantitative risk assessment. *Food Microbiol.* 1995;12:81–90.

10. Corlett DA, Stier RF. Risk assessment within the HACCP system. *Food Control.* 1991;2:71–72.
11. Buchanan RL. The role of microbiological criteria and risk assessment in HACCP. *Food Microbiol.* 1995;12:421–424.
12. van Gerwen SJC, de Bit JC, Notermans S, Zwietering MH. An identification produce for foodborne microbial hazards. *Int J Food Microbiol.* 1997;38:1–15.
13. Notermans S, Mead GC, Jouve, JL. Food products and consumer protection: A conceptual approach and a glossary of terms. *Int J Food Microbiol.* 1996;30:175–185.
14. Buchanan R. The role of microbiological criteria and risk assessment in HACCP. *Food Microbiol.* 1995;12:421–424
15. Panisello PJ, Quantick PC. Application of food micromodel predictive software in the development of hazard analysis critical control point (HACCP) systems. *Food Microbiol.* 1998;15:425–439.
16. World Health Organization. *Applications of Risk Analysis to Food Standard Issues.* Report of the Joint FAO/WHO Expert Consultation, 13–17 March 1995. Geneva, Switzerland.
17. Notermans S, Gallhoff G, Zwietering MH, Mead GC. Identification of critical control points in the HACCP system with a quantitative effect on the safety of food products. *Food Microbiol.* 1995;12:93–98.
18. Foegeding PM. Driving predictive modelling on a risk assessment path for enhanced food safety. *Int J Food Microbiol.* 1997;36:87–95.
19. Holcomb DL, Smith MA, Ware GO, Hung YC et al. *Risk Assessment Models for Food-Borne Pathogens.* II Congreso Venezolano de Ciencia y Tecnología de Alimentos. Dr. Asher Ludin. Caracas, Venezuela, April 24–28, 1999 (Proc. Abstract).
20. Farber M, Ross WH, Harwig J. Health risk assessment of *Listeria monocytogenes* in Canada. *Int J Food Microbiol.* 1996;30:145–154.
21. Beuchat LRB. *Problems Associated with Pathogenic Microorganisms on Raw Fruits and Vegetables.* II Congreso Venezolano de Ciencia y Tecnología de Alimentos. Dr. Asher Ludin. Caracas, Venezuela, April 24–28, 1999 (Proc. Abstract).
22. Codex Alimentarius Commission, Committee on Food Hygiene. *Draft Principles and Application of Hazard Analysis Critical Control Point (HACCP) System.* 1991; Aliform 93/13 VI. FAO/WHO.
23. ILSI Europe. *A Simple Guide to Understanding and Applying the Hazard Analysis Critical Control Point Concept.* ILSI Europe concise monograph series. Washington, DC: In press.
24. World Health Organization (WHO). *A Proposal for Amendment of the Codex Guidelines for the Application of Hazard Analysis Critical Control Point System.* Geneva, 1995.
25. Stier R. Practical Application of HACCP. In: Pearson MD, Corlett DA, eds. *HACCP Principles and Applications.* New York: AVI Book;1992:127–167.
26. Brackett RE. *Personal Communication.* 1999.
27. FDA/USDA. Guide to minimize microbial food safety hazards for fresh fruits and vegetables. *Federal Register* 63 FR 58055. http://vm.cfsan.fda.gov/~lrd/fr981029.html. Accessed October 29, 1998.
28. Díaz RV. *Enumeración y Caracterización de* Aeromonas *sp. en Productos de Origen Animal y Vegetal.* Caracas, Venezuela: Universidad Central de Venezuela; 1990. Thesis.
29. Bracket RE. Antimicrobial effects of chlorine on *Listeria monocytogenes*. *J Food Prot.* 1987;50:999–1003.
30. Zhang S, Farber JM. The effects of various disinfectants against *Listeria monocytogenes* on fresh cut vegetables. *Food Microbiol.* 1996;13:311–321.
31. Francis GA, O´Beirne D. Effects of gas atmosphere, antimicrobial dip and temperature on the fate of *Listeria innocua* and *Listeria monocytogenes* on minimally processed lettuce. *Int J Food Sci Technol.* 1977;32:141–151.
32. Beuchat LR, Ryu JH. Produce handling and processing practices. *Emerg Infect Dis.* 1977;3:459–465.
33. Moreno I. *Comportamiento de* Aeromonas hydrophila *en Ensaladas con Mínimo Procesamiento Almacenadas a 5°C.* Caracas, Venezuela: Universidad Central de Venezuela; 1997. Thesis.
34. Neilsen HJS, Zeuthen P. Influence of lactic acid bacteria and the overall flora of pathogenic bacteria in vacuum-packed cooked emulsion style sausage. *J Food Prot.* 47:28–34.
35. Zhao T, Doyle MP, Besser RE. Fate of enterohemorrhagic *E. coli* O157:H7 in apple cider with and without preservatives. *Appl Environ Microbiol.* 1993;59:2526–2530.
36. Conner DE, Kotrola JS. Growth and survival of *E. coli* O157:H7 under acidic conditions. *Appl Environ Microbiol.*1995, 61, 3382–385.
37. Seo KH, Frank JF. Attachment of *Escherichia coli* O157:H7 to lettuce leaf surface and bacterial viability in response to chlorine treatment as demonstrated by using confocal scanning laser microscopy. *J Food Prot.* 1999;62:3–9.
38. Steinbruegge EG, Maxcy BR, Liewen MB. Fate of *Listeria monocytogenes* on ready to serve lettuce. *J Food Prot.* 1988;42:79–90.
39. Kallander KD, Hitchins AD, Lancette GA, Schmieg JA, et al. Fate of *Listeria monocytogenes* in shredded

cabbage stored at 5°C and 25°C under a modified atmosphere. *J Food Prot.* 1991;54:302–304.

40. Beuchat LR, Brackett RE, Hao DYY, Conner DE. Growth and thermal inactivation of *Listeria monocytogenes* in cabbage and cabbage juice. *Can J Microbiol.* 1986;32:791–795.
41. Berrang ME, Brackett RE, Beuchat LR. Growth of *Listeria monocytogenes* on fresh vegetables stored under controlled atmosphere. *J Food Prot.* 1989;52:702–705.
42. Odumer JA, Mitchell SJ, Alves DM, Lynch JA et al. Assessment of the microbiological quality of ready-to-use vegetables for health-care food services. *J Food Prot.* 1997;6:954–960.
43. Whiting RC. Microbial modeling in foods. *Crit Rev Food Sci Nutr.* 1995;35:467–494.
44. Buchanan RL. Predictive food microbiology. *Trends Food Sci Technol.* 1993;4:6.
45. Walls I, Scott VN. Use of predictive microbiology in microbial food safety risk assessment. *Int J Food Microbiol.* 1977;36:97–102.
46. U.S. Department of Agriculture. *Pathogen Modeling Program*, Version 6. Wyndmoor, PA: Eastern Regional Research Center; 1998
47. Ministry of Agriculture, Fisheries and Food and Leatherhead Research Laboratory. *MicroModel*. Surrey, England: Leatherhead Food Research Association;1998.
48. Linko S. Expert systems—what can they do for the food industry? *Trends Food Sci Technol.*1998;1:3–12.
49. Adair C, Brigg PA. The concept and application of expert systems in the field of microbiological safety. *J Ind Microbiol.* 1992;12:263–267.
50. Kirby R. *Personal communication*. 1999.
51. Zwietering MH, Wijtzes T, De Wit JC, Vant Riet K. A decision support system for prediction of the microbial spoilage in foods. *J Food Prot.* 1992;55(12):973–979.
52. Wijtzes T, van't Riet K, Huis in't Veld JHJ, Zwietering MH. A decision support system for the prediction of microbial food safety and food quality. *Int J Food Microbiol.* 1998;42:79–90.
53. van Gerwen. *Personal communication*. 1999.
54. Notermans S, in't Veld P, Wijtzes T, Mead GC. A user's guide to microbial challenge testing for ensuring the safety and stability of food products. *Food Microbiol.* 1993;10:145–157.
55. Carlin F, Nguyen-the C. Fate of *Listeria monocytogenes* on four types of minimally processed green salads. *Lett Appl Microbiol.* 1994;18:222–226.
56. Marconi A. *Incidencia y Comportamiento de* Listeria monocytogenes *en Escarola Lisa* (Cichorium endivia). Caracas, Venezuela: Universidad Central de Venezuela; 1999. Thesis
57. Carlin F, Nguyen-the C, Abreu da Silva A. Factors affecting the growth of *Listeria monocytogenes* on minimally processed fresh endive. *J Appl Bacteriol.* 1995;78:636–646.
58. Faber JM. Current research on *Listeria monocytogenes*: An overview. *J Food Prot.* 1993;56:640–646.
59. Francis GA, Thomas C, O´Beirne. The microbiological safety of minimally processed vegetables. *Int J Food Sci Technol.* 1999;34:1–22.
60. Finn MJ, Upton ME. Survival of pathogens on modified-atmosphere-packaged shredded carrot and cabbage. *J Food Prot.* 1997;60:1347–1350.
61. Beuchat LR, Bracket R. Inhibitory effect of raw carrots on *Listeria monocytogenes*. *Appl Environ Microbiol.* 1990;56:1734–1742.
62. Nguyen-the, Lund BM. 1991. The lethal effect of carrot on *Listeria* species. *J Appl Bacteriol.* 1991;70: 479–488.
63. Larson AE, Johnson EA, Barmore CR, Huches MD. Evaluation of the botulism hazard from vegetables in modified atmosphere packaging. *J Food Prot.* 1997;60:1208–1214.

Part II

Physicochemical and Structural Aspects

Chapter 6

Chemical and Physicochemical Interactions between Components and Their Influence on Food Stability

Lía Noemí Gerschenson, Carmen A. Campos, Ana M. Rojas, and Guillermo Binstok

INTRODUCTION

When developing a new product or modifying a previously existent one, food technologists must decide the composition and conditions of processing that can provide the adequate microbiologic stability while fulfilling the nutritional and organoleptic standards required.

It is universally known that the quality of a food product is related to the quality of its constituents. A food product contains many classes of components: Some are present in large proportions (water, lipids, carbohydrates, proteins), whereas others are present in small amounts (i.e., minerals, vitamins, flavors, and additives). Chemical and physicochemical interactions can occur among these components during processing and storage, determining that the quality of a food product is also related to the nature of component interactions.

New formulations to meet consumer demands for new and/or healthier foods require the coexistence of components that previously were not jointly present in food products (i.e., fat replacers or artificial sweeteners with other components usually present), broadening the spectra of interactions that must be studied by food scientists to have a thorough understanding of component functionality and final product quality.

It is important that interactions be understood so that beneficial properties arising from ingredient interactions can be optimized, and interactions resulting in detrimental effects on the food can be minimized. However, the complexity of real food products possesses a strong limitation for understanding the exact mechanisms that control different interactions, often requiring the study of interactions in model systems.

This chapter is not expected to be exhaustive but to provide some additional elements to the discussion concerning the importance of component interaction on shelf stability of food products through the analysis of some cases of interest.

ETHYLENEDIAMINETETRAACETIC ACID–SORBATE INTERACTION

Ethylenediaminetetraacetic acid (EDTA) is a known iron complexant widely used as antioxi-

We acknowledge financial support from Universidad de Buenos Aires, Consejo Nacional de Investigaciones Científicas y Técnicas de la República Argentina, STD-3 Program of the European Union, CYTED Program of the Agencia de Cooperación Española, International Development Bank, and Agencia Nacional de Promoción Científica y Técnica de la República Argentina.

dant[1] in lipid-containing food products. It has been also suggested as a chlorophyll degradation inhibitor through the stabilization of Mg^{2+} by chelating action (see Chapter 2). Sorbic acid is a monocarboxylic fatty acid that is frequently used to preserve food products, its inhibitory action being more pronounced against yeasts and molds but acting also against many spoilage and pathogenic bacteria. It is generally recognized as a safe preservative.[2] Sorbic acid and its potassium salt (KS), collectively known as *sorbates*, undergo autoxidative degradation in aqueous solution. This fact can affect microbiologic stability. In addition, the carbonyls formed from the destruction of sorbates can take part in browning reactions, leading to undesirable changes in quality and acceptability. Sorbate degradation was found to be influenced by a number of factors, such as temperature, pH, and system composition.[3,4]

Consumers have shown a general desire to reduce the amount of fat in their diets to reduce calorie intake. Reduced fat intake has been linked to reduced cancer and coronary disease in the population, reinforcing the benefits of that desire. However, to accomplish it, food technologists must perform changes in food formulations through the use of fat substitutes that allow the production of good-quality alternative foods that meet consumer demands. In the case of mayonnaise, EDTA is usually used to prevent lipid oxidation. The change in formulation to provide a "less fat" version creates an increase in pH, a fact that requires the use of preservatives, such as sorbates, to assure microbiologic stability. As a consequence, this dressing is an example of the possible coexistence of EDTA and sorbates.

In our laboratory, we studied the influence of EDTA presence on sorbic acid degradation with the idea of evaluating the shelf stability of food products containing both additives.[5] When model systems of water activity (a_W) depressed to 0.91 with a mixture of 35.0 g/kg of NaCl and enough glycerol as to attain that a_w were studied, the presence of 0.10 g/kg of EDTA in an aqueous solution of pH 5.0 produced an enhancement of sorbate degradation at all the temperatures studied in the range 33–57°C (Table 6–1). For example, at 33°C, a usual storage temperature in subtropical countries, no sorbate destruction was detected (system *a*) but when EDTA was present (system *b*), the half-life of sorbates was reduced to ≅ 257 days.When the same experiment was repeated with 0.50 g/kg of EDTA, the results were similar to the ones obtained with 0.10 g/kg of EDTA.

An analysis of iron content was performed by atomic absorption, observing that an aqueous solution of potassium sorbate (2 g/kg) or one of EDTA (5 g/kg) contained only 2×10^{-6} g/kg of iron. The quantity of phosphoric acid that was necessary to adjust the pH of the system to a value of 5.0 contributed with 28×10^{-6} g/kg of iron. After 42 days at 57°C, the aqueous solution of potassium sorbate with a pH of 5.0 had an iron content of 37×10^{-6} g/kg, but the same system with EDTA had an iron content of 3×10^{-4} g/kg. This result showed that, initially, our system contained iron and its concentration increased significantly during storage when EDTA was present, probably due to the scavenging of iron performed by EDTA from the glass of the flasks used.

Some antioxidants consist of metal sequestrants (i.e., chelating agents that either precipitate the metal or suppress its reactivity by occupying all coordination sites). Iron chelated by EDTA has seven coordination sites, one of which is occupied by water and is, therefore, available for redox reactions. This characteristic, together with the high solubility of Fe (III)-EDTA, renders this chelating agent an efficacious oxidation catalyst.[6] As a consequence, complexation of iron by soluble chelating agents, such as EDTA, does not preclude iron from participating in the Haber-Weiss cycle:

$$Fe^{2+} + O_2 \rightleftarrows Fe^{3+} + O_2^{\bullet -} \qquad \text{I}$$

$$2O_2^{\bullet -} + 2H^+ \rightleftarrows H_2O_2 + O_2 \qquad \text{II}$$

$$Fe^{2+} + H_2O_2 \rightleftarrows \bullet OH + OH^- + Fe^{3+} \qquad \text{III}$$

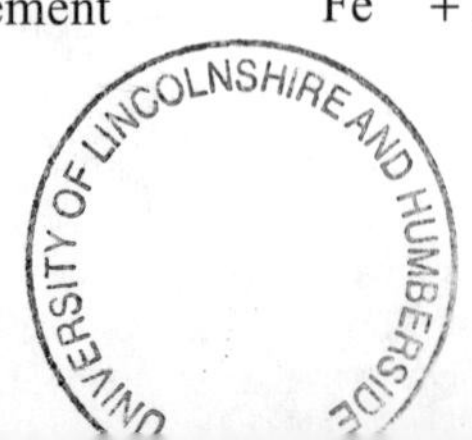

Table 6–1 Influence of EDTA Presence on Sorbic Acid Degradation and Nonenzymatic Browning

System	*Temperature (°C)*	*Sorbic Acid Degradation $(k \pm c) \cdot 10^3$ (day^{-1})*	*Nonenzymatic Browning $(k \pm c) \cdot 10^3$ (ΔAbs/day)*
a	33	NDD	NBD
a	45	NDD	NBD
a	57	1.4 ± 0.4	0.6 ± 0.1
b	33	2.7 ± 0.1	0.9 ± 0.1
b	45	4.5 ± 0.6	3.5 ± 0.3
b	57	10.6 ± 0.9	9.5 ± 0.3

Note: All systems were packaged in glass bottles.
System composition (g/kg):
a: water: 722.5; potassium sorbate: 2.0; NaCl: 35.0; glycerol: 240.5;
b: water: 722.4; potassium sorbate: 2.0; NaCl: 35.0; glycerol: 240.5; EDTA: 0.10.
pH: 5.0 and a_w: 0.91 for both systems.
NDD, no destruction detected; NBD, no browning detected; k, reaction rate constant; c, confidence interval (confidence level, $p = 0.95$).

In this cycle, ferrous ion reduces oxygen to $O_2^{\bullet -}$, which spontaneously disproportionates into H_2O_2 and O_2. The resulting combination of Fe^{2+} and H_2O_2, known as *Fenton reagent*, produces highly reactive hydroxyl radicals (•OH), which indiscriminately oxidize most food constituents. The above metal-catalyzed production of •OH from H_2O_2 requires iron containing at least one free coordination site, a condition met by Fe (III)-EDTA.

Ferric ion alone does not facilitate the generation of oxy-radicals in the presence of oxygen, nor does it promote lipid peroxidation. It is ferrous ion in solution that enhances the production of oxy-radicals, including the •OH, by reactions I, II, and III. Ferrous ion might also initiate a single cycle of lipid peroxidation but the inclusion of a reducing agent with iron provides a continuous source of Fe^{2+}, which, in the presence of oxygen, leads to oxidative damage. Reducing substances such as ascorbic acid, alpha-tocopherol, dopamine, catechols, and reduced glutathione may initiate the Haber-Weiss cycle.[6]

Sorbic acid is a monocarboxylic fatty acid that can be destroyed through an autoxidation mechanism. As Arya[7] stated, heavy metal ions, especially those possessing two or more valence states with a suitable oxidation-reduction potential between them ($M^{(n+1)+} - M^{n+}$), generally accelerate the rate of autoxidation. The following reactions are examples of the mechanism involved:

Decomposition of hydroperoxides into free radicals:

$$ROOH + M^{(n+1)+} \rightarrow ROO^{\bullet} + H^+ + M^{n+}$$

Participation in the initiation reactions:

$$RH + M^{(n+1)+} \rightarrow R^{\bullet} + M^{n+} + H^+$$

We can see that some reactions involved in sorbate autoxidation can promote iron reduction to the valence (II), acting as reducing agents, providing a continuous source of Fe^{2+}, and contributing to more oxidative damage through the Haber-Weiss cycle. According to Mahoney and Graf,[6] the solubility and the oxidation-reduction potential of Fe (III)-EDTA are greater than those of iron alone ($^+$0.177 vs. $^-$0.771), a fact that might be responsible for the higher rates of sorbic acid destruction observed when EDTA was present.

As can be seen in the systems (*a* and *b*) defined in Table 6–1, the presence of EDTA not only increases the sorbate destruction but also the absorbance at 420 nm. Probably the greater production of carbonyls through sorbate degradation when EDTA was present produced the observed enhancement of nonenzymic browning (NEB).[8] To emphasize the magnitude of the influence of EDTA presence on browning development, it can be mentioned that, for example, at 33°C, no browning was detected in system *a,* but when 0.10 g/kg of EDTA was present (system *b*), in ≅ 111 days, an absorbance of 0.1 at 420 nm was reached. Samples with that absorbance show a yellow coloration when visually examined, which can diminish the organoleptic score of some food products.

As the iron-EDTA complex enhances sorbic acid oxidation, the simultaneous presence of EDTA and sorbates might reduce shelf life of food products, showing the importance of taking into account potential interactions between components when formulating a product to have a thorough understanding of a food system's performance.

ASCORBIC ACID–HUMECTANT INTERACTION

Fruits and vegetables are significant sources of dietary vitamin C. The principal biologically active form of this vitamin is L-ascorbic acid (AA), but an oxidation product, L-dehydroascorbic acid (DHA), is also active.[9,10] L-ascorbic acid undergoes aerobic destruction (oxidation) catalyzed by cupric, silver, ferrous, and stannous ions.[11] In general, anaerobic degradation proceeds simultaneously but more slowly than does aerobic degradation, and it is the unique path in the absence of oxygen.[12]

Thermal treatments such as blanching and pasteurization, as well as evaporation, are commonly applied during fruit juice processing, producing degradation of vitamin C and NEB.[13] Nonenzymic browning is considered to be one of the major causes of quality loss in fruit products such as citrus juices during storage, having an important role in flavor, color, and nutritional quality, and is related to vitamin C loss.[14]

Preservation by combined methods controls microbial growth through the application of gentle individual stress factors (i.e., pH control, a_W depression, antimicrobial agents) to fruit products. This combination of inhibiting factors generally assures a product of characteristics more similar to fresh foods than one obtained when applying only one preservation factor.[15]

During the application of combined methods to fruits, different humectants can be used to depress a_W. To understand their effect on vitamin C stability, several experiments[16] were performed in our laboratory with reference to L-ascorbic acid stability in different media of a_W 0.94. We observed that, in glucose and sucrose media (Table 6–2, systems *a* and *b*), AA loss increased with temperature in the range of 24–90°C. NEB showed, in general, the same behavior. However, in the sorbitol system (*c*) and in the control system (*d*, no humectant present) the destruction of sorbic acid increased with temperature but within specific ranges (from 24 to 45°C or from 70 to 90°C), as can be seen in Table 6–2. The system containing sorbitol showed an inverted order between 45 and 70°C concerning destruction rate constants, and the control system showed no significant difference between 24 and 70°C rate constants or between 45 and 80°C constants.

In general, depression of water activity (systems *a–c* vs. *d*) inhibited AA destruction between 24 and 45°C (Table 6–2). However, D-glucose and, to a minor degree, sucrose, were more protective of AA between 24 and 45°C than was sorbitol, although the water activity was the same (a_W ≅ 0.940) in the model systems. Joslyn and Supplee[17] also observed the protective effect of the sugars on anaerobic AA destruction at 38.7°C. As can be seen in Table 6–2, in this range, the rate of browning reactions of humectants and AA was smaller than in the 70–90°C range.

At higher temperatures, the rate of NEB was notoriously lower in systems *c* and *d*, containing sorbitol and no humectant, respectively, than in

Table 6–2 Influence of Humectants on L-Ascorbic Acid Destruction and Nonenzymatic Browning

System	*Storage Temperature (°C)*	*L-ascorbic Acid Destruction (k±c)·10³ (1/hr)*	*Nonenzymic Browning (K±c)·10⁴ (Δabs/hr)*
a	24	0.17 ± 0.02	0.12 ± 0.02 I
	33	0.25 ± 0.07	0.33 ± 0.20 JK
	45	0.52 ± 0.15	0.35 ± 0.14 K
	70	2.46 ± 0.85	2.34 ± 0.41
	80	6.63 ± 0.28	10.19 ± 1.09
	90	18.55 ± 1.81 A	38.70 ± 0.87
b	24	0.30 ± 0.05	0.18 ± 0.10 I
	33	0.59 ± 0.10	0.45 ± 0.18 J
	45	1.03 ± 0.05	0.89 ± 0.38 L
	70	7.85 ± 2.61	22.79 ± 8.88
	80	30.94 ± 4.26	196.26 ± 76.01
	90	60.71 ± 9.64	323.13 ± 91.71
c	24	0.48 ± 0.19	0.20 ± 0.15 IP
	33	0.74 ± 0.05	0.21 ± 0.05 MP
	45	2.61 ± 0.55	0.43 ± 0.12 R
	70	1.34 ± 0.17 B	0.56 ± 0.11 NR
	80	3.65 ± 0.82 C	0.76 ± 0.39 O
	90	13.40 ± 4.14 AD	4.03 ± 1.10
d	24	1.36 ± 0.36 E	0.16 ± 0.04 IS
	33	2.41 ± 0.44	0.17 ± 0.07 MS
	45	6.94 ± 1.81 F	0.67 ± 0.25 LT
	70	1.50 ± 0.40 BE	0.50 ± 0.13 NT
	80	5.00 ± 0.22 CF	1.22 ± 0.32 OU
	90	9.75 ± 0.97 D	1.29 ± 0.45 U

k, reaction rate constant; c, confidence interval ($p = 0.95$); abs/h, absorbance unit per hour.

Aqueous systems composition (g/kg):

a: glucose: 360.0; potassium sorbate: 1.0; L-ascorbic acid: 0.35.

b: sucrose: 520.0; potassium sorbate: 1.0; L-ascorbic acid: 0.35.

c: sorbitol: 340.0; potassium sorbate: 1.0; L-ascorbic acid: 0.35.

d: potassium sorbate: 1.0; L-ascorbic acid: 0.35.

Note: All of the systems had a pH of 3.5 adjusted with citric acid and an a_w of 0.94, with the exception of system d, which had an $a_w \cong 1.00$.

K followed by same letter are not significantly different ($p = 0.95$).

systems with sugars (*a* and *b*), probably due to the lack of carbonyl groups in the humectants of the former systems, a fact that precluded their direct involvement in browning reactions. Sucrose is hydrolized very quickly in acid medium, producing D-glucose and D-fructose,[18,19] the latter being more reactive than glucose in caramelization.[20,21] The contribution of D-fructose to degradation is probably responsible for the observed faster browning reaction rate of the sucrose system (*b*), when compared with glucose-containing system.

The activation energies for AA destruction were different at low and high temperatures for the same model system (i.e., 10 Kcal/mole in the range of 24–45°C and 25 Kcal/mole in the range of 70–90°C for system *a*). The values obtained in the range of 70–90°C were similar to the ones reported for orange juice by Johnson et al.[22] for the 70–98°C range. Nagy and Smooth[23] studied the retention of vitamin C in canned orange juice that was stored for 12 weeks at constant temperature between 4.4 and 48.9°C and observed the same trend: The relation among rate constants and temperature obeyed the Arrhenius model but showed different activation energies in the two ranges of temperatures studied.The break point was at 22–26.7°C: The activation energy of AA loss was larger above than below that temperature.

Curl[24] studied AA retention in aqueous solutions of different composition under anaerobic condition and established a link with the degree of NEB obtained. The sugar presence increased AA loss at 49°C and pH 3.6–3.8, in the order of glucose, sucrose, and fructose. He also studied the use of sorbitol as a humectant and concluded that it showed the same AA destruction rate as that of the system without humectants. In our case, AA destruction showed similar rates for systems *c* and *d* in the range of 70–90°C, and destruction was higher when sugars were present, with sucrose being the more active one. From 70 to 90°C, the pattern of AA destruction paralleled the NEB in all systems: Browning was faster in sucrose (*b*) than in glucose (*a*) media, and color development was very slow in the system with sorbitol (*c*) or without humectant (*d*).

The stability of AA in the model systems might be explained through the different interactions among water and the humectants used. We suppose that between 24 and 45°C, at the same a_w, the structure of water is differentially affected by the humectants, creating a change in the surrounding of the AA molecules and producing a different chemical reactivity. AA seems to be more stable when humectants are "structure makers,"[25,26] such as D-glucose and, to a lesser degree, sucrose, which has a lower strength of association with water molecules. In the case of sucrose, its hydrolysis at pH 3.5 yields glucose, which may partially compensate the "structure breaker" effect produced by D-fructose generated through sucrose hydrolysis.[27] On the other hand, AA destruction is faster in sorbitol medium, as compared to glucose or sucrose systems; this might be ascribed to the shorter time that water molecules stay around this structure breaker humectant and, therefore, stay more available for AA degradation.

Suggett[25] established, through nuclear magnetic resonance studies, that the difference between the hydration properties of structure makers and structure breakers gradually disappears with the increase in temperature above 70°C.

For this reason, the presence of molecules with carbonyl groups (D-glucose and hydrolized sucrose) might accelerate AA destruction at higher temperatures, and the solvent effect might be surpassed in this range by browning reactions that accelerate anaerobic AA destruction through product consumption.

The present results suggest that the rate of anaerobic AA destruction is influenced by the chemical vicinity (water, humectants) of the AA. This effect prevails in the 24–45°C temperature range; in this range, sugars protect ascorbic acid from destruction. At higher temperatures (70–90°C), the solvent effect is surpassed by NEB, which determines the trend observed with the different humectants. These conclusions note the importance of considering component interactions in relation to processing and storage when evaluating stability of food products.

SUGARS–AMINO ACIDS–SORBATES INTERACTION

Simple processes to achieve microbial stability of fruit stored at ambient temperatures have recently been developed for pineapples, bananas, and peaches.[27–29] These processes were based on a combination of lowering of a_W through sugar incorporation, adjustment of pH, and addition of potassium sorbate and sodium bisulfite. Sorbate addition proved to be a key factor in achieving the microbial stability of preserved fruits, assuring shelf stability for periods of around 4 months.

Because it is known that sorbates may be destroyed to a considerable extent during storage of foods,[2] the study of the stability of sorbic acid in model systems comprised of concentrated sugar solutions of acid pH arose as an important task. As a consequence, in our laboratory, that kind of model system (a_W 0.94) at pH 3.5 and 4.5 was studied.[3] Results at pH 3.5 are shown in Table 6–3, where it can be seen that sorbic acid degraded as a function of time with a first-order kinetic. Rate constants increased with temperature and were related to system composition. Rate constants describing the loss of sorbic acid were found to be similar (within the limits of experimental error) for sucrose (*a*) and glucose (*b*) systems with pH 3.5 at 30°C and higher than in aqueous control system (*d*), showing that sugars enhance preservative destruction. It must be remarked that systems *a* and *b* had a lower a_w than did the control system (*d*), a fact that enhances preservative degradation.[30] At 45°C, the model system containing glucose and lysine (*c*) showed two different destruction periods: The rate constant in the first period (≅ 50 days) was somewhat higher than those observed for the model system (*a*) at the same temperature; however, for the second (slower) period, the reverse was observed.

Table 6–3 Rate Constant For Sorbate Degradation In Model Systems Containing Sugars And Aminoacides[3]

System	*Temperature (°C)*	*Sorbic Acid Degradation* $(k \pm c) \cdot 10^3$ *(day^{-1})*
a	20	2.1 ± 0.5
	30	4.8 ± 0.3
b	20	2.5 ± 0.4
	30	4.2 ± 0.7
	45	14.0 ± 1.1
c	20	2.2 ± 0.05
	30	5.3 ± 0.8
	45	20.7 ± 1.0*
		4.7 ± 1.8•
d	30	3.0 ± 0.5

k: reaction rate constant
c: confidence interval ($p = 0.95$).
*first slope.
•second slope.
System composition (g/kg):
a: water: 479.0; sucrose: 520.0; potassium sorbate: 1.0
b: water: 639.0; glucose; 360.0; potassium sorbate: 1.0
c: water: 637.2; glucose: 344.6; lysine: 17.2; potassium sorbate: 1.0.
a_w: 0.94, pH: 3.5
d: water: 999.0; potassium sorbate: 1.0.
a_w: 1.00, pH: 3.5

Visual examination of stored samples revealed that some samples had undergone intensive browning during storage. The glucose plus lysine system developed pronounced browning only at 45°C. The existence of intensive browning at this temperature might help to explain the observed behavior of sorbic acid. It is well known that autoxidized lipids may react with amines, amino acids, and proteins, forming brown molecular products.[31] As a result of oxidation of sorbic acid, carbonyl compounds such as malonaldehyde, crotonaldehyde, and acrolein are formed.[7] These compounds, as well as reducing sugars, form Schiff's bases with free amine groups that are further converted into brown products by aldol reactions. On the other hand, Maillard reaction products were previously found to be capable of retarding the development of rancidity in foods such as cookies and sausages.[32] These two opposite phenomena can explain the behavior of sorbic acid in glucose plus lysine model systems at 45°C. Before 50 days of storage, Maillard reactions probably proceeded to consume autoxidized sorbic acid and enhance the overall rate of degradation. After 50 days, the greater stability of sorbic acid might be due to the inhibitory effect of accumulated Maillard products on sorbic acid degradation. At 20 and 30°C, the formation of Schiff's bases failed to retard or to accelerate sorbic acid degradation, possibly because the degree of Maillard reaction was low in this temperature range.

Present results point out the importance of taking into account the influence of amino acid presence on stability of sugar-containing food products from the point of view of sorbate degradation and consequent food shelf stability.[33]

CONCLUSION

Some examples showing the importance of component interaction on the quality of food products have been presented:

a. the influence of EDTA presence on sorbate degradation.
b. the influence of humectants on ascorbic acid stability, and
c. the influence of sugars and amino acids on sorbate stability.

Even if the nature of the mechanisms involved in these interactions was studied, in general, only for model systems, we consider that results reported here make a valuable contribution to the discussion of the topic. Of course, we must remember that the complexity of real systems superimposes additional effects, such as competition between different interactions, which must be considered when extrapolating results obtained with model systems to food products and evaluating its consequences on food shelf stability.

REFERENCES

1. Shahidi F, Rubin LJ, Diosady LL, Kassam N, Li Sui Fong JC. Effect of sequestering agents on lipid oxidation in cooked meats. *Food Chem.* 1986;21:145–152.
2. Sofos JN. Antimicrobial activity. In *Sorbate Food Preservatives*. Boca Raton, FL: CRC Press; 1989:33–48.
3. Gerschenson LN, Alzamora SM, Chirife J. Stability of sorbic acid in model food systems of reduced water activity: Sugar solutions. *J Food Sci.* 1986;51:1028–1031.
4. Gerschenson LN, Alzamora SM, Chirife J. Effect of sodium chloride and glycerol on the stability of sorbic acid solutions of reduced water activity. *Lebensm-Wiss u-Technol.* 1987;20:98–99.
5. Campos CA, Rojas AM, Gerschenson LN. Studies of the effect of ethylene diamine tetraacetic acid (EDTA) on sorbic acid degradation. *Food Res Int.* 1996;29:259–264.
6. Mahoney JR, Graf E. Role of alpha-tocopherol, ascorbic acid, citric acid and EDTA as oxidants in model systems. *J Food Sci.* 1986;51:1293–1296.
7. Arya SS. Stability of sorbic acid in aqueous solutions. *J Agric Food Chem.* 1980;28:1246–1249.
8. Campos CA. *Estabilidad del Ácido Sórbico en Alimentos Preservados y/o Almacenados*. Buenos Aires, Argentina: Universidad de Buenos Aires; 1995. Ph.D. Thesis.
9. Villota R, Karel M. Prediction of ascorbic acid retention during drying. I. Moisture and temperature distribution in a model system. *J Food Preserv.* 1980;4:111–134.
10. Levine M, Morita K. Ascorbic acid in endocrine systems. In: Aurbach GD, McCormick DB, eds. *Vitamins and Hormones,* Vol. 42. London: Academic Press; 1985:1–64.
11. Bisset OW, Berry RE. Ascorbic acid retention in orange juice as related to container type. *J Food Sci.* 1975;40:178–180.
12. Kurata T, Sakurai Y. Degradation of L-ascorbic acid and mechanism of nonenzymic browning reaction. I. *Agric Biol Chem.* 1967;31:101–105.
13. Massaioli D, Haddad PR. Stability of the vitamin C content of commercial orange juice. *Food Technol Austr.* 1981;33:136–138.
14. Lee HS, Nagy S. Quality changes and nonenzymic browning intermediates in grapefruit juice during storage. *J Food Sci.* 1988;53:168–172, 180.
15. Leistner L. Use of hurdle technology in food processing: Recent advances. In: Barbosa-Cánovas G, Welti-Chanes J, eds. *Food Preservation by Moisture Control.* Lancaster: Technomic Publishing Co.; 1995:377–396.
16. Rojas AM, Gerschenson LN. Ascorbic acid destruction in sweet aqueous model systems. *Lebens-Wiss u-Technol.* 1997;30:567–572.
17. Joslyn MA, Supplee H. Solubility of oxygen in solution of various sugars. *Food Res.* 1949;14:209–215.
18. Schoebel T, Tannenbaum SR, Labuza T. Reaction at limited water concentration. I. Sucrose hydrolysis. *J Food Sci.* 1969;34:324–330.
19. Montes de Oca C, Gerschenson LN, Alzamora SM. Effect of the addition of fruit juices on water activity of sucrose-containing model systems during storage. *Lebens-Wiss u-Technol.* 1991;24:375–377.
20. Van Dam HE, Kieboom APG, Bekkum H. The conversion of fructose and glucose in acidic media: formation of hydroxymethylfurfural. *Stark/Stärke.* 1986;38:95–101.

21. Buera MP, Chirife J, Resnik SL, Wetzler G. Nonenzymatic browning in liquid model systems of high water activity: Kinetics of color changes due to Maillard's reaction among different sugars with caramelization browning. *J Food Sci.* 1987;52:1063–1067.

22. Johnson JR, Braddock RJ, Chen CS. Kinetics of ascorbic acid loss and nonenzymatic browning in orange juice serum. Experimental rate constants. *J Food Sci.* 1995;60:502–505.

23. Nagy S, Smooth JM. Temperature and storage effects on percent retention and percent U.S. recommended dietary intake of vitamin C in canned single-strength orange juice. *J Agric Food Chem.* 1977;25:135–138.

24. Curl AL. Ascorbic acid losses and darkening on storage at 49°C of synthetic mixtures analogous to orange juice. *Food Res.* 1949;14:9–14.

25. Suggett A. In: Frank P, ed. *Water: A Comprehensive Treatise*. New York: Plenum Press; 1949:519–534.

26. Mathlouthi M, Seuvr AM. Solution properties and the sweet taste of small carbohydrates. *Trans Faraday Soc.* 1988;84:2641–2650.

27. Alzamora SM, Gerschenson LN, Cerrutti P, Rojas AM. Shelf-stable pineapple for long-term nonrefrigerated storage. *Lebens-Wiss u-Technol.* 1989;22:233–236.

28. Guerrero S, Alzamora SM, Gerschenson LN. Development of a shelf-stable banana purée by combined factors: Microbial stability. *J Food Prot.* 1994;57:902–907.

29. Sajur SA. *Preconservación de Duraznos por Métodos Combinados*. Mar del Plata, Argentina: Universidad Nacional de Mar del Plata; 1985. M.Sc. Thesis.

30. Gerschenson LN, Campos C. Sorbic acid stability during processing and storage of high moisture foods. In: Barbosa-Cánovas G, Welti-Chanes J, eds. *Food Preservation by Moisture Control*. Lancaster: Technomic Publishing Co.; 1995:761–791.

31. Pokorny J. Browning from lipid-protein interactions. *Prog Food Nutr Sci.* 1981;5:421–423.

32. Lingnert H, Lundgren B. Antioxidative Maillard reaction products. IV. Application in sausage. *J Food Process Preserv.* 1980;4:235–237.

33. Rojas AM, Gerschenson LN. Influence of system composition on ascorbic acid destruction at processing temperatures. *J Sci Food Agric.* 1997;74:369–378.

Chapter 7

Color of Minimally Processed Fruits and Vegetables as Affected by Some Chemical and Biochemical Changes

Lidia Dorantes-Alvarez and Amparo Chiralt

INTRODUCTION

The consumer's color perception of foodstuffs is decisive when buying a product. Color, apart from hedonic connotations, can inform us about many other properties, such as ripeness degree in fruits, product alterations, etc. Other sensory attributes, such as texture or flavor, may play a somewhat less important role in consumer decision on purchase.

The radiant energy that is perceived by the human eye by means of the stimulation of the retina gives rise to color vision. This part corresponds to the wave length range of 380–700 nm, and it is known as *visible region*. The perceived color depends on the spectral composition of the light source, the chemical and optical characteristics of the product, and the eye sensitivity. The human color perception is made up of three attributes: hue (i.e., red, green, blue, etc.), saturation or intensity of the color (i.e., pastel, deep, medium, etc.), and lightness or clarity. To characterize the color of foodstuffs, the chromatic and nonchromatic color attributes must be quantified. This can be accomplished by analyzing the visible spectra distribution in terms of color coordinates, taking into account a standard illuminant and observer (i.e., illuminant D65, observer 10º, as recommended by the Comision International de l'Eclairage (CIE). The visible spectra can be obtained, depending on the product's optical properties, by analyzing the diffused reflected light of opaque samples (i.e., fruits and vegetables minimally processed), the diffused reflected and/or diffused transmitted light of translucent samples (i.e., milk whey), or the regular transmitted light of transparent samples (i.e., apple juice, wines).[1]

Tristimulus colorimeters or spectrophotometers have been widely used to obtain (by direct measurement or calculation) color coordinates, such as *X Y Z* or CIE L*a*b*, from light reflected by the surface of the commodities, precisely where most of the changes of color may take place. These coordinates can be interpreted in terms of clarity, hue, and saturation values, which will correlate with the consumer color perception and give objective criteria to establish color tolerances to control production and storage of foods. For example, in minimally processed fruits and vegetables, browning processes have been monitored, employing several color coordinate analyzers from different manufacturers.[2–6]

One of the aims in minimal processing of fruits and vegetables is the preservation of original color to ensure consumer acceptance. Nevertheless, minimal processing includes unit operations such as washing, peeling, bruising, and size reduction that allow enzymes (chlorophyllase, peroxidase, polyphenoloxidase) and substrates to come into contact, mainly at the surface of the products, bringing about enzymatic reactions related to color deterioration. Polyphenoloxidase, for instance, causes enzymatic browning and affects dramatically the original color of some products. Other operations that may produce changes in color are immersion of the fruits or vegetables in solutions with chemicals and acids, thermal treatments, and exposure of cut fruits to air and light because, in these operations, compounds responsible for color may be altered.

Fruits and vegetables contain natural substances in cells and tissues that are responsible for their characteristic color. Linear correlation between the concentration of several characteristic food pigments and mathematical combinations of food color coordinates has been reported.[7] These substances correspond to different groups of pigments that, in turn, are important in color changes of minimally processed products. The main groups are carotenes and carotenoids, anthocyanins, chlorophylls, and phenolic compounds. These compounds may change throughout different chemical and biochemical reactions occurring during the product processing and storage. Table 7–1 shows the effect that some unit operations of minimal processing may produce in fruits and vegetables pigment. This chapter aims to address the main transformations of fruits and vegetable pigments during processing, with special emphasis on minimally processed products and procedures to control color alterations. Also, focus has been placed on the polyphenoloxidase (PPO) browning action, due to its relevant role in color degradation of minimally processed fruits and vegetables.

CAROTENOIDS

Carotenoids constitute one of the most important groups of pigments and are present in all plant and animal families. As colorants, they are well tolerated because they are naturally present in foodstuffs, are easily metabolized, and their metabolites are beneficial for health. The most notable property of these compounds at the physiologic and dietary level in humans and other animals is that some of them have provitamin A activity. This important characteristic has led some authors to suggest a classification on the basis of their nutritional and biologic activity (antiulcer, anticancer, immunologic regulators, etc.).[8] A convenient grouping of the carotenoids has been to separate them into carotenes (hydrocarbons) and xanthophylls (oxygen-containing compounds). Carotenoids can be altered or partially destroyed by acids and are generally stable in bases, sensitive to light exposure, and prone to enzyme attack. Carotenoids are stable during heating but highly sensitive to oxidation, due to changes in water activity (a_w).

Degradation of Carotenoids

Although carotenoids are fairly stable in their natural environment, on heating or when they are extracted with oil or organic solvents, they become much more labile. In this group of pigments, there are very few structural alterations that promote the formation of other colored compounds, the most significant being the *cis-trans* transformation. In addition, the formation of 5,8 furanoids from 5,6 epoxides may occur, mediated by an acid medium. Such a transformation only slightly affects the color intensity. The most generalized alteration occurring in carotenoids, however, is oxidative destruction, which has a considerable effect on color intensity of foodstuffs. The oxidation of carotenoids is due to oxygen and is catalyzed by enzymes, such as lipoxigenase. Oxidation is also accelerated by metal ions, chemical oxidants, and light, and it is slowed down by the addition of antioxidants, such as ascorbic acid. The high temperatures and/or low moisture levels reached during the processing of vegetables may provoke losses of up to 50% in the concentration of carotenoids with epoxide groups such as some xanthophylls.

Postharvest Storage Effect on Carotenoids

In several fruits, such as *Capsicum annum*, a period of carotenogenesis favored by light after

Table 7–1 Effect of Some Unit Operations in Minimal Processing on Fruit and Vegetable Pigments

Unit Operation	*Effect on Carotenoids*	*Effect on Anthocyanins*	*Effect on Chlorophylls*	*Effect on Phenolics*
Peeling	—	—	—	Promotion of enzymatic browning
Size reduction	—	—	—	Promotion of enzymatic browning
Blanching	Beneficial when peroxidase is inactivated	Protection against coupled oxidation. Leaching when boiling water is used to blanch	Prevention of enzyme bleaching action. Promotion of transformation to pheophytins	Prevention of enzymatic browning
Acidification	Some xanthophyll transformation	Changes of pigment hue and chrome	Transformation to pheophytins	Partial inhibition of PPO activity
Immersion in antibrowning solutions	Protection from oxidation	Leaching of soluble anthocyanins. Sulfites may cause discoloration	Protective action of some antibrowning compounds	Protection from oxidation
Immersion in antimicrobial solutions	—	Leaching of soluble anthocyanins	—	Sorbates and benzoates may reduce enzymatic browning[29]
Radiation	Gamma irradiation has no effect in red capsicums and mangos[47]	—	—	Promotion of browning in some cases
Modified-atmosphere packaging	—	Destabilization in carbon dioxide (>73%).[48]	Increase of chlorophylls in broccoli florets[19]	—

harvesting has been observed.[9] This biosynthetic period may also occur during storage of minimally processed vegetables because they have active enzymatic systems (Figure 7–1).

During lactic fermentation in green table olives, Minguez-Mosquera and Gandul-Rojas[10] found that the only carotenoids that were transformed were those with molecular structure more reactive in the acid medium. A first-order kinetic model described the transformation of these xanthophylls with 5,6 epoxide groups (violaxanthin and neoxanthin) into their corresponding 5,8 furanoid derivatives (auroxanthin and neochrome). However, the overall carotenoid content remained throughout processing, demonstrating the absence of other types of oxidative reactions that degrade them to colorless products.

In general, thermal processing, where products retain a high moisture content (e.g., blanching processes in vegetables), protects the carotenoids from discoloration. Therefore, a minimal processing of fruits and vegetables that includes a mild thermal processing and acidification

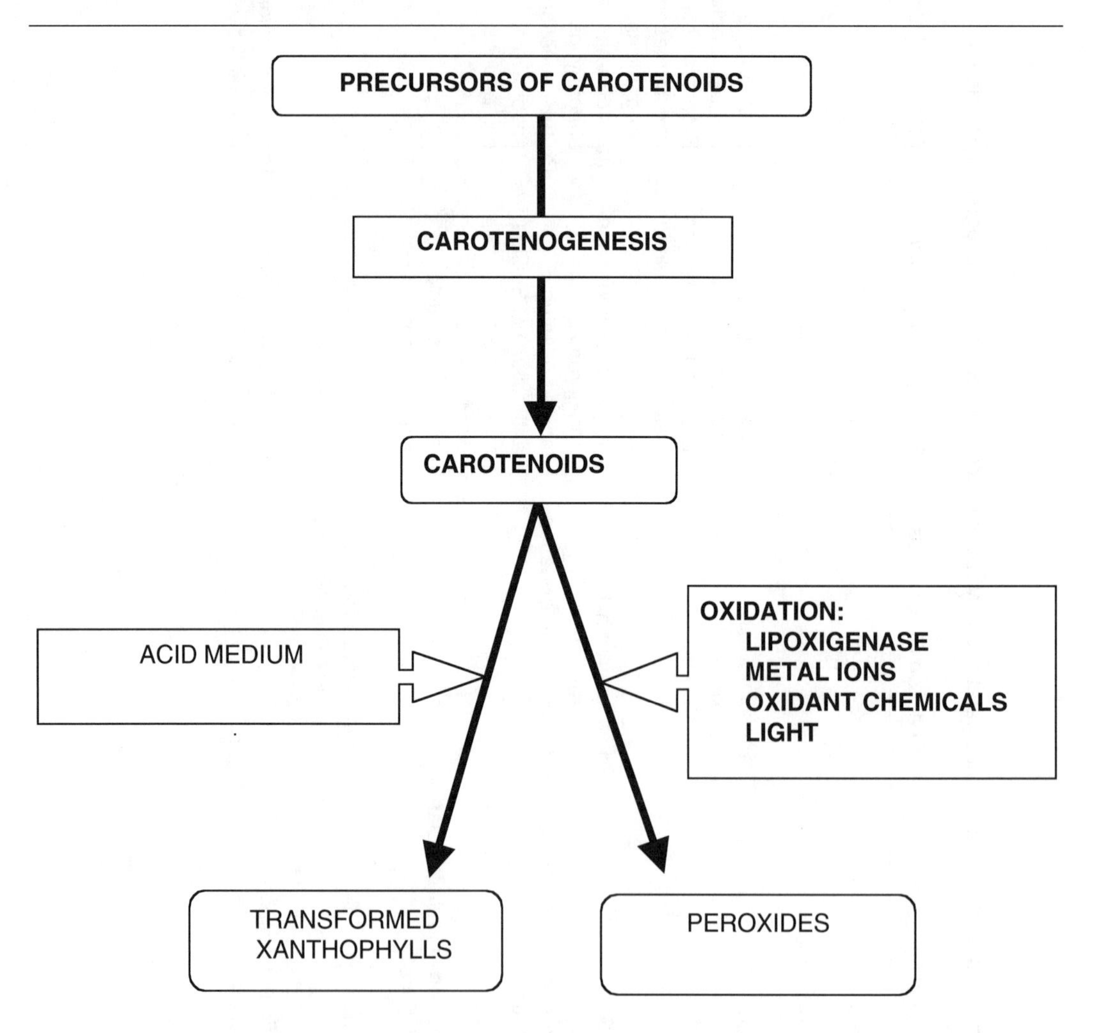

Figure 7–1 Changes of carotenoids in postharvested fruits and vegetables that may also occur in minimal processing.

would affect the color intensity and the total carotenoid content only slightly.

ANTHOCYANINS

The anthocyanins are a group of water-soluble plant pigments that are usually dissolved in the cell fluids, rather than in the lipoidal bodies. They greatly differ from the fat-soluble pigments in their stability and degradation pathways. There has been much interest in anthocyanins because of their importance in fruit and vegetable color, their sensitivity to degradation through chemical and condensation reactions, and their potential use as natural colorants.

Anthocyanins are natural colorants belonging to the flavonoid family and are widely distributed in fruit (particularly in berries and cherries). They are responsible for bright colors, such as orange, red, and blue. The anthocyanins are glycosides and acylglycosides of anthocyanidins. There are over 250 naturally occurring anthocyanins, and all are *o*-glycosylated with different sugar substitutes. The most prevalent sugars substituted on the aglycon, in order of occurrence in nature, are glucose, rhamnose, xylose, galactose, arabinose, and fructose. In addition to their colorful characteristics, anthocyanins possess potent antioxidant properties. Wang et al.[11] determined the antioxidant capacity of 14 anthocyanins among these, cyanidin 3-glucoside had the highest oxygen radical absorbing capacity, which was 3.5 times stronger than trolox (vitamin E analogue), whereas pelargonin had the lowest antioxidant activity but was still as potent as trolox.

The most important anthocyanidins (3-glucosides) and their maximum absorption wave length (nm) of their visible spectrum are pelargonidin (505), cyanidin (525), peonidin (523), delphinidin (535), petunidin (535), and malvidin (535). An increase in the degree of hydroxylation results in an absorption maximum shift toward the blue color (pelargonidin > cyanidin > delphinidin), whereas glycoside formation and methylation result in a shift toward the red color. The color of anthocyanins changes with the pH of the medium: upon dissolution, a complex equilibrium is quickly established between two colored species, flavylium cation (AH^+) and quinoidal base (A), and two colorless ones, carbinol pseudobase (B) and chalcone (C), resulting from the hydration of AH^+. Thus, the final color of the solution at equilibrium is the direct consequence of the equilibrium rate constants controlling the ionization—hydration and tautomeric reactions. Massa and Brouillard[12] calculated these equilibrium constants for cyanidin 3,5-diglucoside and for 3-deoxyflavylium salts, concluding that the latter could be of advantage in the coloring of foods and beverages at neutral pH. In general, the color of anthocyanins at low pH is red, at pH close to 7.0 is purple, and above 7.0 is deep blue.

Transformations of Anthocyanins Due to Processing

Anthocyanins show different responses to changes in pH, depending on their structure, also reflecting differences in stability during processing and storage. These water-soluble pigments, especially the anthocyanidins, change hue in response to pH changes. They are most stable at low pH and are destroyed in the presence of oxygen at higher pH. Anthocyanin breakdown is also pH-dependent and directly related to oxygen concentration, whereas other factors contribute to quality of anthocyanin-containing fruits and vegetables; the color stability of them depends mainly on pH. Rommel et al.[13] found that cyanidin 3-glucoside was the most unstable anthocyanin of red raspberry, disappearing completely during fermentation; on the contrary, cyanidin 3-sophoroside was the most stable pigment. The depectinizing of the juice, combined with its fermentation, caused a great deal of anthocyanin loss (38.7%). However, high-temperature, short-time heating treatment (85–90°C for 1 minute) did not produce an important anthocyanin loss (< 5%) in red raspberry juice. The effect of heat on the stability of anthocyanins is well known; degradation follows first-order kinetics, and pigments change with temperatures above 60ºC; thermal degradation increases as pH decreases, and the exclusion of oxygen is important to anthocyanin stability.

Polyphenoloxidase activity has been associated with anthocyanin loss; Raynal and Moutounet[14] showed that anthocyanins of d'Ente plums can be transformed by a coupled oxidation mechanism involving quinones formed from chlorogenic acid under polyphenoloxidase dependence. This transformation did not occur when PPO was inactivated by heating (90ºC for 5 minutes). The use of sulfites in fruits may produce bleaching of anthocyanins.

Anthocyanin Stability in Fruits

Wrolstad et al.[15] revealed that sucrose addition had a significant protective effect on anthocyanin pigment content in minimally processed and frozen strawberries. There is also evidence that stability of anthocyanins is increased by copigmentation with other polyphenolics coexisting in the same system. Molecules that can copigment with anthocyanins include flavonoids, polyphenols, alkaloids, amino acids, and organic acids. The copigmentation effect exists only in aqueous solutions and is sensitive to pH, temperature, and composition of the solution. Copigmentation protects anthocyanins against hydration, thus preserving their red color. Francia-Aricha et al.[16] detected in model solutions new anthocyanin pigments formed after condensation with flavanols. This was achieved with equimolar solutions of malvidin-3-monoglucoside plus procyanidin B2 in a medium containing acetaldehyde, tartaric acid at 5% and 10%, ethanol, and pH adjusted to 3.2 or 4.0 with sodium hydroxide. Another anthocyanin stabilization procedure is reported by Chandra et al.,[17] in which anthocyanin pigments from tart cherry were stabilized with phosphoric acid, maltodextrin, and α- and β-cyclodextrins.

We may conclude that stability of anthocyanins in minimally processed fruits will depend on operations and additives used in their preparation. Especially important are those that affect the original fruit pH, PPO activity, and other factors that may give rise to the formation of new anthocyanin pigments and copigmentation effect. Storage temperature of processed fruit will play a very important role in the anthocyanin stability because it affects reaction kinetics.

CHLOROPHYLLS

Chlorophylls are the functional pigments of photosynthesis in all green plants. They are the green pigments of leafy vegetables and give the green color to the peel of several fruits, particularly when unripe. They occur alongside a range of carotenoid pigments in the membranes of the chloroplasts, the organelles that carry out photosynthesis in plant cells. The pigments of the chloroplast are intimately associated with other lipophilic components of the membranes, such as phospholipids, as well as the membrane proteins. Fruits and vegetables contain chlorophylls a and b, in the approximate ratio of 3:1. They are essentially porphyrins, similar to those in hæm pigments, such as myoglobin, except for the different ring substituents, the coordination of magnesium rather than iron, and the formation of a fifth ring by the linkage of position 6 to the gamma methine bridge. The porphyrin ring associates readily with the hydrophobic regions of chloroplast membrane proteins. The extended phytol side chain similarly facilitates close association with carotenoids and membrane lipids.

Degradation of Chlorophylls

The degradation of chlorophyll in processed foods and senescent tissues causes a shift in color from brilliant green to olive brown in processed foods and to yellow, brown, or colorless compounds during senescence of fruits. Chlorophyll (blue-green) degradation pathway includes the loss of phytol to form chlorophyllide (blue-green) or the loss of Mg^{2+} to form pheophytin (olive brown), the loss of Mg^{2+} or phytol to form pheophorbide (olive brown), then the accumulation of different intermediates, some of them colorless. A symptom of senescence in harvested green vegetables is the loss of greenness due to degradation of chlorophyll. Quantitative changes in chlorophyll and its degradation products and/or hydrolyzing enzymes have been monitored in vegetables during senescence. Nevertheless, the mechanisms of degradation are not clear. Baardseth and Von Elbe[18] studied the chlorophyll degradation in two spinach varieties in the presence or absence of ethylene. In both varieties, chlorophylls were degraded during storage. Acceleration of degradation due to the presence of ethylene could not be demonstrated. Gradual disappearance did not coincide with activation of chlorophyllase, nor did it produce any new colored compounds. Catalase activity decreased during storage, whereas peroxidase activity increased. Addition of linoleic acid and various enzymes to model systems containing chlorophyll accelerated chlorophyll degradation. The rate of degradation was reduced by addition of antioxidants, catalase, or superoxide dismutase, suggesting involvement of both singlet oxygen and molecular oxygen in the bleaching process.

When processing of green vegetables does not include blanching, degradation of chlorophylls usually takes place throughout storage, because enzymes remain active. Storage conditions of green vegetables, as well as presence of antioxidants and organic acids, will influence chlorophyll retention. Zhuang et al.[19] investigated the effects of modified-atmosphere packaging, automatic misting, and vent packaging on total chlorophyll content of broccoli florets during simulated retail handling and storage at 5°C over 6 days. They found that modified-atmosphere packaging resulted in increased chlorophyll content. Chlorophyll remained near initial levels in vent-packaged products. When green vegetables are blanched prior to freezing or during canning, there is evidence that chlorophyll loses the phytol side chain, which gives the corresponding chlorophyllide a or b, but the most important event is the loss of the magnesium. This occurs most readily in acid conditions, with the magnesium ion being replaced by protons to give pheophytins a and b, with the resulting dirty brown color. One approach to control this phenomenon is to keep the medium slightly alkaline by the addition of a small quantity of sodium bicarbonate. The attempts to retain chlorophyll content during blanching have been addressed by modifying the pH of the heating media or by adding Mg salts in nonacid medium. Robertson[20] evaluated changes in the concentration of chlorophylls and pheophytins in kiwi purée at its normal pH (3.2) and at pH 7.5 adjusted with sodium hydroxide. The purée was frozen and stored at −18°C. After storage, evaluation of chlorophylls showed a better retention in the samples at pH 7.5. Alkaline conditions have an undesirable effect on other green vegetables, which contain chlorophylls and anthocyanins, because the latter turn brown in alkaline media. Unfavorable changes in texture and flavor may also be obtained in alkaline media, and losses of vitamin C are enhanced.

Retention of green color and chlorophyll during storage of minimally processed avocado purée has been studied by Sanchez et al.[21] These authors found that samples with ethylenediaminetetraacetic acid (EDTA) retained the green color better on determination of reflectance and sensory evaluation. After 40 days of storage at 4°C, the total chlorophyll content was higher in these samples, and the pheophytin content was lower in samples with EDTA than in those without it. This may be due to stabilization of chlorophyll magnesium ion by the chelating action of EDTA.

It has been demonstrated that increasing CO_2 concentration (up to 40%) for 3–6 days delayed yellowing, retarded ethylene production and loss of ascorbic acid, and retained chlorophyll content in broccoli florets during storage at 5°C. The effect of CO_2 and chlorophyll retention was attributed to the inhibition of ethylene production.[22] Short-term applications of high CO_2 concentration could be beneficial during storage of green fruits and vegetables. This aspect should be looked into with interest in the field of minimal processing because packaging with increased CO_2 levels may retain color and chlorophyll level. Product a_w affects the rate of chlorophyll degradation. The lower that the a_w is, the better will be the chlorophyll retention,[22] according to the role of the available water in the pigment deterioration mechanisms.

PHENOLIC COMPOUNDS

The families of phenolic compounds commonly found in both fruits and vegetables are benzoic acids, cinnamic acids, flavonols, tannin precursors, and anthocyanidins.[23] Phenolics are involved in one of the most important reactions that affect color of minimally processed fruits and vegetables. When they are sliced, brushed, or damaged in any way, they turn brown or even black. The browning is the result of an enzyme-catalyzed reaction, which is usually highly undesirable and which corresponds to the oxidation of phenols catalyzed by polyphenoloxidases. The Enzyme Nomenclature Commission has classified this group of enzymes according to its specificity. This has designated monophenol mono-oxygenase, cresolase, or tyrosinase as EC 1.14.18.1, *o*-diphenol oxidase, catechol oxidase, or diphenol oxygen oxidoreductase as EC 1.10.3.1, and laccase or *p*-diphenol oxygen oxidoreductase as EC 1.10.3.2.[24] Most PPOs involved in fruit browning are catecholoxidases, except in peach and apricot, where the simultaneous presence of *o*- and *p*-diphenoloxidases has been reported. Both enzymes are also found in mango. Cresolase activity has been detected in several fruits, but this activity is often lost during the purification steps, and its role is controversial.

Cathecholases contain copper as the prosthetic group. For the enzyme to act on its phenolic substrate, the Cu^{2+} must first be reduced to Cu^{+}, a state in which the enzyme can then bind O_2. The phenolic substrates bind only to the oxy-PPO moiety, and, as a result of this binding, hydroxylation of the monophenol or oxidation of the diphenol occurs. The major expression of *o*-diphenol oxidase activity in plants is the oxidation of a range of *o*-dihydroxy phenols to the corresponding *o*-quinones, as shown in

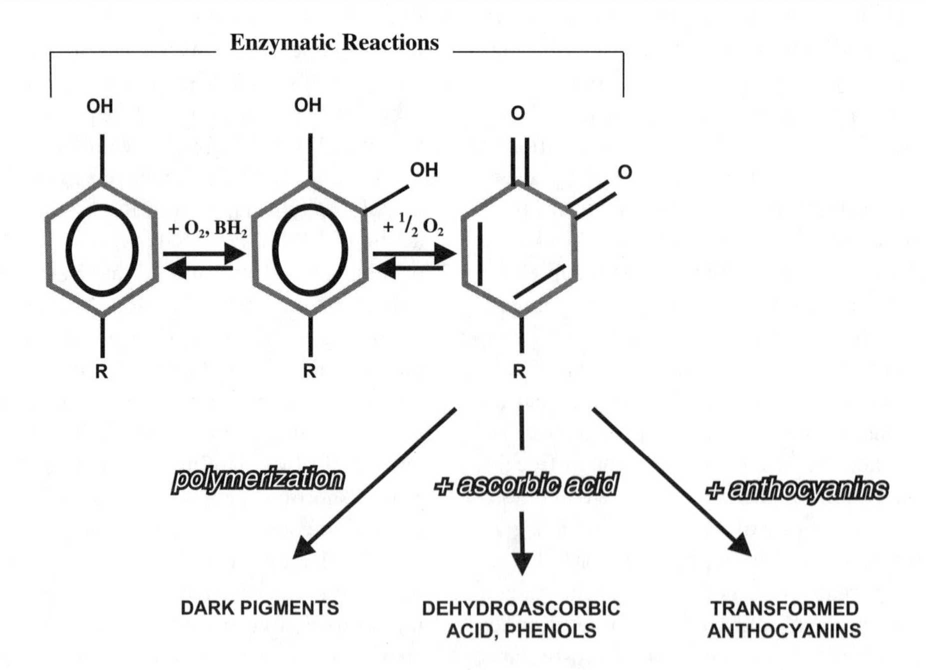

Figure 7–2 Some enzymatic browning reactions in minimally processed fruits and vegetables.

Figure 7–2. Note that the first stage of the reactions is reversible and it is for this reason that ascorbic acid and other reducing agents can prevent enzymatic browning by reducing the colorless *o*-quinones back to their parent phenols. Unfortunately, once all the ascorbic acid is oxidized, the *o*-quinones are no longer reduced, so they then can undergo oxidative polymerization to yield brown-black melanin pigments.

The most common natural substrates for PPO-catalyzed enzymatic browning are chlorogenic acid and its isomers plus catechin and epicatechin. However, some fruit polyphenoloxidases use other phenols as substrates; for example, 3,4-dihydroxyphenylethylamine is the major substrate in bananas, and dihydroxyphenylalanine is the natural substrate in the leaves of broad beans. Grape cathecolase acts on *p*-coumaryl and caffeoyl-tartaric acids whereas dates contain an unusual combination of diphenol oxidase substrates, including a range of caffeoyl-shikimic acids; these are analogous to the ubiquitous isomers of chlorogenic acid. PPO activity has been detected in all parts of fruits, including the peel, flesh, and cortex. When twelve apple cultivars were analyzed, the activity was mainly concentrated in the peel or the cortex, although the levels in the cortex were higher than those in the peel in seven cultivars, similar in four others, and lower in one cultivar.[25] The intracellular location of fruit PPO has been shown to be in chloroplasts, particularly on the inner face of thylakoids. In some cases, PPO activity has also been obtained from mitochondria in olive and apple, microbodies in avocado or partly associated with the cell wall in banana. There is generally a change from bound to more soluble enzyme forms during ripening, but activity of the soluble form is always lower than that observed in young fruits (about 17% in apple and 4–5% in grape).

Most of the unit operations that usually are carried out during minimal processing affect the degree of browning. Operations such as peeling, slicing, and cutting produce physicochemical modifications of membranes of the fruit cells, causing subcellular decompartmentation and leading to enzyme-substrate contact and browning. Likewise, these operation increase fruit respiration rates and, thus, their susceptibility to browning. One of the difficulties in selecting a method to prevent browning is the need to comply with safety and regulatory standards while maintaining the sensory attributes of the fruit or vegetable. The principles on which prevention of browning techniques are based are the following: inhibition or inactivation of the enzyme; elimination or transformation of the substrates; and transformation of the intermediate products of the reaction. Inhibition techniques generally involve two of the above and may be divided in three groups: thermal treatments, addition of chemicals, and combinations.

Thermal Treatments To Control Browning

Inactivation of polyphenoloxidase may be achieved by processing fruit and vegetables at temperatures above 80°C. Svensson[26] reported that inactivation of peroxidase, polyphenoloxidase, and lipoxigenase of potatoes followed a first-order reaction kinetics. To compare thermal inactivation of enzymes with thermal death curves for microorganisms, the heat treatment data for enzymes should be reduced to D-values and z-values. The z-values for lipoxigenase, PPO, and peroxidase were 3.6°, 7.8°, and 35°, respectively.

If we consider the definition of *minimal processing* suggested by Rolle and Chism,[27] that is, "minimal processing includes all unit operations (washing, sorting, peeling, slicing, etc.) that might be used prior to blanching on a conventional processing line," then thermal treatments are not included in minimal processing. On the other hand, Wiley[28] defines *minimally processed refrigerated* (MPR) fruits and vegetables as "those prepared by a single or any number of appropriate unit operations such as peeling, slicing, shredding, juicing, etc., given a partial but not an end-point preservation treatment including use of minimal heat, use of preservatives, or irradiation." It seems from this point of view that blanching should fit into the definition. If blanching is necessary, as is true in some fruits and vegetables, high-temperature short-time processes should be preferred to long-time lower-temperature treatments that may, in turn, result deleteriously to some commodities such as avocado.

Addition of Antibrowning Compounds

Many antibrowning agents have been tested; however, a limited number of these are permitted for use in fruits and vegetables. Chemicals exhibit different modes of action on preventing enzymatic browning; therefore, the following groups may be considered: reducing agents, chelating agents, and inorganic salts, among others.

Reducing Agents

These compounds react with quinones, reducing them to phenols, or act on the enzyme itself, linking irreversibly the copper of the "met" or "oxi" forms of the PPO. Sulfites, sulfur dioxide, and bisulfites inhibit enzymatic and nonenzymatic browning and are also effective against microbial infections. However, due to their harmful effect on sensitive and asthmatic people, their use has been restricted or even banned in some countries. In the United States, the Food and Drug Administration (FDA) has limited their use in salad bars.[23]

Ascorbic acid is a moderate reducing compound, acidic in nature, forms neutral salts with bases, and is water soluble. The product may be added to foods as tablets or wafers, dry premixes, liquid sprays, or as a pure compound. It is important to add the ascorbic acid as late as possible during processing to maintain high levels during the expected shelf life of the food commodity. Ascorbic acid (vitamin C) and its various neutral salts and other derivatives have been leading generally recognized as safe (GRAS) antioxidants for use in fruits and vegetables and their juices to prevent browning and other oxidative reactions.

Erythorbic acid is the D isomer of ascorbic acid but has no vitamin C activity. Most research suggests that L-ascorbic acid and erythorbic acid have about equal antioxidant properties; thus, L-ascorbic acid might be used only where vitamin C addition is a need, because L-ascorbic acid is about five times more expensive than erythorbic acid. Ascorbic acid derivatives have been studied to prevent browning reactions in fruits and vegetables[29]; among them are to be found ascorbic acid-2-phosphate, ascorbic acid-3-phosphate, and ascorbic acid-6-fatty acid ester. These substances have not yet been approved by the FDA. Ascorbic acid phosphates can act as ascorbic acid reservoirs because ascorbic phosphatase present in plants may gradually liberate ascorbic acid.

Another reducing agent is cysteine. The action of this amino acid is complex. It forms additional compounds with phenolic substrates, such as acid 5,S, cysteinil-3,4-dihydroxytoluene, which is produced when cathecol reacts with cysteine; or acid 2,S cysteinilchlorogenic, formed with cysteine and chlorogenic acid. Also, cysteine reduces quinones and forms thiol adducts, thus preventing the formation of pigments. Friedman and Bautista[30] proposed that the mechanism of action of cysteine includes the breakage of the copper–nitrogen from a hystidine link in the active center of PPO. They suggested that there is a strong affinity of the thiol group for the enzyme copper, displacing the hystidine residues that link copper to the rest of the protein, thus producing changes in the conformation of PPO. When kinetics of activity of avocado polyphenoloxidase were studied,[31] the curves of enzyme activity in system PPO-cathecol with different concentrations of cysteine showed two inhibitory effects: first, a lag period in the curves that was enhanced upon increasing the concentration of cysteine, and second, the diminution of the slope of the curves of activity. This behavior was also observed by Kahn[32] and agrees with the mechanism of inhibition described above. Cysteine has shown to be effective as an antibrowning agent in red delicious apples,[33] in Bosc pears,[34] and in avocado,[6] and looks promising in minimal processing of such fruits.

Chelating Agents

The mechanism of preventing enzymatic browning using chelating agents is the formation of a complex between these inhibitors and copper through an unshared pair of electrons in their molecular structures, which provides the complexing or chelating action. The best-known chelating agents for use on fruits and vegetables that are GRAS are citric acid and EDTA. In addition, derivatives of phosphoric acid also sequester metal ions. The short-chain polyphosphates, particularly sodium acid pyrophosphate, show the best sequestering ability, and their efficiency decreases as the pH increases.

Inorganic Salts

Salts of calcium, zinc, and sodium have been tested as antibrowning agents. Pizzocaro et al.[35] studied the inhibition of apple polyphenoloxidase by the action of sodium chloride. They found that dipping apple slices in NaCl (0.2–1.0 g/L range) solutions for 5 minutes increased the PPO activity. However, it was proven that ascorbic acid–sodium chloride mixtures inhibited 90–100% of the PPO activity. Van Rensburg and Engelbrecht[36] studied the effect of calcium salts on susceptibility to browning of avocados. Calcium treatments suppressed both respiration and polyphenol oxidation, and reduced the content of total phenolics, leucoanthocyanins, and flavonols. Of the calcium treatments used, calcium arsenate gave the best results. Zinc chloride has also been tested to control browning in minimally processed mushrooms[2] and avocado,[6] but results indicated no effective antibrowning action on such products.

Some Novel Inhibitors of Enzymatic Browning

Different compounds have been investigated as inhibitors of enzymatic browning as substitutes for sulfites; some of them have not yet been approved as food additives. The characteristics of some of them are discussed, based on the description of Vámos-Vigyázó.[37] Tropolone (2-hydroxy-2,4,6-cicloheptatrien-1-one) is a copper chelator slowly binding to the "oxy" form of the enzyme. It was found to be a substrate of horseradish peroxidase in the presence of hydrogen peroxide and is, therefore, assumed to be helpful in distinguishing this enzyme from PPO. Kojic acid (5-hydroxy-2-hydroxymethyl-pyrone) interacts with *o*-quinone formation from *o*-diphenols by decreasing oxygen uptake by the enzyme. (+)-Catechin 3′O-α-D-glucopyranoside proved to be a strong inhibitor of mushroom PPO; because other PPO substrates can be transformed by glucosidation into inhibitors, as well; this group of compounds might be of interest for future research. Carrageenans, a group of naturally occurring sulfated polysaccharides, as well as amylose sulfate and xilan sulfate were reported to inhibit browning of unpasteurized apple juice and diced apples at low concentrations (less than 0.5%). A peptide of MW 600 Da present in honey showed PPO inhibition in model solutions. Carbon monoxide gas atmosphere was found to inhibit mushroom PPO reversibly, whereby it prevented self-inactivation of the enzyme. However, carbon monoxide as a substance harmful to human health does not seem to be of any practical use in preventing browning of food materials.

Control of Enzymatic Browning through Hurdle Technology

In minimally processed fruits and vegetables, combinations of several factors can inhibit browning by different mechanisms. This technology is considered more promising than the use of individual inhibitors because of the possibility of synergistic interactions. Table 7–2 gives some combinations reported by several authors that were effective in prevention of browning in some fruits.

Kahn[32] studied the effect of proteins, protein hydrolysates, and amino acids on *o*-dihydroxyphenolase activity of mushroom, avocado, and banana. L-cysteine 0.65 mM prevented brown-

Table 7–2 Some Reported Combinations of Antibrowning Agents in Browning Control in Minimal Processing of Fruits and Vegetables

Product	*Reducing Agent*	*Acid*	*Chelating Agent*	*Inorganic Salt*	*Other Antibrowning Compound*	*Remarks*	*Reference*
Mushroom	Erythorbic acid	Citric acid	Ethylenediamine tetraacetic acid	Zinc chloride	4-hexyl resorcinol		2
Apple	Ascorbic acid derivatives	Citric acid	Complexing agents				29
Apple	Ascorbic acid				4-hexyl resorcinol		38
Potato	Ascorbic acid	Citric acid					39
Avocado	L-cysteine	Citric acid	Sodium pyrophosphate			Reduced a_w	6
Banana, avocado	L-cysteine				L-hystidine L-lysine L-phenylalanine Triglycine		32
Potato	Ascorbic acid	Citric acid	Sodium acid pyrophosphate	Calcium chloride	Ascorbic acid phosphate		40
Garlic (Chopped)		Citric acid					41
Prepeeled Potato	Ascorbic acid					Combined with lye surface digestion	40

ing in homogenates of avocado, and the same amino acid at 10 mM controlled browning in banana for 40 hours. Also, he found that 1-cm-thick slices of ripe avocado and ripe banana dipped in a solution containing 230 mM cysteine kept the original color (light green and light cream) six hours after cutting the slices.

Considering the results obtained by Kahn, Dorantes et al.[6] employed a combination of factors to prevent browning in minimally processed avocados. They found that two combinations of antibrowning compounds were more effective in preventing browning—a mixture containing L-cysteine, sodium erythorbate, disodium ethylenediamine tetracetate (EDTA sodium salt), and another mixture with sodium erythorbate, disodium ethylene diamine tetracetate, pyrophosphate tetrasodium salt, and L-cysteine (Table 7–2). On the basis of these results, experiments were carried out to control browning in avocado products. A constant concentration (0.2%) of L-cysteine was used. Various concentrations of pyrophosphate, a_w, and pH levels of the immersing solution were tested. A surface response analysis showed that one of the best combinations to prevent browning was 1% tetrasodium pyrophosphate, 0.2% cysteine at a_w 0.80, and pH 5.5. The selected formulation gave acceptable results for color, flavor, and texture of minimally processed avocado slices after 192 hours of storage.

Treatments to control discoloration of minimally processed mushrooms were investigated by Sapers et al.[2] Whole or sliced mushrooms were immersed in browning inhibitor solutions and evaluated for color changes during storage. Washing sometimes induced purple discoloration associated with bacterial attack. Other discolorations were induced by hypochlorite, 4 hexylresorcinol, and acidic dips. The most effective treatment was a combination of sodium erythorbate, cysteine, and EDTA at pH 5.5. Addition of preservatives to browning inhibitor dips did not improve storage life. However, dipping in 5% hydrogen peroxide prior to application of browning inhibitors significantly increased shelf life.

Sapers et al.[42] studied the browning in cut potato and apple, as affected by vacuum and pressure infiltration. Samples were cut in different geometrical forms. A solution containing 4% ascorbic acid plus 1% citric acid was employed. The soluble solid content of treated samples (plugs) was increased significantly by pressure infiltration, compared with dipping at atmospheric pressure. However, some darkening occurred during pressure infiltration, as indicated by the decrease in L* value. Potato samples (plugs) infiltrated at 108 kPa gained 2–4 days of storage, compared with dipping, but this treatment was ineffective with potato dices. The contrasting response of potato plugs and dices to pressure infiltration was attributed to the lower surface/volume ratio of the plugs. Although there was no advantage in applying browning inhibitors to dices (3/8 inches) by pressure infiltration, this technique may be advantageous with larger pieces or even prepeeled tubers. Improvements in the storage life of pressure-infiltrated potatoes might be achieved by vacuum packaging of the treated product.

Sapers et al.[43] controlled enzymatic browning in potato plugs with ascorbic acid-2-phosphate (AA-2-phosphate). Treatment effectiveness was greatly improved by reducing pH to 2.0 with phosphoric acid to inhibit endogenous acid phosphatase, because it appears that potatoes have a high level of acid phosphatase activity, as compared with apple, which results in the rapid hydrolysis of AA-2-phosphates and premature consumption of the ascorbic acid generated thereby. It is possible that pH as low as 2.1 could also inhibit polyphenoloxidase.

Another method investigated by Sapers and Miller[40] was the control of enzymatic browning in prepeeled potatoes by surface digestion with 17% NaOH for 4 minutes at 49°C. This lye digestion, in conjunction with conventional browning inhibitors (4% ascorbic acid, 1% citric acid, 1% sodium acid pyrophosphate, and 0.2% calcium chloride), represents a viable alternative to sulfiting prepeeled potatoes.

Amiot et al.[44] studied the effect of bruising on browning in apples of eleven cultivars. The degree of browning was determined by the simultaneous measurements of soluble (absorbance at 400 nm: A_{400}) and insoluble (lightness: L*) brown pigments. For the eleven cultivars, good correlations were obtained between A_{400} and the amount of hydroxycinnamics degraded, and between L* and the amount of flavan-3-ols degraded. According to this work, cultivars could be divided into two classes, those with a weak browning capacity (Elstar, Florina, Mutsu, and Golden), characterized by a low absorbance values ($A_{400} < 0.4$) and high lightness ($L^* > 53$) and those with a high browning capacity (Charden, Canada, McIntosh, Gala, Fuji, and red delicious), characterized by a high absorbance ($A_{400} > 0.4$) and low lightness ($L^* < 52$).

Browning of apples was also studied by Luo and Barbosa-Cánovas,[45] who evaluated the effect of 4-hexylresorcinol (HR) on the browning inhibition of delicious apple slices during cold storage (0.5 or 4.4°C). Significant ($p < 0.001$) inhibition in browning of apple slices was obtained with HR solution at concentrations as low as 0.005%. Discoloration was observed on the vascular bundles of the fruit 2 days after slicing. Combining 0.5% ascorbic acid with HR eliminated vascular discoloration and synergistically enhanced the browning inhibition with partial-vacuum (20-inch Hg vacuum) packaging and low-temperature storage. More than 50 days of browning-free storage life of the apple pieces were obtained with 5-minute dipping in a solution containing 0.01% HR plus 0.5% ascorbic acid and 0.2% calcium chloride.

Tronc et al.[46] studied the enzymatic browning inhibition in cloudy apple juice (pH = 3.8), acidified to pH 2.7, at 16–24°C and at a constant current density of 40 mA/cm^2 in an electrodialysis unit composed of an AB electrocell with three compartments—a bipolar membrane and two ion exchange membranes (anion and cation exchange). The electrodialysis acidification of apple juice and its subsequent storage with decreased pH resulted in a significant decrease in PPO activity (81% at pH 2.7), compared with the control. After 6 hours of storage, the a* value of control of weakly acidified juices was around 6, whereas that of acidified juice was around 4.5, which means that the acidified samples retained the original color of apple juice much better.

The preparation of minimally processed lettuce includes selection, washing, disinfection, and reduction of size. Enzymatic browning particularly appears in this last operation and throughout storage, and its control is important. Some commercial antibrowning solutions in the prevention of browning in minimally processed lettuce include citric acid, ascorbic acid, and sodium chloride. Castañer et al.[5] evaluated treatments to control browning of lettuce (cv. iceberg) stem discs, measuring L*, a*, and b* parameters, as well as hue angles. Results at 24 hours showed that a* was the most suitable parameter to measure browning, due to changes observed in the stem disks that led to development of reddish and brownish colors. Some browning inhibitors were tested on lettuce stem discs. Browning was decreased ($p < 0.05$) when dipped in water solutions with 4 g/L cysteine, 100 g/L citric acid, 5 g/L EDTA, or 0.1 g/L resorcinol. Additionally, some organic acids were tested. Acetic acid solutions (pH 2.3–2.81) resulted in better browning inhibition than citric acid solutions (pH 1.67–2.25), although the pH was higher for acetic acid. The authors gave a possible explanation to these results in terms of the differences in acid diffusion through the lettuce slices. If acetic acid diffuses better than citric acid, it would reach PPO more efficiently and exert its action earlier. They concluded that both vinegar and 50 ml/L acetic acid solutions could be very useful in preventing browning in cut lettuce stem during cold storage and commercial handling.

CONCLUSION

Color of minimally processed fruits and vegetables can be mainly affected by:

1. oxidation of phenolics catalyzed by poly–phenoloxidase,
2. conversion of chlorophylls to pheophytins by acidification, and
3. modification of anthocyanins by oxidation and acidification of the medium.

Methods that are most successful in preserving color in minimally processed fruits and vegetables include combinations of factors or hurdle technology. Looking to the near future, the modification of fruits and vegetables by genetic engineering—suppressing the genes that express polyphenoloxidase and other enzymes that affect color—is going to be very important.

REFERENCES

1. Francis FJ, Clydesdale FM. *Food Colorimetry: Theory and Applications.* Westport, CT: AVI Publishing Co.; 1975.
2. Sapers GM, Miller LR, Miller CF, Cooke PH, Sang-Won Ch. Enzymatic browning control in minimally processed mushrooms. *J Food Sci.* 1994;59:1042–1046.
3. Lozano JE, Drudis-Biscarri R, Ibarz-Ribas A. Enzymatic browning in apple pulps. *J Agric Food Sci.* 1994;59:564–567.
4. Alzamora SM, Cerruti P, Guerrero S, Lopez-Malo A. Minimally processed fruits by combined methods. In: Barbosa-Cánovas GV, Welti-Chanes J, eds. *Food Preservation by Moisture Control. Fundamentals and Applications.* Lancaster, PA: Technomic Publishing Co; 1995:463–492.
5. Castañer M, Gil MI, Artes F, Tomas-Barberan FA. Inhibition of browning of harvested head lettuce. *J Food Sci.* 1996;61:314–316.
6. Dorantes AL, Parada DL, Ortiz MA, Santiago PT, Barbosa-Cánovas VG, Chiralt A. Effect of antibrowning compounds on the quality of minimally processed avocados. *Food Sci Technol Int.* 1998;4:107–113.
7. Clydesdale FM. Analysis and processing of colorimetric data for food materials. In: Rha Ch, ed. *Theory, Determination and Control of Physical Properties of Food Materials.* Dordrecht: Reidel Publishing Co; 1975:297–309.
8. Olson JA. Biological actions of carotenoids. *J Nutr.* 1989; 119:94–97.
9. Minguez-Mosquera MI, Hornero-Méndez D. Changes in carotenoid esterification during the fruit ripening of *Capsicum annuum* cv. Bola. *J Agric Food Chem.* 1994;42:640–644.
10. Minguez-Mosquera MI, Gandul-Rojas B. Mechanism and kinetics of carotenoid degradation during the processing of green table olives. *J Agric Food Chem.* 1994;42:1551–1554.
11. Wang H, Guohua C, Prior RL. Oxygen radical absorbing capacity. *J Agric Food Chem.* 1997:45:303–309.
12. Massa G, Brouillard R. Color stability and structural transformations of cyanidin 3,5 diglucoside and four 3-deoxyanthocyanins in aqueous solutions. *J Agric Food Chem.* 1987;35:422–426.
13. Rommel A, Heatherbell DA, Wrolstad RE. Red raspberry juice and wine: Effect of processing and storage on anthocyanin pigment composition, color and appearance. *J Food Sci.* 1990;55:1011–1017.
14. Raynal J, Moutounet M. Intervention of phenolic compounds in plant technology. 2. Mechanisms of anthocyanin degradation. *J Agric Food Chem.* 1989;37: 1051–1053.
15. Wrolstad RE, Skrede G, Per L, Enersen G. Influence of sugar on anthocyanin pigment stability in frozen strawberries. *J Food Sci.* 1990;55:1064–1072.
16. Francia-Aricha EM, Guerra MT, Rivas-Gonzalo JC, Santos-Buelga C. New anthocyanin pigments formed after condensation with flavanols. *J Agric Food Chem.* 1997;45:2262–2266.
17. Chandra A, Muraleedharan GN, Iezzoni AF. Isolation and stabilisation of anthocyanins from tart cherries (*Prunus cerasus* L.). *J Agric Food Chem.* 1993;41: 1062–1065.
18. Baardseth P, Von Elbe JH. Effect of ethylene, free fatty acid and some enzyme systems on chlorophyll degradation. *J Food Sci.* 1989;54:1361–1363.
19. Zhuang H, Barth MM, Hildebrand DF. Packaging influenced total chlorophyll soluble protein, fatty acid composition and lipoxygenase activity in broccoli florets. *J Food Sci.* 1994;59:1171–1174.
20. Robertson G. Changes in the chlorophyll and pheophytin concentration of kiwi fruit during processing and storage. *Food Chem.* 1985;17:25–31.
21. Sanchez PME, Ortíz MA, Dorantes AL. The effect of ethylene diamine tetracetic acid on preserving the color of an avocado puree. *J Food Process Preserv.* 1991;15:261–271.
22. Simpson KL. Chemical changes in natural food pigments. In: Richardson T, Finley JW, eds. *Chemical Changes in Food During Processing.* Westport, CT: AVI Publishing Company; 1985:409–442.
23. Martínez M, Whitaker JR. The biochemistry and control of enzymatic browning. *Trends Food Sci Technol.* 1995;6:195–200.
24. Lee CY, Whitaker JR. *Enzymatic Browning and Its Prevention.* Washington, DC: ACS Symposium Series; 1995:600.
25. Janovitz-Klapp H, Richard C, Goupy M, Nicolas J. Inhibition studies on apple polyphenoloxidase. *J Agric Food Chem.* 1990;38:1437–1441.
26. Svensson S. Inactivation of enzymes during thermal processing. In: Hoyen T, Kvale O, eds. *Physical, Chemical and Biological Changes in Food Caused by Thermal Processing.* London: Applied Science; 1977:202.

27. Rolle RS, Chism GW. Physiological consequences of minimally processed fruits and vegetables. *J Food Qual.* 1987;10:157–177.
28. Wiley RC. *Minimally Processed Refrigerated Fruits and Vegetables.* New York: Chapman & Hall; 1994.
29. Sapers GM, Hicks KB, Phillips JG, et al. Control of enzymatic browning in apple with ascorbic acid derivatives, polyphenol-oxidase inhibitors, and complexing agents. *J Food Sci.* 1989;54:997–1002.
30. Friedman M, Bautista FF. Inhibition of polyphenol-oxidase by thiols in the absence and presence of potato tissue suspensions. *J Agric Food Chem.* 1995;43:69–76.
31. Dorantes AL, Ramirez JS, Davila OG. Influence of water activity, cystein and papain on the activity of avocado polyphenoloxidase. In: Roos YH, ed. *Proceedings of International Symposium on Properties of Water.* Helsinki: ISOPOW-7; 1998:127–130.
32. Kahn V. Effect of proteins, protein hydrolyzates and aminoacids on *o*-dihidroxyphenolase activity of polyphenol-oxidase of mushroom, avocado and banana. *J Food Sci.* 1985;50:111–114.
33. Richard-Forget F, Goupy M, Nicolas J. Cystein as an inhibitor of enzymatic browning. 2. Kinetic studies. *J Agric Food Chem.* 1992;40:2108–2113.
34. Siddiq M, Cash N, Sinha K, Akhter P. Characterization and inhibition of polyphenol oxidase from pears (*Pyrus communis* L. Cv. Bosc and Red). *J Food Biochem.* 1994;17:327–337.
35. Pizzocaro F, Torreggiani D, Gilardi G. Inhibition of apple polyphenol oxidase by ascorbic acid, citric acid and sodium chloride. *J Food Process Preserv.* 1993;17:21–30.
36. Van Rensburg E, Engelbrecht AHP. Effect of calcium salts on susceptibility to browning of avocado fruit. *J Food Sci.* 1986;51:1067–1068.
37. Vámos-Vigyázó L. Prevention of enzymatic browning in fruits and vegetables: A review of principles and practice. In: Lee CHY, Whitaker JR, eds. *Enzymatic Browning and Its Prevention.* Washington, DC: ACS Symposium Series 600; 1995:49–62.
38. Monsalve-Gonzalez A, Barbosa-Cánovas GV, Cavalieri RP, Mc Evily AJ, Yengar R. Control of browning during storage of apple slices preserved by combined methods. 4-hexylresorcinol as antibrowning agent. *J Food Sci.* 1993;58:797–800.
39. Langdon TT. Preventing of browning in fresh prepared potatoes without the use of sulfiting agents. *Food Technol.* 1987;5:64–67.
40. Sapers GM, Miller RL. Control of enzymatic browning in pre-peeled potatoes by surface digestion. *J Food Sci.* 1993;58:1076–1078.
41. Bae RN, Lee SK. *J Korean Soc Hortic Sci.* 1990; 3:213–218. Re: *J Food Sci Technol Abstracts.* 1991;23: 4T52.
42. Sapers GM, Garzarella L, Pilizota V. Application of browning inhibitors to cut apple and potato by vacuum and pressure infiltration. *J Food Sci.* 1990;55: 1049–1053.
43. Sapers GM, Miller RL, Douglas Jr FW, Hicks KB. Uptake and fate of ascorbic acid-2-phosphate in infiltrated fruit and vegetable tissue. *J Food Sci.* 1991;56:419–422.
44. Amiot MJ, Tacchini M, Aubert S, Nicolas J. Phenolic composition and browning susceptibility of various apple cultivars. *J Food Sci.* 1992;57:958–962.
45. Luo Y, Barbosa-Cánovas GV. Inhibition of apple-slice browning by 4-hexylresorcinol. In: Lee ChY, Whitaker JR, eds. *Enzymatic Browning and Its Prevention.* Washington DC: ACS Symposium Series 600; 1995: 240–250.
46. Tronc JS, Lamarche F, Makhlouf J. Enzymatic browning inhibition in cloudy apple juice by electrodialysis. *J Food Sci.* 1997;62:75–78.
47. Mitchell GE, McLauchlan RL, Beattie TR, Banos C, Gillen AA. Effect of gamma irradiation on the carotene content of mangos and red capsicums. *J Food Sci.* 1990;55:1185–1186.
48. Lin TY, Koehler PE, Shewfelt RL. Stability of anthocyanins in the skin of star krimson apples stored unpackaged under heat shrinkable wrap and in package modified atmosphere. *J Food Sci.* 1989; 54:405–407.

CHAPTER 8

Mathematical Modeling of Enzymatic Reactions as Related to Texture after Storage and Mild Preheat Treatments

C. van Dijk and L.M.M. Tijskens

INTRODUCTION

The invention of the sterilization process by Nicolas Appert almost two centuries ago was a major breakthrough in food processing. This process offered mankind the possibility to extend the keeping quality of perishable agricultural products during substantially longer periods than was ever optional for the raw materials, as such.

However, in time, it was realized that this process strongly affects quality attributes of the product, such as texture, color, and taste. Furthermore, this process negatively affects (hidden) product properties, such as, for example, the vitamin C content.

During the last three decades, prolonging keeping quality has turned from art into science. The underlying reasons for this transition are related to:

1. an improved basic understanding about plant physiology, plant genetics, and (some of) their relations;
2. raw materials with improved properties: These properties are related to quality attributes such as texture, taste, flavor, and color;
3. improved logistics and management of the raw material in a chain; and
4. developments in process technologies, ranging from the improvement of existing technologies to the industrial implementation of technologies not yet applied.

The results of the research presented here were obtained within the framework of the EU-AIR project, under project number AIR1-CT92–0278, entitled: "The (bio)chemistry and archestructure of fruit and vegetable tissue as quality predictors for optimising storage and processing regimes: Basic research leading to applicable models and rules," partly financed by the European Commission. The authors wish to express their gratitude to the participants of this project:

K.W. Waldron, Institute of Food Research, Norwich, UK; A.M.M. Stolle-Smits, K. Recourt, and C. Boeriu, ATO, Wageningen, Netherlands; P.S. Rodis, Agricultural University of Athens, Athens, Greece; B.E. Verlinden and J. DeBaerdemaeker, Catholic University Leuven, Leuven, Belgium; T. de Barsy and R. Deltour, University of Liege, Liege, Belgium; I. Zarra, University of Santiago de Compestella, Santiago de Compestella, Spain. C. van Dijk was coordinator of this project.

In conjunction with these scientific and technologic developments, the perception of product quality acquired by the consumer is semi-continuously influenced by the mutually dependent and interacting parameters product, consumer, and market.[1] At this very moment, the consumer is very concerned about the fresh image of agricultural products, as related to a healthy image. In addition, the consumer is also spending less time to produce a meal. This implies an increase in the demand of characteristics such as "easy to use" and convenience. As a consequence of these developments, one can observe activities focused on the extension of the keeping quality of agricultural raw materials by processing these raw materials minimally. On the other hand, efforts are undertaken to reduce the impact of the sterilization process on both product properties and quality attributes of the perishable products by adapting this process. In essence, these two approaches are followed to deliver products to the market with an optimal "fitness for use" for the consumer.[2,3]

The contents of this chapter are focused on the effects of both temperature and time on the important quality attribute of "texture" of fruits and vegetables. Information was gathered about specific enzymes, which are either known or assumed to affect texture during storage or heat treatments. A set of chemical reactions was defined, based on this information, information from literature, and generally accepted theories. In essence, these sets of chemical reactions are the core of the models. Using the well-known rules of chemical kinetics,[4–6] these models can be developed further into their analytical solutions. In other words, underlying processes affecting texture were modeled, rather than texture itself. So far, two main items have been mentioned: texture and modeling. To obtain a better insight of the modeling approach chosen in relation to texture, a verbal description of both texture and modeling will be given. Examples will be worked out and discussed, making use of this "chemical kinetic" approach during either storage or preheating treatments of fruits and vegetables.

TEXTURE AND TEXTURE-GENERATING FORCES: A VERBAL MODEL

Global Origin of Texture

Considering the origin of texture, three levels of abstraction can be discerned: the molecular level, the cellular level, and the organ level. At the *molecular level*, the key determinants of texture are the chemical nature of the plant cell wall and the interactions between the constituting biopolymers. During the development and senescence of plant organs, modification of cell wall polymers results in a change in the contribution of texturally perceived properties. Compare, for example, the perceived texture between an immature, a mature, and an overly mature apple. The majority of these changes in the raw material have been attributed to cell wall metabolism during either ripening[7–9] or storage.[10] Such changes can lead to a deterioration in texture and palatability of plant organs, as a result of oversoftening or toughening. Even in the most-studied example of a pectin degrading enzyme, polygalacturonase, no firm conclusions have been reached on its role in the softening of fruit during ripening.[11] It is, however, generally accepted that these cell wall modifying enzymes contribute to the texture of the final product, whether or not this product has been stored or given a heat treatment. However, little information is available on the effect on texture of either the storage or the heating process, nor is there much information on the rate and contribution of the chemical and biochemical alterations taking place in the cell wall.

At the *cellular level*, the key determinant of texture is the tissue archestructure, which comprises cell wall thickness, cell size, cell shape, cell adhesion, and tissue organization. For example, when cells become larger, the surface/volume ratio decreases. Assuming that the surface relates to the cell wall, with its major contribution to texture, it can easily be anticipated that this increase in cell volume will have an effect on the perceived texture. When, on the other hand, the cell-cell adhesion decreases, this will also have consequences for the perceived texture. By applying a rupture force on a tissue with low cell-cell adhesion forces (the cells are not strongly glued together), the tissue will break between the cells and the cell contents are retained within the cell upon fracturing of the tissue. When the cell-cell adhesion forces are high (the cells are strongly glued together), the tissue will break through the cells, and the cell contents are liberated. Thus, the cell-cell adhesion not only has consequences for the perceived texture, but also for the perceived taste and flavor. The main locus of the cell-cell adhesion is the middle lamellae region.

Textural properties are measured at the *organ level* by rheologic measurements and/or perceived by sensory analysis. Nevertheless, the way that textural properties change ultimately depends on effects exerted at the molecular level, affecting tissue archestructure at the cellular level and finally having consequences at the organ level. Little information is available on the effects of either storage or heating on the molecular level and how these effects at the molecular level are translated to the cellular and organ levels and, finally, to the perceived texture. The relations between (bio)chemical composition, physical appearance, and mechanical properties are largely unclear. It is obvious that the textural behavior of a product, either during storage or a heat treatment, cannot be described exclusively, based on information from only one of these levels. Going from a molecular level through the cellular level to the organ level, a hierarchical interaction pattern is described. In addition to this hierarchical interaction pattern, it is obvious that the texture of agricultural products is caused by a combination of physical forces originating from the following processes or properties:

1. turgor pressure inside intact living cells and the associated tissue tension,
2. special compounds inside cells possibly generating strength (e.g., starch),
3. cohesive forces within a cell: chemical properties of the cell wall,
4. adhesive forces between cells: chemical properties of the pectin,
5. overall structure and shape of separate cells,
6. overall structure and shape of tissue: strength and distribution of, for example, stronger shaped vascular tissue.

In this summation, items 1–4 represent the chemical and physical based forces, and items 5–6 represent the histologic and morphologic ones (archestructure).

Depending on relative occurrence and relative importance of one of the mentioned items, a very diverse range of textural behavior can be depicted, e.g.:

- With only tissue tension (turgor) as a major item, products such as fresh strawberries are soft and juicy, losing texture upon processing.
- With pectin forces overruling, products such as fresh apples are essentially crispy and juicy (rupture through cells; juice with contents becomes liberated from disrupted cells).
- With cell wall forces overruling, products are essentially mealy and dry (rupture along cells; juice with contents remains inside intact cells), such as some times in overripe apples.
- With vascular tissue important, products are essentially tough and fibrous, such as sometimes in asparagus.

Cohesive Forces within a Cell and Adhesive Forces between Cells

In the schematic model representation of the primary plant cell wall by Carpita and Gibeaut,[12] the cellulose microfibrils are embedded in or surrounded by hemicelluloses. This cellulose-hemicellulose network, chemically more or less inert, provides the basis of the plant cell wall. Interwoven with this network are the pectic polymers. Pectin is the description for a family of heteropolysaccharides, rich in (methylated)

D-galacturonic acid, including polysaccharide side chains.[13] Pectins are block polymers[14] that contain both linear homogalacturonan blocks ("smooth regions") and branched blocks of rhamnogalacturonan ("hairy regions") with neutral side chains.[15] The knitwear of these three distinct polysaccharides forms the basis of the cohesive forces within a cell. Furthermore, it is assumed that the cellulose-hemicellulose matrix forms the stretch-resistant, load-bearing part and that the pectin matrix forms the compression-resistant part of this network.[16] In addition to their contribution to the cohesive forces, pectins are thought to be mainly responsible for the adhesive, cementing forces between cells. This cementing function of pectins is assumed to be mainly performed in the middle lamellae region between adjacent cells. However, it is not yet clear whether this cementing function is caused by "egg-box" structures, induced by calcium complexing of unsubstituted homogalacturonans[17] or by neutral side chains of the pectic polysaccharide entangled in and with the cellulose-hemicellulose region, causing intercellular adhesion.[18]

Relation between Chemical Properties of (Bio) Polymers and Physical Forces

The stereo configuration of biopolymers (curly like starch or linear like cellulose) has a major impact on their physical behavior (strength and elasticity). This stereo configuration, however, depends on polymer type only. For the cellulose-hemicellulose network, the stereo configuration is considered to be constant for agro-products, either when put into storage or during a heat treatment. The rupture strength of linear polymers is generally accepted to be proportional (for the chemical aspect) with the degree of polymerization. Based on the assumption that pectin and cellulose are long linear polymers, decay of their chain length explains the loss of rupture force with a first-order reaction: For each cut in each chain the remaining length is half its original.[19] For linear polymers with side chains, such as pectins, the situation is somewhat more complicated. Although it is not quite clear whether short side chains increase the intrinsic rupture strength of the complete chain, the embedding factor (e.g., root system in soil) is of major importance. The effect of side chains on texture will depend less on the degree of polymerization (DP) than on spatial distribution and slipping inhibition.

For real three-dimensional polymers, the situation is even more complicated: In addition to the increase in embedding force (exponential toward limit), for each cross-link existing between the same two polymer chains, one cut is necessary for an effective chain shortening to occur. Thus, decay can take place a long time before any signs of decay can be observed on a physical basis (rupture). On the other hand, one new cross-link between unlinked chains generates a huge increase in the degree of polymerization and, hence, in rupture strength. The spatial structure and the distribution of side chains and cross-links will, therefore, define the type of behavior: Too many cross-links on a small volume does not really increase strength but improve brittleness; too few cross-links generate linear polymers with high elasticity and low strength. Thus, for nonlinear polymers, the relative position of the cross-links is of major importance.

Enzymes and Texture

Like any part of living plants, cell walls are continuously prone to orchestrated, enzyme-catalyzed alterations of their chemical composition during growth, maturation, and senescence. For green beans, during these sequential stages of growth, maturation, and senescence, the chemical composition of the cell walls changes, as does the amount of enzymes capable of reacting with cell wall polymers.[20] Tremendous efforts have been made to study the role of pectins and pectolytic enzymes in the softening of fruit tissue during ripening.[21] With regard to these texture-modifying enzymes, it has to be realized that the enzyme activity at a given time is not the most determining factor for texture behavior. It is the total exerted action of the enzyme activity over a given period of time that determines the chemical changes that have taken place in the cell wall polysaccharides and the subsequent physical consequences.

MATHEMATICAL MODELING OF ENZYMES AND TEXTURE

Models and Model Levels

Models are the mathematical description of a part of a real-world situation. Two approaches with regard to modeling can be distinguished, leading to either empirical models or more fundamental models. Empirical models are based on the (statistical) analysis of a relation between input and output data, a dose-response relation. Examples are exponential or logistic functions describing the softening of fruit, or Near Infra Red calibration curves predicting either chemical and physical parameters or perceived texture attributes of agro-products. More fundamental models are based on kinetic mechanisms and fundamental laws, e.g., Arrhenius. Of main importance for these types of models is the basic understanding of the processes underlying the observed phenomena one wants to model, rather than the phenomena themselves.[22]

With empirical models, one tries to integrate domain knowledge of specific products with a measuring technique to generate statistically based prediction models. It has to be realized that the empirical models do not add to the understanding of product behavior. These empirical models are, however, generally well suited for practical applications because they have a high predictive power most of the time. In general, these empirical models are limited to the measuring situation.

The information provided within this chapter focuses on the more fundamental models. These models can be divided into three distinct levels of dynamism. In increasing hierarchical order, one can distinguish:

- The *static model* gives a state description at a certain constant time. The static model states in fact the situation of the product, together with the relational functions between state and observed properties. It relies primarily on data from chemical analysis.
- The *dynamic model* describes the product behavior with time, but with otherwise constant environment (external factors). The dynamic model describes state changes in time. State changes in time refer, for example, to changes taking place during a heat treatment or during storage. This type of model can (possibly) be formulated with algebraic functions (analytical solutions). In most cases, however, differential equations are necessary. The model needs information, not only about chemical compounds (type and amount), the enzymes acting upon them (type and amount), but also the generation and degeneration in time of the enzymes involved. This dynamic model relies on the static model, extended with enzyme data.
- The *supra-dynamic model* describes product behavior with time and with changing environment (external factors); in other words, it describes changes in state during time under changing external factors. It is impossible to formulate this type of model with algebraic functions (analytical solution), and one has to make use of differential equations. It needs the same kind of information as the previous (dynamic) type of model. In addition to all of this, the influence of the changing external factors upon these activities has to be known (estimated). It relies on the previous type of models, augmented with data of processing and its influence, e.g., with a temperature process. This implies that activation energies are required to be known, based on the assumption of an Arrhenius-type behavior.

Model Development and Statistical Analysis

The models were developed using a system of problem decomposition.[1] This system is oriented toward modeling of the underlying processes that cause the observed phenomena, rather than the modeling of the observed phenomena themselves. The models are based on kinetic mechanisms describing the particular process. The models were developed further by using the well-known rules of chemical kinetics. The mathematical development and statistical analysis were carried out according to Tijskens et al.[22] No transformations were applied to the data to prevent errors during the estimation.[23] The data were analyzed as one integral set, using time and temperature simultaneously as explaining variables.[24–26]

Most of the experiments are conducted at constant conditions of external factors such as, for example, temperature. To analyze the experimental data, analytical solutions of the model formulation at constant external conditions are required. These analytical solutions will be deduced from the differential equations but are applicable only at constant conditions. In practice, constant conditions are very rare. However, the model formulations applicable at any time and temperature are the differential equations. The formulation of the differential equations is the core of the model, rather than the resulting analytical solutions. These analytical solutions are a logical consequence of the differential equations. The boundary conditions for the differential equations are defined by the experimental setup.

Symbols and Notation Used

Within the notation used to describe the underlying chemical and biochemical processes affecting the firmness of agro-products, the following has to be realized. With reference to the enzymes, the activities of these enzymes were determined in vitro, under standardized conditions with respect to the amount of substrate, pH, buffer, temperature, etc. These enzymes were extracted from the agro-products under study, which were either stored or heat-treated under defined conditions. Thus, the increase or decrease in activity was related to either formation or inactivation of enzyme. Because all of the modeling work was based on the information supplied by the in vitro assays, these activities served as input information for the models and not the enzyme amount or concentration. For this reason, in the modeling part, the notation referring to enzymes was not put in brackets, to distinguish it from real concentrations.

As a consequence of enzymatic reactions, pectin is either demethylated (PE action) or depolymerized enzymatically (PG action) or chemically (β-degradation). Within the study presented, neither the degree of esterification (DE) nor the DP of pectin was determined. The consequences of either chemical or enzymatic breakdown of the pectic polymer are mathematically described, and these effects are related to changes in firmness. Because neither the DP nor the DE are concentrations, but rather relative amounts, these notations were not put in brackets either.

MODELING APPROACH

The Mode of Action of the Enzyme Activities To Be Modeled

In this chapter, three enzymes are discussed. Their function and activity are assumed to be (strongly) related to the texture of fruits and vegetables. Two of these enzymes have pectin as their main substrate. These enzymes are, respectively, pectin methyl esterase (PE; *EC 3.1.1.11*) and endopolygalacturonase (PG; *EC 3.2.1.15*). PE removes methanol from methylated pectin. The plant enzyme works as a zipper and removes the methyl groups blockwise, in contrast to the fungal enzyme, which demethylates pectins more at random.[27,28] The consequences of the PE action are threefold. First, due to its blockwise demethylating action, the enzyme increases the probability that two adjacent polygalacturonic polymer chains form "egg-boxes" in the presence of calcium ions. The formation of a three-dimensional network, also caused by egg-box structures, results in an apparent increase in the chain length of the pectic polymers. As a consequence, the firmness of the plant tissue is increased. In addition, this demethylation process enhances both shielding and repulsion forces by the electric charges within the pectic polymer matrix of the cell wall and middle lamellae. Second, demethylated pectin is not vulnerable to eliminative breakdown, in contrast to methylated pectin.[29–31] β-Eliminative breakdown is virtually absent at room temperature (rate constant $k \approx 0$). It starts to contribute to observable changes in firmness above about 90°C. Therefore, it is assumed that preheating of plant tissue at moderate temperatures (50–80°C) reduces the softening of the plant tissue, compared with no preheating treatment. Third, demethylated pectin forms the substrate for PG. Due to its mode of action, PG action is assumed to decrease firmness, because this enzyme depolymerizes the pectin, thereby decreasing its DP.

The second enzyme studied is PG. This enzyme is observed in many fresh fruits and vegetables.[27] It depolymerizes pectic polysaccharide chains, preferably at those locations where the methyl groups have been removed. Due to its mode of action, the DP of the pectic polysaccharide chains decreases. This decrease in DP is assumed to be at least (partly) responsible for the frequently observed softening of plant tissue.[32,33] However, conflicting information exists about the observed PG activity and fruit softening.[11,34–37]

The third enzyme studied is peroxidase (POD; *EC 1.11.1.7*). Peroxidase is one of the plant's most heat-resistant enzymes and, therefore, it is often used as marker enzyme to assess the effectiveness of a heat treatment.[38] Peroxidase is also assumed to be involved in the oxidative cross-linking of cell wall polymers by catalyzing the formation of, for example, diferulic acid and isodityrosine cross-links.[39–42]

Two remarks are in order. With regard to the enzymes mentioned, it has to be realized that these enzymes form only a small part of all enzymes involved in cell wall metabolism. However, these enzymes are assumed to have the highest contribution to enzyme-catalyzed, texture-modifying actions. The second remark refers to the texture of plant materials. During either storage or heat processing, the firmness of the plant material decreases to a certain minimal value. In other words, the firmness of plant material consists of two components, a fixed part ($Firm_{fix}$) and a variable part ($Firm_{var}$). The firmness generated by the cellulose-hemicellulose domain of the plant cell wall will hardly be affected during heat processing,[43] let alone during storage, and, thus, represents the fixed part of the firmness. The variable part of the firmness refers to the pectic fraction of the cell wall. Its chemical composition and, as a consequence thereof, its physical properties are affected by either the enzymes, such as described above, or enhanced temperatures.

Firmness Decrease during Storage, with Peaches as Example

In Figure 8–1, the measured PG activity of peaches harvested at two sequential years is shown.[44]

The first observation is that the form of the curves at each storage temperature used is similar; at increasing storage temperatures, the PG activity first increases in time, followed by a decrease in activity. The second observation is that the total observed activity differs with a factor of about 20 between years. The observed increase in PG activity during the early stages of storage and the observed decrease in activity upon prolonged storage can be explained with the following mechanism. An increase in enzyme activity is due to the formation of active enzyme from an inactive, latent form of the enzyme; a decrease in activity can be caused by inactivation of the enzyme. At storage temperatures, it is difficult to imagine that enzyme denaturation occurs; therefore, inactivation (senescence) is assumed. This behavior of PG can be described as a set of consecutive reactions (see Equations 1a and 1b).

$$PG_{pre} \xrightarrow{k_f} PG \qquad \textbf{(1a)}$$

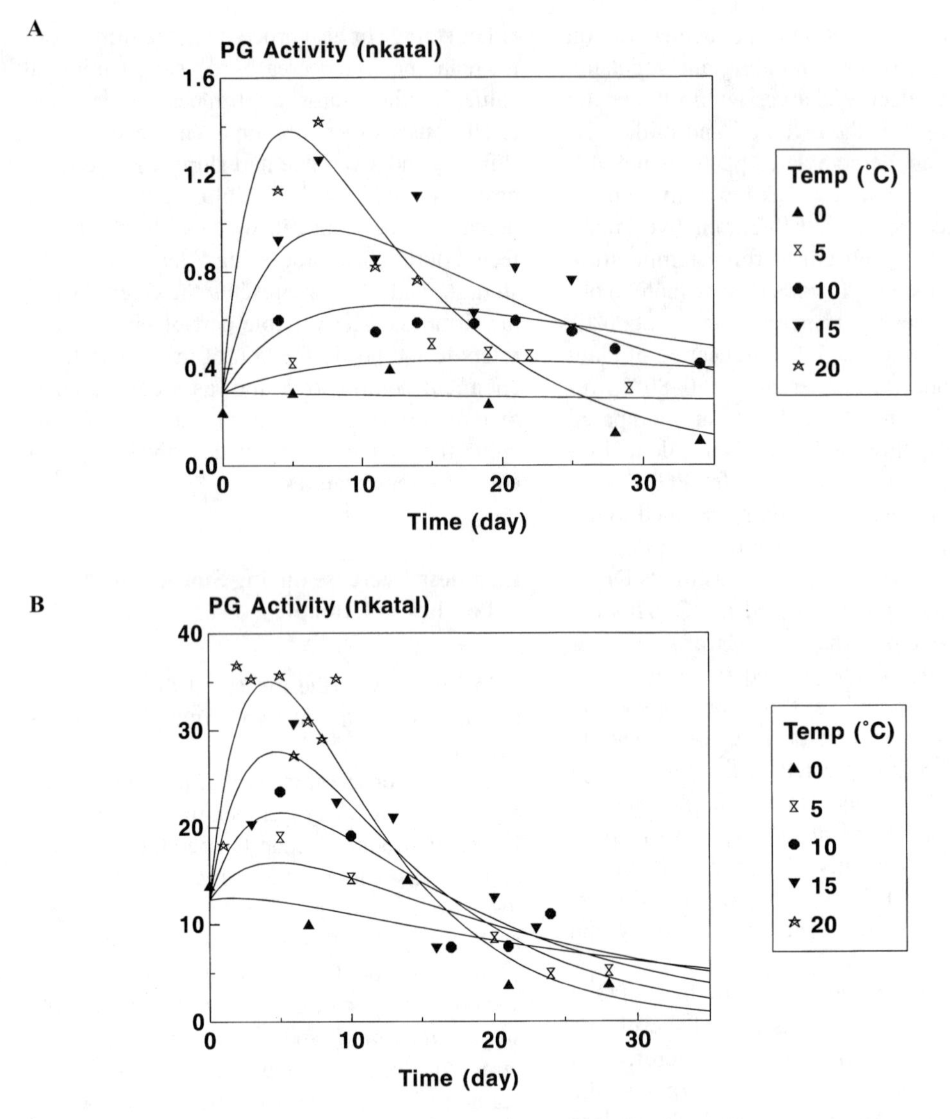

Figure 8–1 Measured (symbols) and simulated (solid lines) PG activities of peaches during storage at different temperatures for season 1 (A) and season 2 (B).

$$PG \xrightarrow{k_d} PG_{na} \quad \textbf{(1b)}$$

This set of chemical reactions can be converted into a set of differential equations (Equations 2a and 2b) and solved analytically for constant temperatures (Equation 3).

$$\frac{\partial PG_{pre}}{\partial t} = -k_f \cdot PG_{pre} \quad \textbf{(2a)}$$

$$\frac{\partial PG}{\partial t} = k_f \cdot PG_{pre} - k_d \cdot PG \quad \textbf{(2b)}$$

$$\mathrm{PG}(t) = PG_{\mathrm{pre},0} \cdot k_{\mathrm{f}} \left(\frac{\exp(-k_{\mathrm{d}} \cdot t) - \exp(-k_{\mathrm{f}} \cdot t)}{k_{\mathrm{f}} - k_{\mathrm{d}}} \right) + \mathrm{PG}_0 \cdot \exp(-k_{\mathrm{d}} \cdot t) \tag{3}$$

Equation 3 describes the total PG activity at any time at a given temperature. In these equations, t is the time of storage, PG is the activity of the enzyme measured under standard conditions, and k is the reaction rate constant. The index *pre* refers to the inactive precursor, *f* refers to the formation process of active enzyme, *d* refers to the denaturation process, *na* refers to the inactivated enzyme, and *0* refers to the initial condition ($t = 0$; with reference to the start of the measurements). The temperature dependence of chemical reactions is described by Arrhenius' law. Therefore, both reaction rate constants k_f, and k_d (see Equation 1) are assumed to depend on temperature, according this law (see Equation 4).

$$k_{\mathrm{i}} = k_{\mathrm{i,ref}} \cdot \exp\left[\frac{\mathrm{Ea_i}}{R} \left(\frac{1}{T_{\mathrm{ref}}} - \frac{1}{T_{\mathrm{abs}}} \right) \right] \tag{4}$$

In this equation T_{ref} refers to a chosen reference temperature (K), T_{abs} is the storage temperature (K), $k_{i,ref}$ is the reaction rate constant i of a chemical reaction at T_{ref}, Ea_i is the energy of activation, and R is the universal gas constant.

The above series of differential equations describe the observed phenomena, as depicted in Figure 8–1, and their relation with temperature dependence of the underlying rate constants according to Arrhenius' law.

The data on the PG activities, measured in peaches stored at different constant temperatures, were analyzed statistically using Equation 3 for the time dependence and Equation 4 for the temperature dependence of all reactions and their rate constants involved (see Equations 1 and 2). The results of this statistical analysis for the two seasons, studied independently, and for the combined information of the two seasons together, are given in Table 8–1.

Based on the results shown in Table 8–1, the following conclusions can be made. First, the model is capable of describing the apparent different PG activity between seasons (see Figure 8–1) with the same estimates for the kinetic parameters. This strongly suggests that the classification of the parameters used for the kinetic parameters (cultivar-specific) and batch parameters (batch-/season-specific) is allowed and valid. Second, it also proves that the model formulation, in conjunction with the assumptions made, are valid, irrespective of growth and harvest conditions. In other words, the model is capable of dealing with stochastic behavior during successive seasons and growing areas. The next question to be answered is, of course, how this information, contained in a set of differential equations, can be related to the observed changes in firmness (see Figure 8–2).

Thus, the basic activity of PG, decreasing the DP of the pectin matrix, has also to be described in differential equations. The peach variety studied for the storage experiments (cv. Red Haven) contained virtually no PE activity. If PE would be present, in time, increasing amounts of pectin would be demethylated, resulting in an increase in the amount of substrate for PG. Because PE is absent, this reaction does not have to be included in the model formulation, and DE can be considered constant throughout the storage period. The amount of active PG affects the DP of pectin. This can be described according to the following mechanism and differential equation:

$$\mathrm{DP} + \mathrm{PG} \xrightarrow{k_{\mathrm{PG}}} \mathrm{PG} \tag{5}$$

$$\frac{\partial \mathrm{DP}}{\partial t} = k_{\mathrm{PG}} \cdot \mathrm{PG} \cdot \mathrm{DP} \tag{6}$$

In these equations, PG is the activity of polygalacturonase, DP is the degree of polymerization of pectin, and k_{PG} is the reaction rate constant of the depolymerization reaction. The chemical reaction (Equation 5) describes that PG depolymerizes PG-accessible pectin and that pectin becomes depolymerized without chang-

Table 8–1 Results of the Statistical Analysis for the PG Activity Measured in Stored Peaches during Two Seasons and of the Combined Analysis

Parameter	*Season 1 Estimate*	*Season 2 Estimate*	*Season (1+2) Estimate*
Kinetic parameters			
$k_{f,ref}$	$1.73\ 10^{-2}$	$6.64\ 10^{-2}$	$6.38\ 10^{-1}$
Ea_f/R	$1.36\ 10^{4}$	$6.27\ 10^{3}$	$1.17\ 10^{4}$
$k_{d,ref}$	$1.79\ 10^{-1}$	$2.69\ 10^{-1}$	$1.02\ 10^{-1}$
Ea_d/R	$5.67\ 10^{3}$	0 (fixed)	0 (fixed)
Batch parameters			
$PG_{0,season\ 1}$	$3.00\ 10^{-1}$	n.a.	$2.35\ 10^{-1}$
$PG_{pre,0,season\ 1}$	8.37	n.a.	2.28
$PG_{0,season\ 2}$	n.a.	$1.25\ 10^{1}$	$1.82\ 10^{1}$
$PG_{pre,0,season\ 2}$	n.a.	$1.22\ 10^{2}$	$4.09\ 10^{1}$
Administrative information			
T_{ref}	10	10	10
R^2_{adj}	81.8	85.2	81.6
N_{obs}	31	30	61

n.a., not applicable.

ing the amount of active enzyme. Equation 5 can be expressed in its differential equation (Equation 6) by again applying the fundamental rules of chemical kinetics. Substituting the analytical solution for PG (Equation 3) into this differential equation and solving for constant

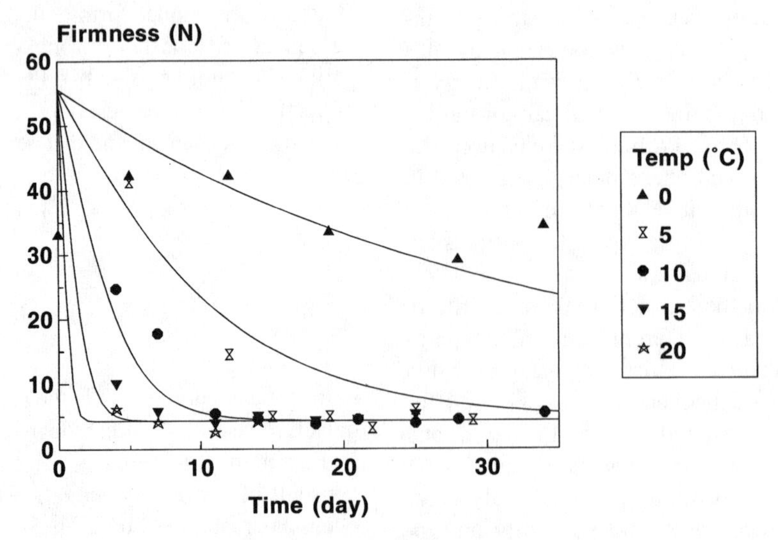

Figure 8–2 Measured (symbols) and simulated (solid lines) firmness of peaches during storage at different temperatures for season 1.

conditions (constant temperatures), one obtains Equation 7.

$$\frac{\mathrm{DP}(t)}{\mathrm{DP}_0} = \exp\left[k_{\mathrm{PG}} \cdot \mathrm{PG}_0 \frac{\exp(-k_{\mathrm{d}} \cdot t) - 1}{k_{\mathrm{d}}} - k_{\mathrm{PG}} \cdot \mathrm{PG}_{\mathrm{pre},0} \frac{k_{\mathrm{d}} - k_{\mathrm{f}} + k_{\mathrm{f}} \cdot \exp(-k_{\mathrm{d}} \cdot t) - k_{\mathrm{d}} \cdot \exp(-k_{\mathrm{f}} \cdot t)}{k_{\mathrm{d}} \cdot (k_{\mathrm{d}} - k_{\mathrm{f}})}\right] \quad (7)$$

Equation 7 describes the degree of polymerization at a given time, at constant temperature conditions, due to the action of PG. As indicated earlier, the firmness of plant products consists of a variable part (pectin-based) and a fixed part (cellulose-hemicellulose-based). Because all changes in firmness are ascribed to changes in the DP of pectin, Equation 7 is accordingly converted into an equation describing the changes in observed firmness on the basis of the enzyme-catalyzed depolymerization of the pectin matrix:

$$\mathrm{Firm} = \mathrm{Firm}_{\mathrm{fix}} + \mathrm{Firm}_{\mathrm{var}} \cdot \frac{\mathrm{DP}(t)}{\mathrm{DP}_0} \quad (8)$$

The firmness data (see Figure 8–2) of stored peaches (season 1) were analyzed using Equations 8 and 4 simultaneously. The values of the kinetic parameters $k_{f,ref}$, Ea_f/R, and the batch parameters PG_0 and $PG_{pre,0}$ were used, such as determined (see Table 8–1, season 1). The results of this analysis are given in Table 8–2.

The fact that the firmness behavior can be explained by the activity of the appropriate enzyme, estimated on separate enzyme data assessed in vitro with an R^2_{adj} of about 90%, strongly indicates and confirms the reliability of the modeling approach used. The simulated data using the estimated parameters of Tables 8–1 and 8–2 are represented by the solid lines in Figure 8–2.

Table 8–2 Results of Statistical Analysis of Firmness (Season 1) of Stored Peaches Based on the Measured PG Activity

Parameter	*Estimate*
Kinetic parameters	
$k_{PG,ref}$	$6.70\ 10^{-1}$
Ea_{PG}/R	$1.48\ 10^{4}$
Batch parameters	
$Firm_{var}$	$5.13\ 10^{1}$
$Firm_{fix}$	4.32
Administrative information	
T_{ref}	10
R^2_{adj}	89.2
N_{obs}	32

The Activity of Pectin Methyl Esterase during Mild Heat Treatments of Potatoes, Carrots, and Peaches

The main aims of a heat treatment of agroproducts are twofold. First, a heat treatment decreases the microbial load of these products, thereby extending their keeping quality or shelf life. The second aim is the activation and/or inactivation of enzymes present in the plant tissue.[27] The apparent activity of enzymes exhibits a well-known behavior. At temperatures below about 50°C, a continuous increase in activity is observed with increasing temperatures. This increase in activity can generally be described by Arrhenius' law. At still higher temperatures, a rather steep decline in the apparent activity is observed due to denaturation.[5,6,45]

One of the enzymes presumed to have an effect on the firmness of heat-treated plant products is the enzyme pectin methyl esterase (PE). As such, PE has a long record of confusing effects on the contribution to firmness during heat treatments. Obviously, the relation between measured PE activity and observed firmness is complex. One of the factors that might add to this confusion is the existence of different PE-isoenzymes[46,47] and the observation that PE can be either soluble or bound to the cell wall polysaccharides.[48]

To obtain a better insight into the effect of mild heat treatments for two vegetables (carrots and potatoes) and one fruit (peaches), the effect of a range of temperatures was used to determine the effect at a given temperature on the observed PE activities in time. A mathematical model was developed that describes the dynamic changes in PE activity.[26,49]

In Figure 8–3, the measured PE activities for carrots (A), potatoes (B), and peaches (C), as functions of heating time and temperature, are shown, respectively. For carrots and potatoes, at temperatures up to 70°C, an (initial) increase in the measured PE activity can be observed. Obviously, below this temperature, additional PE activity is formed. Above this temperature, the decrease in activity seems to be exponential, indicative of a first-order inactivation process. The formation process is apparently completely overruled by the inactivation process at temperatures higher than 70°C. For peaches, a different behavior is observed. At temperatures below 70°C, in contrast to carrots and potatoes, no (initial) increase in measured PE activity can be observed; a heat-resistant PE activity seems to exist. An exponential decrease in activity in time seems to occur only at temperatures above 70°C. To account for these observations in the three products studied, a conversion from a bound to the soluble PE activity is assumed to occur. This assumed behavior can be represented in a set of chemical reactions given in Equation 9.

$$PE_{bnd} \xrightarrow{k_c} PE_{sol} \tag{9a}$$

$$PE_{bnd} \xrightarrow{k_{d,bnd}} PE_{na} \tag{9b}$$

$$PE_{sol} \xrightarrow{k_{d,sol}} PE_{na} \tag{9c}$$

The total activity of the PE enzyme comprises both the bound and the soluble configuration:

$$PE_{tot} = PE_{bnd} + PE_{sol} \tag{10}$$

The index *sol* indicates the soluble fraction, *bnd* the bound fraction, *c* the conversion reaction, and *d* the denaturation of the enzyme. This set of chemical reactions can be converted by fundamental kinetics into a set of differential equations:

$$\frac{\partial PE_{bnd}}{\partial t} = -k_{d,bnd} \cdot PE_{bnd} - k_c \cdot PE_{bnd} \tag{11a}$$

$$\frac{\partial PE_{sol}}{\partial t} = k_c \cdot PE_{bnd} - k_{d,sol} \cdot PE_{sol} \tag{11b}$$

At constant temperature, the solution of this set of differential equations for the sum of both active configurations of the enzyme (PE_{tot}) is given in Equation 12a.

$$PE_{tot} = PE_{sol,0} \cdot \exp(-k_{d,sol} \cdot t) + PE_{bnd,0} \cdot \left(\frac{(k_{d,bnd} - k_{d,sol}) \cdot \exp\{-(k_c + k_{d,sol}) \cdot t\} + k_c \cdot \exp(-k_{d,sol} \cdot t)}{k_{d,bnd} + k_c - k_{d,sol}} \right) \tag{12a}$$

In case the conversion of bound enzyme too far precedes its denaturation (the reaction rate for the inactivation process, $k_{d,bnd}$, is much smaller than that of the formation process k_c), as is the case for carrots and potatoes, Equation 9b is not of relevance. As a consequence, Equation 12a simplifies to:

$$PE_{tot} = PE_{sol,0} \cdot \exp(-k_{d,sol} \cdot t) + PE_{bnd,0} \cdot \left(\frac{k_c \cdot \exp(-k_{d,sol} \cdot t) - k_{d,sol} \cdot \exp(-k_c \cdot t)}{k_c - k_{d,sol}} \right) \tag{12b}$$

The first term in Equations 12a and 12b describes the denaturation of the initially present soluble PE activity. The second term describes the unbinding of PE and the denaturation of the enzyme in its unbound form (Equation 12b; carrots and potatoes) or denaturation of both its bound and its unbound form (Equation 12a; peaches).

Again, by applying Arrhenius' law to the reaction rate constants in combination with the equations for denaturation (Equations 12a and b), the general pattern of enzyme denaturation at any constant temperature can be described. The data were analyzed with Equation 12a for peaches and Equation 12b for carrots and potatoes, together with Equation 4 (Arrhenius' law), using nonlinear regression, with time and temperature simultaneously as dependent variables.

For the three products analyzed, the initial level of PE activity was slightly different for each temperature-time activity curve (Figure 8–3). This can be caused by differences between batches and/or experimental errors. For this reason, the initial PE activity levels were estimated separately for each temperature, allowing each temperature series to have its own initial value; the kinetic parameters were estimated in common. The results of this analysis for the individual data for carrots, potatoes, and peaches (results of two seasons combined) are given in Table 8–3.

The results for peaches (Figure 8–3) suggest two active configurations of the PE; the bound form (PE_{bnd}) prevailing at the lower temperature region and the soluble form (PE_{sol}) prevailing at the higher temperatures by a rapid temperature-dependent conversion (see Equation 11a). The reaction rate constant of the denaturation at the reference temperature for the soluble PE form ($k_{d,s,ref}$) is smaller than for the bound PE configuration ($k_{d,b,ref}$), which indicates a better heat stability at the reference temperature for the soluble PE. The activation energy for the denaturation of the bound configurations ($PE_{d,bnd}$) is much smaller than for the soluble form ($PE_{d,sol}$). The consequence of this difference is that the dena-

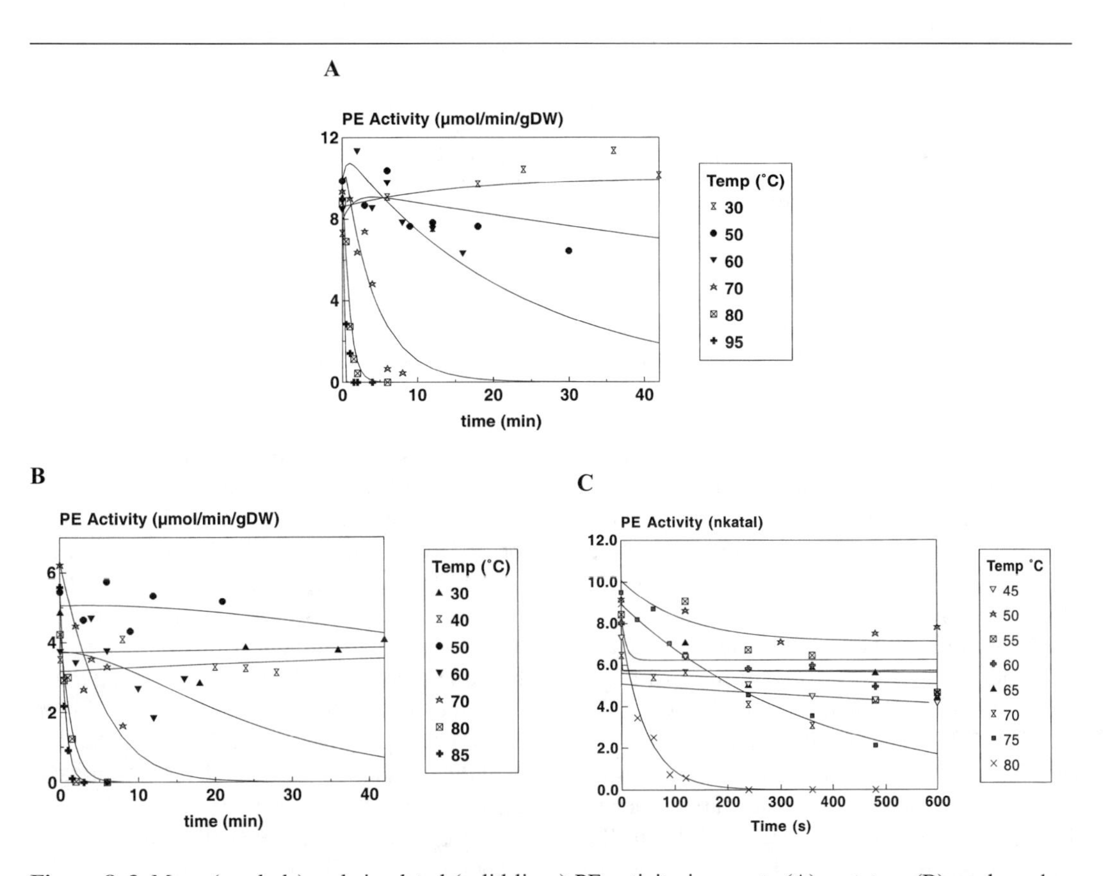

Figure 8–3 Mean (symbols) and simulated (solid lines) PE activity in carrots (A), potatoes (B), and peaches (C), as function of time and preheating temperature. Reproduced from Tijskens et al.,[26,49] with permission of Elsevier Science Limited.

Table 8–3 Results of the Statistical Analysis for the PE Activity of Carrots, Potatoes, and Peaches (Two Seasons Combined)

	Estimate for		
Parameter	*Carrots*	*Potatoes*	*Peaches*
Kinetic parameters			
$k_{c,ref}$	2.35	9.42 10^{-2}	1.03 10^{-2}
Ea_c/R	1.40 10^{4}	1.30 10^{4}	8.03 10^{3}
$k_{d,bnd,ref}$	n.a.	n.a.	8.0 10^{-3}
$Ea_{d,bnd}/R$	n.a.	n.a.	5.00 10^{2}
$k_{d,sol,ref}$	4.25 10^{-2}	6.18 10^{-2}	3.91 10^{-4}
$Ea_{d,sol}/R$	1.96 10^{4}	1.72 10^{4}	6.81 10^{4}
Batch parameters			
PE_{sol}; Activity range	9.01–10.00	3.17–6.29	n.a.
$PE_{bnd,0}$	1.32	2.48	n.a.
PE_{bnd}; Activity range	n.a.	n.a.	0.84–10.8
$PE_{sol,0}$	n.a.	n.a.	4.88
Administrative information			
T_{ref}	89.1	84.4	91.5
R^2_{adj} f	60	60	60
N_{obs}	159	160	90

n.a., not applicable.

turation of the soluble configuration is more affected by a temperature increase than the denaturation of the bound form.

The fact that both the solubilization reaction and the two inactivation reactions (see Equation 11) are required to explain the observed behavior signifies that both reactions can precede the other at some situation in time and/or temperature. The three-dimensional simulation of the PE activity of carrots and peaches as a function of time and (constant) temperature is shown in Figure 8–4, based on the results of analysis (Table 8–3). The striking difference in this simulation between carrots and potatoes (the latter is not shown, because it is almost similar as for carrots) on one hand and peaches on the other hand is that the PE of carrots and potatoes almost obeys a first-order inactivation mechanism and that the inactivation of PE of peaches occurs in two sequential steps. For peaches the explained parts, R^2_{adj}, for the first and second season, analyzed independently, are, respectively, 93% and 90%. For the analysis of the combined information this value is 91.5%, showing no loss of explaining power. The major advantage of combined analysis lies in the increase in explaining and predicting power for different situations and different batches. Furthermore, the fact that the activity of PE in peaches of different seasons and, therefore, different properties, can be analyzed together with the same model and the most important parameters in common, constitutes a major validation of the principles, assumptions, and deduction techniques underlying this model. In Figure 8–4 the three dimensional simulation of the PE activity of carrots (A) and peaches (B) is respectively shown as a function of time and temperature. For carrots it is obvious that the enzyme activity decays (almost) exponentially in time at any temperature.

A

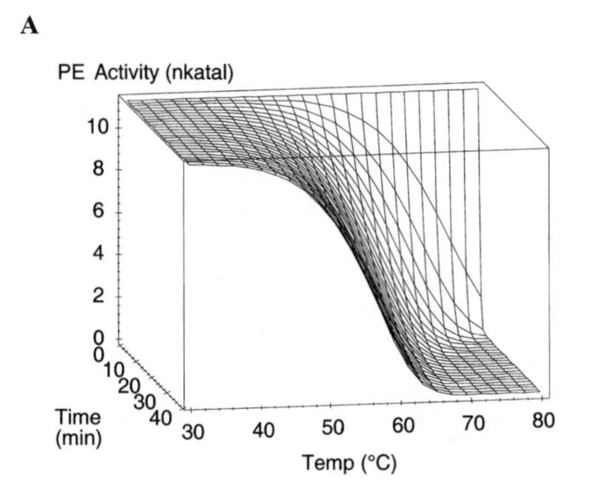

B

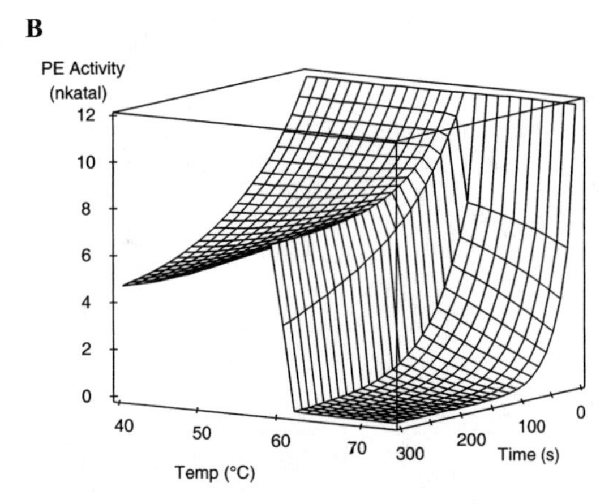

Figure 8–4 Three-dimensional simulation of the PE activity in carrots (A) and peaches (B), as function of time and temperature, based on the parameter estimates of Table 8–3.

The Activity of Peroxidase during Mild Heat Treatments of Potatoes, Carrots, and Peaches

The enzyme POD is found in almost all living organisms. The primary action of this enzyme is to control the level of peroxides, generated in oxygenation reactions, to avoid excessive formation of radicals, which and harmful to all living organisms. Due to its relatively high heat stability, POD serves in the processing of fruits and vegetables as a marker enzyme.[38] It is assumed that if a tissue is POD-negative, all enzymes are denaturated. With regard to texture, this enzyme regained attention, due to its potential to establish phenolic cross-links between the neighboring polymers in relation to cell-cell adhesion and the consequences for thermal stability of texture.[39,40] To gain insight into the temperature-time behavior of this enzyme, as was the case for PE (discussed earlier), two vegetables (carrots and potatoes) and one fruit (peaches) were studied.[25] The plant enzyme consists of a complex spectrum of isoenzymes.[50] Furthermore, the existence of both soluble and bound isoenzymes has been reported, each with different susceptibilities to heat denaturation.[51] In other words, from the enzymatically active bound form of POD, the enzymatically active soluble form is generated, both catalytically active forms being susceptible to heat denaturation. Comprising this information into a mathematical formulation leads to:

$$POD_{bnd} \xrightarrow{k_c} POD_{sol} \tag{13a}$$

describing the solubilization of the bound POD,

$$POD_{bnd} \xrightarrow{k_{bnd}} POD_{na} \tag{13b}$$

$$POD_{sol} \xrightarrow{k_{d,sol}} POD_{na} \tag{13c}$$

and Equations 13b and c, describing the thermal denaturation process of both the bound and soluble enzyme. This reaction mechanism is the same as described for the PE activity in peaches (see earlier and Tijskens et al.[49]).

The total activity of the POD enzyme comprises both the bound and the soluble enzyme configurations:

$$POD_{tot} = POD_{bnd} + POD_{sol} \tag{14}$$

This set of chemical reactions can be converted into a set of differential equations:

$$\frac{\partial POD_{bnd}}{\partial t} = -k_{d,bnd} \cdot POD_{bnd} - k_c \cdot POD_{bnd} \tag{15a}$$

$$\frac{\partial POD_{sol}}{\partial t} = -k_{d,sol} \cdot POD_{sol} + k_c \cdot POD_{bnd} \tag{15b}$$

As can be observed in Figure 8–5, at prolonged times and elevated temperatures, POD activity is observed for both the bound as well as the soluble form. This fixed part of the activities, of course, is not included in these differential

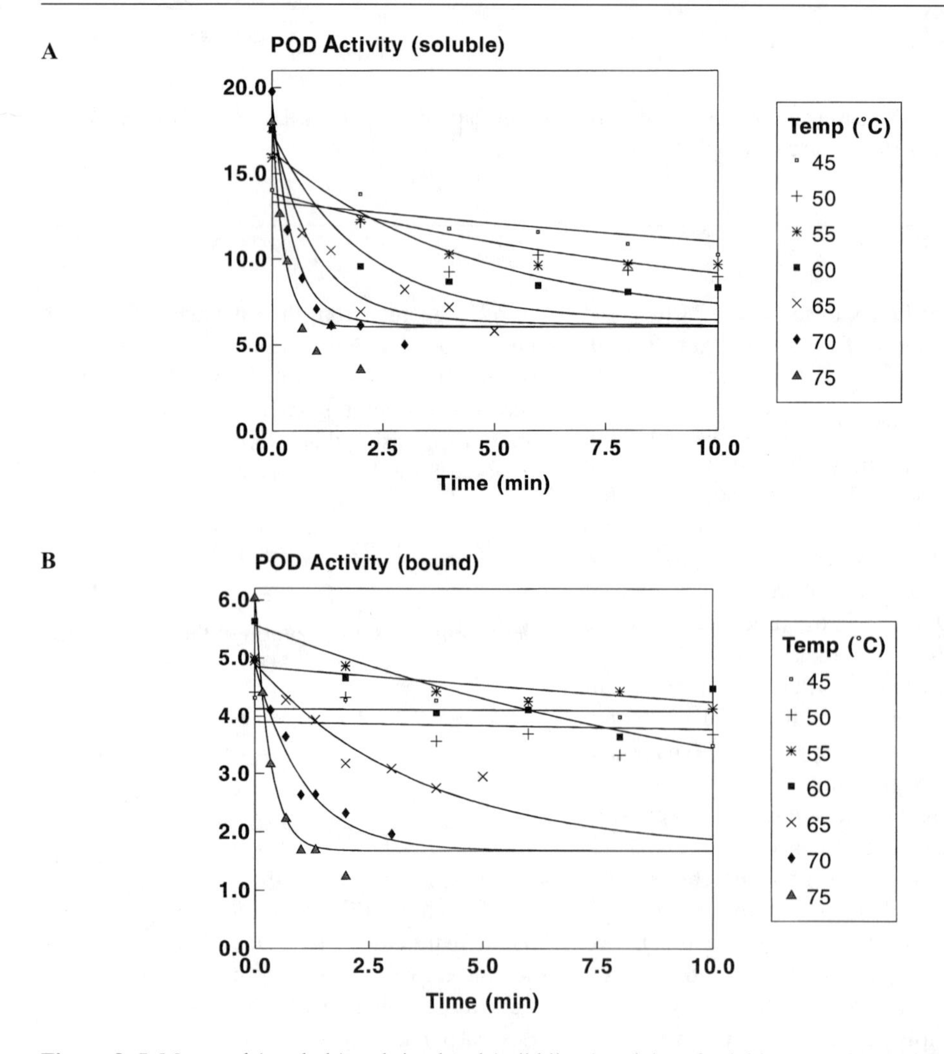

Figure 8–5 Measured (symbols) and simulated (solid lines) activity of soluble (A) and bound (B) POD activity in peaches as function of preheating temperature and time.

equations but is accounted for in the analytical solutions of Equation 16.

The analytical solution of this set of differential equations for the sum of both the bound and soluble form of this enzyme, taking the fixed part into consideration, results in Equation 16.

$$POD_{bnd} = POD_{bnd,\,var} \cdot \exp\{-(k_{d,\,bnd} + k_c) \cdot t\} + POD_{bnd,\,fix} \quad \textbf{(16a)}$$

$$POD_{sol} = POD_{sol,var} \cdot \exp(-k_{d,sol} \cdot t) + POD_{sol,fix} + POD_{bnd,var} \cdot \left(\frac{k_c \cdot [\exp(-k_{d,sol} \cdot t) - \exp\{-(k_{d,bnd} + k_c) \cdot t\}]}{k_{d,bnd} + k_c - k_{d,sol}} \right) \quad \textbf{(16b)}$$

The first term in Equation 16a describes the denaturation of the initially present bound POD activity. The second term describes the invariable part of the bound activity for the observed activity remaining after heat treatment. In Equation 16b, the first term describes the denaturation of the initially present soluble POD activity. The second term describes an invariable part of the soluble POD activity for the observed activity remaining after heat treatment. The third term describes the combination of heat denaturation of the bound POD activity and the formation of the soluble POD activity from the bound POD activity. By applying Arrhenius' law (Equation 4) to the reaction rate constants, in combination with the equations for denaturation (Equation 16), the general pattern of enzyme denaturation at any constant temperature can be described.

For carrots and potatoes, only the bound enzyme was assayed; for peaches, both the soluble and bound enzymes were assessed separately. Consequently, the analysis of the enzyme data on peaches is more elaborate but also more comprehensive and reliable than that for potatoes and carrots. In Figure 8–5, both the soluble and bound activity levels of POD in peaches as a function of heating time and temperature are shown. In Figure 8–6, the bound POD activity as a function of heating time and temperature is shown for carrots and potatoes. As with the PE activity in peaches,[49] carrots, and potatoes,[26] the initial POD activity was slightly variable on a day-to-day basis. Therefore, the initial POD activity levels (bound and soluble) were estimated separately, and the kinetic parameters were estimated in common (see earlier).

During data analysis, $k_{d,bnd,ref}$ consistently approached zero. Thus, the denaturation of the bound form is not supported by the data. Apparently, the conversion of the bound form into the soluble configuration is faster and more important than the denaturation of the bound configuration. In other words, all of the heat-labile bound POD will already be solubilized by the time direct denaturation of the bound POD occurs. During further statistical analysis of the data, $k_{d,bnd,ref}$ was fixed at zero. A compilation of the estimated values for peaches, carrots, and potatoes is given in Table 8–4.

The explained part, R^2_{adj}, for both carrots and peaches is about 95%, indicating the correctness of the assumptions being made. For potatoes, the explained part is lower (70%). This could be caused by the limited number of objects studied. The results of the statistical analysis strongly indicate that the bound form of POD solubilizes and is subsequently denaturated. The heat-labile bound POD will already be solubilized by the time direct inactivation of the bound POD occurs.

The simulated three-dimensional activity–time–temperature behavior of the POD activity in peaches, both bound and soluble, is shown in Figure 8–7. The values of the parameters to perform this simulation were from Table 8–4, the value for $POD_{sol,var}$ was set at 10.6, and the value for $POD_{bnd.var}$ was set at 3.23. Viewed along the time axis, the POD activity, both bound and soluble configurations, behave roughly according to an exponential decay process at constant temperature. Viewed along the temperature axis, the difference in susceptibility to disappear, either by conversion or denaturation, can be observed.

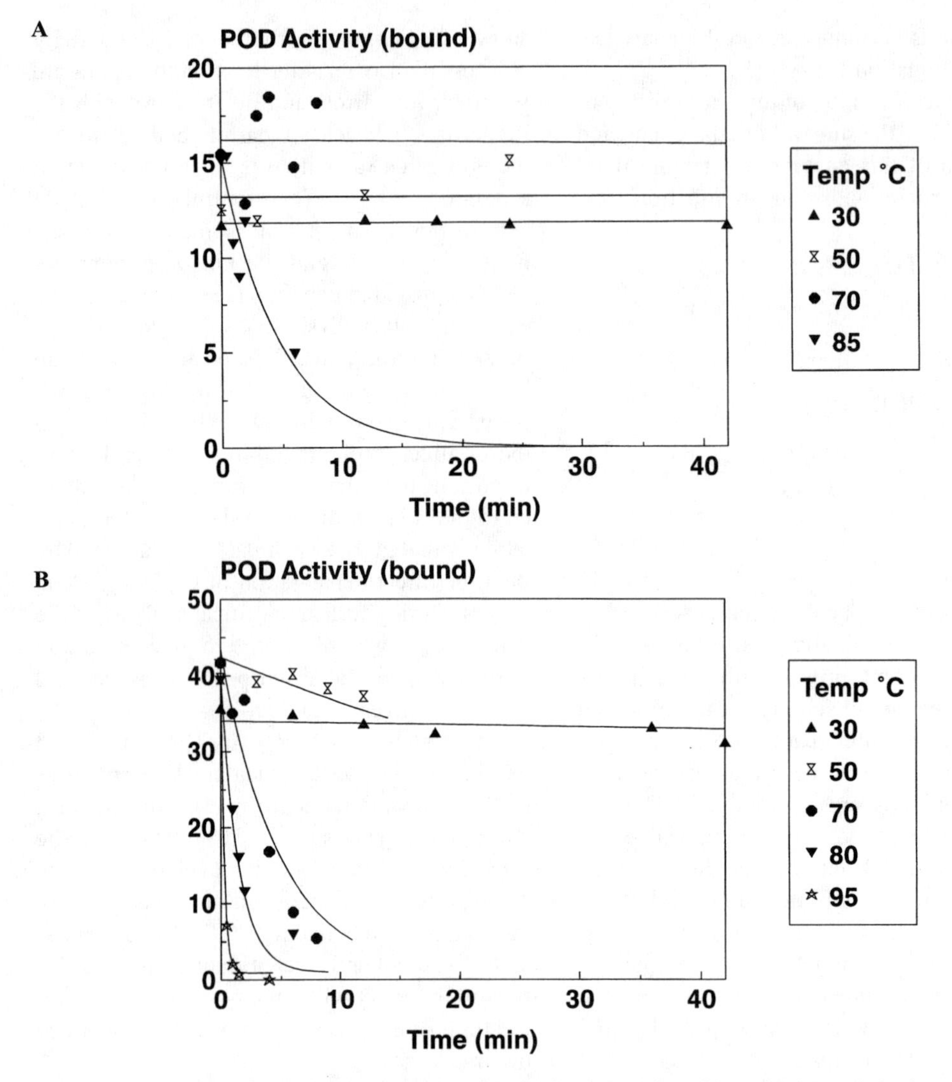

Figure 8–6 Measured (symbols) and simulated (solid lines) activity of bound POD activity in carrots (A) and potatoes (B) as function of preheating temperature and time.

A Preheating-Cooking Model: PCM

In previous sections, attention has been given to the formulation of the kinetic reactions assumed to underlie the observed enzyme activity and their time- and temperature-dependent behavior in the form of differential equations and analytical solutions. By proper merging of these analytical solutions with Arrhenius' law, estimates were made about the rate constants and activation energies. For stored peaches, the PG activity was directly related to the change in firmness of this fruit.

Preheating is, as mentioned earlier, one of the means to decrease the microbial load of agroproducts, thereby extending their keeping quality or shelf life. However, preheating, as such, is also assumed to have an effect on the firm-

Table 8–4 Results of the Statistical Analysis for the POD Activity of Carrots, Potatoes, and Peaches

Parameter	*Estimate for*		
	Carrots	*Potatoes*	*Peaches*
Kinetic parameters			
$k_{c,ref}$	9.61 10^{-4}	8 10^{-7}	1.30 10^{-3}
Ea_c/R	1.42 10^{4}	5.10 10^{4}	2.85 10^{4}
$k_{d,bnd,ref}$	n.a.	n.a.	0 (fixed)
$E_{ad,bnd}/R$	n.a.	n.a.	0 (fixed)
$k_{d,sol,ref}$	n.a.	n.a.	8.44 10^{-3}
$E_{ad,sol}/R$	n.a.	n.a.	1.80 10^{4}
Batch parameters			
$POD_{bnd,fix}$	0 (fixed)	0 (fixed)	1.68
$POD_{bnd,var}$;Activity range	34.1–43.4	0.53–1.10	2.22–4.44
PODsol,fix	n.a.	n.a.	6.03
PODsol,var; Activity range	n.a.	n.a.	7.34–12.3
Administrative information			
T_{ref}	94.5	70.0	96.4
R^2_{adj}	60	60	60
N_{obs}	216	21	90

n.a., not applicable.

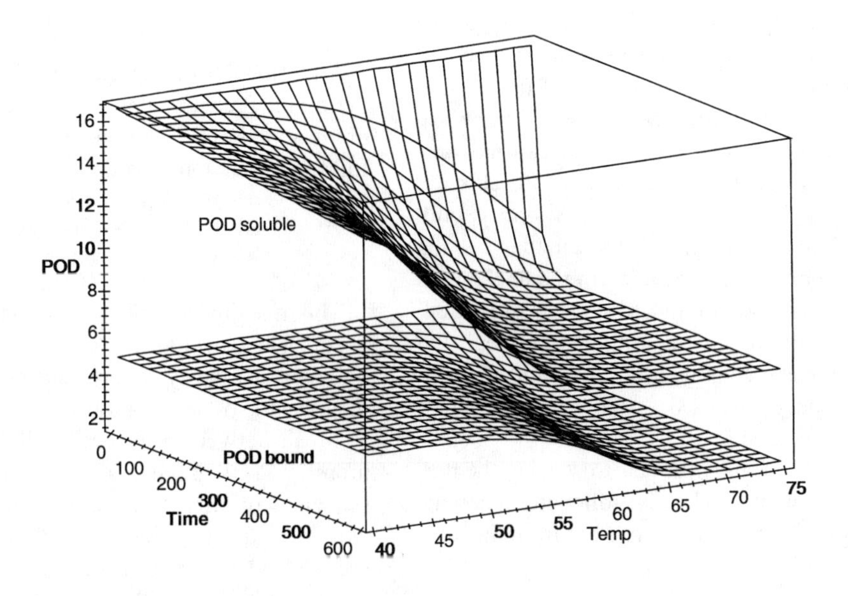

Figure 8–7 Three-dimensional of bound and soluble POD activity of peaches as function of time and temperature, based on the parameter estimates of Table 3–4.

ness of the cooked or sterilized products. In this section, the effect of preheating on the firmness of the cooked products will be formulated in a set of reaction mechanisms. These mechanisms will be developed into mathematical equations according to the strategy discussed in the previous sections. Emphasis is focused on the observed firming effect of PE during preheating and the consequences for the texture of the cooked product. Here it has to be realized that the total process is divided into two subprocesses, preheating and cooking. During preheating, PE exerts its action, demethylation of pectin. The preheating is performed at temperatures below 90°C. Virtually no β-degradation of pectin takes place below this temperature. However, it has to be emphasized again that the number of sites vulnerable to β-degradation decreases due to the PE action.

During cooking, all PE is inactivated very quickly (a 10^3-fold decrease in activity within 20 seconds), so the cooking process virtually does not continue to change the degree of methylation. However, the sites not affected by the PE action are vulnerable to β-degradation during cooking. In Figure 8–4A, the three-dimensional behavior comprised of the PE activity of carrots is shown as a function of (preheating) time and temperature, based on the results of the statistical analysis of the PE activity (see Table 8–3). The bound form of PE (PE_{bnd}) of carrots comprises only about 10% of the total PE activity, as defined in Equation 11. To simplify both the mathematics and the statistical analysis, the bound form was not included in the analysis to simplify the system and to make a statistical analysis possible. To distinguish between the preheating and cooking times, the indices *p* and *c* are, respectively, used to indicate the different types of processing.

For the preheating process to develop the preheating-cooking model, the following reactions and time-dependent equations are formulated:

$$\mathrm{PE} \xrightarrow{k_d} \mathrm{PE}_{na} \tag{17a}$$

$$\frac{\partial \mathrm{PE}}{\partial t_p} = -k_d \cdot \mathrm{PE} \tag{17b}$$

In Equation 17, the bound form of PE was neglected. The reaction of methylated pectin with PE is described with Equation 18.

$$\mathrm{DE} + \mathrm{PE} \xrightarrow{k_s} \mathrm{PE} \tag{18a}$$

$$\frac{\partial \mathrm{DE}}{\partial t_p} = -k_s \cdot \mathrm{DE} \cdot \mathrm{PE} \tag{18b}$$

This reaction states that the PE-accessible pectin is demethylated in time, without changing the amount of active enzyme. However, the preheating process leaves the demethylated pectin backbone with the same DP as before the action of PE.

Substituting Equation 17 into Equation 18 results in:

$$\mathrm{DE}(t_p) = \mathrm{DE}_0 \cdot \exp\left[\frac{k_s \cdot \mathrm{PE}_0 \cdot \{\exp(-k_d \cdot t_p) - 1\}}{k_d}\right] \tag{19}$$

Equation 19 describes the DE at any time during the preheating process, as a consequence of the action of PE. For the cooking process, the following reaction is formulated:

$$\mathrm{DP} + \mathrm{DE} \xrightarrow{k_\beta} \mathrm{WSP} + \mathrm{DE} \tag{20}$$

This reaction describes the β-degradation of pectin. The galacturonic acids within the pectin polymer are partly methylated and partly unmethylated. β-Degradation causes the polymer to break between adjacent methylated galacturonic acids within the polymer chain. The consequence of this reaction is that, during the course of the reaction (cooking), the DP decreases and results in pectic fragments, which are water soluble (WSP) and, therefore, do not contribute to the texture of the product. This equation also describes that, during this β-degradation, the DE is not affected. Equation 20 can be converted into a differential equation (Equation 21a) and solved at constant temperature to give Equation 21b.

$$\frac{\partial \mathrm{DP}}{\partial t_c} = -k_\beta \cdot \mathrm{DP} \cdot \mathrm{DE} \qquad \textbf{(21a)}$$

$$\mathrm{DP}(t_c) = \mathrm{DP}_0 \cdot \exp\{-k_\beta \cdot \mathrm{DE}(t_p) \cdot \mathrm{t_c}\} \qquad \textbf{(21b)}$$

Equation 21 describes the decrease of DP due to β-degradation as a function of cooking time, given the DE resulting as a consequence of the preheating time, such as described in Equation 19.

Combining Equations 19 and 21 integrates the effect of both the preheating and cooking time and results in:

$$\mathrm{DP}(t_c) = \mathrm{DP}_0 \cdot \exp\left(-k_\beta \cdot t_c \cdot \mathrm{DE}_0 \cdot \exp\left\{\frac{\mathrm{k_s} \cdot \mathrm{PE}_0 \cdot [\exp(-k_d \cdot t_p) - 1]}{k_\mathrm{d}}\right\}\right) \qquad \textbf{(22)}$$

This triple exponential function is the mathematical description of the combined preheating process (at a given temperature during a given time t_p) and cooking process (at a given temperature during a given time t_c), and the consequences of these combined processes on the DP at the end of the cooking process.[22]

Described earlier, the firmness of peaches decreases, due to the action of PG. All changes in firmness were ascribed to changes in the DP of pectin. These changes in DP were accordingly converted into Equation 8, describing the changes in observed firmness on the basis of the PG-catalyzed depolymerization of the pectin matrix. The results of this analysis, given in Table 8–2, strongly suggest the validity of this approach. In other words, the firmness of plant tissue consists of a fixed (cellulose-hemicellulose-based) and variable (pectin-based) part. Ascribing the changes in DP to the variable part, due to the combined preheating and cooking process, results in:

$$\mathrm{Firm} = \mathrm{Firm_{fix}} + \mathrm{Firm_{var}} \cdot \frac{\mathrm{DP}(t_c)}{\mathrm{DP}_0} \qquad \textbf{(23a)}$$

In combining Equations 22 and 23, a mathematical description (Equation 24) is formulated to describe how the variable part of the firmness is affected by both the preheating and the cooking process and, consequently, the firmness of the product.

$$\mathrm{Firm} = \mathrm{Firm_{fix}} + \mathrm{Firm_{var}} \cdot \exp\left(-k_\beta \cdot t_c \cdot \mathrm{DE}_0 \cdot \exp\left\{\frac{k_\mathrm{s} \cdot \mathrm{PE}_0 \cdot [\exp(-k_\mathrm{d} \cdot t_\mathrm{p}) - 1]}{k_\mathrm{d}}\right\}\right) \qquad \textbf{(23b)}$$

The result of this analysis is compiled in Table 8–5. In Figure 8–8, the actually measured firmness of carrots after cooking, preceded by a preheating treatment, is given in a three-dimensional representation. Each point in this figure gives the firmness of a sample after cooking at a given preheating time and temperature. The lines in this figure are the simulated behavior based on Equation 23 and the parameter values from Table 8–5. This figure clearly shows that virtually no firming effect is observed by preheating at temperatures either below 40°C or above 75°C. The absence of an observed firming effect below 40°C is caused by the absence of a noticeable total exerted effect due to PE functioning; the enzyme is not sufficiently active below this temperature. Above 75°C, the temperature-dependent inactivation is fast; the PE is already inactivated before the consequence of its enzymatic action becomes apparent.

On the other hand, if the preheating time is short, the highest effect is observed at relatively high temperatures; the temperature-dependent enzyme activation process prevails over the temperature-dependent inactivation process. For example, after 30 minutes at 70°C at this preheating temperature, the highest firmness is observed. After this period of time at this temperature, only 0.1% of the PE is still active.

Table 8–5 Results of the Statistical Analysis for the PE Activity in Carrots, Based on Either PE Activity Measurements (PE Model) or Firmness Modulation during Preheating and Cooking (PCM Model)

Parameter	*PE Model**		*Preheating-Cooking Model (PCM)*	
	Estimated Parameters			
	Estimate	*s.e.*	*Estimate*	*s.e.*
Kinetic parameters				
$k_{d,sol,ref}$	4.25 10^{-2}	4.65 10^{-3}	4.87 10^{-2}	2.68 10^{-3}
$Ea_{d,sol}/R$	1.93 10^{4}	9.56 10^{2}	1.95 10^{4}	4.64 10^{2}
k_s	n.a.	n.a.	2.42 10^{-1}	5.71 10^{-3}
Ea_s/R	n.a.	n.a.	1.37 10^{4}	2.18 10^{2}
Batch parameters				
PE_{sol}; Activity range	9.01–10.0	0.24–0.62	n.a.	n.a.
$Firm_{var}$ ($\equiv DP_0$)	n.a.	n.a.	4.73	1.59 10^{-1}
	Fixed Parameters			
Kinetic parameter			Fixed at	
k_β	n.a.	n.a.	2.3 10^{-2}	n.a.
Batch parameters				
PE_{sol}	n.a.	n.a.	1	n.a.
DE_0	n.a.	n.a.	6 10^{1}	n.a.
$Firm_{fix}$	n.a.	n.a.	1.5	n.a.
Administrative information				
R^2_{adj}	89.5		95.4	
T_{ref}	60		60	
t_c	n.a.		30	
N_{obs}	159		294	

*Data derived from Table 8–3.
n.a., not applicable.

Increasing the preheating time results both in an increase in firmness and a shift in the maximal measured firmness to lower preheating temperatures. Obviously, at about 55°C, a balance is reached between the total exerted activity and enzyme denaturation; for example, at this temperature after 2 hours, about 13% of the enzyme is still active.

The information shown in Figure 8–8 was used as input information for Equation 23 to estimate the values of the rate constants and activation energies. The results of this analysis are compiled in Table 8–5.

Several conclusions can be derived from the information given in Tables 8–3 and 8–5. Based on the temperature and time-dependent studies of PE in carrots (PE model in Table 8–3; see earlier), estimates could be made for the kinetic value $k_{d,sol,ref}$ and its activation energy $Ea_{d,sol}/R$. Based on firmness measurements of carrot samples preheated at different times and temperatures, a model was formulated (preheating-cooking model [PCM), including both the enzymatic action of PE and the chemical depolymerization of pectin due to β-degradation (see Equation 23), describing how the variable part of the firmness is affected by both the preheating and cooking times. Making use of this last model, the parameters $k_{d,sol,ref}$ and $Ea_{d,sol}/R$ were also estimated. Comparing the values for $k_{d,sol,ref}$ and

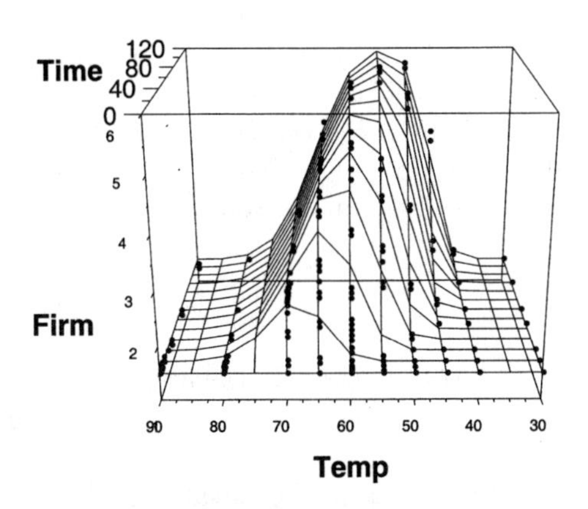

Figure 8–8 Three-dimensional representation of the measures (symbols) and simulated (solid lines) firmness of carrots as function of preheating time and temperature followed by a cooking process.

$Ea_{d,sol}/R$ for the PE model and for PCM, it can be concluded that these values are identical. This conclusion is supported by the values of the explained parts R^2_{adj} being 89.1% and 95.4% for the PE model and PCM, respectively.

In conclusion, based on different starting points—respectively, structured enzyme (in)activation studies and structured firmness studies, using the same kinetic formulations to describe underlying processes—identical results are obtained. This altogether strongly supports the validity of the approach presented.

A Histologic, Stochastic Approach To Relate the Mode of Tissue Rupture with Rupture Stress

The sensory perceived texture of agro-products comprises several elements, e.g., firmness, crispiness, mealiness, etc. The value of some of the product properties relating to firmness can be measured, e.g., making use of compression or tensile measurements. In some cases, the results of the instrumentally determined firmness relate well with the sensory perceived firmness, such as is the case in carrots. In other cases, this relation is completely absent. For example, the potato varieties Nicola and Irine represent with regard to their sensory perceived texture characteristics two extremes. Nicola represents a waxy potato and Irene a very crumbly, mealy potato.[52] Analyzing the tissue softening during cooking by compression measurements shows that the softening process of both varieties can be described by an exponential decay process, with slightly different rate constants, representing the variable part of the firmness.[53] The fixed part of the firmness is almost identical for both varieties. Nevertheless, this fixed part is perceived by the consumer and is assessed to be completely different between these varieties. In other words, no relation exists between instrumentally determined firmness and sensory perceived texture. This difference can be explained on the basis of differences between these varieties at the cellular level. Apparently, the crumbly, mealy potato variety tissue fractures through the cells, and the waxy variety fractures along the cells, through the middle lamellae region of the cell walls. The fracture force measured is, however, in both cases identical. In other words, sensory perceived texture attributes are, in this example, strongly related to the mode of tissue fracture, rather than to the required fracture force.

To address this problem, Verlinden et al.[54–57] modeled the relationship between the macroscopic tissue strength and the strength of both the cell wall and the middle lamellae, using a stochastic approach. The macroscopic (tissue) strength of a vegetable, in this case carrots, was determined by measuring the mode of rupture and the rupture stress, using a specially designed ring-shaped device.[54–57] The measured macroscopic values represent average tissue properties resulting from the interactions between individual cells of the tissue. The strength of a tissue is determined by the sum of the normalized cell–cell interactions (normalized either per cell or surface rupture area) and the cell wall strength. Again, the cell–cell interactions are determined by pectin, confined to the middle lamellae region of the cell walls, and represent the variable part of the firmness. The cell wall strength is determined by the cellulose-hemicellulose matrix determining the fixed part of the firmness. The middle lamellae strength can be defined as the middle lamellae breaking force, F_l, and the strength of the cell wall as cell-wall-breaking force, F_w. The value of the breaking force of either middle lamellae or cell wall depends on its intrinsic strength and on the dimensions of the middle lamellae or cell wall, respectively. The geometric arrangement of the cells and their relative orientations (the tissue archestructure) with respect to the applied forces on the tissue also affect the rupture force of a particular cell. Confining this information in a mathematical description gives:

$$F_l = S_l \cdot A_l \tag{24a}$$

$$F_w = S_w \cdot A_w \tag{24b}$$

In Equation 24, S stands for the intrinsic strength, defined as the rupture stress of the material, and A is the dimensional and geometric (archestructural) factor. Each individual cell has its own dimension and geometric orientation. Furthermore, the cell wall thickness and the properties of the pectin in the middle lamellae have their own biologic variability. Taking these considerations into account, the cell wall strength, F_w, and the middle lamellae strength, F_l, can be considered as stochastic (probably) independent variables, each with its own mean value (μ_w and μ_l) and standard deviation (σ_w and σ_l). Whenever the middle lamellae strength, F_l, is higher than the strength of the cell wall, F_w, the cell will rupture through the cells, and vice versa, when F_w is higher than F_l, the cells will rupture along the middle lamellae. This is mathematically written as follows:

$$F_d = F_l - F_w \tag{25a}$$

$$\text{If } F_d > 0 \Rightarrow \text{cell wall break} \tag{25b}$$

$$\text{If } F_d < 0 \Rightarrow \text{middle lamellae break} \tag{25c}$$

On the rupture surface of the tissue, the percentage of cell wall breaks equals the proportion of cell for which $F_d > 0$. This proportion can be calculated using the cumulative distribution function of F_d. In other words, the mode of surface rupture is related to a stochastic process.

The next question to be addressed is the relation between the percentage cell wall breaks on the rupture surface and the rupture stress as measured with the tensile test. Verlinden et al.[57] worked this problem out into a statistical but dynamic distribution model. With these models, it becomes possible to estimate stresses acting on single cells from the macroscopic tissue strength (organ level) properties and the histologic properties (cellular level), based on cell sizes and cell wall break data. The mathematics are, however, too complex to include the complete deduction and validation of these models in the framework of this chapter.

REFERENCES

1. Sloof M, Tijskens LMM. Problem decomposition: Application in experimental research, statistical analysis and modelling. *Presented at International Symposium on Intelligent Data Analysis* IDA-95, August 1995. Baden-Baden, Germany: 1995.
2. Kramer A, Twigg BA. *Quality Control for the Food Industry.* Westport, CT: AVI Pub Co; 1970.
3. Steenkamp JBEM. *Product Quality: An Investigation into the Concept and How It is Perceived by Consumers.* Wageningen, Netherlands: Agricultural University Wageningen; Ph.D. Thesis: 1989.
4. Cabral JMS, Best D, Boross L, Tramter J, eds. *Applied Biocatalysis.* Working Party on Applied Biocatalysis of the European Federation of Biotechnology. Chur, Switzerland: Harwood Academic Publishers; 1993.
5. Segel IH. *Enzyme Kinetics. Behavior and Analysis of Rapid Equilibrium and Steady-State Enzyme Systems.* New York: John Wiley & Sons; 1993.
6. Whitaker JR. *Principles of Enzymology for the Food Sciences,* 2nd ed. New York: Marcel Dekker; 1994.
7. Martin-Cabrejas MA, Waldron KW, Selvendran RR, Parker ML, et al. Ripening related changes in the

cell walls of Spanish pear (*Pyrus communis*). *Physiol Planta.* 1994:91;671–679.

8. Luri S, Levin A, Greve LC, Labavitch JM. Pectic polymer changes in nectarines during normal and abnormal ripening. *Phytochemistry.* 1994;29:725–731.
9. Seymour GB, Colquhoun IJ, DuPont MS, Parsley KR, et al. Composition and structural features of cell wall polysaccharides from tomato fruits. *Phytochem.* 1990;29:725–731.
10. Harker FR, Hallett IC. Physiological changes associated with development of mealiness of apple fruit during cool storage. *Hort Sci.* 1992;27:1291–1294.
11. Smith CJS, Watson CF, Ray J, Bird CR, et al. Antisense RNA inhibition of polygalacturonase gene expression in trangenic tomatoes. *Nature.* 1988;334:724–729.
12. Carpita NC, Gibeaut DM. Structural models of primary cell walls in flowering plants: consistency of molecular structure with the physical properties of the walls during growth. *Plant J.* 1993;3:1–30.
13. Fry SC. Cross-linking of matrix polymers in the growing cell walls of angiosperms. *Ann Rev Plant Physiol.* 1986;37:165–186.
14. Jarvis MC. Structure and properties of pectin gels in plant cell walls. *Plant Cell Environ.* 1984;7:153–164.
15. De Vries JA, Rombouts FM, Voragen AGJ, Pilnik W. Enzymic degradation of apple pectins. *Carb Pol.* 1982;2:25–33.
16. Mutter M. *New rhamnogalacturonan degrading enzymes from* Aspergillus aculeatys. Wageningen, Netherlands: Agricultural University Wageningen; Ph.D. Thesis: 1997.
17. Grant GT, Morris ER, Rees DA, Smith PJ-C, et al. Biological interactions between polysaccharides and divalent cations: the egg-box model. *FEBS Lett.* 1973;32:195–198.
18. Kikuchi A, Edashige Y, Ishii T, Fujii T, et al. Variations in the structure of neutral sugar chains in the pectic polysaccharides of morphological by different carrot calli and correlations with the size of cell clusters. *Planta.* 1996;198:634–639.
19. Saedt APH, Homan WJ, Reinders MP. A finite state Markov model with continuous time parameter for physical and chemical cutting processes. *Eur J Oper Res.* 1991;55:279–290.
20. Stolle-Smits AAM. *Effects of Thermal Processing on Cell Walls of Green Beans: A Chemical and Ultrastructural Study.* Nijmegen, Netherlands: Catholic University Nijmegen; Ph.D. Thesis: 1998.
21. Atherton JG, Rudich J, eds. *The Tomato Crop.* London: Chapman & Hall; 1986.
22. Tijskens LMM, Hertog MLATM, van Dijk C. Generic modelling and practical applications In: Nicolaï BM, De Baerdemaeker J, eds. Food quality modelling. *Cost Action.* 1997;915:145–151.
23. Ross GJS. *Nonlinear Estimation.* New York: Springer Verlag; 1990.
24. Tijskens LMM. Modelling color of tomatoes. Advantage of multiple nonlinear regression. *Proceedings COST94 Workshop Post-Harvest Treatment of Fruit and Vegetables.* September 14–15, 1994:175–185.
25. Tijskens LMM, Rodis PS, Hertog MLATM, Waldron KW, et al. Activity of peroxidase during blanching of peaches, carrots and potatoes. *J Food Engr.* 1997;34:355–370.
26. Tijskens LMM, Waldron KW, Ng A, Ingham L, et al. The kinetics of pectin methyl esterase in potatoes and carrots during blanching. *J Food Engr.* 1997;34:371–385.
27. Pilnik W, Voragen AGJ. The significance of endogenous and exogenous pectic enzymes in fruit and vegetable processing. In: Fox PF, ed. *Food Enzymology.* Amsterdam: Elsevier Science; 1991:303–336.
28. Burns JK. *The Chemistry and Technology of Pectin.* Walter RG, ed. London: Academic Press; 1991:165–188.
29. Van Buren JP, Peck NH. Effect of K fertilization and addition of salts on the texture of canned snap bean pods. *J Food Sci.* 1981;47:311–313.
30. Sajjaanantakul T, Van Buren JP, Downing DL. Effects of methyl ester content on heat degradation of chelator-soluble carrot pectin. *J Food Sci.* 1989;5:1273–1277.
31. Keybets MJH, Pilnik W. β-Elimination of pectin in the presence of several anions and cations. *Carb Res.* 1974;33:359–362.
32. Shewfelt AL. Changes and variations in the pectic constitution of ripening peaches as related to product firmness. *J Food Sci.* 1965;30:573.
33. Shewfelt AL, Paynter VA, Jen JJ. Textural changes and molecular characteristics of pectic constituents in ripening peaches. *J Food Sci.* 1971:36:573.
34. Awad M, Young RE. Postharvest variation in cellulase, polygalacturonase and pectinmethylesterase in avocado (*Persea americana Mill*, cv. Fuerte) fruits in relation to respiration and ethylene production. *Plant Physiol.* 1979;79:635–640.
35. Grierson D, Fray R. Control of ripening in trangenic tomatoes. *Euphytica* 1994;79:251–263.
36. Christopher JSS, Watson CF, Morris CP, Bird CR, et al. Inheritance and effect on ripening of antisense polygalacturonase gene in trangenic tomatoes. *Plant Mol Biol.* 1990;14:369–379.
37. Giovannoni JJ, Della Penna D, Bennett AB, Fischer RL. Expression of a chimeric polygalacturonase gene in trangenic rin (ripening inhibitor) tomato fruit results in polyuronide degradation but not fruit softening. *Plant Cell.* 1989;1:53–63.

38. Robinson DS. Peroxidases and catalases. In: Robinson DS, Eskin NAM, eds. *Peroxidases and Catalases in Foods*. London: Elsevier Science; 1991:1–47.
39. Parker ML, Waldron KW. Texture of Chinese water chestnut: involvement of cell-wall phenolics. *J Sci Food Agric.* 1995;68:337–346.
40. Parr AJ, Waldron KW, Ng A, Parker ML. The wall-bound phenolics of Chinese water chestnut (*Eleocharis dulcis*). *J Sci Food Agric.* 1996;71:501–507.
41. Brett C, Waldron KW. *Physiology and Biochemistry of Plant Cell Walls.* London: Unwin Hyman; 1990.
42. Fry SC. *The Growing Plant Cell Wall: Chemical and Metabolic Analysis.* Harlow: Longman Scientific & Technical; 1988.
43. Hinton DM, Pressey R. Cellulase activity in peaches during ripening. *J Food Sci.* 1973:39:783–785.
44. Tijskens LMM, Rodis PS, Hertog MLATM, Kalantzi U, van Dijk C. Kinetics of polygalacturonase activity and firmness of peaches during storage. *J Food Engr.* 1998;35:111–126.
45. Wiley RC, ed. *Minimally Processed Refrigerated Fruits and Vegetables*. New York: Chapman & Hall; 1994.
46. Recourt K, Stolle-Smits T, Laats JM, Beekhuizen JG, et al. Pectins and pectolytic enzymes in relation to development and processing of green beans. In: Visser, Voragen, eds. *Pectins and Pectinases, Progress in Biotechnology 14*. Amsterdam: Elsevier Science; 1996:399–404.
47. Ebbelaar CEM, Tucker GA, Laats JM, van Dijk C, et al. Characterization of pectinases and pectin methylesterase cDNAs in pods of green beans (*Phaseolus vulgaris* L.). *Plant Mol Biol.* 1996;31:1141–1151.
48. Laats MM, Grosdenis F, Recourt K, Voragen AGJ, et al. Partial purification and characterization of pectin methyl esterase from green beans (*Phaseolus vulgaris* L.). *J Agric Food Chem.* 1997;45:572–577.
49. Tijskens LMM, Rodis PS, Hertog MLATM, Proxenia N, van Dijk C. Activity of pectin methyl esterase during blanching of peaches. *J Food Engr.* 1999;39:167–177.
50. Shannon LM. Plant isoenzymes. *Annu Rev Plant Physiol.* 1968;19;187–210.
51. Gkinis AM, Fenema OR. Changes in soluble and bound peroxidase during low temperature storage of green beans. *J Food Sci.* 1978;43:527–531.
52. Van Marle JT, Van der Vuurst de Vries R, Wilkinson EC, Yuksel D. Sensory evaluation of the texture of steam-cooked potatoes. *Potato Res.* 1997;40:79–90.
53. Van Marle JT. *Characterization of Changes in Potato Tissue during Cooking in Relation to Texture Development.* Wageningen, Netherlands: Agricultural University Wageningen; Ph.D. Thesis: 1997.
54. Verlinden BE. *Modeling of Texture Kinetics during Thermal Processing of Vegetative Tissue.* Belgium: Katholieke Universiteit Leuven; Ph.D. Thesis: 1996.
55. Verlinden BE, de Barsy T, DeBaerdemaeker J, Deltour R. Modeling the mechanical and histological properties of carrot tissue during cooking in relation to texture and cell wall changes. *J Texture Studies.* 1996;27;15–18.
56. Verlinden BE, DeBaerdemaeker J. Modelling low temperature blanched carrot firmness based on heat induced processes and enzyme activity. *J Food Sci.* 1997;62:213–229.
57. Verlinden BE, Nicolaï BM, DeBaerdemaeker J. Modelling the relation between macroscopic vegetable strength and the strength of cell walls and middle lamellae: a stochastic approach. *ASEA Annual International Meeting*. Minneapolis, MN: No. 976025; 1997b.

CHAPTER 9

The Role of Tissue Microstructure in the Textural Characteristics of Minimally Processed Fruits

Stella M. Alzamora, María A. Castro, Susana L. Vidales, Andrea B. Nieto, and Daniela Salvatori

INTRODUCTION

Texture is one of the most important but least understood quality determinants in fruits and vegetables. On a global basis, there is an ever-increasing demand for improved texture and high quality in fruits. The attractive features of fruits contribute not only to their aesthetic qualities, but also to their mineral and vitamin content and antioxidant properties.

Mechanical properties of biologic tissues depend on contributions from the different levels of structure and their chemical and physical interactions.[1] As van Dijk and Tijskens pointed out in the previous chapter, three levels can be distinguished:

1. the molecular level (i.e., the chemicals and the interactions between the constituting polymers)
2. the cellular level (i.e., the architecture of the tissue cells and their interaction)
3. the organ level (i.e., the arrangements of cells into tissues), whose rheologic properties or sensory characteristics are analyzed.

Because of the hierarchical components in the study of the textural characteristics of plant tissues, we are faced with the problem of trying to integrate a range of different approaches. Biochemists describe the polymers present in the cell wall, their cross-links, and the effect of the enzymatic action; microscopists describe the microanatomy of the tissue; physiologists detail the physical behavior of the cell walls; food material scientists study mechanical and rheologic properties; and sensory analysis experts examine the perceived textural properties.

At the macroscopic level, fruits are viscoelastic products, i.e., products that combine the effects of an elastic solid and a viscous fluid in response to applied loads. Thus, their mechanical properties are time-dependent.[2] Quasi-static, dynamic, creep, and stress-relaxation properties of fruits, as well as of a number of viscoelastic models, have been largely reported in the literature.[2–4]

Research has also concentrated on the chemistry and biochemistry of the wall, i.e., on the molecular level. On the contrary, histologic and fine-structural studies remain an underworked area of fruit and vegetable research.

In this chapter, after a brief description of key structural elements, we present some of our recent attempts to correlate the cellular parameters (plasma membrane and tonoplast integrity, wall organization, adhesion between cells) within the changes in the mechanical properties of some minimally processed fruits. The understanding they will give of how fresh tissues respond to various processes will make possible improvements in quality assessment of commercial fruits.

KEY STRUCTURE FACTORS DETERMINANT OF TEXTURE

The edible portion of most plant foods is predominantly composed of parenchymatous tissue. The parenchyma cells, ≅ 50–500 μm across and polyhedral or spherical in shape, show, from out to inner, the middle lamella that glues adjacent cells; the primary cell wall within the plasmodesmata; the plasma membrane; a thin layer of parietal cytoplasm containing different organelles (mitochondria, spherosomes, plastids, chloroplasts, endoplasmic reticulum, nucleus, and so on); and, bound by the tonoplast membrane, one or more vacuoles that contain a watery solution of organic acids, salts, pigments, and flavors that are responsible for the osmotic potential of the cell (Figure 9–1). Cells and intercellular spaces are arranged into tissues and these last into the final organ.

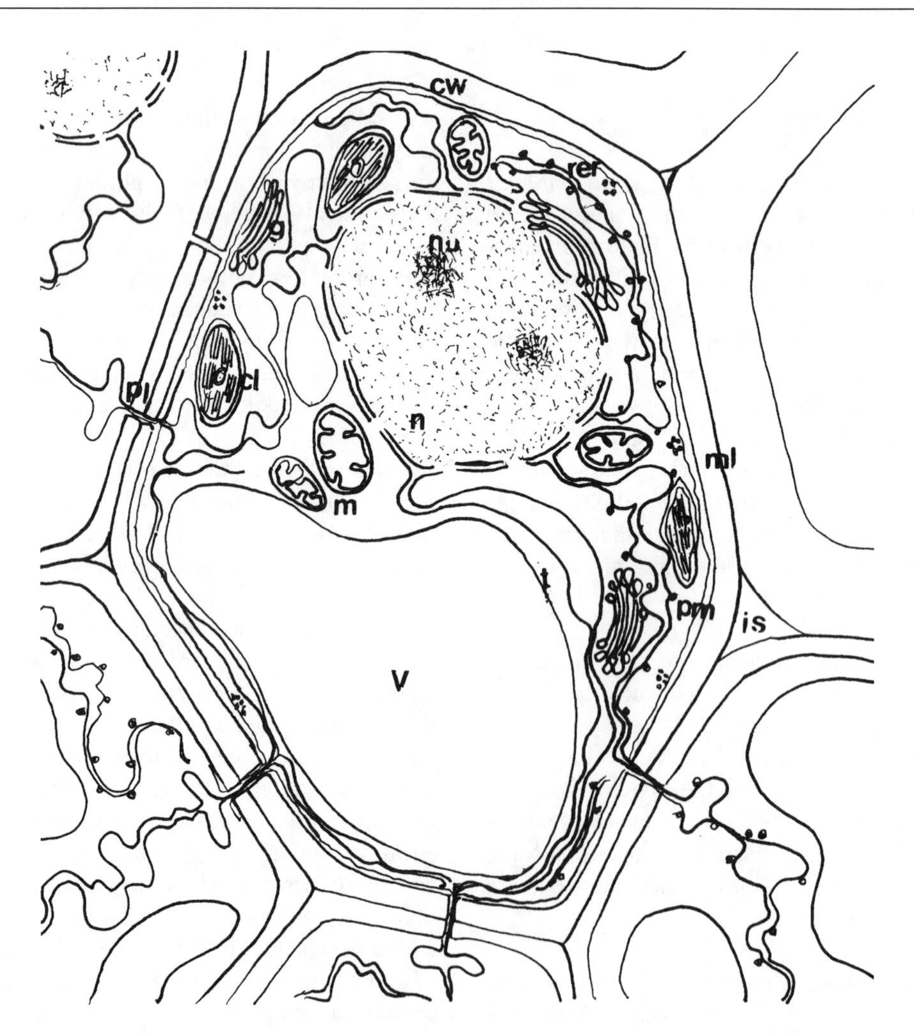

Figure 9–1 Diagrammatic representation of a parenchyma cell. cl = chloroplast; cw = cell wall, g = Golgi; is = intercellular space; nu = nucleolus; n = nucleus; ml = middle lamella; pl = plasmodesmata; pm = plasmalemma; rer = rough endoplasmic reticulum; v = vacuole.

At the cellular and tissue level, the three major structural factors that contribute to textural properties of plant-based foods are *turgor*, the force exerted on the cell membrane by intracellular fluid, *cell wall rigidity,* and *cell–cell adhesion,* determined by the integrity of the middle lamella and the plasmodesmata.[1,5] In addition to these major structural elements, the relative percentage of the different tissues, size and shape of the cells, ratio of cytoplasm to vacuoles, volume of intercellular spaces (which may contain either fluids or interstitial air), type of solutes present, and presence of starch and its state are also important.[6]

Turgor Pressure

Cell turgor arises because sugars and salts in the cell's vacuolar fluid cause an osmotic pressure gradient across the semipermeable plasma membrane lining the cell's structural wall. Water flows into the cell until the osmotic gradient is balanced by a static pressure difference. Because it is a component of water potential (ψ), turgor (P) is a measure of water status.

Relationship between wall properties and turgor is different for nongrowing and growing cells.[7] In most nongrowing, primary-walled plant cells, the outward pressure produced by turgor is approximately equaled by the inward counterforce on the cell contents exerted by the wall.[7] As water enters the cell, the wall stress increases, and the elastic energy stored in the strained polymer bonds compresses the protoplasm, resulting in turgor and preventing changes in volume. At this point the static pressure in the fully turgid cell may be 3–4 atmospheres. For mature, turgid cells of higher plants, changes of water potential are mainly reflected in changes of turgor (i.e., $\Delta P/\Delta\psi \cong 1$) because of the high rigidity of walls.[8] However, in growing cells, after exceeding a threshold turgor pressure, the wall begins to expand irreversibly, resulting in a permanent increase in cell volume and, hence, in plant cell growth, as we will analyze in the next section. According to Bourne,[9] the rigidity and perceived crispness of fresh fruits and vegetables can be attributed to cell turgor, which distends the cell vacuoles against the partially elastic cell walls.

Plasmodesmata

Plasmodesmata are minute channels that traverse the plant cell wall to provide a cytoplasmic pathway for communication between neighboring cells (Figure 9–1). Although they maintain protoplasmic continuity (symplasm), the pores are small enough to prevent organelle movements between cells.[10] Plasmodesmata are anatomically and taxonomically widespread and occupy anywhere from 0.08 to 0.002 times the area in common between adjacent cells.[11,12] The typical radii are 4×10^{-6} cm (40 nm), ranging from 1.2×10^{-6} cm to 8.8×10^{-6} cm for cells in quite a wide range of genera. The pore frequencies range from an unusually high value of 6×10^9 pores/cm^2 to 1.5×10^8 pores/cm.[2,11,12] Recently, these connections have been shown to be involved in two types of intercellular transport: the passive movement of small molecules and ions, and the active movement of specific macromolecules.

Knowledge of the structure of plasmodesmata has mostly relied on the interpretation of images from electron microscopy involving chemical fixation. They are bounded by plasma membrane that lines the surface of the wall and the desmotubule—the cylinder of tightly furled membrane running through a plasmodesma—is continuous with the endoplasmic reticulum (ER) in neighboring cells. There is a layer of proteinaceous material, either integral or closely associated with the desmotubule, that helps to maintain the structure of the plasmodesma.[7,12] In mature walls, single plasmodesmata often are distended at the level of the middle lamella, the diameter of the lumen increases, and the desmotubule may become so tortuous in its path that continuity cannot be seen. In addition, branched plasmodesmata are common: Plasmodesmata may extend for many micrometers in the center of the wall, developing numerous branches leading out to the surrounding cell.[13] Sometimes, they are branched on one side only, sometimes on both sides. Walls surrounding these sides show removed pectin and hemicellulose fraction but intact cellulose microfibrillar skeleton.[14]

Overall and Blackman[12] have suggested a dynamic model of a plasmodesma where actin and myosin are helically arranged around the desmotubule, connecting the plasma membrane to the desmotubule. Contractile proteins link the plasma membrane to the ER via anchoring proteins at the neck of the plasmodesma, presumably acting as a sphincter and limiting movement through the plasmodesma. An electron-lucent sleeve appears surrounding the plasmodesma. This sleeve may contain sphincter particles encircling the neck regions, spirals of electron-dense materials, and plasma-membrane wall connections. The size exclusion limit for passive transport between cells would depend on the dimensions of the gap between the desmotubule and the plasma membrane, as determined by contractile proteins. Transport of specific macromolecules might occur due to a transient increase in the size exclusion limit of plasmodesmata via the actin–myosin mobile system or via a dilated ER.[7,12]

Plasmolysis stretches and can rupture plasmodesmata but most of them reform when the tissue is returned to an isotonic or hypotonic medium, although with a higher exclusion limit. Moreover, desmotubules are not lost during plasmolysis but form a stable structure, connected to the plasmodesma through "spokes" or connections around the neck.[7,12,13]

The Cell Wall or Extracellular Matrix

Despite being external to the plasma membrane, the wall is a very much active metabolic compartment. The cell wall is of considerable functional importance for plants, having diverse roles in cell expansion and, hence, plant growth, cell differentiation, cell-to-cell adhesion, and plant defense.[15]

The cell wall is a complicated structure, and the organization and functional interactions of its components are not yet fully understood. However, advances in cell wall research are impressive, and our view of the cell wall and its role is evolving rapidly.[7,15–25]

Components of the Cell Wall

Cells of fruit pulp contain primary cell walls. Primary walls are composed of approximately 90% polysaccharides ($\cong$ 35% pectic polysaccharides; $\cong$ 30% cellulose; $\cong$ 25% hemicelluloses) and 10% protein. However, the structures of many of the cell wall polysaccharides are so complex that even the most sophisticated technologies available today may not be capable of completely delineating the primary structure of these polymers.[23] The content and structural features of the wall polymers vary, depending on the species, development stage, and differentiation of the tissue.[16] Modern cytochemical studies have also revealed differences in cell wall composition between different cells in a tissue and even between different areas of the wall of a given cell.[26]

Cellulose composes linear chains of β-1,4-linked glucose residues, which aggregate ($\cong$ 30–100 such chains) together by intra- and inter-chain hydrogen bounds to form insoluble structures ("microfibrils") that are $\cong$ 5–15 nm in diameter and several thousand units long. The major structural variables in the cellulose of different tissues are in the degree of polymerization and in the degree and type of crystallinity of the glucan chains.[23]

Hemicelluloses of primary cell walls are principally xyloglucans, alongside a smaller amount of xylan. Xyloglucan backbone consists of β-(1 → 4)-linked glucose residues that are highly substituted by xylose residues linked via α-(→6) bonds. Some glucose residues can be substituted with disaccharide or trisaccharide chains (i.e., xylose, galactose, and fucose or arabinose). All of the xyloglucans are very structurally similar, but there are some differences in the side chains.[16,23] More of the cell wall xyloglucan is hydrogen-bonded to cellulose, probably limiting the self-association of cellulose fibers and providing sites for covalent cross-linking of cellulose.[22,23]

The phenolic components of primary cell walls are principally polymer-esterified ferulic and p-coumaric acids. Recent evidence suggests that the feruloyl esters are subjected to peroxidase-catalyzed coupling to yield diferuloyl bridges, thereby cross-linking the polysaccharide molecules and contributing to cell adhesion.[1,23]

The pectic polysaccharides may be diverse in their structural composition, containing rhamnogalacturonan I and II, arabinans, homogalacturonans, galactans, and arabinogalactans. The building blocks of pectins present a high proportion of D-galacturonic acid residue linear chains in α-1,4-glycosidic linkages. The carboxylic acid groups may be esterified with methanol. Numerous L-rhamnose residues are normally interspersed within the chain. Pectic substances also contain variable proportions of neutral polysaccharides. In particular, the earliest-formed wall layer deposited at cell division or middle lamella (which acts as an adhesive between adjacent cells) is composed of heat-labile pectic substances. These pectins have fewer rhamnose residues, fewer and shorter branches, and a higher degree of esterification than do the pectins of the primary cell wall.[7]

The major types of proteins in the cell wall are hydroxyproline-rich glycoproteins (extensin and potato lectin), proline-rich glycoproteins, glycine-rich proteins, arabinogalactan proteins, and enzymic proteins (peroxidases, phosphatases, glycosyl hydrolases, dehydrogenases).[16] The quantitatively dominant amino acid is extensin, which is found in an insoluble form within the cell wall. It has been suggested that proteins could provide the cell wall with combined elasticity and strength that carbohydrates and polyphenolics could not impart to the cell wall.[16] Because extensin is tightly cross-linked, it may play a role in restricting cell expansion.[23]

Cell Wall Architecture

How do individual cell wall components interact to determine the mechanical properties of the cell wall? Here we must recognize that a temporal sequence of genetically controlled changes in cell wall composition and structure typically takes place from plant development to fruit ripening, influencing significantly the physical properties. Growing walls behave as viscoelastic liquids (i.e., they do not recover original size and shape after removal of external force), whereas nongrowing cell walls behave as viscoelastic solids (i.e., their size and shape are restored in a greater or lower degree upon removal of force).[20]

Recently, several models about the interactions between components have been proposed.[7,17,21,22,26] In general, they postulate that three networks, intertwined and perhaps covalently linked with each other, are responsible for bearing tensile stresses within the wall[20]:

1. a stretch-resistant, load-bearing cellulose/xyloglucan network, where the hemicelluloses are thought to form a hydrogen-bonded surface coat over the microfibrils and may bridge between microfibrils;
2. a compression-resistant pectic polysaccharide network, where the junction zones are originated by Ca^{2+} cross-linking and by ester linkages with dihydroxy-cinnamic acids, such as diferulic acid; and
3. a third network consisting of the structural proteins covalently linked by oxidative phenolic cross-bridges and other linkages.

During growth, plant cells usually enlarge 10- to 1,000-fold in volume by massive vacuolation and irreversible cell wall expansion.[19] Growing plant cells have turgor pressures ranging from 3 to 10 bar. Because its cross-sectional area is about 1/100 that of the cell, the growing wall must bear a very much higher tensile stress ($\cong$ 1,000 bar).[20] It is believed that each of the three polymeric networks cited above bears a fraction of the wall stress.

During growth, cells regulate specific loosening processes that result in wall stress relaxation. Wall relaxation progressively reduces cell turgor and water potential, and water is drawn into the cell (a reversible process). Thus, the cell enlarges by water uptake and the irreversible expansion of the cell wall. That is, the growing cell walls are "extensible": They deform irreversibly in a time-dependent manner under tensile forces in the wall generated by cell turgor.[20] Wall creep and stress relaxation would be induced by expansins in a pH-dependent manner. Expansins would transiently displace short stretches of hemicelluloses that are bonded to the surface of the microfibril. If the wall is in tension, the polymers creep. If the wall is relaxed, no polymer movement will occur. In the meantime, other wall enzymes (i.e., hydrolases, endoglucanases, transglycosylases) would cut matrix glucans and restructure the wall, reducing their viscous resistance to wall creep. When cells mature, cross-linking of the matrix by peroxidases and/or pectin methyl esterases may increase the resistance to creep, and the wall is no longer extensible by expansions.[18]

On the other hand, softening in ripening is associated with disassembly of the primary cell wall. Modifications of the structure and composition of the constituent polysaccharides have been correlated with the expression of a range of enzymes (i.e., hydrolases, transglycosylases) and the alteration of covalent and noncovalent interactions between different polysaccharides. Pectins and hemicelluloses typically undergo solubilization (i.e., from the loss of galactosyl residues) and depolymerization, contributing to wall loosening and disintegration. As expected, this dynamic of the cell wall architecture and the heterogeneity between species, between tissues, and even between domains within a single wall have profound implications on the rheologic behavior and textural response of fruits.

EXAMINATION OF PLANT TISSUE STRUCTURE

Tissue structure is usually studied using microscopy (optical, electron, and atomic microscopy) and other imaging techniques (e.g., magnetic resonance imaging), together with localization techniques (e.g., immunolocalization, X-ray microanalysis, energy-dispersive detection system). Such techniques have proved most advantageous for the examination of food structure because they generate data in the form of an image.[27,28] Recent advances in microscopy and imaging techniques, as well as their application to the field of food science, have been reviewed by Kaláb et al.[27] The visualization of the true tissue structure is extremely difficult because the preparation of a specimen for microscopy alters the sample to some extent.[27] The best way is to subject the sample to several imaging techniques to compare and confirm the results. On the other hand, all of these techniques are complementary. Each one reveals a partial structure aspect (i.e., scanning electron microscopy [SEM] is used to examine surfaces, transmission electron microscopy [TEM] reveals the internal structure). If the various microscopy techniques are combined and their results integrated, we will reach a better understanding of the structural features of these highly heterogeneous plant materials.

As an example of the different microscopic approaches, Figure 9–2 presents some microstructural and ultrastructural observations in fresh fruits, obtained using microscopy techniques with different resolution level. We can observe, from a general aspect of mango tissue, intact starch granules inside the cells, obtained by light microscopy (LM); the topography of strawberry tissue, as seen in SEM; and a detail of the cellular wall of mango with a nitid middle lamella and intact cellular membranes or a detail of a plasmodesma of papaya, densely stained, obtained by TEM.

DIVERSITY WITHIN AND BETWEEN FRUITS

Whatever the approach or level emphasized, an important disadvantage exists in studying textural characteristics—the variability within and between fruits originating from the following phenomena:

a. Plant samples are composed of different tissues, with different types of cells exhibiting quite different turgidity, osmotic pressure, elasticity, size, and composition, so that some kind of averaged value of turgor or mechanical wall resistance would be obtained. Even in fairly ho-

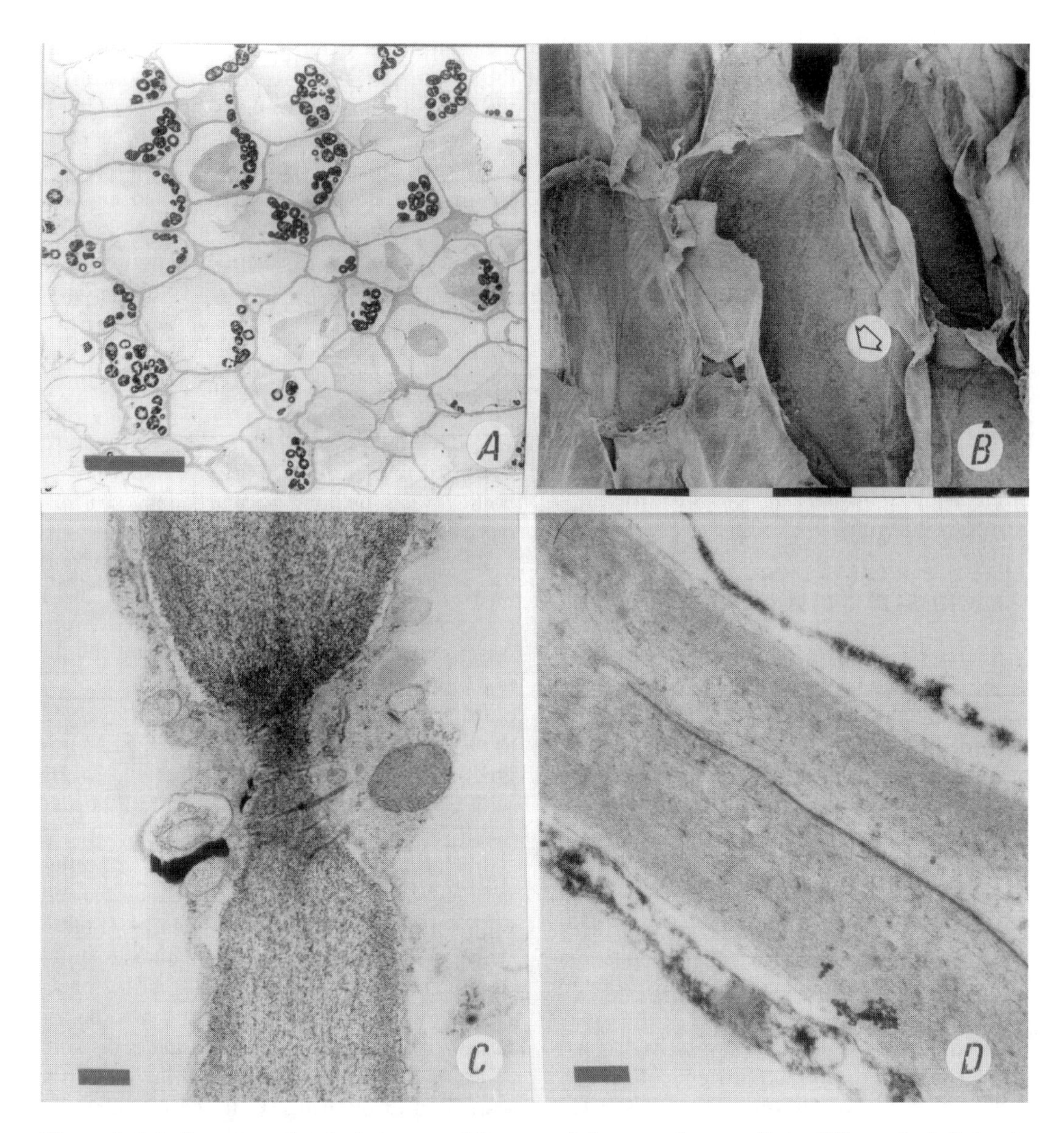

Figure 9–2 A–D, structural and ultrastructural features of the parenchyma cells in different fresh fruits. A: Mango, light micrograph (tissue general aspect); B: Strawberry, SEM micrograph (tissue general aspect); C–D, TEM micrographs; C: papaya (plasmodesmata and cell wall detail); D: mango (cell wall detail). Scale: A–B = 100 μm; C–D: 500 nm.

mogeneous tissue, turgor may vary considerably from cell to cell.[8] Moreover, cytochemical studies have demonstrated the structural diversity of individual wall polysaccharides: At least some structures of the noncellulose polysaccharides vary from cell to cell, and even among the different wall faces of a single cell.[29]

b. Physical and chemical properties depend on time of harvest in the field. As previously mentioned, growing cell walls differ from mature walls in many ways; in particular, growing cell walls show distinctive rheologic properties.[18] On the other hand, differences between fruits in the canopy of the same tree have been detected long ago in most fruit crops. As an example, after individually picking 1,800 fruits from a single Valencia orange tree, a clear relation between radiation intensity and fruit composition was found—the percentage of fully colored fruits, juice, and ascorbic acid being higher in the top outside than in the inside fruit, whereas oranges from other tree portions were intermediate.[30] Due to this biologic material variance, a sufficient number of replications are necessary to obtain an acceptable level of confidence in microscopic, instrumental, and sensory tests.

TEXTURAL CHANGES: A MICROSTRUCTURAL VIEW

Bourne[31] has described in detail what happens at the microscopic level when a compressive test is applied to a vegetable tissue. The cell is deformed in the direction of the applied load, and, as the cell content is incompressible, the cell surface/volume ratio increases, resulting in a distension of the cell wall and an increase in cell wall stresses. These stresses are resisted by the turgor pressure, which also increases, provoking a water efflux until internal and external water potential (Ψ) are equal. The middle lamella is deformed and, in consequence, the cell-to-cell contact area is modified. Removal of the external load allows the cell to return to its original shape (depending on the elasticity of its cell wall) and to recover its strain. Because of the water efflux and the time-dependent properties of the cell wall, the mechanical properties of the whole tissue show time dependence. Also, because pectin has plastic characteristics, cell reorientation will not be totally reversible and, in consequence, there will be the same degree of recoverable strain at the macroscopic level.[2]

Thus, if we conclude that initial turgor pressure, viscoelasticity of the cell wall and the middle lamella, and plasmalemma hydraulic permeability are determinant of macroscopic mechanical characteristics, the modification of these parameters due to processing will affect the textural properties of the tissues.

Next, we present some examples of the relationship of microstructure–mechanical properties in fruit tissues subjected to some operations that are usually involved in minimal processing of fruits (blanching, osmotic dehydration, and calcium incorporation) and their combinations.

Blanching

Strawberry

Fresh whole strawberries (cv. Pájaro) were subjected to blanching in saturated vapor or in boiling water for 2 minutes and cooling in drinking water at 20ºC.[32]

Structural changes were recorded by TEM and SEM. In the raw strawberries (Figure 9–3a,d), cell walls, plasmalemma, and tonoplast were intact, and a large vacuole limited the cytoplasm to a thin peripheral layer between tonoplast and plasmalemma. The middle lamella was observed as a markedly electron-dense region. SEM micrographs of fresh fruit showed individual cells with good definition and arranged in a honeycomb pattern.

In blanched tissues (Figure 9–3b,c,e,f), cellular membranes were broken, decreasing the cell turgor pressure. Hot-water blanching (Figure 9–3c,e,f) resulted in a very severe ultrastructural disorganization of the cell walls: Fibrillar organization was lost, the cell wall appeared perforated (Figure 9–3c, *arrow*), and the cells were difficult to identify because of the structural disruption of the walls. Cell walls exhibited a severe loss of material and extreme swelling. Vapor blanching, on the contrary, maintained the original arrangements of cells, as seen in SEM (Figure 9–3b), although cell walls appeared more porous (*arrow*). Examination of tissues with TEM (microphotographs not shown) indicated in the vapor-treated sample electron-dense cell walls, a network of microfibrils, and a pectin matrix very similar to those of the fresh fruit, although the middle lamella had been partially lost.

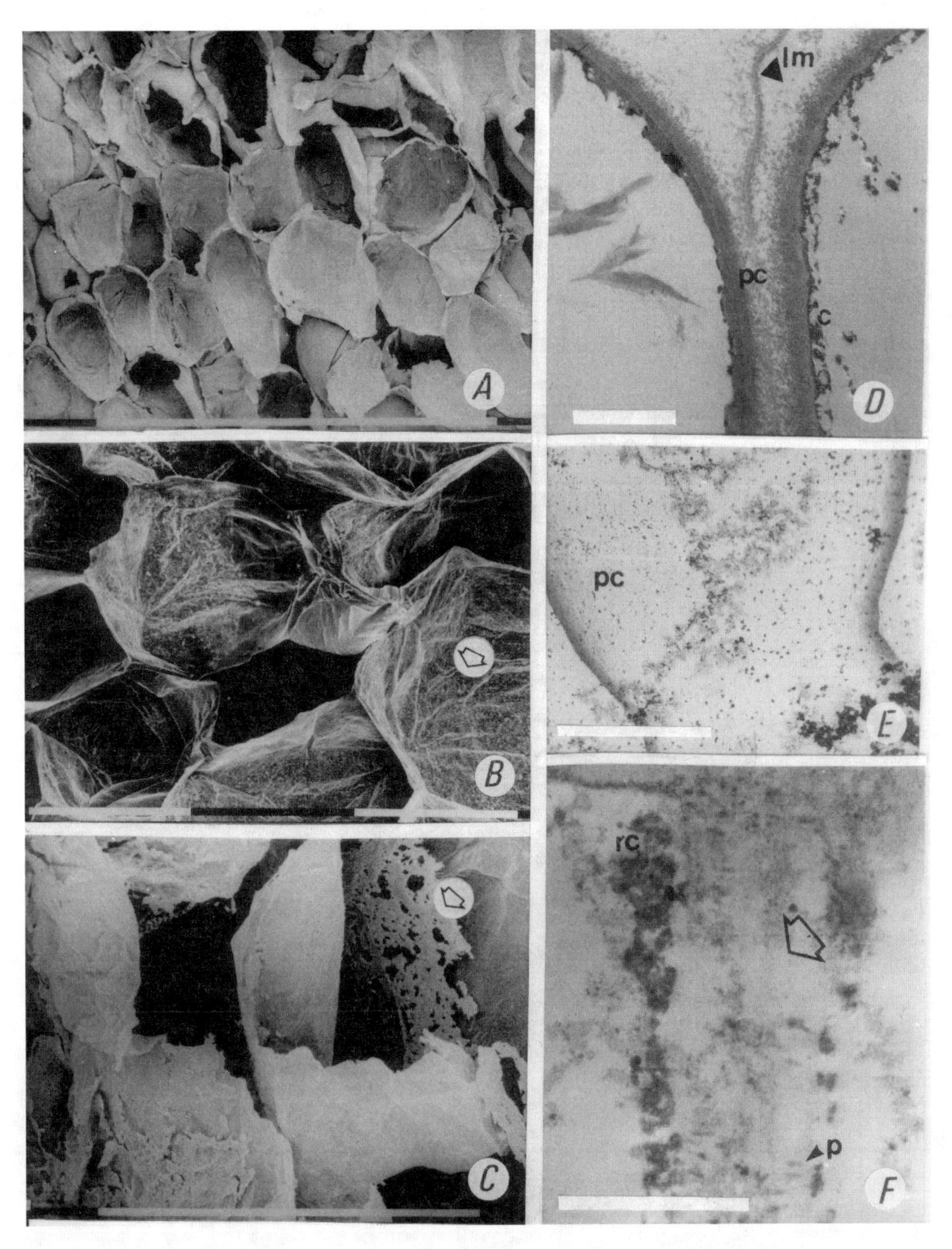

Figure 9–3 A–F, Blanched strawberry fruit, ultrastructural features. A–C, SEM micrographs: A, control; B, vapor treatment; C, water treatment. Note the partially eroded or perforated cell wall. D–F, TEM micrographs: D, control; E, water treatment: cell walls separated along middle lamella; F, water treatment: substantial negative effect at the plasmodesmata level (arrow). pc = cell wall; lm = middle lamella; p = plasmodesmata; rc = cytoplasmatic rests. Scale: A = 1 mm; B–C = 0.1 mm; D = 5 μm; E = 10 μ; F= 1 μ.

These ultrastructural changes in blanched fruits correlated well with the instrumental changes in texture. Textural properties were less modified by vapor treatment. A slight decrease in force occurred due to vapor blanching but water blanching resulted in a significant softening. Average maximum extrusion force for fresh fruit was 11.5 kg. Fruits blanched in vapor or in water exhibited an average extrusion force equal to 9.3 kg or 6.1 kg, respectively.[32]

Apple

Cylindrical samples of apple (cv. Granny Smith) of diameter 1.9 cm and length 0.6 cm were exposed to steam for 1.5 minutes at atmospheric pressure, then cooled in water at 5°C.[33–35] Before blanching, LM observations (micrographs not shown) indicated cell and intercellular space loosely arranged in a netlike pattern that was inhomogeneous and anisotropic. As seen in TEM (Figure 9–4a), fresh tissues exhibited densely stained, tightly packed fibrillar material in the walls and a conspicuous middle lamella in between. Also, in some areas, it was observed that the middle lamella was degraded, with formation of empty regions in the wall (micrographs not shown). Heating resulted in tonoplast and plasmalemma disruption

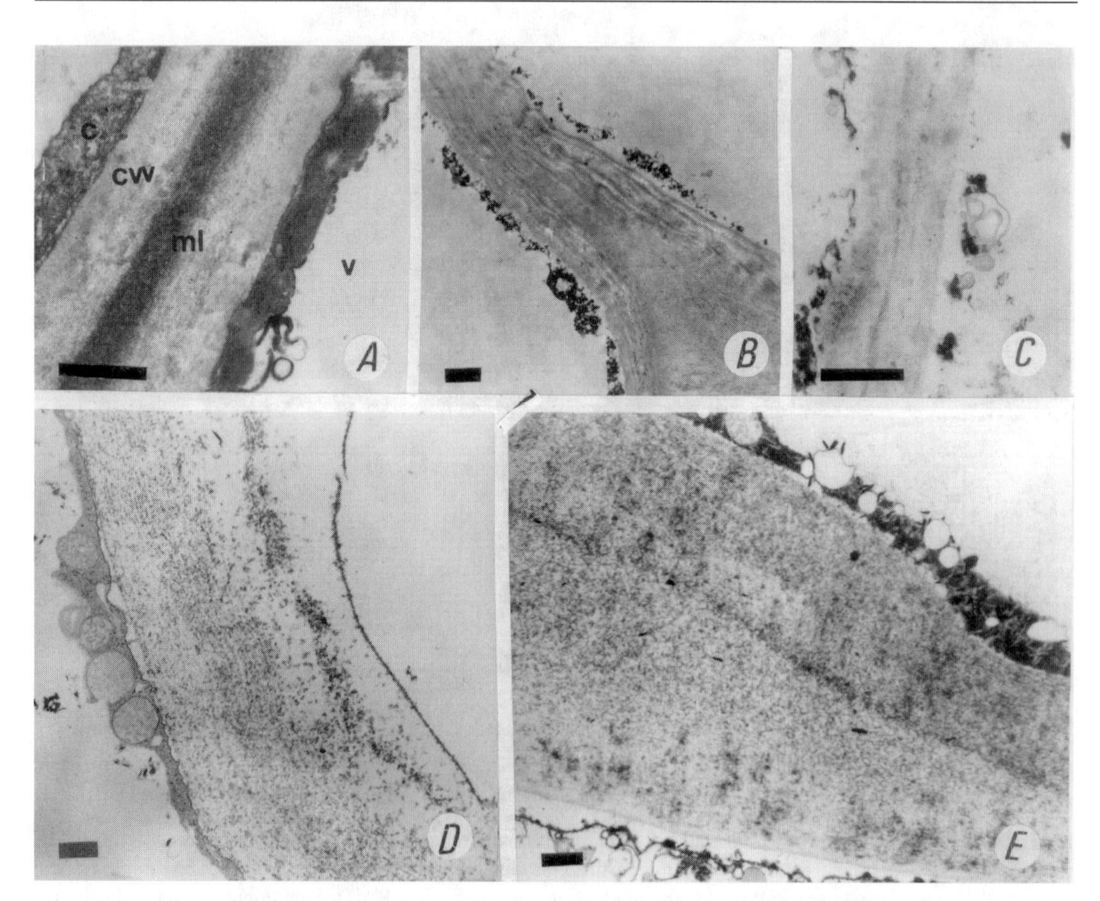

Figure 9–4 Vapor-blanched apple and kiwifruit. A–C, apple: A, control; B–C, blanched. D–E, kiwifruit: D, control; E, blanched. c = cytoplasm; cw = cell wall; ml = middle lamella; v = vacuole. Scale: 1 μm.

and appearance of vesicles in the cytoplasm, as well as some degradation of cell wall (Figure 9–4b,c). In some areas, the middle lamella was still visible. Although microfibrillar disorganization was obvious, the optical density of cell wall was only slightly lower than the one of the fresh fruit. Separation of adjacent cells was not observed, but the treatment resulted in shrinkage (≅ 23%) of the tissues. In agreement with the moderate structural modifications of the cell wall, tissue failure for vapor-blanched apple was slightly lower than the value for the fresh fruit (12.5 vs. 14.6 kg), and the resistance to deformation also decreased slightly.

Kiwifruit

Kiwifruits (cv. Hayward), sliced into halves, were blanched in saturated vapor for 2 minutes, then cooled in water at 10°C.[35,36] Fresh kiwifruit tissues (flesh zone) showed parenchyma cells intact, with thin walls and numerous plasmodesmata connecting adjacent cells (also see Figure 9–7a). The blanching treatment produced slight shrinking and elongation of cells (LM microphotographs not shown). Although the cell walls were swollen, they showed in TEM (Figure 9–4e) an electron density similar to that of raw cells (Figure 9–4d).

Compression force for the blanched specimen was not different from the value exhibited by the fresh fruit, in concordance with the little damage encountered through microscopic analysis. It is interesting to note that, when other lots were analyzed for mechanical properties, the effect of blanching appeared to be dependent on the ripeness of the fruit, the compression force varying from 64% to ≅ 100% of the value of the fresh fruit.

Blanching and/or Osmotic Dehydration

Apple

For glucose impregnation at atmospheric pressure, apple cylinders were immersed into 22.0% w/w (a_w = 0.97) glucose solution, with forced convection at 25°C until water and solids contents were almost constant. For vacuum glucose impregnation, fruit samples were immersed in a 59.0% w/w (a_w = 0.84) glucose solution at 25°C, and a pressure equal to 600 mm Hg was applied to the system for 10 minutes, then the atmospheric pressure was restored and the samples drained.[33–35] For both treatments, the a_w value at the end of the osmotic dehydration was ≅ 0.97. Some samples were previously blanched, as described before.

TEM micrographs for samples dehydrated at atmospheric pressure (Figure 9–5b,c,f) did not demonstrate a great damage in the cell walls, as compared with the fresh apple (Figure 9–5a). Cell walls appeared with good electronic density, and the network of microfibrils was similar to the one of the control. As seen in LM (micrographs not shown), cytoplasm separated from the wall, and cells looked more rounded but well defined. Although heated samples showed broken membranes with formation of vesicles (Figure 9–5b), staining of the cell walls was slightly darker or at least equal to the corresponding nonheated tissue (Figure 9–5c). In environmental scanning electronic microscopy (ESEM), cells appeared more rounded than in the control (Figure 9–5f vs. Figure 9–5e), and the intercellular spaces decreased. Cells of vacuum-impregnated tissues appeared also rounded, and the intercellular contact did not decrease (LM micrographs not shown). TEM observations indicated a very densely packed fibrillar material (Figure 9–5d). Vacuum-treated apples exhibited a greater shrinkage than did those impregnated under atmospheric pressure (37% vs. 23%), as well as a lower moisture content (72% w/w vs. 76% w/w), probably due to the loss of native liquid that occurred by the expansion of gas during vacuum application and to the larger mass transfer potential gradient from the inside of the cell to the walls and intercellular spaces filled with concentrated solution when the atmospheric pressure was restored.

According with the structural observations, texture analysis demonstrated that there were no significant differences between the failure force values for blanched or for blanched and osmotically dehydrated apples under atmospheric pres-

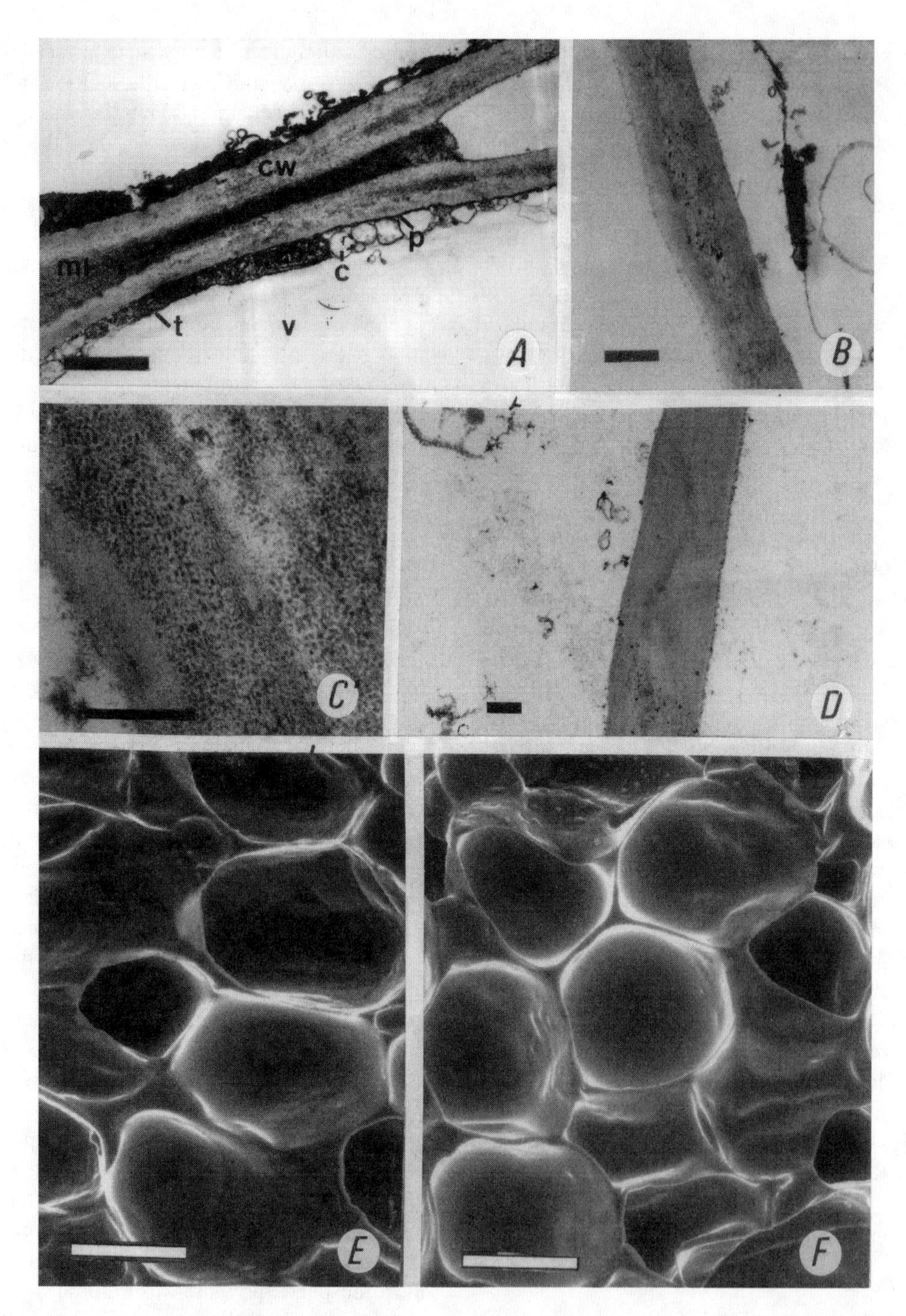

Figure 9–5 A–F, Glucose-impregnated apple fruit. A–D, TEM micrographs: A, control; B, glucose impregnated at atmospheric pressure with previous vapor blanching; C, glucose impregnated at atmospheric pressure without vapor treatment; D, vacuum infiltrated: note the conspicuous packed cell wall. E–F, ESEM micrographs: E, control; F, glucose impregnated at atmospheric pressure. ml = middle lamella, cw = cell wall, c = cytoplasm, t = tonoplast, p = plasmalemma, v = vacuole. Scale: A–D = 1 μm; E–F = 400 μm.

sure, both values being slightly lower than the one corresponding to the fresh fruit (12.5 kg and 12.7 kg, respectively, vs. 14.6 kg for raw apple). Glucose impregnation without previous heating and vacuum treatment slightly increased the failure force (16.2 kg and 15.4 kg, respectively), although both treatments significantly reduced the resistance to deformation.

Blanching and/or Osmotic Dehydration and/or Calcium Addition

Strawberry

Fresh, whole strawberries were subjected to osmotic dehydration at room temperature to attain a_w 0.93 or 0.95, with or without previous vapor blanching (saturated vapor, 2 minutes) and with or without calcium lactate addition (0.1% w/w). Glucose and calcium infiltration was carried out by mixing the whole fruit, glucose, and calcium lactate in the required proportions.[32]

Cell walls of glucose-impregnated strawberries dehydrated to a_w 0.93 without previous blanching were not degraded, appearing electron-dense and, in some areas, with neat presence of the middle lamella (Figure 9–6e). TEM observations also demonstrated the existence of cell wall–plasmodesmata complexes, very well maintained (micrographs not shown). In contrast, contraction of membranes from the cell wall or lysis of plasmalemma and tonoplast with formation of numerous vesicles occurred (micrographs not shown). Tissues dehydrated to a_w 0.95, with calcium addition and previously blanched showed cell walls with an optical density lower than the one of the fresh fruit (TEM micrographs not shown) but, as seen in SEM, the cells retained the original shapes and arrangements (Figure 9–6b). Tissues with the same treatment except calcium infiltration exhibited, on the contrary, a more severe collapse, and cells were difficult to identify (Figure 9–6c).

No significant differences were detected in maximum extrusion force between fresh and nonheated glucose-impregnated strawberries with a_w 0.95 or 0.93 after processing and/or storage. A slight decrease in force occurred in fruit previously blanched after processing, and loss of firmness continued during storage, principally in those fruits without calcium addition. However, calcium addition to nonheated glucose-impregnated fruits did not appear to improve textural characteristics. These macroscopic measurements appeared to be in good agreement with the structural observations.

Kiwifruit

Kiwifruits, sliced into halves, were immersed at 25°C under atmospheric pressure in glucose aqueous solutions during 6 days, to reach $a_w \cong$ 0.97, with or without 0.1% w/w calcium lactate addition during the infusion step.[36] LM observations revealed in fresh tissue thin-walled parenchymatous cells, with a large vacuole limiting the cytoplasm to a thin peripheral layer pressed against the cell wall between tonoplast and plasmalemma (Figure 9–7a). Cell walls appeared densely stained and exhibited small, localized areas of intense staining that were plasmodesmatal connections. Glucose treatment caused extensive plasmolysis of cellular membranes, severe degradation of cell walls and middle lamella, and decreasing cell-to-cell contact (Figure 9–7b).

When calcium was added, swelling of cell walls occurred but cell-to-cell cohesion was maintained through the middle lamella in some areas, and principally by the plasmodesmatas (Figure 9–7c) in others. The staining of the cell walls was not so reduced as in the absence of calcium, indicating a greater integrity of the cell wall. These microstructural characteristics were in good correlation with the mechanical strength of the tissues subjected to the different treatments.

Osmotic dehydration with glucose significantly decreased the failure force (7 N vs. 54 N for the fresh fruit). The residual relaxation force and the relaxation time of raw fruit were reduced after atmospheric infusion to $\cong$ 10% of the corresponding values for raw fruit, indicating an important loss of the elastic component. Addition of 0.1% w/w calcium lactate not only increased mechanical strength of the tissues in more than 40% of the value for the sample without calcium, but the force-deformation curve showed a similar pattern to that of the fresh fruit. However, the parameters of the relaxation curve demonstrated the same trend as did the one ob-

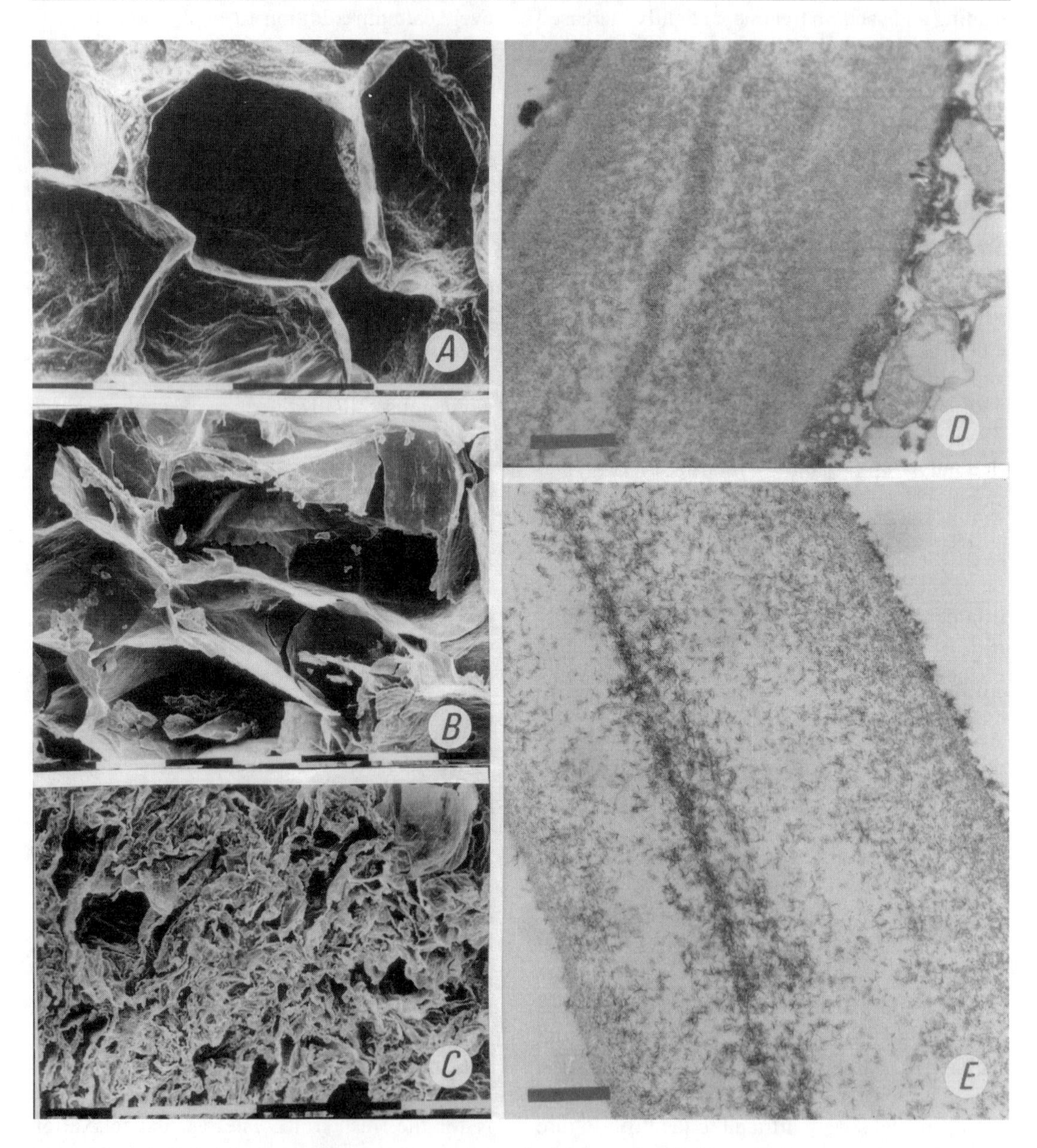

Figure 9–6 A–E, Calcium effect in glucose-impregnated strawberry fruit, with and without previous blanching. A–C, SEM micrographs: A, control; B–C: vapor blanched and glucose impregnated (a_w 0.95), B: with Ca^{2+}; C, without Ca^{2+}. D–E, TEM micrographs: D, control; E: glucose impregnated with Ca^{2+} and without blanching (a_w 0.93). Note the conservation of cell wall in E, as compared with D (control). Scale: A–C = 0.1 mm; D–E = 0.5 μm.

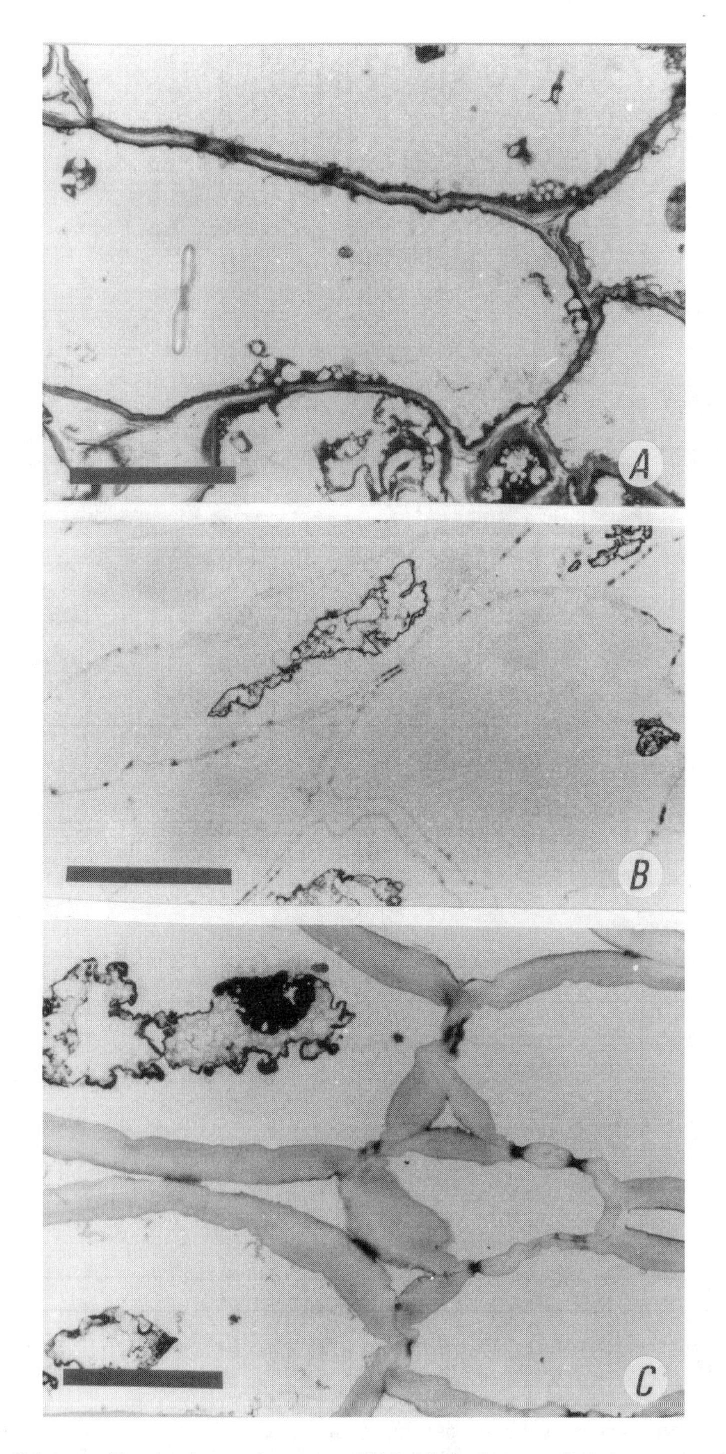

Figure 9–7 A–C, Calcium effect in glucose-impregnated kiwifruit. A, control; B, impregnated without Ca^{2+}; C, impregnated with Ca^{2+}. Note in C a more stained tissue than in B and the particular substantially positive effect at the plasmodesmata level. Scale: 50 μm.

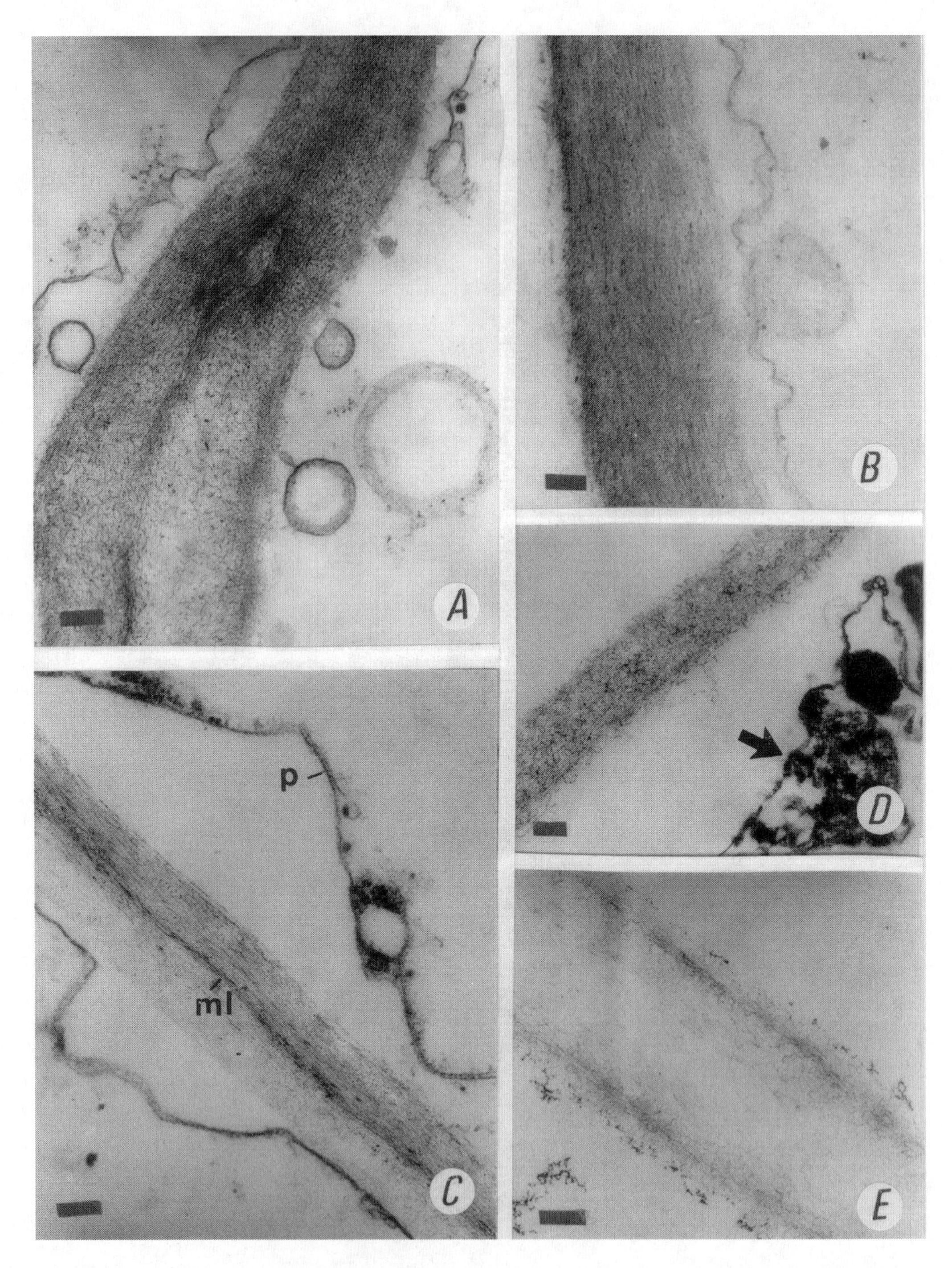

Figure 9–8 A–E, Effect of calcium and glucose impregnation (atmospheric or in vacuum) on melon fruit. A, control; B–C, vacuum infiltrated: B, with Ca^{2+}; C, without Ca^{2+}. D–E, glucose impregnated at atmospheric pressure: D, with Ca^{2+}; E, without Ca^{2+}. Note the significant difference in staining between E and the rest. p = plasmalemma, ml = middle lamella. Scale: 200 nm.

served when calcium was not added, indicating tissue damage in both cases. In conclusion, calcium infiltration increased failure forces, due to enhanced cell cohesion (through plasmodesmata and middle lamella) and increased cell wall integrity.

Melon

Melon (honeydew var.) was cut into cylindrical specimens (30.0 mm in diameter and 14.5 mm in length) and subjected to vacuum (immersion in a 55% w/w glucose aqueous solution for 10 minutes at a pressure of 213 mbar, followed by 10 minutes at atmospheric pressure, final a_w 0.98) or atmospheric impregnation (immersion in a 0.98 a_w glucose aqueous solution for 4 days, final a_w 0.98). In some cases, calcium lactate was added to the impregnation solution.[37]

Ultrastructural studies of the melon flesh in the fresh fruit demonstrated darkly stained cell walls with greater intensity toward the margin and in the central zone of the middle lamella (Figure 9–8a). The arrangement of cytoplasm was marginal, with numerous invaginations of the plasma membrane. Vacuum treatment caused slight plasmolysis of cellular membranes, but cell walls appeared with good electron density and a middle lamella very nitid (Figure 9–8c). Addition of calcium during vacuum provoked a very dense longitudinal staining fiber pattern (Figure 9–8b). Atmospheric infusion caused folding and rupture of cell walls (micrographs not shown), and breakage of cellular membranes with formation of coarse granules that were assembled along cell walls (Figure 9–8e). Cell walls appeared with very much reduced staining in the central zone. When the osmotic dehydration was performed in presence of calcium, the cells showed cell walls with good electron density, a clear reticulate pattern, and broken membranes with vesicle formation (*arrow*) (Figure 9–8d).

In concordance with these observations, the effect of the osmotic dehydration on textural properties depended on the pressure treatment. Impregnation performed in vacuum produced a similar resistance of melon to puncture than did the one exhibited by the fresh fruit. However, atmosphere-treated samples showed a smaller maximum puncture force (≅ 66% of the value for the raw fruit). For both treatments, particularly in the vacuum process, the presence of calcium significantly increased the puncture force. Whatever the treatment, residual relaxation force and relaxation time were very much reduced.

CONCLUSION

Viscoelastic properties of fruits change vastly, not only during maturation, ripening, and storage, but also during processing because of the alterations of their structural components. The present results clearly demonstrate the close correlation between structural cell changes and the mechanical properties of the tissues.

Although all the processes studied decreased residual relaxation force and relaxation time, indicating a significant reduction of the elastic component of the tissue, mainly by the loss of turgor pressure, the extent of cell wall modifications differed between fruits, probably because of the differences at molecular level. Thus, for a complete understanding of the mechanical behavior, we must address simultaneously the study at the micro-, intermediate, and macroscopic levels. Unfortunately, most of the research on the subject did not present an integral approach, in part because of the necessity for various disciplines to best resolve the problem.

REFERENCES

1. Waldron KW, Smith AC, Parr AJ, Ng A, et al. New approaches to understanding and controlling cell separation in relation to fruit and vegetable texture. *Trends Food Sci Technol.* 1997;8:213–221.

2. Pitt RE. Viscoelastic properties of fruits and vegetables. In Rao MA, Steffe JF, eds. *Viscoelastic Properties of Foods*. Amsterdam: Elsevier Science; 1992:49–76.
3. Peleg M. Compressive failure patterns of some juicy fruits. *J Food Sci.* 1976; 41:1320–1324.
4. Peleg M. Characterization of the stress relaxation curves of solid foods. *J Food Sci.* 1979;44:277–281.
5. Jackman RL, Marangoni AG, Stanley DW. The effects of turgor pressure on puncture and viscoelastic properties of tomato tissue. *J Text Stud.* 1992;23:491–505.
6. Ilker R, Szczesniak AS. Structural and chemical bases for texture of plant foodstuffs. *J Text Stud.* 1990;21:1–36.
7. Brett CT, Waldron KW. *Physiology and Biochemistry of Plant Cell Walls*, 2nd ed. London: Chapman & Hall; 1996.
8. Steudle C. Pressure probe techniques: Basic principles and applications to studies of water and solute relations at the cell, tissue and organ level. In: Smith JAC, Griffiths H, eds. *Water Deficits. Plant Responses from Cell to the Community.* Oxford: BIOS Scientific Publishers; 1993:5–36.
9. Bourne MC. Texture of fruits and vegetables. In: de Man JM, Voisly PW, Rasperr VF, Stanley DW, eds. *Rheology and Texture in Food Quality.* Westport, CT: AVI Publishing Co.; 1976:275–307.
10. Dey PM, Brownleader MD, Harborne JB. The plant, the cell and its molecular components. In: Dey PM, Harborne JB, eds. *Plant Biochemistry*. Bath: Academic Press; 1997:1–48.
11. Tyree MT. The symplastic concept. A general theory of symplastic transport according to the thermodynamics of irreversible processes. *J Theor Biol.* 1970;26:181–214.
12. Overall RL, Blackman LM. A model of the macromolecular structure of plasmodesmata. *Trends Plant Sci.* 1996;1:307–311.
13. Gunning BES, Robards AW. Plasmodesmata and symplastic transport. In: Wardlaw IF, Passioraaa JB, eds. *Transport and Transfer Processes in Plants.* New York: Academic Press; 1976:15–41.
14. Juniper BE. Some speculations on the possible roles of the plasmodesmata in the control of differentiation. *J Theor Biol.* 1977;66:583–592.
15. Casero PJ, Knox JP. The monoclonal antibody JIMS indicates patterns of pectin deposition in relation to pit fields at the plasma-membrane-face of tomato pericarp cell walls. *Protoplasma.* 1995;188:133–137.
16. Brownleader MD, Jackson P, Moabsheri A, Pantelides AT, et al. Molecular aspects of cell wall modifications during fruit ripening. *Crit Rev Food Sci Nutr.* 1999;39: 149–164.
17. Carpita NC, Gibeaut DM. Structural models of primary cell walls in flowering plants: consistency of molecular structure with the physical properties of the walls during growth. *Plant J.* 1993;3:1–30.
18. Cosgrove DJ. Cell wall loosening by expansins. *Plant Physiol.* 1998;118:333–347.
19. Cosgrove DJ. Relaxation in a high-stress environment: The molecular bases of extensible cell walls and cell enlargement. *Plant Cell.* 1997;9:1031–1041.
20. Cosgrove DJ. Tansley Review No. 46. Wall extensibility: Its nature, measurement and relationship to plant cell growth. *N Phytol.* 1993;124:1–23.
21. McCann MC, Roberts K. Changes in cell wall architecture during cell elongation. *J Exp Botany.* Special Issue 1994;45:1683–1691.
22. McCann MC, Wells B, Roberts K. Direct visualization of cross-links in the primary plant cell wall. *J Cell Sci.* 1990;96:323–334.
23. McNeil M, Darvill AG, Fry SC, Albersheim P. Structure and function of the primary cell walls of plants. *Annu Rev Biochem.* 1984;53:625–663.
24. Varner JE, Lin LS. Plant cell architecture. *Cell.* 1989;58:231–239.
25. Taiz L. Plant cell expansion. Regulation of cell wall mechanical properties. *Annu Rev Plant Physiol.* 1984; 35:585–657.
26. Grant Reid JS. Carbohydrate metabolism: Structural carbohydrates. In: Dey PM, Harborne JB, eds. *Plant Biochem*. Bath: Academic Press; 1997:205–236.
27. Kaláb M, Allan-Wojtas P, Miller SS. Microscopy and other imaging techniques in food structure analysis. *Trends Food Sci Technol.* 1995;6:177–186.
28. Aguilera JM, Stanley DW. *Microstructural Principles of Food Processing and Engineering.* Amsterdam: Elsevier Science; 1990:2–18.
29. Albersheim P, An J, Freshour G, Fuller MS, et al. Structure and function studies of plant cell wall polysaccharides. *Biochem Soc Trans.* 1994;22:374–378.
30. Monselise SP. Citrus. In: Monselise SP, ed. *CRC Handbook of Fruit Set and Development.* Boca Raton, FL: CRC Press; 1990:87–108.
31. Bourne MC. Physical properties and structure of horticultural crops. In: Peleg M, Bagley EB, eds. *Physical Properties of Foods.* Westport, CT: AVI Publishing Co.; 1983:207–228.
32. Vidales SL, Castro MA, Alzamora SM. The structure–texture relationship of blanched and glucose-impregnated strawberries. *Food Sci Technol Int.* 1998;4:169–178.
33. Vidales SL, Nieto AB, Alzamora SM. *Personal communication*, 1999.
34. Nieto A, Castro MA, Salvatori D, Alzamora SM. Structural effects of vacuum solutes infusion in mango and apple tissues. In: Akritidis CB, Marinos-Kouris D, Saravacos GD, eds. Mujumdar AS, series ed.

Drying'98, Vol. C. Thessaloniki, Greece: ZITI Editions; 1998:2134–2141.

35. Alzamora SM, Gerschenson LN, Vidales SL, et al. Structural changes in minimal processing of fruits: Some effects of blanching and sugar impregnation. In: Fito P, Ortega-Rodríguez E, Barbosa-Cánovas GV, eds. *Food Engineering 2000.* New York: Chapman & Hall; 1997.

36. Muntada V, Gerschenson LN, Alzamora SM, et al. Solute infusion effects on texture of minimally processed kiwifruit. *J Food Sci.* 1998;63:616–620.

37. Rojas AM, Castro MA, Gerschenson LN, et al. *Firmness and Structural Characteristics of Glucose Impregnated Melon.* Proceedings of the Poster Session, ISOPOW 7, Helsinki, Finland, June 30–July 4, 1998.

Part III

Preservation Technologies

Chapter 10

Improved Drying Techniques and Microwave Food Processing

Constantino Suárez, Pascual E. Viollaz, Clara O. Rovedo, Marcela P. Tolaba, and Mónica Haros

INTRODUCTION

The term *food processing* covers a wide range of technologies and methods for preserving and/or transforming the product from the site of agricultural production to the consumer. Most food processing involves procedures that change the inherent freshlike quality attributes. For example, heat processing, often beneficial to reduce bacterial load, inactivates enzymes, softens tissues, and, applied in excess, may provoke deteriorative reactions, such as the loss of vitamins, color, texture, and appearance.

The removal of water from foods by drying is probably the oldest technology for preservation. The sun drying of fruits is a well-known process that originated in antiquity. A dried food product has the advantage of decreased weight and the consequent reduced cost for transportation. However, it is well known that current dehydration by air causes irreversible changes in the dehydrated product, particularly when high temperatures are used. Flavor is one of the most important quality factors for acceptability of a food product by the consumer. Water markedly influences the affinity of the volatile compounds for proteins, because it affects the conformation of the proteins and the form of the binding sites. Thus, aroma retention during drying is considerably affected by the moisture content of the product.

As the consumers demand finished products with little or no loss in sensory characteristics, new techniques are required that use lower temperatures and/or decreased drying times. The new techniques must conveniently combine technologies and hurdles to obtain a quality product. For instance, several novel dehydration methods have been proposed in the literature in the last decade, even though most of them are not well known by food processors. Although some of the innovative techniques have been commercialized successfully, there is still potential to improve and exploit them further. In this chapter, several traditional and novel methods of drying preservation, including microwave drying, will be discussed, mainly focusing on quality changes. Other microwave applications are also discussed.

DRYING OF FOODSTUFFS

The problem of drying hygroscopic materials is one of long-standing interest and of increasing importance.[1] Drying of solids has been studied from the beginning of this century at an increasing rate. For instance, the *Drying Technology Journal*, which started in 1980 with only two issues per year, is now editing 10 issues per year of nearly 500 pages each issue.

The authors acknowledge the financial support from the University of Buenos Aires and from the Consejo Nacional de Investigaciones Científicas y Técnicas of Argentina.

This profuse publication may be indicating that drying is a very active research topic and/or that, as yet, there are no adequate models to describe the drying process in the full range of drying conditions. Another difficulty from the experimental point of view is the strong interaction between heat and mass transfer mechanisms. Thus, the drying of solids of high moisture content under isothermal conditions can be performed only by operating with radiant energy. In this way, the coupling between heat and mass transfer coefficients, given by the Lewis number, can be eliminated. Furthermore, solids with high initial moisture content may present during drying zones in the rubbery state, coexisting with others in the glassy state and in zones with elastic characteristics. In this way, the mechanical properties of the solid must also be considered, together with the drying kinetics of the product, to assess its quality.

Achanta[2] has made a thorough analysis of the drying process and its effect over the obtained product. If the material is above the glass transition temperature, Tg, it shrinks during drying, the product behaves as a supercooled liquid, and the volume reduction is nearly equal to the evaporated water. For temperatures below Tg (low-temperature air drying), a small volume change caused by capillary forces occurs. Besides, the moisture profile formed at the beginning of drying, with low moisture content at the solid surface, may conduct to a vitreous state, i.e., below Tg. At this condition, the skeletal Brownian movements of the polymer disappear, only the microbrownian movement of the lateral chains of the polymer remains; the molecular diffusivity slows down; and the moisture of the interior of the solid cannot migrate rapidly to the surface, allowing case hardening to occur. To avoid the formation of such impervious zones at the surface of the solid and to maintain the surface at the rubbery condition, mild drying conditions must be used, i.e., low air velocity, high relative moisture content, or low temperature.

Another method to avoid the formation of a crust is to diminish the size of the sample, given that diffusion is a slow phenomenon for big solids, but it is a very fast phenomenon at a cellular scale. Because the Biot number is small for a small body, there is no moisture gradient in the solid, and the migration of moisture from the center of the body to the surface can avoid the formation of the crust.

Case hardening or crust formation is very frequent when drying solid foods. However, for products such as cereal foods and puffed corn, where crispiness and low velocity of rehydration are required, such a phenomenon is desired. A rapid formation of a crust is also necessary for microencapsulated flavors to avoid the loss of flavors after processing. In other cases, such as dried vegetables, case hardening is not desirable because it does not allow a rapid rehydration of the product.

In drying of foods with a high moisture content (50% and higher), the entire solid is in a rubbery condition, i.e., above the Tg temperature. In this case, the contraction of the solid is equal to the volume of evaporated water. Because the loss of moisture is more important at the surface, the crust formed at the surface would be in a vitreous state and, therefore, under a tensile condition, whereas the nucleus of the solid would be compressed. The mathematical approach of this phenomenon is complicated, even though some investigations have been performed in the area of wood and gel drying.[3]

Some Considerations on Drying Models

Among various drying models found in the literature, many of them do not take into account the energy balance. These models are definitely inappropriate for predicting the drying kinetics of food materials with high moisture content. Also, because the evolution of temperature as a function of time cannot be simulated from these models, they are not appropriate for predicting deteriorative reactions.

Sun and Meunier[4] found the following criteria for a simplified analysis of drying. When the Lewis number (Le, ratio of thermal diffusivity to mass diffusivity) is higher than 10, the heat transfer inside the solid is fast, and a uniform temperature profile within the solid can be assumed. The value of the temperature changes as a function of time, but the temperature gradient in the solid is negligible. In this case, it is necessary to know the diffusion coefficient and the film heat transfer coefficient to simulate the process. On the other hand, if the product Le Bi > 100, an isothermal condition can be assumed for the entire drying process, i.e., there is no temperature gradient in the solid, and its temperature is constant during the process (the energy balance is not necessary). In this case, it is necessary to know only the value of the diffusion coefficient.

These criteria allow the simplification of the treatment of the drying kinetics, because when uniform temperature is assumed across the solid, the energy balance reduces to a simple ordinary differential equation. Also, for predicting the drying kinetics of food in the rubbery state (moisture content higher than 50%), it is absolutely necessary to take into account the shrinkage phenomenon to obtain diffusion coefficients with physical significance.

Predrying Treatments To Hasten Drying Rate

Suárez and Viollaz[5] found that it is possible to improve the quality of dehydrated foods by means of a pressurized gas freezing pretreatment before air drying. In short, the method consists of the pressurization (20–60 bar) with nitrogen at room temperature of solid foods, such as fruit and vegetables, for no more than 2 hours. This is followed by total freezing of the sample at 253°K. After that, the pressure is released and the sample let to thaw at room temperature before drying. The procedure was used satisfactorily for the drying of carrots, apples, onions, and potatoes. Such pretreatment not only increases the drying rate about 100 times, in comparison with the untreated samples, but also reduces shrinkage, improving the rehydration rate and color of dehydrated products.

An interesting chemical pretreatment that reduces the drying rate of some waxy fruits and vegetables consists of the use of long-chain fatty acids. Radler[6] and, recently, Ponting and McBean[7] found that when some waxy fruits, such as cherries and prunes, are dipped for a few minutes in a cold aqueous emulsion of ethyl oleate, the drying rate is considerably increased. A similar treatment was successfully used by Wieghart et al.[8] to hasten the drying rate of cut alfalfa. In that work, the authors use a mixture of methyl esters of long-chain fatty acids and a surfactant (isopropanol) to spray the alfalfa prior to the drying process.

Also, in Australia, predrying treatments are used to hasten drying of grapes. Grapes are dipped in or sprayed with an alkaline oil-water emulsion, known to growers as *cold dip*, which is a mixture of ethyl esters of fatty acids and free oleic acid.[9] The mixture is emulsified in a solution of potassium carbonate in water. By this procedure, drying times are reduced from about 4–5 weeks for untreated grapes to 8–14 days for treated ones, when dried on racks. This acceleration of drying leads to a rapid rise in sugar concentration, which inhibits the action of the enzyme polyphenol oxidase, responsible for darkening in untreated fruits.

Suarez[10] investigated the use of an emulsion dipping procedure to increase the drying rate of sweet corn and grain corn with lower moisture content (field corn). In both cases, the author found a beneficial effect of dipping procedure on the drying rate of corn; however, this beneficial effect was less marked as the harvested corn moisture was reduced.

The importance of cuticular resistance on drying rate of waxy materials is not well known. Harris et al.[11] reported that the waxy cuticle is the main barrier to water loss during drying. Formerly, it was suggested that the emulsion treatment provoked a disruption of the cuticle wax platelets, removing the waxy bloom. However, it has been shown that almost no wax is removed and that the effect is reversible by washing. The mode of action seems to be a modification of the structure of the outer wax layer, with the consequent increase of water permeability. It has been observed that the skin of the emulsion-treated berries appears to be more transparent to infrared rays, which would allow a better radiant heat energy uptake.[12]

Effect of Drying on Nutrient Loss

An increase of the drying temperature increases the drying velocity, but also increases the rate of undesirable reactions (vitamin loss, enzymatic and nonenzymatic browning, oxidation reactions) and unfavorable texture changes. Nutrient loss is a complex function of temperature, moisture content, and process duration.

Enzyme-catalyzed reactions generally occur at low temperatures because, at high temperatures, enzymes become denaturalized and are not functional as a catalyst. These enzymes can be present in the product, produced by microorganisms, or purposely added to the product. In some processes, such as the posttreatment of tea, coffee, cocoa, and grapes, certain enzymes, such as phenoloxidase, must be functional to impart flavors to the product.[13] However, the extent of those reactions must be controlled to obtain an acceptable product.

Living organisms and foods are stable because enzymes and substrates are separated. When they are put in contact because of mechanical damage, the enzymatic reaction can occur at a high rate. For example, the action of the lipase oxidase is very fast in rice after milling. Hard drying conditions can produce the rupture of the cells, putting in contact enzymes with substrates, promoting in this way deteriorative reactions.

The activation energy for drying is generally less than 10 Kcal/mole and, consequently, is lower than the value of 14 Kcal/mole reported by Rojas[14] for ascorbic acid destruction under aerobic conditions. The activation energies of enzyme-catalyzed reactions are within the 4–20 Kcal/mole range (mostly about 11 Kcal/mole). Deactivation energies vary between 40 and 120 Kcal/mole (mostly about 70 Kcal/mole). Thus, an increase of the drying temperature initially produces an increase of vitamin loss and a higher rate of the enzyme-catalyzed reactions. A further increase of temperature will produce an increase in the rate of the enzyme destruction and, therefore, a decrease of the reactions catalyzed by enzymes.

Freeze Drying

Freeze drying is a sublimation process, i.e., water in a solid state goes directly to a vapor state. Spieles et al.[15] recognize four stages in freeze drying. Briefly, these are the following:

a. Preparation. During this stage, a phase separation process occurs in which a crystalline phase of pure ice and an amorphous phase, consisting of biologic materials, stabilizers, and a small amount of noncrystalline water, is formed.
b. Primary drying. During this stage, all of the crystalline water sublimates.
c. Secondary drying. This stage consists of a partial removal of water of the amorphous phase. The time for this secondary drying can be a substantial share of the total drying time.
d. Rehydration.

If the sample is not homogeneous, stages may overlap. Some regions of the solid could be in a primary drying process, whereas others could be evolved to a secondary drying process. In fact, if the rate of cooling is very fast, all of the water is in an amorphous state, and there are not pure ice crystals present in the solid. Samples with low water content generally give a product with high mechanical stability.

Freeze drying produces a product of superior quality, in comparison with other drying processes. An important fact is that there is no liquid present in the sample, so the capillary forces, which are mainly responsible for the shrinking process, are absent; the porosity of the material is very high and its rehydration very fast. Additional advantages are the low temperature of the freeze-drying process and the low amount of liquid water in the sample, preventing the loss of flavors and limiting the extent of deteriorative reactions (enzymatic and nonenzymatic).

On the other hand, freeze drying is an expensive process because drying kinetics at low temperature is slow, and vacuum pump and equipment for freezing the material are required. Most of freeze dryers operate at temperatures of –10ºC or lower.

The heat of sublimation necessary for freeze drying can be supplied by conduction through the iced material, by radiation, or by convection through the dried layer. As a consequence, the heat-transfer resistance increases during the drying process, not only because the thickness of the dried layer increases, but also because the thermal conductivity of the dry material is low, due to the low gas pressure in the pores. The use of microwaves seems to be an interesting alternative to supply heat for sublimation because a temperature gradient is not needed for heat flux. However, owing to the low pressure involved in freeze drying, there is a great tendency for ionization of the gases, causing plasma discharge that can burn the product.[16] According to Liapis,[16] the problem of discharge can be minimized when working at pressures below 50 microns. The operation at high vacuum pressures is, however, quite expensive, owing to the need for a condenser at very low temperatures, aside from the low drying rate that must be expected when operating under such conditions. Microwave freeze drying is a process very difficult to control. Because the loss factor of water is higher than the dielectric factor of ice, any localized melting may produce a rapid chain reaction that results in runaway overheating. There are, however, some indications that these difficulties can be solved and that the market will provide units for microwave freeze drying similar to those already manufactured for coffee.[17]

Finally, it can be concluded that, even given the evolution of freeze drying in last 20 years, this process has been restricted to a very selective group of food products (coffee, mushrooms, diced chicken) whose relatively high cost of production is balanced by an increase in quality. Its potential evolution is still considerable and will depend on successive progress in basic research and creativity level in the design and operation of plants and instruments.[16]

Fluidized-Bed Drying

This process, initially used on large scale by petroleum companies, became very popular over the past three decades in the chemical industry, owing to its versatility and adaptability to various unit operations. The technique, very simple in its principle, consists of the mobilization of solid particles in an upward-flowing gas stream, producing an intimate contact between the solids and the hot gas carrier. A typical commercial fluidized-bed dryer consists of a cylindrical column, with the hot gas entering from the bottom of the preloaded bed and exiting at the top. This basic structure has been successfully modified during the last two decades to increase the applicability of the process.

In food processing, it is generally recognized that fluidized-bed drying requires relatively small, uniform, and discrete particles that can be readily fluidized. Thus, the process has been successfully applied to process small vegetable pieces, whereas powders would be inappropriate because they clog up the cyclone.

The variants introduced in the fluidized conventional process are becoming applied in industrial food drying, even though some users are unaware of some of these innovative techniques. In that way, fluidized-bed dryers are now used to dehydrate pastes, slurries, suspensions, continuous webs, sheet-form materials, and mixtures of large and small particles. Mujumdar[18] reviewed the diversity of operating conditions, applications, and possible designs of fluidized-bed dryers.

An interesting modification is the pulsofluidized bed with relocated gas stream.[19] In this new drying process, hot air is pumped by a high-pressure blower through a rotary valve-distributor. The distributor interrupts the air flow periodically and directs it to the respective sections in a gas chamber located below a supporting grid of a standard fluidized-bed dryer. The air fluidizes only that segment of the bed is located above the acting section. In this way, the entire bed is fairly fluidized, and a reduction of the minimum fluidization velocity of 12% is attained. This type of fluidized bed permits the fluidization of shallow beds (about 0.1 m) and nonisometric particles such as carrot, onion, potato, and celery dried into 1.5- to 3.5-mm discs.[20]

Drying of vegetables takes place largely during the falling rate period, where heat and moisture transfer occurs in the interior of the product, becoming limiting factors and, therefore, increasing the drying time. Microwaves can accelerate the drying process by generating positive vapor pressures that push moisture out. The combination of both processes was performed at the Department of Biological System Engineering at Washington State University. A variation of fluidized bed, known as *spouted bed*, was used. In this kind of bed, only the central cone of the bed is fluidized. This provokes a vigorous fluidization of the particles in the cone that are forced upward, falling downward in the periphery of the bed. The residence time of the particles in the cone is very short, allowing the use of drying air at elevated temperatures and increasing the rate of heat transfer from the air to the particles. Feng and Tang[21] used the combination of microwaves and spouted bed to dehydrate apple dices, with satisfactory results.

The new procedure not only reduced the drying time, compared with the classic air drying, but also improved texture and color. Using the same procedure to dry blueberries from 85% to 10% moisture content, Feng and Tang[21] found that the drying time is reduced 1/6 and 1/20 with respect to only spouted bed or microwave drying, respectively.

Superheated Steam Drying

Energy costs in drying represent a significant part of the total manufacturing cost. For example, in milk drying, about 25% of manufacturing costs are energy costs.[22] New technologies that can reduce this energy input are of great interest. The "airless" drying, i.e., drying with superheated steam instead of air, has the potential of making a large reduction in dryer energy input.[23] This idea is at least a century old, even though its commercial use appeared on the scene only in the last two decades.

Superheated steam drying uses a recycle-and-purge mode, whereby a quantity of exhaust of superheated steam equivalent to the water evaporated from the product is purged, and the rest is recycled. Energy-saving potential is due to eliminating the need for heating from ambient temperature and to the ability to recover energy from the purged superheated steam. Energy can be obtained from purged superheated steam by heat transfer to a process fluid or by compressing a portion of purged steam to a pressure such that its latent heat of evaporation can be transferred to the recycled steam, increasing its temperature to drier inlet superheated steam temperature.

In comparison with the classic air-heated dryers, drying rates using superheated steam are 2–3 times higher. This is the result of the higher values of the heat and mass-transfer coefficients than when heated air is used. There are several other advantages. The higher heat capacity of steam, as compared with air, reduces the mass flow rate required to deliver a given amount of thermal energy. High heat transfer coefficients imply small heating surfaces and low investment costs. Another advantage of steam drying is the quality of the dried product. There is a potential reduction for oxidative damage of products, such as discoloration and browning reactions, improving the quality of the product. As water evaporates inside the product, the internal pressure tends to puff the product, providing a porous product of low density and better rehydration characteristics than products dried in hot air.

For heat-sensitive products, lower pressure operation appears to have an interesting potential. At least one Japanese company uses steam for drying and deodorizing soy products, sterilizing the material at the same time. Other applications have been successful at pilot plant scale levels for spray drying of whey products.

Beet pulp is dried in Europe with superheated steam, with better results than when dried in hot air.[18] Whereas air-dried beet pulp is brown and used for cattle feed, the new technique produces a bright product that can be used for human consumption.

Steam drying is successfully used to dewater and dry citrus by-products.[24] The by-products of citrus fruit processing are valuable as cattle feed. The new method delivers higher yields than does conventional drying and recovers more citrus oil.

Mujumdar[25] has reviewed published results on industrial and/or pilot scale steam drying of diverse products such as bark, paper, wood particles, beet pulp, etc. It must be pointed out that any direct (convection) dryer could be adapted to superheated steam operation. The thermal energy consumption in direct dryers could be reduced by supplying a part of the thermal energy requirements indirectly, such as, for example, incorporating heated panels or tubes immersed within a bed of particulate solids. Because the 300- to 500-kJ/kg water evaporated,[18] steam drying should be examined carefully in the near future. It is attractive from energy and environmental standpoints, and results in a better product than does air drying. For all of these reasons, it might be expected that steam drying can be the drying technology of the next century.

MICROWAVE FOOD PROCESSING

The benefit of microwave technology has been realized over the past decades with the growing acceptance of microwave ovens in homes. Since the 1950s, microwave heating has been gaining acceptance by the food industry to cover various types of processes. Some well-defined advantages over other food processing are recognized. Among them, speed is an important factor: The heating time is one-quarter of the time used in conventional heating. Uniformity of heating is generally achieved, due to internally generated heat, and overheating of the surface can be avoided. This fact can improve the quality of the product by avoiding case hardening.

Although cost represents a major barrier to wider use of microwaves in the industry, an equally important barrier is still the lack of understanding of how microwaves interact with materials. Effects such as resonance within the material, as well as large variations in field patterns at the material surface, can occur and affect the efficiency of microwave processing.[26,27] Even though the most widely accepted microwave operations are the precooking of meat products and the tempering or thawing of frozen foods, other processing operations are gaining acceptance in food industry.

Decareau[28] has pointed out six major classes of possible microwave processing: tempering, dehydration, blanching, cooking, pasteurization, and sterilization. Other potential uses of microwave heating were listed by Gerling,[29] based on patents such as baking, coagulation, coating, gelatinization, puffing, and roasting.

It can be said, in general, that microwave use offers an interesting potential, given its rapid and uniform heating of food products, reduced thermal degradation of essential nutrients, increasing retention of food quality factors in comparison with conventional thermal processes, and usefulness for defrosting food.

Another important microwave use is saving energy. Heating with microwaves is more efficient than conventional heating because heat is generated in the food and not in the air, container, or oven (40% vs. 10% for conventional ovens), according to Trub.[30] Among the disadvantages that have been described are the inability to brown food, nonuniform cooking of foods of irregular shape, and excessive drying in some foods, such as breads.[31] On the other hand, the work of Rosen[32] seems to be quite conclusive about any possibility that microwave energy can induce chemical reactions. To estimate which types of radiation could cause chemical changes in a material, respective quantum energies[32] were compared with the energies of chemical bonds. Rosen concluded that the quantum energies of microwaves fall short by several orders of magnitude of breaking chemical bonds and, consequently, the statements about observed chemical effects of microwaves must be viewed with skepticism. Before a review of some of the applications of microwaves to food processing, a brief discussion of the fundamentals of microwave heating will be presented.

Heat Generation of Microwaves

It is important to point out that microwaves are a form of energy and not a form of heat. The microwaves, which are waves of very short wave length (centimeter order), are manifested as heat only upon interaction with a material. In biologic materials, the polar molecules, such as water, are primarily responsible for such interaction. Because the microwave field reverses its polarity 2.45×10^9 times per second, the polar molecules attempt to line up with the field, and friction will occur between adjacent molecules. In doing so, considerable kinetic energy is extracted from the microwave field, and heating takes place.

Ionic conduction is another microwave heating mechanism. When ionic solutions are in a microwave field, the ions are obliged to flow first in one direction, then in the opposite one with the rapidly alternating field.

The fundamental equation for microwave power absorption by a substance is expressed as:

$$P = 2\pi\, E_0\, E''\, f\, E^2,$$

where P = the power developed in a volume of material; E_0 = the dielectric constant; f = the frequency of the energy source; E'' = the dielectric loss factor of the substance, and E = the electric field gradient (volts per centimeter). Using $E_0 = 8.85 \times 10^{-1}$ farads/m, it results:

$$P(W/m^3) = 55.61 \times 10^{-20}\, f\, E''\, E^2$$

The loss factor or dielectric loss, E'', and the dielectric constant, E', are related by the loss tangent:

$$\tan \alpha = E'' / E',$$

which is a measure of the material's ability to be penetrated by an electric field and to dissipate electrical energy into heat.[33] From these considerations, it can be said that a "loose" material is one that heats well, whereas a "low-loss" material is one that heats poorly and, consequently, is more transparent to microwave radiation.

The dielectric properties of biologic materials at microwave frequencies are mainly determined by two factors: salt content (ionic concentrations) and moisture.[34] Water is the major adsorbent of microwaves in moist materials and, consequently, the higher the moisture content, the better will be the heating. In contrast, the organic components of a given material are dielectrically inert ($E' < 3$ and $E'' < 0.1$) and, compared to aqueous ionic fluids or water, may be considered transparent to microwaves.[33] At very low moisture contents, the traces of bounded water are not affected by the rapid changing of the microwave field, and the components of low specific heat become the major factors in heating. Mudgett[33] has found that, in high-carbohydrate foods and syrups, the dissolved sugars are the main microwave susceptors.

Given that the dielectric properties provide an intangible link between the electromagnetic waves and the material being processed, they are very important in microwave processing. For many food materials, the dielectric properties have been reported in the literature.[33,35]

Industrial Application of Microwaves To Process Fruit and Vegetables

Dehydration

The first large industrial use of microwaves was in the finish drying of potato chips.[36] Because Maillard reaction takes place during frying of potato chips, these are conditioned for several weeks to convert glucose to starch in order to avoid an overbrowning of the product. The use of microwave processing reduces browning, and chips can be fried to the proper color. With such modification, water is partially removed and not replaced with oil, with the result that the final moisture content is higher than in conventional chips. This extra water may act as a barrier to oxygen and protect oil from oxidative rancidity, increasing the shelf life of the chips.[37] Although successful in the 1960s, the process did not survive, not because of equipment shortcomings, but rather for industry-related reasons.[38] Other products dried under atmospheric pressure using microwaves include rough rice and tomato paste.[33]

Decreased time and temperature for vacuum microwave drying of orange juice concentrate into powder results in increased retention of ascorbic acid and volatile compounds, compared with other drying processes.[36] The cost per kilogram is claimed to be less than that for either spray-dried or freeze-dried products.

A combination of the classic drying procedure and microwave was utilized by Tulasidas et al.[39] They used pulsed mode of microwave power at a constant air velocity to investigate the drying rate of grapes. The product was dried faster than with conventional air drying, and good quality of the dehydrated product was obtained. Kaensup et al.[40] found that a combined microwave with fluidized bed dryer appears to exert considerable effect on the drying rates of pepper seeds, particularly at low inlet air temperature. Significantly improved drying rates were obtained by the combined procedure, compared with a conventional fluidized bed.

Orange juice concentrate can be transformed into powder by vacuum microwave drying with lower time and temperature than other drying processes, with increasing retention of ascorbic acid and volatile compounds.[28] According to Decareau,[28] the cost per kilogram is less than that of spray drying or freeze drying.

Little information exists in the literature concerning the effect of microwaves on starches. The physicochemical properties of potato and tapioca starches subjected to microwave radiation were recently investigated by Lewandowicz et al.[41] Both starches showed a rise in temperature gelatinization and a change in their pasting properties, from that typical of tuber to that more characteristic of cereal starches. Other changes observed were a drop in the solubility of starch granules and a change in the X-ray diffraction patterns from type B to type A, also characteristic of cereals.

The influence of microwave power, inlet air temperature, and inlet air velocity on the drying properties of corn was investigated by Shivhare et al.[42] The authors concluded that, at higher levels of microwave, power drying time is reduced considerably, whereas the inlet air temperature has no effect on drying behavior. In particular, the authors recommended the use of microwave for corn drying when used as seed. Greater than 92% germination was obtained when absorbed power at 0.25 W/g was used for corn drying. Also, Decareau[28] has pointed out the advantages of microwave drying, compared with hot-air drying of wheat and corn, to be increased germination capability of the grain, uniform drying, decreased fuel usage, and drying speed (an important factor at harvest).

To conclude, it can be said that microwave drying is more adequate as finish drying because initial drying, particularly for products of very high moisture content, is more economic and efficient when done conventionally. As generally occurs, drying is limited by heat transfer to the core of the product and mass transfer of water out of the material. It would be expected that microwave drying will occur at more uniform conditions than air drying, given that energy is mainly absorbed by those areas of the material with highest moisture content, exactly where the most drying is needed, thus reducing the possibility of case hardening. With reference to flavor, very little conclusive evidence exists that there are real differences in conventionally versus microwave-heated vegetables. Considering the patterns and times of heating in microwave and conventional drying, flavor differences are conceivable.

Blanching

This process is used in the industry to inactivate enzymes in fruits and vegetables. Recent studies have demonstrated that enzyme inactivation by microwave heating may offer the advantage that heat-sensitive nutrients and flavor compounds can be preserved. Combining microwave radiation and blanching, it is possible to inactivate pectin-methylesterase in orange juice concentrate in only 4–5 minutes.[28]

Collins and McCarty[43] used microwave energy and boiling water methods to inactivate peroxidase in white potatoes. The inactivation pattern was similar for both methods; however, microwaves caused denaturation at a faster rate. The authors found that at least 13 minutes of heating in boiling water were required to inactivate enzymes in the core (2.27 cm), whereas, with microwave energy, this was accomplished in 4.7 minutes.

Low-temperature evaporation of orange juice in the frozen concentrate industry does not inactivate pectin-methylesterase, which catalyzes the hydrolytic removal of methoxyl groups from pectin. It was observed that the activity of this enzyme correlates well with product deterioration. Copson[44] designed an apparatus to inactivate the enzyme in the orange juice concentrate by microwaves, using a counter flow system for exposing the concentrate to radiation from a helical director. The procedure not only provided a means for inactivation of enzymes, but also facilitated the recovery of heat, reducing the volume of material to be irradiated.

Microwave blanching was shown to have little effect on the ascorbic acid content of a large variety of vegetables, as compared with the traditional steam or boiling water blanching (Figure 10–1). The blanching studies were performed at 3,000 MHz and a power level of 2 kW. Proctor and Goldblith[45] suggested that microwave blanching might have practical advantages in food processing if continuous microwave equipment became a reality.

Brussels sprouts were blanched by conventional and microwave methods, and a combination of microwaves and steam.[46] Findings were that the combined method was very effective to inactivate the peroxidase enzyme and that such a procedure was effective in retention of chlorophyll and ascorbic acid, as well as in stabilization of flavor. The results obtained by Eheart[47] also indicate that microwave energy minimizes

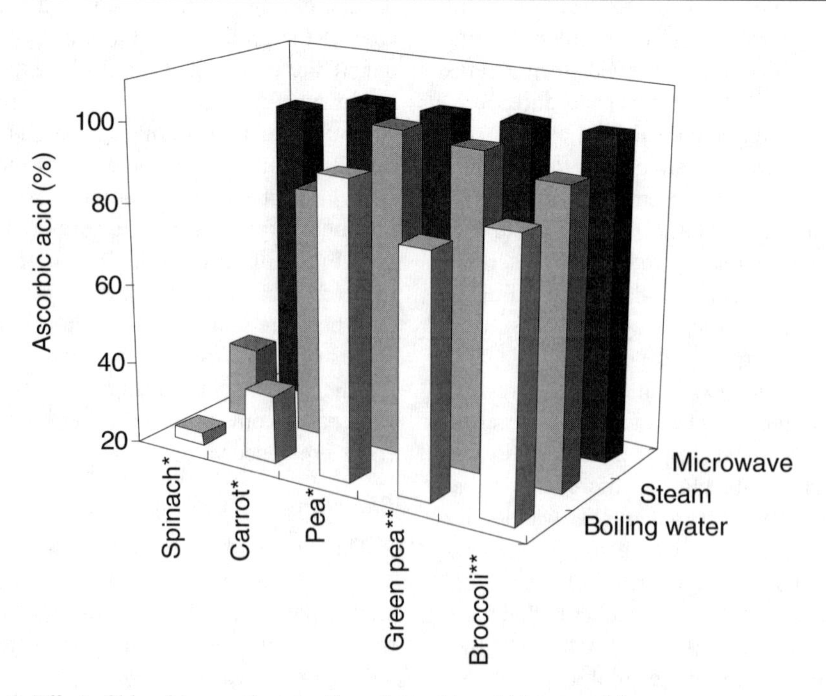

Figure 10–1 Effect of blanching on the retention of ascorbic acid in vegetables.

losses in chlorophyll during home and commercial blanching processes.

There is, however, little indication that enzyme inactivation is a direct result of the radiation treatment. The combination of time and tempering when a sample is exposed to microwave radiation is responsible for the loss of enzyme activity in foods. When using microwave heating, the inactivation takes place in a shorter period of time, compared with conventional blanching, and without detrimental effect on most foods. However, given the great variety of fruits and vegetables, different treatment times will be required that will depend on maturity, size, texture, shape, and type of enzyme to inactivate. Microwave energy produced comparable softening in about one-third the time required in boiling water. Collins and McCarty[43] demonstrated that microwave energy inactivates potato peroxidase in 4.7 minutes, whereas the time required in boiling water is about 13 minutes. Similarly, polyphenol oxidase was inactivated more rapidly with microwaves than in hot water.

Chen et al.[48] have reported that microwave heating, combined with boiling water treatment, reduced blanching time of potato, compared with hot water alone, also improving the texture of the product; the use of hot water cooked the outer tissues before sufficient heat had reached the core of potato.

Cooking and Pasteurization

Microwaves have several applications that facilitate cooking procedures. However, the information available in the literature concerning fruits and vegetables is relatively poor and, in some cases, not very recent. Comparison among different cooking methods was carried out by Gordon and Noble,[49] who studied three cooking methods for some vegetables: boiling, pressure saucepan, and waterless microwaves. The results are reported in Figure 10–2. It can be observed that, in all cases, microwave cooking increases ascorbic acid retention. Eheart and Gott,[50] however, did not arrive at the same conclusion. Ascorbic acid retention in peas, spinach, broccoli, and potatoes cooked with and without water in the microwave oven and by conventional methods showed no significant differences. The amount of water used was evidently not sufficient to cause a significant leaching. Thus, waterless cooking did not improve ascorbic acid retention in that study. In general, shorter heating times in microwave ovens serve to retain heat-labile nutrients.[51]

On the other hand, Lanier and Sinstrunk[52] did not find differences in the content of niacin, pantotenic acid, and total carotenoids when sweet potatoes were cooked by boiling, steaming, canning, or microwave. It must be pointed out that the authors did not report details of microwave power or water amount used during cooking. Eheart and Gott[50] used selected peas for determining carotene retention after cooking them in water with a conventional oven and in a microwave oven, with and without added water. In the three cases, carotene content was completely retained. Complete retention or even increases of carotene content during microwave cooking have also been reported by Bender.[53]

Retention of chlorophylls in green vegetables has an important bearing on their acceptability. Panel scores for color and percentage of unchanged chlorophyll in cooked broccoli correlated significantly. Greater losses in chlorophylls have been reported in frozen compared with freshly cooked vegetables.[54] However, how the heat generated by microwaves affects chlorophylls retention is not available.

It is clear that there is as yet no standardized cooking method for vegetables. However, market indicators support the conclusion that high-quality vegetables result by microwave heating.[55] This cooking procedure may provide quality vegetables with maximum nutritional value. As has already been pointed out, microwave cooking achieves a more rapid enzyme inactivation than do conventional methods. Because the flavors of many foods are produced enzymatically, considerable differences between microwave and conventionally cooked vegetables must be expected. Unfortunately, little is known about the effect of microwaves on flavor of different

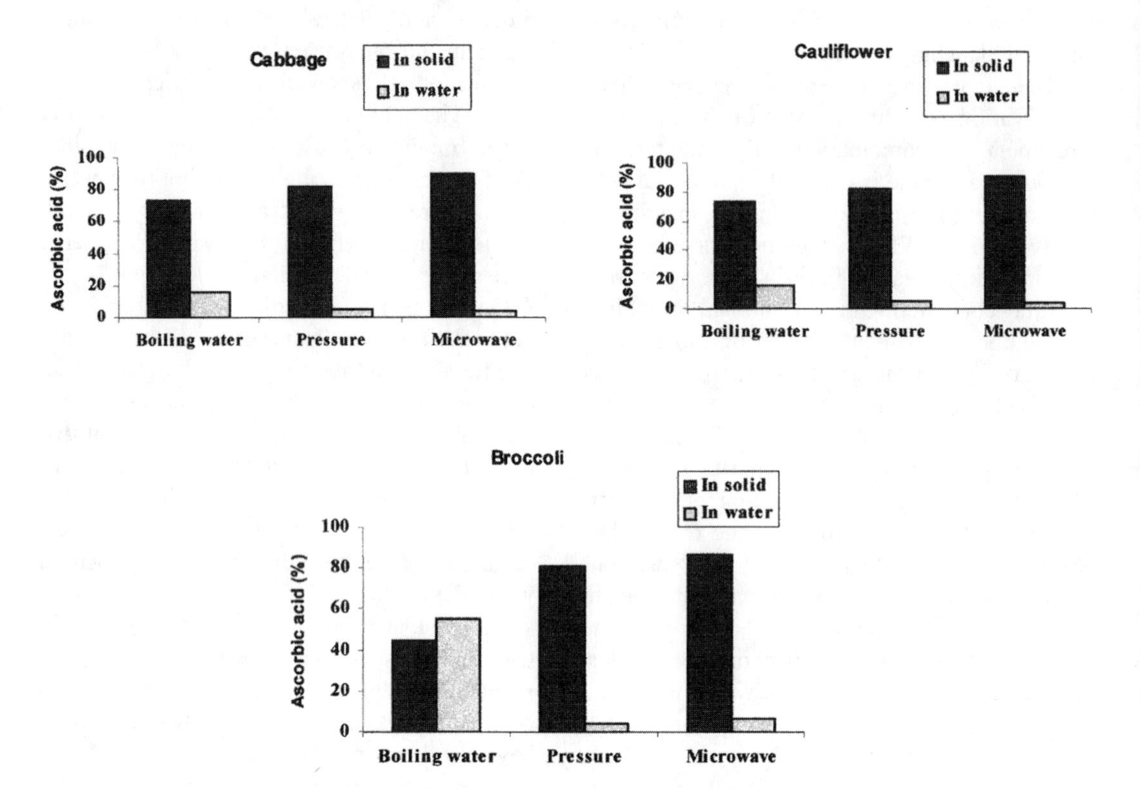

Figure 10–2 Comparison of traditional and microwave (no water added) cooking methods of vegetables on ascorbic acid retention.

food products. The increasing use of this technology will surely encourage research in this area of knowledge.

Pasteurization by radio frequency was first investigated in the 1940s to destroy mold inoculated into slices of bread.[56] Slices inoculated with *Aspergillus niger* were microwave-processed at 2,450 MHz to destroy mold spores.[57] The samples that received microwave treatment did not show mold growth, and the shelf life of the product was extended, without need of chemical preservatives. In Sweden, the first large-scale use of microwave energy for pasteurization of packaged sliced breads has been built.[58]

Microwaves were also used for pasteurization of peeled, vacuum-packed potatoes for 6 minutes; after that, the product was quickly cooled and stored at 8°C for 6 weeks.[59] Pasteurization of corn-soy milk blend for infants to destroy *Salmonella* can yield nondetectable levels of the microorganism with 6.7–7.0 minutes of microwave treatment to a temperature of 61–67°C.[60]

REFERENCES

1. Kolhapure NH, Venkatesth KV. An unsaturated flow of moisture in porous hygroscopic media at low moisture contents. *Chem Engr Sci.* 1997;52:3383–3392.
2. Achanta J. *Moisture Transport in Shrinking Gels during Drying.* Purdue University; Ph.D. Thesis; 1995.
3. Majumdar P, Marchetas A. Heat, moisture transport,

and induced stresses in porous materials under rapid heating. *Num Heat Trans.* Part A, 1997;32:11–130.

4. Sun LM, Meunier F. A detailed model for nonisothermal sorption in porous adsorbents. *Chem Engr Sci.* 1987;42:1585–1593.
5. Suárez C, Viollaz PE. Effect of pressurized gas freezing pre-treatment of carrot dehydration in air flow. *J Food Technol.*1982;17:607–613.
6. Radler F. The prevention of browning during drying by the cold dipping treatment of Sultana grapes. *J Sci Food Agric.* 1964;20:1149–1153.
7. Ponting JD, McBean DM. Temperature and dipping treatment effects on drying rates and drying times of grapes, prunes and other waxy fruits. *Food Technol.* 1970;24(3):85–89.
8. Wieghart M, Thomas JW, Tesar MB. Hastening drying rate of cut alfalfa with chemical treatment. *J Anim Sci.* 1980;51:1–9.
9. Grncaveric M, Radler F. A review of the surface lipids of grapes and their importance in the drying process. *Am J Enol Vitic.* 1971;22:80–86.
10. Suárez C. Effectiveness of ethyl oleate dipping, steam blanching and other pretreatments on drying of corn kernels. *Lebensm-Wiss u-Technol.* 1987;20:123–127.
11. Harris CE. The effect of organic phosphates on the drying rate of grass leaves and dry matter losses during drying. *J Agric Sci.* 1978;83:353–358.
12. Scholefield PB, May P, Neale TF. Harvest pruning and trellising of sultaner vines. *Sci Hort.* 1977;7:115–122.
13. Ashie INA, Simpson BK, Smith JP. Mechanisms for controlling enzymatic reactions in foods. *Crit Rev Food Sci Nutr.* 1996;36:1–30.
14. Rojas A.M. *Destrucción de Vitamina C en Alimentos Procesados y Almacenados.* Universidad de Buenos Aires; 1995. Doctoral Thesis.
15. Spieles G, Marx T, Heschel Y, Rau G. Analysis of desorption and diffusion during secondary drying in vacuum freeze-drying of hydroexyethyl starch. *Chem Engr Process.* 1995;34:351 357.
16. Liapis AI. Freeze drying. In: Mujumdar A, ed. *Handbook of Industrial Drying.* New York: Marcel Dekker; 1987;295–326.
17. Schiffman PJ. Food products develop want for microwave processing. *Food Technol.* 1987;40(6):94–98.
18. Mujumdar AS. Drying technologies of the future. *Drying Technol.* 1991;9:325–347.
19. Kudra T. Novel drying technologies for particles, slurries and pastes. In: Mujumdar AS, ed. *Drying'92.* Amsterdam: Elsevier Science; 1992:224–239.
20. Glaser R. On possibility of drying vegetable slices in a pulsofluidized bed. *Proc VII Drying Symposium*, Lodz, Poland; 1991:147–154.
21. Feng H, Tang J. Microwave finish drying of diced apples in spouted bed. *J Food Sci.* 1998;4:679–683.
22. Robertson LJ, Baldwin AJ. Process integration study of a milk powder plant. *J Dairy Res.* 1993;60:327–338.
23. Stubbing TJ. Airless drying: its invention, method and applications. *Chem Engr Res Des.* 1993;70:488–495.
24. Covington RO. Steam dryer aimed at byproducts. *Food Engr.* 1983;102–103.
25. Mujumdar AS. *CEA Report.* No. 817 U671. 1990. Contact: Exergex Corp., 3795 Navarre, Brossard, Quebec, Canada, J4Y 2H4.
26. Kritikos HN, Foster KR, Schwam HP. Temperature profiles in spheres due to electromagnetic heating. *J Microw Pow.* 1981;16:327–344.
27. Metaxas RC, Meredith RJ. *Industrial Microwave Heating.* London: Peter Peregrinus; 1983.
28. Decareau RV. *Microwaves in the Food Processing Industry.* Orlando, FL: Academic Press; 1985.
29. Gerling JE. Microwaves in the food industry: Promise and reality. *Food Technol.* 1986;82–83.
30. Trub L. 1979. Microwave ovens: Who's using them and why. *Natl Food Rev.* Winter, 1979;24–25.
31. Murray BX. Microwave oven usage and ownership investigated in consumer study. *Food Prod Dev.* 1977;11:72–76.
32. Rosen CG. Effects of microwaves on food and related materials. *Food Technol.* 1972;26(4):36–41.
33. Mudgett RE. Developments in microwave food processing. In: Schwartzberg HE, Rao HA, eds. *Biotechnology and Food Process Engineering*, Basic Symp. Series. New York: Marcel Dekker; 1990:359–403.
34. Mudgett RE. Microwave properties and heating characteristics of foods. *Food Technol.* 1986;30(5):84–98.
35. Nelson SO, Kraszewski AW. Dielectric properties of materials and measurement techniques. *Drying Technol.* 1990;8:1123–1142.
36. Decareau RV. *Microwave Foods: New Product Development.* Trumbull: Food and Nutrition Press; 1992.
37. Quast DG, Karel M. Effects of environmental factors on the oxidation of potato chips. *J Food Sci.* 1972;37: 584–588.
38. O'Meara JP. Why did they fail? A backward look at microwave applications in the food industry. *J Microw Pow.* 1973;8:167–171.
39. Tulasidas TN, Raghavan GSV, Kudra T, Gariepy Y, et al. Microwave drying of grapes in a single mode resonant cavity with pulsed power. *ASAE Meeting Presentation.* No. 94–6547. St. Joseph, MI: 1994.
40. Kaensup W, Wongwises S, Chutina S. Drying of pepper seeds using a combined microwave/fluidized bed dryer. *Drying Technol.* 1998;16:853–862.

41. Lewandowicz G, Fornal J, Walkwoski A. Effect of microwave radiation on physico-chemical properties and structures of potato and tapioca starches. *Carbohydr Polym.* 1997;34:213–230.
42. Shivhare US, Raghavan GS, Bosisio RG. Microwave drying of corn. II. Constant power, continuous operation. *Trans ASAE.* 1992;35:951–957.
43. Collins JL, McCarty IE. Comparison of microwave energy with boiling water for blanching whole potatoes. *Food Technol.* 1969;23(6):1337–1342.
44. Copson DA. Microwave irradiation of orange juice concentrate for enzyme inactivation. *Food Technol.* 1954;8(2):397–399.
45. Proctor BE, Goldblith SA. Radar energy for rapid cooking and blanching and its effect on vitamin content. *Food Technol.* 1948;2(3):95–104.
46. Dietrick WC, Huxell CC, Guadagni DG. Comparison of microwave, conventional and combination blanching of Bruselles sprouts for frozen storage. *Food Technol.* 1970;24(5):105–109.
47. Eheart MS. Effect of microwave—vs. water—blanching on nutrients in broccoli. *J Am Diet Assoc.* 1967;50: 207–211.
48. Chen SC, Collins JL, McCarty IE, Johnston MR. Blanching of white potatoes by microwaves followed by water. *J Food Sci.* 1971;37:742–745.
49. Gordon J, Noble I. Comparison of electronic and conventional cooking of vegetables. *J Am Diet Assoc.* 1959;35(3).
50. Eheart MS, Gott C. Conventional and microwave cooking of vegetables. *J Am Diet Assoc.* 1964;44:116–119.
51. Ang CY, Chang CM, Fray AE, Livingston GE. Effects of heating methods on vitamin retention in six fresh and frozen preparated food products. *J Food Sci.* 1975;40:997–1003.
52. Lanier NJ, Sinstrunk WA. Influence of cooking method on quality attributes and vitamin content of sweet potatoes. *J Food Sci.* 1979;44:374–376.
53. Bender AE. Nutritional effects of food processing. *Rev Nutr Food Sci.* 1968;13:6–9.
54. Sweeney JP, Gilpin GL, Martin ME, Dawson EH. Palatability and nutritive value of frozen broccoli. *J Am Diet Assoc.* 1960;36:122–125.
55. Snyder OP Jr. Increasing the quality of vegetables in foodservice operations. *Microw En Appl Newsl.* 1993;9:3– 7.
56. Bartholomew JW, Harris RG, Sussex F. Electronic preservation of Boston brown bread. *Food Technol.* 1948;2(3):91–95.
57. Olsen CM. Microwave inhibits bread mold. *Food Engr.* 1965;37:51–56.
58. Bengtsson NE, Ohlsson T. *Food Process Engineering*, Vol. 1. In: Linko P, Malkki Y, Olkko J, Larinkari J, eds. London: Applied Science Publishers; 1980.
59. Decareau RV. Microwaves in food processing. *Food Technol Austr.* 1984;36:81–86.
60. Bookwalter GN, Shulka TP, Kwolek WF. Microwave processing to destroy salmonellae in corn-soy-milk blends and effect on product quality. *J Food Sci.* 1982;47:1683–1686.

Chapter 11

Vacuum Impregnation of Plant Tissues

Pedro Fito and Amparo Chiralt

INTRODUCTION

Vacuum impregnation (VI) of a porous product consists of exchanging the internal gas or liquid occluded in open pores for an external liquid phase by the action of hydrodynamic mechanisms (HDMs) promoted by pressure changes.[1,2] The operation is carried out in two steps after the product immersion in the tank containing the liquid phase. In the first step, vacuum pressure is applied in the closed tank, thus promoting the expansion and outflow of the product internal gas. The releasing of the gas takes the product pore native liquid with it. In the second step, the atmospheric pressure is restored in the tank, and compression leads to a great volume reduction of the remaining gas in the pores and, thus, the subsequent in flow of the external liquid in the porous structure. Compression can also reduce the pore size, depending on the mechanical resistance of the solid matrix.

VI operation can allow us to incorporate any ingredients in a porous product to adapt its composition to certain stability or quality requirements in a quick and simple way. Structured foods such as fruits and vegetables have a great number of pores (intercellular spaces) that are occupied by gas or native liquid to quite an extent and that offer the possibility of being impregnated by a determined solution to improve composition: incorporation of acids, preservatives, sugars, or other water activity (a_w) depressors, etc. In this sense, VI can be considered as a tool in minimal processing of fruits or vegetables that is much more effective than the conventional soaking of the fruit in the active component solutions.

Recently, extensive studies have been carried out to determine the feasibility of VI in several fruits;[3,4] the influence of VI treatments on their physical, chemical, and structural properties;[5–7] and the effectiveness of VI in the incorporation of different kinds of components into the fruit tissue.[8] Quality parameters (such as texture and color) and stability of VI fruit and vegetable samples have been compared with those achieved by simple conventional immersion until the same stabilizing factor levels (pH, a_w, antimicrobial concentration, etc.) were reached.[9–11]

The aim of this chapter is to compile fundamental aspects of VI in terms of a mathematical model that permits operation control, as well as the more relevant features associated with VI of plant tissues in relation to changes in product quality or process advantages.

MODEL OF VACUUM IMPREGNATION: EQUILIBRIUM AND KINETICS

Vacuum impregnation is one of the more practical applications of controlled HDMs promoted by pressure changes. The pressure gradi-

We thank the Comision Interministerial de Ciencia y Tecnología, the U.E. (STD3 programme), and the CYTED Program for financial support.

ent in the system is the driving force in bulk mass transfer if the product has pores and is immersed in a liquid phase. Pore internal gas can be expanded or compressed, in line with the external solution inflow while pressure gradients persist. This phenomenon is HDM. The model of this mechanism predicts the entry of an external liquid in a porous product when high pressure is applied in the system; nevertheless, liquid will be released when the product returns to normal pressure. The advantage of the HDM action after a vacuum period in the system lies in the partial gas release throughout this period while the mechanical equilibrium is being achieved; the restoring of atmospheric pressure implies compression, mainly of the residual gas and flow into the pores of the external liquid. The product thus remains filled with the liquid phase at normal pressure. From the HDM model, it is possible to predict the amount of the external liquid that can be introduced into a porous food as a function of the product-effective porosity and the applied compression ratio.

In a first approach, HDM was modeled for rigid products with homogeneous pores of diameter D and length z, which were immersed in a liquid.[1,2] The interior of the pore is assumed to be occupied by gas at an initial pressure p_i, whereas the pressure in the gas-liquid interface inside the pore (p_e) is greater than p_i, even when the system is at constant pressure, due to the capillary effects. Pressure difference $\Delta p = p_e - p_i$ is equal to capillary pressure (p_c), given by the Young-Laplace equation as a function of capillary diameter and liquid surface tension (σ).

The amount of liquid flowing into the pores by pressure gradients has been estimated[1] from the application of the Hagen-Poiseulle equation in differential form to the system (Equation 1). In Equation 1, the liquid penetration was expressed as the pore volume fraction (x_v) occupied by the liquid at each time (t), i.e., a function of the liquid viscosity (μ). The x_v value increases with time as internal gas is compressed until Δp equals zero and mechanical equilibrium is reached; then $(dx_v/dt) = 0$. By assuming an isothermal compression and taking into account the relationship between Δp and x_v, Equation 1 becomes Equation 2 if mechanical equilibrium is reached. In Equation 2, p_1, is the initial gas pressure in the pores and p_2 the external system pressure.

$$-\Delta p + \frac{32\mu z^2}{D^2} x_v \frac{dx_v}{dt} = 0 \quad (1)$$

$$x_v = \frac{p_2 + p_c - p_1}{p_2 + p_c} \quad (2)$$

Equation 2 shows that the volume fraction of a pore penetrated by external liquid in stiff products depends only on the promoted pressure changes and capillary effects. If no pressure changes are imposed on the system ($p_1 = p_2$), only capillary effects cause liquid entry, and x_v value at equilibrium can be estimated by Equation 3.

$$x_v = \frac{p_c}{p_2 + p_c} \quad (3)$$

Equation 2 is usually written in terms of compression ratio r (Equation 4), given by Equation 5. In most cases, p_c is difficult to estimate because the pore diameter is unknown. Nevertheless, from the mathematical analysis of Equation 4, it can be deduced[1] that the capillary force contribution to liquid penetration may be neglected when the apparent compression ratio is higher than 4 or 6, depending on the pore diameter range. Nevertheless, in VI process, if $p_1 > 200$ mbar, a great capillary contribution can be expected for products with small pore diameter ($\approx 10\ \mu m$ or less).

$$x_v = 1 - \frac{1}{r} \quad (4)$$

$$r = \frac{p_2 + p_c}{p_1} \quad (5)$$

From x_v, the sample volume fraction impregnated by the liquid (X) can be easily obtained by multiplying Equation 4 by the product's effective porosity (ε_e) or pore volume fraction available to the HDM action. Equation 6 predicts the amount of sample volume that will be occupied by an external liquid of a determined composition at the mechanical equilibrium status, in terms of the product porosity and the compression ratio, when porosity remains constant during compression.

$$X = \varepsilon_e \left(1 - \frac{1}{r}\right) \tag{6}$$

The HDM model was extended for viscoelastic porous products, where pressure changes cause not only gas or liquid flow but also solid matrix deformation-relaxation phenomena (DRP).[12,13] In viscoelastic solid matrices, expansion–compression processes lead to changes in pore volume, which will be time-dependent. During the first VI step, product volume usually swells, associated with gas expansion, and, afterward, the solid matrix relaxes; capillary penetration or expelling of internal liquid also occurs in this period. In the second step, compression causes volume deformation and subsequent relaxation, coupled with the external liquid penetration in the pores. Mechanical properties of the solid matrix and flow properties of the penetrating liquid in the pores will define characteristic penetration and deformation-relaxation times responsible for the final impregnation and deformation status of the samples at equilibrium. Equation 7 describes the equilibrium relationship between the compression ratio (r), initial sample porosity (ε_e), final sample volume fraction impregnated by the external solution (X), and sample volume deformations at the end of both the process (γ) and the vacuum step (γ_1), all of these being referred to the sample initial volume.

$$\varepsilon_e(r - 1) = (X - \gamma)\, r + \gamma_1 \tag{7}$$

Mathematical modeling of kinetics of VI can be deduced for stiff porous products by integration of Equation 1, taking into account the function $\Delta p = f(x_v)$ (Equation 8), and assuming that x_v takes the x_{v0} and x_v values at $t = 0$ and t, respectively. Equation 9 gives the pore volume penetration at a time t, in a reduced way with respect to its equilibrium (x_{ve}) value (Equation 10), in terms of B and k parameters (Equations 11 and 12). These are, respectively, time dimension, dimensionless, and constant for a determined system and determined operation conditions. In B, three terms can be distinguished: the liquid viscosity (μ); a product structural number, depending on the sample characteristic depth (e), the pore radius (r_p), and tortuosity factor (F_t); and a pressure number, depending on the equilibrium pressures in the system before (p_1) and during (p_2), the compression step. The x_{ve} can be estimated by Equation 4 from pressure values.

$$-\Delta p = p_2 \left(\frac{1 - x_{ve}}{1 - x_v} - 1\right) \tag{8}$$

$$\frac{t}{B} = -\frac{1}{2}(x_r^2 - x_{r0}^2) - k\left[(x_r - x_{r0}) + \ln\left(\frac{1 - x_r}{1 - x_{r0}}\right)\right] \tag{9}$$

$$x_r = \frac{x_v}{x_{ve}} \tag{10}$$

$$k = 1 + \frac{1}{x_{ve}} \tag{11}$$

$$B = 8\mu \left(\frac{eF_t}{r_p}\right)^2 \frac{p_2}{(p_2 - p_1)^2} \tag{12}$$

Figure 11–1 shows the development of x_r as a function of time and liquid viscosity for samples with two different pore diameters, 160 (in the order of apple pores) and 5 µm, considering a tortuosity factor $F_t = 2$, and a pore length (≈

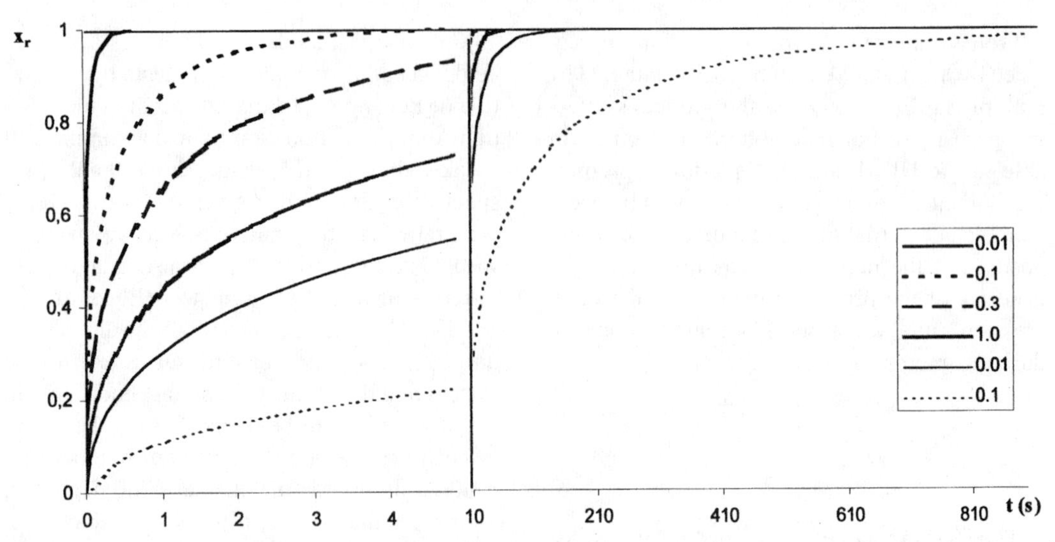

Figure 11–1 Vacuum impregnation kinetics in an ideal stiff porous product. 1 cm characteristic dimension, with pore diameter of 160 µm (*thick lines*) and 5 µm (*thin lines*), as a function of impregnating solution viscosity (Pa s). Predicted values using Equation (9) with a pore tortuosity factor $F_t = 2$ and a pore length of ≈ sample characteristic dimension.

---

sample depth) of 1 cm. Zero value was taken for x_{v0}. A very short time was required to achieve the x_{ve} value ($x_r = 1$) when viscosity ranges between the usual values of aqueous solution used in VI process. Even for more viscous solutions (hydrocolloids, concentrated syrups, etc.), the impregnation time is relatively short, when compared with other process times. For the lowest pore radius and low viscosity, the equilibrium time is still close to 1 minute. This time range is on the order of that required to achieve stationary pressure conditions in the tank. In fact, in different attempts to analyze VI kinetics in a great number of fruits, using sugar isotonic solutions ($\mu \sim 0.006$–0.008 Pa s) and varying the length of the compression period from 5 to 30 minutes, no significant differences between the x_v values at the different times were found.[3,4]

When matrix exhibits a viscoelastic mechanical response to compression, a fast mechanical pseudoequilibrium is achieved as a first step, with a notable reduction of product porosity, coupled with partial liquid penetration. Afterward, the true equilibrium is reached throughout relaxation of deformed porous matrix, which leads to a progressive liquid entry, in line with sample volume recovery, while the product remains immersed.[14] Kinetics of impregnation in this second step will be dependent on relaxation time of the matrix. This two-step behavior will be especially promoted for long penetration times associated with high liquid viscosity or/and small pore diameter.

VACUUM IMPREGNATION IN FRUIT PROCESSING

VI Feasibility in Fruits

Plant tissue shows intercellular spaces that may contain a gas or liquid phase and that are susceptible to impregnation with an external solution and diverse objectives. Figure 11–2 shows the parenchymal tissue of some fruits and vegetables observed by cryogenic scanning electron microscopy (cryoSEM). In this technique, samples were cryofixed by immersion in slush nitrogen, cryofractured, and directly observed

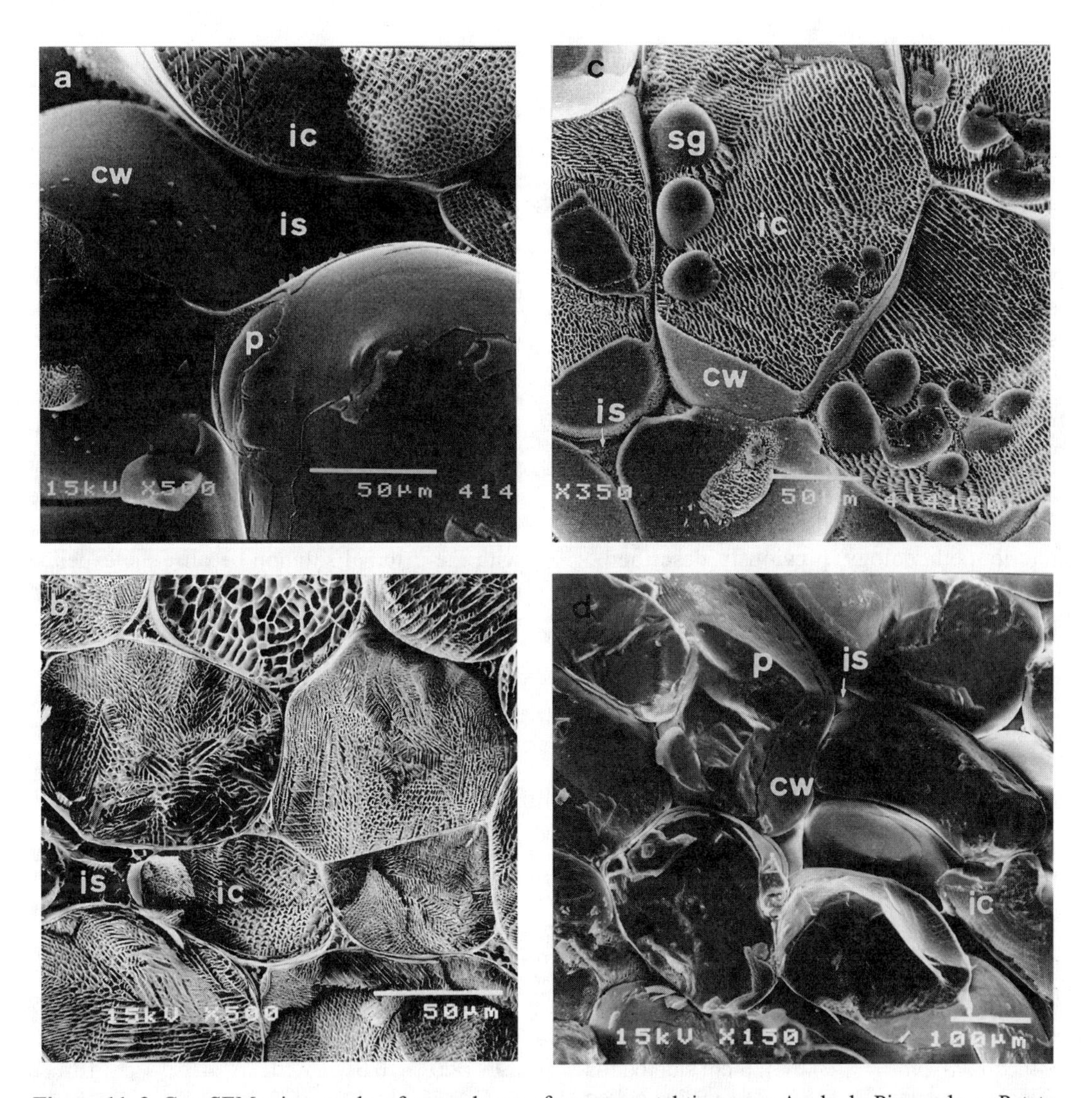

Figure 11–2 CryoSEM micrographs of parenchyma of some vegetal tissues. a: Apple. b: Pineapple. c: Potato. d: Strawberry, showing intercellular spaces with different sizes, with and without native liquid phase. is: intercellular spaces, cw; cell wall, p; plasmalemma, ic; intracellular content, sg; starch granules.

---

after surface etching and coating with gold. According to Bomben and King,[15] the bright regions in the micrographs are the solute-water glass of the cell sap, cell membranes, and cell walls. The darker regions correspond to the ice microcrystal sites or to sample voids. Differences in electron density are due to differences in height between the ice (which sublimes during etching), the solute-water glass, and insoluble structures of the tissue. Solute-water glass appears as a dentritic zone because of ice microcrystal sublimation and can be observed in the intracellular zone and in some intercellular spaces containing native liquid. In some cases, these appear completely empty. The total volume fraction of empty intercellular spaces constitutes the tissue porosity, which greatly varies in different fruits and vegetables. In Figure 11–2, the differing compactness of cellular arrangement can be observed, as can the differences in size of intercellular spaces. Apple tissue shows large voids among cells, whereas strawberry cells are densely packed, showing very small pores. Pineapple and potato show a great part of intercellular spaces filled with native liquid, which is revealed by their dentritic aspect.

The response of many fruits to VI process (impregnation and deformation levels at equilibrium in the two process steps) has been characterized experimentally by means of a gravimetric methodology, previously described.[3,13] Table 11–1 gives the obtained VI parameters for a great number of analyzed fruits. Positive volume deformations at the end of the vacuum step (γ_1) were obtained, in most cases due to the solid matrix deformation associated with depression and gas expansion. At the end of the compression period, deformation (γ) was negative or positive, depending on the fruit; e.g., eggplant shows a great volume reduction after VI, whereas orange peel and mango swell. With reference to the liquid phase fluxes, most of the fruits show negative values of X_1, which quantify the net gain of liquid at the end of the vacuum step, due to loss of native liquid (released in line with gas outflow), coupled with capillary entry. Negative values of X_1 suggest that pore volume initially occupied by native liquid will also be available to be impregnated with the external solution because the expelled native liquid will be replaced by the external one throughout the compression step.

Table 11–1 Vacuum Impregnation Response of Several Fruits Applying 50 mbar in the Vacuum Step

Fruit, Variety	*Sample**	X_1	γ_1	X	γ	ε_e†	*Reference*
Apple *Granny Smith*	2×2 cm cylinders	–4.2	1.7	19.0	–0.6	21.0	3
Apple *Red Chief*	2×2 cm cylinders	–5.0	2.1	17.9	–2.4	20.3	3
Apple golden	2×2 cm cylinders	–2.7	2.8	11.2	–6.0	17.4	3
Mango *Tommy Atkins*	1 cm slices	0.9	5.4	14.20	8.9	5.9	3
Strawberry *Chandler*	Whole fruit	–2.1	2.9	1.9	–4.0	6.4	3
Kiwifruit *Hayward*	Fruit quarters	–0.2	6.8	1.09	0.8	0.66	3
Peach *Miraflores*	2.5 cm side cubes	–2.29	2.0	6.5	2.1	4.7	16
Peach *Catherine*	2.5 cm side cubes	–1.4	0.5	4.4	–4.2	9.1	16
Apricot *Bulida*	Fruit halves	–0.2	1.5	2.1	0.11	2.2	16
Pineapple *Española Roja*	1 cm slices	–6.5	1.8	5.7	2.3	3.7	16
Pear *Passa Crassana*	2.5 cm side cubes	–1.3	2.8	5.3	2.2	3.4	16
Prune *President*	Fruit halves	–1.0	0.6	1.0	–0.8	2.0	16
Banana *Giant Cavendish*	0.5–2 cm slices	–6.1	3.6	10.6	1.3	10.1	4
Melon *Inodorus*	2×2 cm cylinders	–4.0	2.0	5.0	–0.4	6.0	16
Eggplant	1×1 cm cubes	–9.6	2.4	15	–37	53.8	‡
Zucchini	2.5 cm cubes	–2.47	3.2	5.85	3.27	2.6	‡
Orange peel *navel late*	2.5×6 cm	–6	5	40	14	20	17

* Size and shape of processed samples.
† Effective porosity to VI of the initial sample, estimated by HDM model (Equation 7).
‡ Unpublished results (Grass, M. *personal communication*).
X: impregnated sample volume fraction; *γ*: sample relative volume of deformation.

The level of final impregnation (X) was greatly affected by the coupling of penetration-deformation phenomena throughout the action of pressure gradients in the system, due to the viscoelastic response of plant tissue.[3,13] This fact makes the analysis of the pressure influence on each parameter difficult because pressure changes affect both liquid fluxes and matrix deformation. The estimated porosity values (Table 11–1) of the fruit on the basis of the HDM model (Equation 7) are very close to those values obtained by other techniques.[3] This validates the theoretical model. On the basis of the analyzed VI response of fruit as a function of the applied vacuum pressure, the vacuum level given by most common industrial vacuum pumps (~50 mbar) is recommended for practical use to carry out VI process efficiently.[16]

Table 11–1 presents the relative feasibility of a great number of fruits and vegetables to VI. With the exception of strawberry, kiwifruit, peach, and apricot, notable levels of final impregnation (X) are reported. Even in these fruits, it is possible to introduce a certain volume fraction of the external solution, which, at the adequate concentration, can be the vehicle to introduce a selected active compound. On the other hand, VI process provokes an extensive release of any internal gas phase, which can contribute to improve sample stability, as in the case of some oxygen-dependent enzymatic reactions.

Compositional Changes in Fruit Promoted by VI

VI may promote fast compositional changes in fruit, this being useful in many cases to assure processed fruit stability (decrease of pH or a_w, incorporation of antibrowning agents or antimicrobials)[8,9] or quality enhancement (improvement of the sweet-sour taste relationship, fortification with specific nutrients, etc.).

Prediction of compositional changes in short VI treatments can be easily obtained by applying Equations 13 and 14 if the value of impregnated volume fraction (X) and densities of the initial product (ρ^0) and the impregnating solution (ρ^{IS}) are known. The mass fraction of any component (water or solutes) in the impregnated product x_i^{VI} can be estimated by Equation 13, deduced from a mass balance in the system, in terms of the sample initial composition and the mass ratio of the impregnating solution in the initial product (x_{HDM}) and the solution composition (y_i). From these equations, the required solution concentration of a determined component (water, sugar, acid, additive) can be calculated to achieve a desired level in the final product.

$$x_i^{VI} = \frac{x_i^0 + x_{HDM} y_i}{1 + x_{HDM}} \quad (13)$$

$$x_{HDM} = X \frac{\rho^{IS}}{\rho^0} \quad (14)$$

where x_i = mass fraction of component i before and after VI (0 and VI superscripts) in the product and y_i = mass fraction of component i in the impregnation solution.

Composition changes promoted by concentration gradients between external solution and product were not taken into account in Equation 13, due to the short length of VI processes. Nevertheless, when driven forces dependent on concentration are high (or long treatment times were applied), deviations from the predicted final concentrations can be found. Figure 11–3 shows the relationship between experimental and predicted values of the mass fraction of soluble solids reached in Granny Smith apple slices (1 cm thick) after VI (5 minutes at 180 mbar plus 10 minutes at atmospheric pressure) with sucrose syrups of different concentrations (between 25 and 65°Brix). Despite the expected concentration effects, experimental values of x_{ss} are slightly lower than those predicted by Equation 13, especially when the more concentrated syrups were used. This can be due to the sample pore collapse in the compression step because of the higher syrup viscosity and the subsequent reduction of the impregnation level.

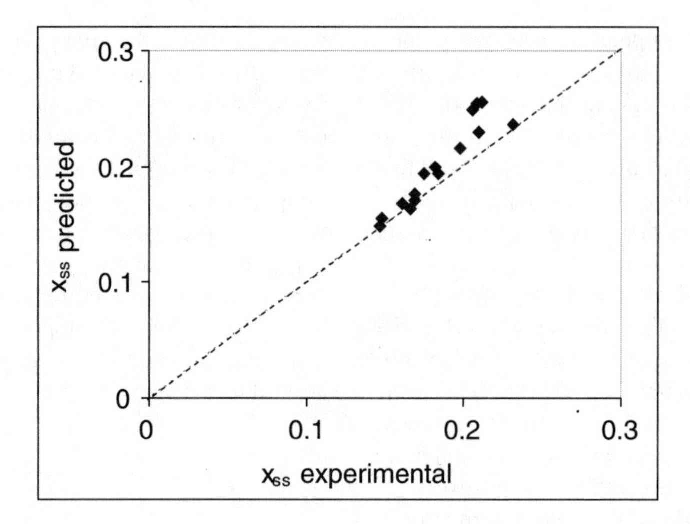

Figure 11–3 Relationship between experimental and predicted (Equations 13 and 14) soluble solid concentration of apple (var. Granny Smith) after VI (5 min at 180 mbar plus 10 min at atmospheric pressure) with sucrose solution of different concentration.

Changes in Mechanical and Structural Properties of Fruit Due to VI

Throughout VI, the exchange of internal gas with an external liquid, as well as sample volume changes, may imply variations in sample mechanical, structural, and other physical properties that will affect the overall fruit quality. Several works[5–7,11] have been developed to analyze some of these changes.

The effect of VI with hypotonic, isotonic, and hypertonic solutions on mechanical properties of apple tissue has been studied through stress relaxation tests (8% constant strain, 200 mm/minute deformation rate).[7] When isotonic solutions were used (cell turgor unaltered), no significant differences in the initial elasticity modulus between fresh and VI apples were found. Nevertheless, the relaxation rate and the total relaxation level increased in VI samples, in line with the impregnation level (X).[5] These changes in sample viscoelastic behavior have been attributed to the exchange of compressible gas phase by a viscous one (noncompartmented liquid in the intercellular space), which will flow out from the pores throughout compression.[5,7] Impregnation with hypotonic solution implied a greater level of stress relaxation than did the isotonic treatments, which was explained by outflow of the intracellular liquid, in line with cell rupture and promoted by an excessive turgor.[7]

The promoted structural changes by VI with isotonic solutions have been observed by cryo-SEM in apple,[7] strawberry, pineapple, and other fruits.[17,18] CryoSEM observations of samples did not show cellular alterations or debonding but showed the sample intercellular spaces completely flooded by the external solution, with a similar aspect to that observed in fresh pineapple or potato with free native liquid (Figure 11–2).[7]

Impregnation with hypertonic solutions promotes sample osmotic dehydration that contributes to changes in chemical and physical properties of the product. Due to dehydration, cells lose turgor and elastic properties, provoking a sharp increase in the viscous character of the sample.[19] However, advantages of VI treatments, conducted to reduce fruit water activity, have been observed in texture retention in several fruits.[11,20–21] Mechanical response in high deformation compression/puncture tests of VI a_w-depressed fruits (kiwi and melon) was closer to that of fresh fruit than to that obtained for fruit at the same a_w level but prepared by conventional soaking in osmotic solutions.[9–11] These advantages have been correlated with microstructure features in terms of alterations of cell wall and membranes, as observed by transmission electron microscopy (TEM).[9,11,21] Cell integrity was better preserved in VI treatments than in conventional soaking. Differences in treatment length could be greatly responsible for distinct structural features. Whereas 15 minutes were used in VI treatments of kiwi with a 59°Brix glucose syrup to reach $a_w = 0.97$, 5 days were needed for the sample to reach the same a_w value after immersion in a diluted glucose solution.[11] During immersion time, soluble constitutive components can leach to quite an extent, thus greatly affecting cell integrity and fruit texture.

In addition to the reported ultrastructural differences in osmosed fruit tissues promoted by VI, cryoSEM analysis also allows us to detect a different cellular response to the osmotic stress in terms of cell wall–plasmalemma interactions.[7,22–23] In VI hypertonic treatments, osmotic dehydration of the tissue promotes plasmolysis but no significant shrinkage of the cellular wall, whereas the space between plasmalemma and cell wall appeared to be completely full of solids.[7,24] On the contrary, osmosis of the tissue at atmospheric pressure, when intercellular spaces are occupied by gas, provokes cell wall shrinkage without plasmalemma separations,[25] even at very high osmotic stress.[24] This different behavior has been explained in terms of the pressure gradients generated in the tissue, due to the loss of water volume in the cell and the different pressure drop of gas and liquid phases into the intercellular spaces during their flux toward the generated volumes in the tissue.[22] The force balance on both sides of the plasmalemma-cell wall layer during cell water volume loss leads to the separation of the double layer and to the flux of the external liquid through the cell wall or to the cell wall deformation, together with the plasmalemma.[22] Hydrodynamic mechanisms are generated in line with the osmotic-diffusional mechanisms, due to the mass transport, which promote different fruit tissue developments, depending on the viscosity of the fluid phase in the intercellular spaces. In osmosed samples without VI, the capillary penetration of the external solution near the interface leads to similar behavior in the more external cells to those observed in vacuum-impregnated samples.[22,25] Likewise, fruit tissue, where native liquid floods the intercellular spaces, behaves like VI samples.[26]

Changes in Color and Thermal Properties of Fruit Due to VI

Other physical properties, such as optical or thermal properties, also change due to VI. The gas-liquid exchange in the fruit implies a more homogeneous refraction index throughout the sample and, consequently, an increase in the product transparency. Thus, when color was measured by diffuse reflection, a decrease in the reflection percentages was obtained for VI samples, as compared with fresh samples. Figure 11–4 shows the reflection-visible spectra of a mango sample surface, before and after VI (5 minutes at 50 mbar plus 10 minutes at atmospheric pressure) with a 55°Brix sucrose solution. A sharp decrease in reflectance percentages after VI treatment can be observed. Changes in the visible spectra have the corresponding repercussion on color coordinates. Sample lightness and color saturation usually decrease, whereas hue is slightly modified. Table 11–2 presents the color, in terms of CIE L*a*b* coordinates, for a considerable number of fruits with different porosity levels (Table 11–1), before and after VI (5 minutes at 50 mbar plus 10 minutes at atmospheric pressure) with different solutions. Figure 11–5 shows the development in the chromatic plane of hue and chrome of the different fruits. The chromatic trajectory during impregnation leads the points to the center of the plane, in agreement with the mentioned re-

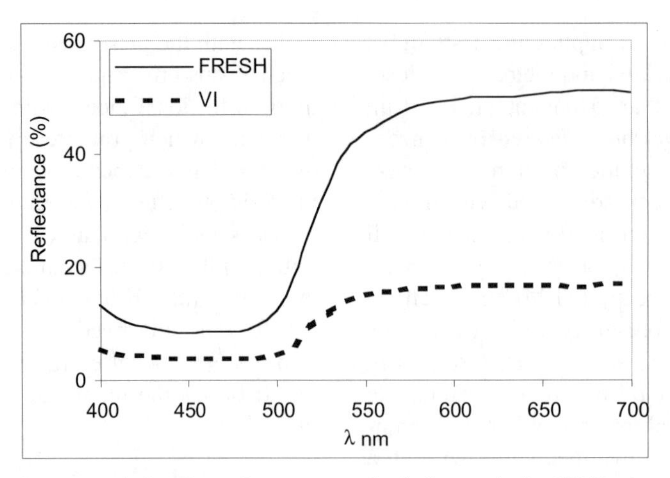

Figure 11–4 Reflection spectra of surface of mango samples before and after VI (5 min at 50 mbar plus 10 min at atmospheric pressure) with a 55°Brix sucrose solution.

duction of chrome. Total color difference between fresh and impregnated samples is shown in Table 11–2. Apple, strawberry, and papaya show the biggest color changes. In general, this change did not have a negative effect on consumer perception.

VI promotes changes in product thermal properties, especially in thermal conductivity with highly porous matrices. Changes are dependent on total porosity and pore distribution, in relation with the heat flow sense and impregnating solution composition.[14,31] Figure 11–6 shows the predicted relative changes, as referred to the initial sample, of thermal conductivity and diffusivity as a function of sample porosity and pore distribution in line with heat flow, without

Table 11–2 CIE L*a*b Color Coordinates* of Sample Surface before and after Vacuum Impregnation with Different Impregnating Solutions (IS)

	Fresh Fruit			*VI Fruit*					
Fruit (variety)	*L**	*a**	*b**	*L**	*a**	*b**	*ΔE*†	*IS*	*Reference*
Apricot *(Canino)*	52.1	11.5	36	34.1	8.7	26.1	21	77°Brix CS‡	27
Papaya *(Solo)*	64.2	11.2	44.9	34.5	5.6	23.5	37	77°Brix CS‡	28
Banana *(Cavendish)*	47.8	4	21.7	64.9	3.7	27.5	18	Orange juice	29
Apple *(Granny Smith)*	69	–2.4	14	36	0	7	34	Isotonic RGM§	14
Strawberry *(Chandler)*	47	36	27	18	12	2.1	45	65°Brix sucrose	5
Kiwifruit *(Hayward)*	34.8	–8.8	21.1	25.7	–7.5	15.8	11	63°Brix RGM§	30
Mango *(Tommy Atkins)*	69.8	11.9	59.5	46.8	8.1	43.6	28	60°Brix sucrose	6

* Mean values of almost five replicates.
† Color difference between fresh and VI fruit; $\Delta E = (\Delta L^{*2} + \Delta a^{*2} + \Delta b^{*2})^{0.5}$
‡ CS: corn syrup
§RGM: Rectified grape must.
VI conditions: t_1 = 5 min; P_1 = 50 mbar; t_2 = 10 min; P_2 = atmospheric pressure.

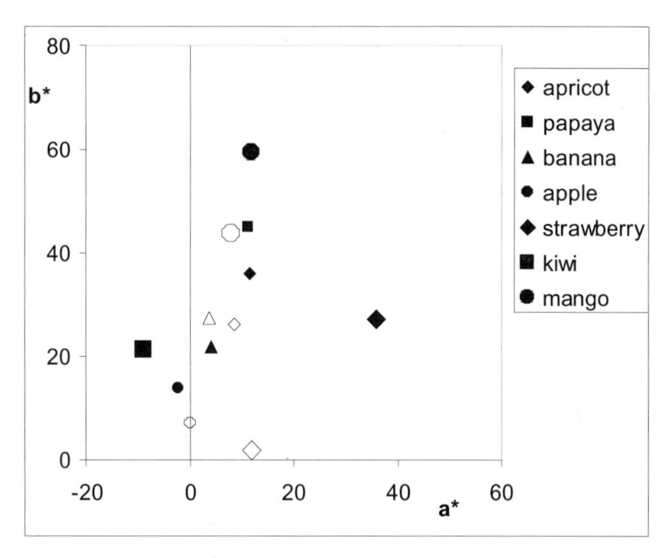

Figure 11–5 Changes in chromatic attributes (a*, b*) promoted by vacuum impregnation in different fruits. Open symbols show the respective location of each fruit in the chromatic plane after VI with solutions specified in Table 11–2.

changes in product composition during VI.[31] In the predictive model, a pore distribution factor (a) is included, which corresponds to the pore volume fraction perpendicular to the heat flow, being (1–a) the volume fraction of parallel pores. Lesser modifications were observed in thermal diffusivity because of the simultaneous sample density increase during VI. An increase in the impregnating solution concentration promoted the expected reduction in all thermal properties,

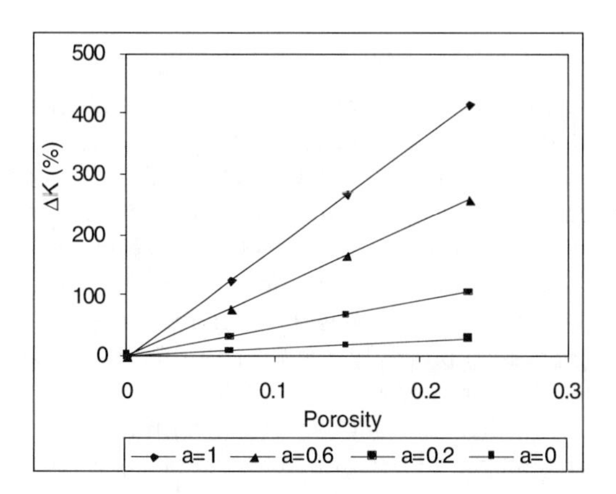

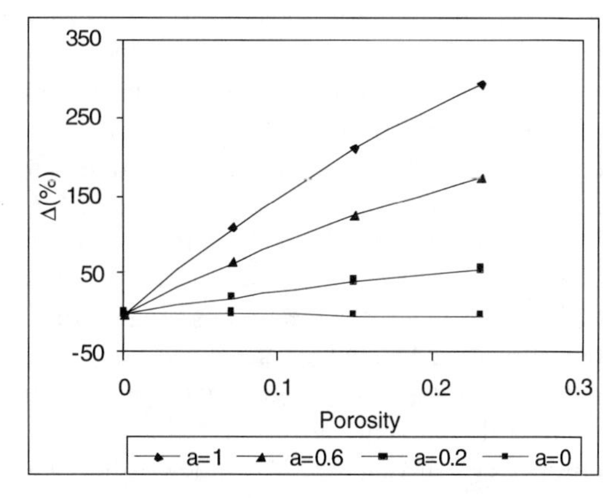

Figure 11–6 Relative changes in thermal conductivity (K) and diffusivity (α) promoted by VI (referred to the value of nonimpregnated sample) as a function of sample porosity and pore volume fraction perpendicular to heat flow (a = 1 perpendicular orientation, a = 0 parallel orientation). Values estimated on the basis of HDM model for a product with 85% moisture content and 15% of carbohydrates. VI conditions: isotonic impregnating solution, p_1 = 50 mbar, p_2 = 1,030 mbar.

as compared with impregnation without compositional changes.[14,22]

Influence of VI on Osmotic Processes

Osmotic dehydration (OD) is the usual process to decrease product a_w in minimally processed fruits and vegetables or in some deeply processed fruits, such as candied fruits or jam.[32,33] The a_w decrease is based on osmotic water flow through cell membranes when the product is immersed in sugars and/or salt concentrated solutions (osmotic solution [OS]). Vacuum treatment leads to a faster osmotic process, due to the coupling of osmotic/diffusional mechanism and HDM.[2,32–34] Three kinds of osmotic treatments have been defined, depending on the pressure applied to the system: OD (at atmospheric pressure), VOD (at vacuum pressure), and PVOD (pulsed vacuum osmotic dehydration).[12,35] In PVOD, VI with the OS takes place during the first 5–10 minutes of the process by the action of a vacuum pulse. This implies a fast compositional change in the product that will affect the osmotic driving force and mass transfer kinetics.[36,37]

A generalized approach to model osmotic (OD, VOD, and PVOD) dehydration kinetics has been reported[37] on the basis of decoupling of HDM and pseudodiffusional (PD) mechanisms. This hypothesis has its basis in the faster HDM kinetics, as compared with slow diffusion/osmotic transport. Another aspect to be considered in kinetic analysis is that mass transport occurs mainly in the food liquid phase (FLP: water plus soluble solids); therefore, reduced driven force of the process would be referred to the composition changes in this product fraction (Equation 15). Taking these considerations into account, Equation 16 gives the value of the reduced driven force at time *t* of the osmotic process, in terms of water or soluble solid concentration ($Y_t^w = Y_t^{ss}$), as a function of a term that depends on sample response to VI and the applied pressure ratio ($Y^w_{t,HDM,\,t=0}$), and a time-dependent term ($Y^w_{t,PD,\,t>0}$). The first represents the initial change in Y_t^w, due to sample pore impregnation after the vacuum pulse in PVOD process or by capillary forces in VOD or OD processes. The second one gives the change in Y_t^w, due to pseudodiffusional mechanisms. This can be expressed on the basis of the Fickian approach as a function of the effective diffusion coefficient.

$$Y_t^w = Y_t^{ss} = \frac{z_t^w - z_e^w}{z_0^w - z_e^w} = \frac{z_t^w - y^w}{z_0^w - y^w} \tag{15}$$

$$Y_t^w = Y^w_{tHDM,t=0}\; Y^w_{tPD,t>0} \tag{16}$$

By substituting in Equation 16 the expression of Y_t^w and $Y_t^w{}_{HDM,\,t=0}$, a pseudodiffusional equation with decoupled HDM was obtained (Equation 17). The $z_t^w{}_{HDM}$ value can be estimated if the sample VI response is known on the basis of a mass balance.

$$\frac{\dfrac{z_t^w + y^w}{z_0^w + y^w}}{\dfrac{z^w_{t\,HDM} + y^w}{z_0^w + y^w}} = \frac{z_t^w + y^w}{z^w_{t\,HDM} + y^w} = Y^w_{t_1\,PD} \tag{17}$$

Table 11–3 presents the values of the effective diffusivity (D_e) obtained by fitting the described model in terms of changes of composition of the FLP, other than those promoted by HDM mechanism, for OD and PVOD processes carried out in identical conditions for several fruits. For most of the cases, D_e is higher in PVOD processes, probably due to the different structural changes provoked in the fruit tissue.[7,25]

Another relevant aspect concerning the VI advantages in osmotic dehydration of porous fruit is the lesser weight loss of the product throughout the process. Because an initial mass of the external solution is introduced into the pores, the weight loss-versus-time curve develops always with lesser values than in the nonimpregnated samples, which implies a better process yield.[24,38] On the other hand, the filling of the intercellular spaces during VI greatly promotes the solute diffusion in these pores and, thus, the solute gain in the product, therefore contributing to mitigate the volume and mass losses of the impregnated product.[39]

VI in PVOD processes also affected the development of osmosed samples to the final equilibrium in terms of the mass, volume, density, and structural changes occurring in the samples.[24,38] These aspects are very important in long-term osmotic processes, such as those performed in the candy fruit processes.[40,41]

CONCLUSION

Vacuum impregnation of porous fruit may be a useful tool in fruit processing. Adequate modifications of composition can be carried out with short process times and without high temperature requirements. Vacuum impregnation feasibility increases in line with fruit porosity, but the mechanical properties of solid matrices also play an important role in the process efficiency. In terms of process yield, VI allows us to reduce the weight loss of osmotically dehydrated fruits; the higher is the product porosity, the greater will be the VI advantages.

Table 11–3 Effect of VI and the Osmotic Solution on Effective Diffusion Coefficient (D_e) in Several Fruit Liquid Phase during Osmotic Dehydration

Fruit (variety)	*T (°C)*	*$D_e \times 10^{10}$ (m/s²)*		*Osmotic Solution*	*Reference*
		OD	*PVOD*		
Pineapple (*Cayena lisa*)	40	3.43	4.55	65°Brix sucrose	37
Pineapple (*Cayena lisa*)	50	5.13	5.11	65°Brix sucrose	37
Pineapple (*Cayena lisa*)	60	4.68	5.95	65°Brix sucrose	37
Banana (*Cavendish*)	25	4.19	5.02	65°Brix sucrose	37
Banana (*Cavendish*)	35	3.82	4.06	65°Brix sucrose	37
Banana (*Cavendish*)	45	3.89	4.51	65°Brix sucrose	37
Papaya (*Solo*)	30	4.0	4.7	77°Brix CS[1] (a_w=0.739)	28
Papaya (*Solo*)	40	6.7	8.4	65°Brix sucrose	28
Apple (*Granny Smith*)	30	1.36	3.36	65°Brix sucrose	36
Apple (*Granny Smith*)	40	2.55	5.61	65°Brix sucrose	36
Apple (*Granny Smith*)	50	4.46	8.80	65°Brix sucrose	36
Mango (*Tommy Atkins*)	30	0.77	1.13	50°Brix glucose	3
Mango (*Tommy Atkins*)	30	1.56	1.91	60°Brix sucrose	3
Kiwifruit (*Hayward*)	25	2.01	2.26	63°Brix RGM(2) (aw=0.789)	30
Kiwifruit (*Hayward*)	35	3.11	3.46	63°Brix RGM(2) (aw=0.789)	30
Kiwifruit (*Hayward*)	45	5.93	6.15	63°Brix RGM(2) (aw=0.789)	30
Kiwifruit (*Hayward*)	25	1.13	1.65	65°Brix sucrose	30
Kiwifruit (*Hayward*)	35	2.84	2.87	65°Brix sucrose	30
Kiwifruit (*Hayward*)	45	3.80	3.97	65°Brix sucrose	30
Apricot (*Canino*)	30	2.05	1.95	65°Brix sucrose	27
Apricot (*Bulida*)	30	2.31	2.41	65°Brix sucrose	27
Strawberry (*Chandler*)	30	0.16	0.21	65°Brix sucrose	4

[1] CS: corn syrup.
[2] RGM: Rectified grape must.
[3] Unpublished results (Talens, P., *personal communication*).
[4] Unpublished results (Ayala, A., *personal communication*).

NOTATION

p_i: internal pressure in the pores.
p_e: pressure at the gas-liquid interface in the pore.
p_c: capillary pressure.
p_1: vacuum pressure in the first VI process step.
p_2: pressure in the system in the second VI process step.
μ: solution viscosity.
x_v: pore volume fraction impregnated by the solution.
x_{ve}: pore volume fraction impregnated by the solution at mechanical equilibrium.
x_r: reduced pore volume fraction impregnated by the solution (x_v / x_{ve})
ε_e: sample effective porosity.
r: compression ratio in the VI process.
X_1: sample volume fraction impregnated by the solution at the end of the first VI step.
X: sample volume fraction impregnated by the solution at the end of the VI process.
γ_1: relative volume deformation of the sample due to pressure change at the end of the first VI step
γ: relative volume deformation of the sample due to pressure change at the end of the VI process.
e: sample characteristic dimension.
r_p: pore radius
F_t: tortuosity factor of the sample pores.
k: dimensionless parameter of the model of VI kinetics.
B: time dimension parameter of the model of VI kinetics.
ρ^0: density of the initial product.
ρ^{IS}: density of the impregnating solution.
x_{HDM}: mass ratio of the impregnated solution in the initial product.
y_i: mass fraction of the component *i* in the impregnating solution.
x_i^{IV}: mass fraction of component *i* in the impregnated product.
x_i^0: mass fraction of component *i* in the initial product.
K: thermal conductivity
α: thermal diffusivity.
Y_i: reduced driven force referred to component *i*.
z_t^i: mass fraction of component *i* in the food liquid phase at time *t* of the process.
$z_t^i{}_{HDM}$: value of z_t^i reached in the sample after VI with the osmotic solution.

REFERENCES

1. Fito P. Modelling of vacuum osmotic dehydration of food. *J Food Engr.* 1994;22:313–328.
2. Fito P, Pastor R. On some non-diffusional mechanism occurring during vacuum osmotic dehydration. *J Food Engr.* 1994;21:513–519.
3. Salvatori D, Andrés A, Chiralt A, Fito P. The response of some properties of fruits to vacuum impregnation. *J Food Proc Eng.* 1998;21:59–73.
4. Sousa R, Salvatori D, Andrés A, Fito P. Analysis of vacuum impregnation of banana (*Musa acuminata* cv. Giant Cavendish). *Food Sci Technol Int.* 1998;4: 127–131.
5. Martínez-Monzó J, Martínez-Navarrete N, Fito P, Chiralt A. Cambios en las propiedades viscoelásticas de manzana (*Granny Smith*) por tratamientos de impregnación a vacío. In: Ortega E, Parada E, Fito P, Matos-Chamorro AR, Sobral PJ, Chiralt A, Alzamora SM, eds. *Equipos y Procesos para la industria de alimentos, Vol. II, Análisis cinético, termodinámico y estructural de los cambios producidos durante el procesamiento de alimentos.* Valencia: Servicio de Publicaciones de la Universidad Politécnica; 1996:234–243.
6. Martínez-Monzó J, Martínez-Navarrete N, Fito P, Chiralt A. Effect of vacuum osmotic dehydration on physicochemical properties and texture of apple. In: Jowitt Red. *Engineering and Food at ICEF 7.* Sheffield: Sheffield Academic Press; 1997:G17–G20.
7. Martínez-Monzó J, Martínez-Navarrete N, Fito P, Chiralt A. Mechanical and structural changes in apple (var. Granny Smith) due to vacuum impregnation with cryoprotectants. *J Food Sci.* 1998;63:499–503.
8. Tapia MS, López-Malo A, Consuegra R, Corte P, Welti-Chanes J. Minimally processed papaya by vacuum osmotic dehydration (VOD) techniques. *Food Sci Technol Int.* 1999;5:43–52.

9. Muntada V, Gerschenson LN, Alzamora SM, Castro MA. Solute infusion effects on texture of minimally processed kiwifruit. *J Food Sci.* 1998;63:616–620.
10. Alzamora SM, Gerschenson LN, Vidales SL, Nieto A. Structural changes in minimal processing of fruits: Some effects of blanching and sugar impregnation. In: Fito P, Ortega-Rodríguez E, Barbosa-Cánovas GV, eds. *Food Engineering 2000*. New York: Chapman & Hall; 1997:117–139.
11. Alzamora SM, Tapia MS, Leúnda A, Guerrero SN, Rojas AM, Gerschenson LN, et al. Relevant results on minimal preservation of fruits in the context of the multinational project XI.3 of CYTED, an Ibero-American R&D Cooperative Programme." In: Lozano JE, Barbosa-Cánovas G, Parada Arias E, Añón MC, eds. *Trends in Food Engineering.* Unpublished.
12. Fito P, Chiralt A. An update on vacuum osmotic dehydration. In: Barbosa-Cánovas GV, Welti-Chanes J, eds. *Food Preservation by Moisture Control: Fundamentals and Applications.* Lancaster, PA: Technomic Pub. Co.; 1995:351–372.
13. Fito P, Andrés A, Chiralt A, Pardo P. Coupling of hydrodynamic mechanism and deformation-relaxation phenomena during vacuum treatments in solid porous food-liquid systems. *J Food Engr.* 1996;27:229–240.
14. Martínez-Monzó J. *Cambios físico-químicos en manzana Granny Smith asociados a la impregnación a vacío. Aplicaciones en congelación.* Valencia, Spain: Universidad Politécnica; Thesis. 1998.
15. Bomben JL, King CJ. Heat and mass transport in the freezing of apple tissue. *J Food Technol.* 1982;17: 615–632.
16. Chiralt A, Fito P, Andrés A, Barat JM, Martínez-Monzó J, Martínez-Navarrete N. Vacuum impregnation: A tool in minimally processing of foods. In: Oliveira FAR, Oliveira JC, eds. *Processing of Foods: Quality Optimization and Process Assessment.* Boca Raton: CRC Press; 1999:341–356.
17. Chiralt A, Chafer M, Ortolá MD, Fito, P. *Orange Peel Response to Vacuum Impregnation and Osmotic Dehydration: Role of Microstructure.* Presented at IFT Annual Meeting, Chicago, IL; July 24–28, 1997.
18. Salvatori D. *Deshidratación osmótica de frutas: ambios composicionales y estructurales a temperaturas moderadas.* Valencia, Spain: Universidad Politécnica; Thesis. 1997.
19. Pitt RE. Viscoelastic properties of fruits and vegetables. In: Rao MA, Steffe JF, eds. *Viscoelastic Properties of Foods.* London: Elsevier Science; 1992:49–76.
20. Alzamora SM, Gerschenson LN. Effect of water activity depression on textural characteristics of minimally processed fruits. In: Barbosa-Cánovas GV, Lombardo S, Narsimhan G, Okos MR, eds. *New Frontiers in Food Engineering. Proceedings of the 5th Conference of Food Engineering* New York: AICHE; 1997:72–75.
21. Nieto A, Castro MA, Salvatori D, Alzamora SM. Structural effects of vacuum solute infusion in mango and apple tissues. In: Akritidis CB, Marinos-Kouris D, Saravacos GD, eds. *Drying'98*, vol C. Thessaloniki: Ziti Editions; 1998:2134–2141.
22. Fito P, Chiralt A, Barat JM, Martínez-Monzó J, Fito P. Vacuum impregnation in fruit processing. In: Lozano JE, Barbosa-Cánovas G, Parada Arias E, Añón MC, eds. *Trends in Food Engineering.* Gaithersburg, MD: Aspen Publishers; 1999: in press.
23. Barat JM, Albors A, Chiralt A, Fito P. Equilibration of apple tissue in osmotic dehydration. Microstructural changes. In: Akritidis CB, Marinos-Kouris D, Saravacos GD, eds. *Drying'98*, vol A. Thessaloniki: Ziti Editions; 1998:827–835.
24. Barat JM, Chiralt A, Fito P. Equilibrium in cellular food osmotic solution systems as related to structure. *J Food Sci.* 1998;63:1–5.
25. Salvatori D, Albors A, Andrés A, Chiralt A, Fito P. Analysis of the structural and compositional profiles in osmotically dehydrated apple tissue. *J Food Sci.* 1998;63:606–610.
26. Castro D. *Procesamiento mínimo de piña por deshidratación osmótica.* Valencia, Spain: Universidad Politécnica; Thesis. 1999.
27. García Redón E. *Deshidratación osmótica de albaricoque* (Prunus armeniaca). Valencia, Spain: Universidad Politécnica; Thesis. 1997.
28. Acosta E. *Deshidratación osmótica a vacío de papaya. Desarrollo de un producto mínimamente procesado.* Las Palmas de Gran Canaria, Spain: Universidad de Las Palmas; Thesis. 1995.
29. Sousa R. *Aplicación de la impregnación a vacío y de la deshidratación osmótica al desarrollo de productos de banana.* Valencia, Spain: Universidad Politécnica; Thesis. 1996.
30. García Pinchi R. *Deshidratación osmótica a vacío por pulso de kiwi. Ricardo.* Valencia, Spain: Universidad Politécnica; Thesis. 1997.
31. Barat JM, Martínez-Monzó J, Alvarruiz A, Chiralt A, Fito P. Changes in thermal properties due to vacuum impregnation. In: Argaiz A, López-Malo A, Palou E, Corte P, eds. *Proc of the Poster Session, ISOPOW practicum II.* Puebla: Universidad de las Américas;1994:117–120.
32. Shi XQ, Chiralt A, Fito P, Serra J, Escoín C, Gasque L. Application of osmotic dehydration technology on jam processing. *Drying Technol.* 1996;14:841–847.
33. Shi XQ, Fito P. Vacuum osmotic dehydration of fruits. *Drying Technol.* 1993;11:1429–1442.
34. Shi XQ, Fito P, Chiralt A. Influence of vacuum treatment on mass transfer during osmotic dehydration of fruits. *Food Res Int.* 1995;28:445–454.

35. Fito P, Andrés A, Pastor R, Chiralt A. Vacuum osmotic dehydration of fruits. In: Singh P, Oliveira FAR, eds. *Process Optimization and Minimal Processing of Foods*. Boca Raton, FL: CRC Press; 1994:107–121.

36. Barat JM, Alvarruiz A, Chiralt A, Fito P. A mass transfer modelling in osmotic dehydration. In: Jowitt R, ed. *Engineering and Food at ICEF 7*. Sheffield: Sheffield Academic Press; 1997:G81–84.

37. Fito P, Chiralt A. Osmotic dehydration: An approach to the modelling of solid food-liquid operations. In: Fito P, Ortega-Rodríguez E, Barbosa-Cánovas G, eds. *Food Engineering 2000*. New York: Chapman & Hall; 1996:231–252.

38. Fito P, Chiralt A, Barat J, Salvatori D, Andrés A. Some advances in osmotic dehydration of fruits. *Food Sci Technol Int*. 1998;4:329–338.

39. Martínez-Monzó J, Martínez-Navarrete N, Chiralt A, Fito P. Osmotic dehydration of apple as affected by vacuum impregnation with HM pectin. In: Akritidis CB, Marinos-Kouris D, Saravacos GD, eds. *Drying'98*, vol A. Thessaloniki: Ziti Editions; 1998:836–843.

40. Fito P, Barat JM, Chiralt A. *Applying Vacuum Impregnation to Fruit-Vegetable Candy Processes*. Presented at IFT Annual Meeting; July 24–28, 1997; Chicago IL.

41. Barat JM. *Desarrollo de un modelo de la deshidratación osmótica como operación básica*. Valencia, Spain: Universidad Politécnica; Thesis. 1998.

Chapter 12

High Hydrostatic Pressure and Minimal Processing

Enrique Palou, Aurelio López-Malo, Gustavo V. Barbosa-Cánovas, and Jorge Welti-Chanes

INTRODUCTION

Minimally processed fruits and vegetables must maintain freshlike characteristics while providing a convenient shelf life and assuring safety and nutritional value. Therefore, handling, processing, and storage operations during fruit and vegetable transformation into a minimally processed product must be carefully selected from the available food preservation technologies. Minimal processing includes a wide range of technologies for preserving otherwise short shelf life of vegetable and fruit products while minimizing changes to their freshlike characteristics, as well as improving quality of the long shelf life of minimally processed fruit and vegetable products. The first targets in minimal processing are pathogenic and spoilage microorganisms. According to Gould,[1] food preservation technologies can be classified into those that prevent or slow down microbial growth, those that inactivate microorganisms, and those based on the hurdle technology concept,[2] which could act in inhibiting or inactivating microorganisms, depending on the combination of hurdles applied. From a microbial safety point of view, technologies that inactivate food-borne pathogenic and spoilage flora are preferred over technologies that are based on preventing or slowing down microbial growth. Few techniques act primarily by inactivation, and heat is by far the most used. In heat sterilization and pasteurization, the applied treatment inactivates and considerably reduces the number of microorganisms initially present. However, food sensory and nutritional characteristics are also strongly affected. Nonthermal processes applied to food preservation without the collateral effects of heat treatments are being deeply studied and tested. High hydrostatic pressure (HHP) as a food preservation technique inactivates microorganisms and can be applied without raising temperature; thus, sensory and nutritional characteristics could be maintained.

Some of the current industrial applications of HHP include fruit jams, dressings, jellies and yogurts, grapefruit and mandarin juices, sugar-impregnated tropical fruits, and avocado paste (guacamole). The first pressure processed foods were strawberry, kiwi, and apple jams produced by Meidi-ya Food Co., which had a vivid and natural color and flavor. Hayashi[3] presents a list of pressure-processed foods in the Japanese market, which includes fruit sauces and desserts. Cheftel[4] mentioned that many of these products are acid foods; hence, they have an intrinsic safety factor. Also, some of the products are stored and sold refrigerated; consequently, oxidative reactions are retarded. Several new industrial applications are expected soon. Current and potential applications of HHP treatment of foods have focused on the application of pressure as an alternative technology to heat treatments, especially for heat-sensitive foods, and as nonthermal processes to assure safety and quality attributes in minimally processed foods.

This chapter explores the potential of HHP treatments for minimal processing of foods, with special attention to acid foods, and emphasizes the effects of pressure in some food-borne microorganisms and quality deteriorative enzymes.

HIGH HYDROSTATIC PRESSURE TREATMENT

High pressures can be generated by heating of the pressure medium, direct or indirect compression. High isostatic pressure technology is the application of pressure uniformly throughout a product. The food industry employs the technique of isostatic pressing for applying high pressures to foods. Most industrial isostatic pressing systems utilize the indirect method of pressure generation. The isostatic pressing systems may be operated as cold, warm, or hot systems. Several reviews on HHP[5–7] describe the available methods to generate high pressure in more detail. However, most of the scientific work has been performed using the hydrostatic pressure generation system, in which, once loaded and closed, the vessel is filled with the pressure-transmitting medium, air is removed, and HHP is then generated in a batch processing system. The technical advantage of the batch-type pressure vessel is the simplicity of fabrication, when compared with a continuous flow pressure vessel operating at pressures as high as 500–900 MPa. Operating pressure vessels in sequence with no lag in the processing times so that the system operates sequentially can increase the production rate of the batch process. A semicontinuous system with a processing capacity of 600 L/hour of liquid food at a maximum operating pressure of 400 MPa is used commercially to produce grapefruit juice in Japan.[6] Recently, Flow International has engineered a concept that is compatible with food production lines involving pumpable products (Figure 12–1). This on-line production unit uses a device called an *isolator*. The food is pumped into a

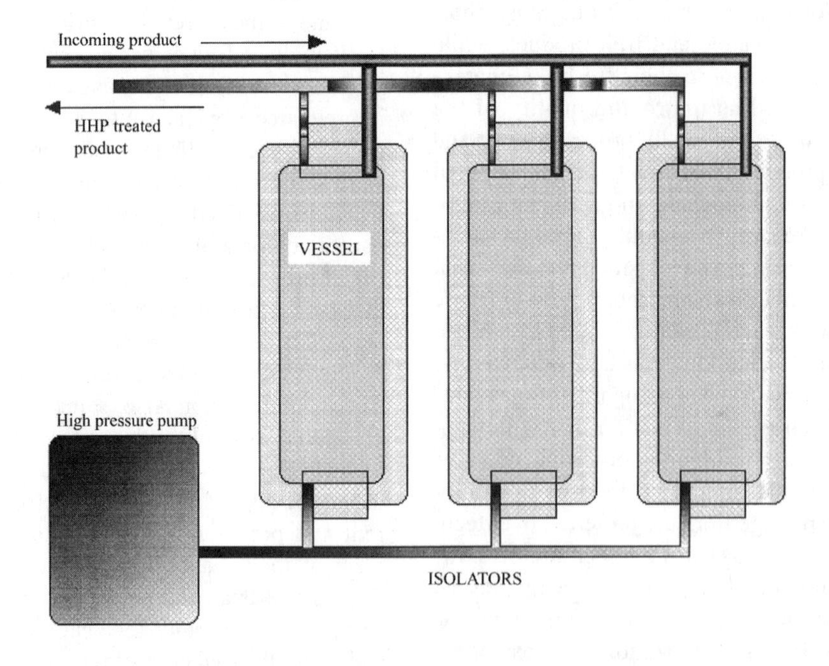

Figure 12–1 Semicontinuous high hydrostatic pressure food system.

vessel, then pressurized, and, after a holding time, the food is discharged from the vessel into a clean filling station. The process then repeats for as long as needed to meet production requirements. Multiple units can be sequenced so that while one unit is being filled others are in various stages of operation to produce an almost continuous output.[8] The engineering challenges of the application of high pressure in the food industry are basically the fabrication of pressure vessels to handle large volumes of food while withstanding the high pressures. Pressure vessels should have a short cycle time, be easy to clean, and be safe to operate with accurate process controls. It is desirable to develop a continuous process of pressurization for industrial purposes at reasonably low capital and operating costs. Most of these challenges are being met to some extent, but research is still being conducted to develop the necessary high-pressure equipment for the modern food industry.[6,7]

Any reaction, conformational change, or phase transition that is accompanied by a decrease in volume will be favored at high pressures while reactions involving an increase in volume will be inhibited, according to the Le Chatelier principle.[4,7] However, due to the complexity of foods and the possibility of changes and reactions that can occur under pressure, predictions of the effects of high-pressure treatments are difficult, as are generalizations for any particular type of food. The tremendous amount of information generated in the last 10 years raises evidence about the effects of HHP on microbial inactivation, chemical and enzymatic reactions, as well as structure and functionality change of biopolymers.[6,7,9–11] HHP processing of food provides an alternative to conventional thermal treatments. However, for the development of HHP processes, it is essential to understand the influence of pressure on quality deteriorative enzymes, microorganisms, and sensory acceptability. Microorganisms and enzymes are affected by HHP treatments, whereas flavor and color quality of certain foods are retained.[12–20] HHP differs from other nonthermal and thermal preservation technologies, in that the mechanical force accompanying pressure generation can deform or noticeably modify fruit and vegetable integrity, especially in porous products. The air confined in the food matrix is subjected to compression and expansion during pressurization and decompression, disrupting food tissues. This drastically affects textural characteristics, promoting other reactions, such as enzymatic browning. Thus, the application of HHP often focuses on fruit and vegetable products such as juices, jams, jellies, and on diced forms, blended or immersed with other ingredients, as in brines, syrups, salsas, yogurt, or dressings.

MICROBIAL RESPONSE TO HIGH HYDROSTATIC PRESSURE TREATMENTS

The sensitivity to pressure depends on the type of microorganisms, and the most important groups are the vegetative and spore forms of microbes. In general, microbial vegetative forms are inactivated at 400–600 MPa, whereas spores of some species may resist pressures higher than 1,000 MPa at ambient temperatures. The relative pressure sensitivity of the vegetative forms of microorganisms has made them the obvious first targets for the preservation of foods by high pressure, particularly for low-pH foods and other foods in which the intrinsic preservation systems already operating ensure that the pressure-resistant food poisoning or spoilage spore-formers that may survive are unable to grow.[1,21]

A characteristic shared by most fruits and low-pH foods is their high acidity, and a pH under 4.0 is not uncommon. pH is the single most important factor, with respect to the type of microorganism, that can spoil these classes of food products. Although most species of bacteria are inhibited by the hydrogen ion concentration, lactic acid bacteria, yeast, and molds are more aciduric, and many find these pH values to be tolerable, if not optimum, for growth. It is because of acidity, therefore, that fungi and lactic acid bacteria are the principal spoilage microorganisms of fruits, fruit products, and low-pH foods. The application of pressure as a preservation technique for acid foods must focus on the effects on deteriorative lactic acid bacteria and food-borne fungi. Although the acid concentration and low pH of fruit juices may be antagonistic toward most pathogenic bacteria, these factors alone do not ensure product safety. Recent outbreaks of food poisoning from *Salmonella* species and *Escherichia coli* O157:H7 were traced to nonpasteurized orange and apple juices, respectively.[22] Nonthermal pasteurization of fruit juices emerges as a necessity to assure safe products. On the other hand, most vegetables provide a good media for the growth of microorganisms and may present some health risks, owing to their relatively high pH. The presence of food-borne pathogens in minimally processed vegetables is of special concern because they are consumed without any further microbial inactivation treatment, such as heating or cooking.[23] *Listeria monocytogenes*, *Vibrio cholerae*, *Salmonella* species, *Shigella* species, *E. coli*, *Bacillus cereus*, *Yersinia enterocolitica*, *Aeromonas* species, and *Campylobacter* species have been linked as possible causes of infections due to minimally processed vegetable consumption.[23,24]

Extensive reviews about the effects of HHP treatment on microorganisms concur that, at least for spoilage and pathogenic microbial vegetative forms, pressure applied with sufficient intensity can inactivate them. Also, these reviews show some general findings: The extent of microbial inactivation achieved at a particular pressure treatment depends on a number of interacting factors, including type and number of microorganisms, magnitude and duration of HHP treatment, temperature, and composition of the suspension media or food.[7,16,25]

The microbial response to a particular pressure treatment can be divided into inactivation and survival responses. The microbial cell damage will determine survival or death, and the extent of damage obviously depends on HHP treatment characteristics, such as pressure level, time of exposure, temperature, and suspension media composition. However, the result—survival or inactivation—is often determined by plate counts, and, usually, dilution and plating are made just after treatment. Survival, as measured immediately after pressure release, may differ from that determined after a repair period in the food or in an enriched medium.[4] Recovery after pressure treatment is a very important consideration for process efficacy and death kinetics assessment.[25] The lack of nutrients in water, peptone water, or phosphate buffers prevents the recovery of the pressure-damaged cells. The possibility of cell recovery exists, and, in many cases, pressure-treated microorganisms may not be detected in plate count methods because of their failure to initiate growth when they are plated immediately after treatment. However, if the repair mechanism remains intact, the microorganisms may be capable of regeneration and growth. Isaacs et al.[26] observed that, for *E. coli* suspensions treated at 200 MPa up to 6 minutes and plated on selective (Mac-Conkey and eosin methylene blue agars) and nonselective (tryptone soy agar) media, the survival fraction was greater for bacteria plated on the nonselective agar. This was attributed to the inhibitory ingredients contained in the selective media, indicating that there is a proportion of microorganisms that, after pressurization, can repair and reproduce, whereas the added stress caused by culturing on selective media inhibits the repair process. Figure 12–2 presents *Zygosaccharomyces bailii* decimal reductions in model systems (water activity [a_w] 0.98, pH 3.5) after pressure treatments at 345 and 517 MPa, determined after a repair period of 48 and 120 hours. For pressure treatments that lead to significant reductions on the initial population of *Z. bailii*, but not enough to inactivate the initially inoculated cells, further incubation reveals that the survivors can grow.[16] Yeast cells subjected to this physical hurdle may become injured or sublethally stressed; their recovery requires generally more incubation time, and the injured survivors are capable of growth. In treatments that render the complete inhibition of 10^5 *Z. bailii* CFU/mL after 48 hours (decimal reduction time [D value, discussed below] of 5), the same result was obtained with a longer

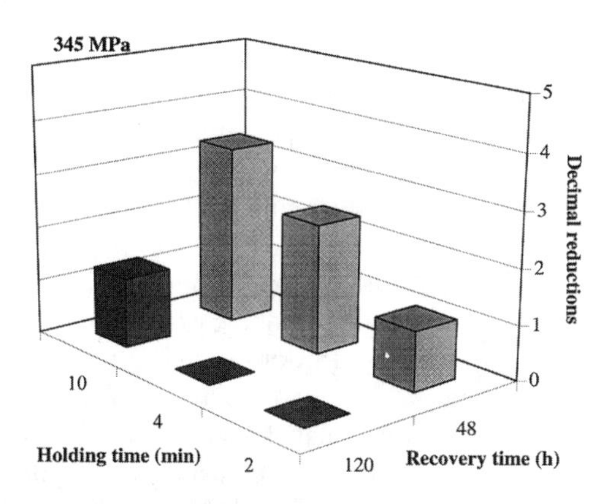

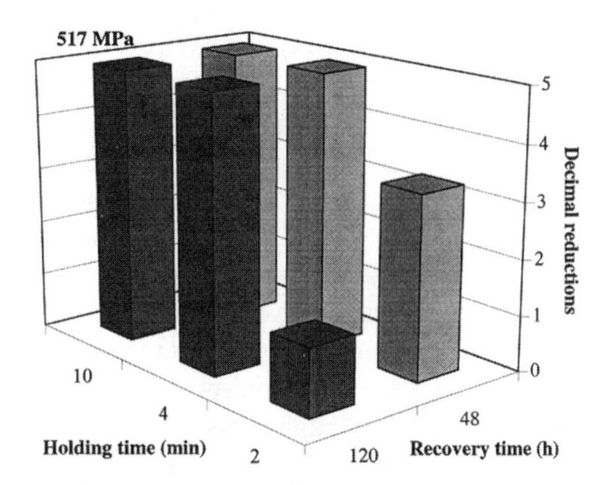

Figure 12–2 High hydrostatic pressure holding time and *Zygosaccharomyces bailii* decimal reductions after recuperation of 48 or 120 h at 25°C.

(120-hour) incubation period.[16] The lag time before reappearance of microbial growth was related to the intensity of the pressure treatment. Two different pressure levels can be defined: a lower one that causes microbial injury and delays growth, and a higher one that induces inactivation of vegetative microorganisms. The ultimate fate of the injured cells will depend on the conditions after pressure treatment. However, the fact that pressure can cause injury may be advantageous when high hydrostatic pressure is combined with other preservation methods.

High Hydrostatic Pressure Mechanism of Action

To explain the response of microorganisms to pressure, cell morphology, cell integrity, and biologic molecules have been studied. Protein denaturation, lipid phase change, and enzyme inactivation can perturb the cell morphology, genetic mechanisms, and biochemical reactions. In the inactivation of microorganisms by HHP, the membrane is the most probable key site of disruption.[9] Inactivation of key enzymes, including those involved in DNA replication and transcription, is also mentioned as a possible inactivating mechanism. The lethal effect of high pressure on vegetative microorganisms is thought to be the result of a number of possible changes taking place simultaneously in the microbial cell.[25]

Shimada et al.[27] suggested that the structural impact of HHP on yeast cells occurred directly in the membrane system, particularly in the nuclear membrane. The outer cell shape of *Saccharomyces cerevisiae*, observed by scanning electron microscopy (SEM), was almost unaffected in high-pressure treatments up to 300 MPa, but at pressures higher than 500 MPa, there was disruption and damage in the cell wall.[27] Transmission electron microscopy (TEM) revealed that the inner structures were damaged, especially the nuclear membrane, even at 100 MPa.[27] Mackey et al.[28] reported that *L. monocytogenes* and *S. thompson* subjected to HHP treatments (250 MPa, 10 minutes) and observed by TEM showed significant changes in the appearance of the nuclear material. The damage profile after HHP treatments reveals pressure sensibility differences among microorganisms.[28,29] HHP treatments can alter the membrane functionality, such as active transport or passive permeability, and, therefore, perturb the physicochemical balance of the cell.[30] The physical state of lipids surrounding membrane proteins plays a crucial role in the activity of membrane-bound enzymes. There is considerable evidence that pressure tends to loosen the contact between attached enzymes and membrane surfaces, as a consequence of the changes in the physical state of lipids that control enzyme activity.[31] Smelt[32] observed that intracellular pH decreased under pressurization and associated the pH drop with the loss of ATPase activity and the reduction on the proton efflux from the cell interior. Knorr[12,13] reported that the reduced Na/K ATPase activity during and after pressurization could be related with a decrease in the bilayer membrane fluidity. Smelt[32] postulated that, to maintain the internal pH homeostasis, membrane-bounded ATPase acts as an ion pump. High pressure can denature the enzyme or cause a dislocation in the membrane; thus, microbial cells could die by internal acidification. Microbial death is also attributed to permeabilization of the cell membrane after an HHP treatment.

Microbial Inactivation

For vegetative cells, it has been reported[4,7] that the increase in holding time can be noticed only when pressures above 200–300 MPa are used. Therefore, a pressure threshold exists for each microorganism and depends on cell growth phase, food composition, and the levels of other variables. The pressure sensitivity of microorganisms varies with the stage of the growth cycle at which the organisms are subjected to HHP treatment. In general, cells in the exponential phase are more sensitive to pressure treatments than are cells in the lag or stationary phases of growth.[25,33] Palou et al.[34] demonstrated pressure sensitivity differences between *Z. bailii* cells from the exponential and stationary growth phases. Yeast cells from the stationary growth phase showed a critical pressure (to reduce 50% the initial inocula) around 300 MPa, whereas, for cells from the exponential growth phase, this critical pressure is reduced to < 172 MPa. Smelt et al.[35] stated that, in the stationary phase, microbial cells are under less optimal conditions than during the exponential phase, due to depletion of the nutrient or formation of toxic metabolites that contribute to expose cells to a stress condition. Cells from the stationary phase will show a stress response and, thus, will be more resistant to adverse environmental conditions. Whereas for formulating a safe process stationary phase cells should be used, exponentially growing cells are particularly suitable for studying mechanistic aspects because the culture conditions are better defined.

Table 12–1 presents a summary of the effects of HHP treatments on selected microorganisms. For each microorganism, there exists a pressure level at which the effect of increasing treatment time causes significant reductions of the initially present or inoculated microorganisms. The intrinsic conditions of the suspension media, such as pH, a_w, and nutrients, may influence the pressure threshold. This pressure threshold can increase or decrease, depending on the microorganism. Suspension media composition drastically influences the microbial response to HHP treatments, showing synergistic or antagonic effects. Baroprotective effects of reduced a_w for organisms that can grow at reduced a_w is reported.[7,16,17] Ogawa et al.[36] reported that, for *S. cerevisiae* inoculated in concentrated fruit juices, the number of surviving microorganisms depends on the juice-soluble solids concentration and observed that the inactivation effect at pressure ≤ 200 MPa decreased as juice concentration increased. Hashizume et al.[37] also reported an increase in *S. cerevisiae* surviving cells with increasing concentrations of sucrose (0–30% w/w), when pressurized at 260 MPa for 20 minutes at 25°C. Oxen and Knorr[38] observed that an HHP treatment at room temperature and 400 MPa for 15 minutes inactivated the yeast *Rhodotorula rubra* when the a_w of the suspension media was higher than 0.96, whereas the number of survivors was higher when the a_w was depressed. The resistance to inhibition at reduced a_w values may be attributed to cell shrinkage, which probably causes a thickening in the cell membrane that reduces membrane permeability and fluidity.[17] The increased baroresistance of microorganisms at low a_w may also be attributed to a partial cell dehydration due to the osmotic pressure gradients between the internal and external fluids, which may render smaller cells, thicker membranes, and an increased pres-

Table 12–1 High Hydrostatic Pressure Treatments and Selected Microorganisms

Microorganism	*Substrate Media*	*Treatment Conditions*		*Decimal Reductions*	*Reference*
Saccharomyces cerevisiae	Satsuma mandarin juice	250 MPa	5 min	≅ 2.0	Ogawa et al.[36]
			10 min	≅ 4.0	
			30 min	6.0	
Aspergillus awamori	Satsuma mandarin juice	250 MPa	5 min	≅ 3.0	Ogawa et al.[36]
			10 min	≅ 4.0	
			30 min	> 4.0	
		300 MPa	5 min	5.0	
Aspergillus niger	Satsuma mandarin juice	400 MPa	10 min	≅ 4.0	Takahashi et al.[44]
Penicillium citrinum	Satsuma mandarin juice	250 MPa	10 min	≅ 4.0	Takahashi et al.[44]
Candida albicans	Satsuma mandarin juice	250 MPa	10 min	≅ 4.0	Takahashi et al.[44]
Zygosaccharomyces bailii	Spaghetti sauce pH 4.0	305 MPa	10 min	≅ 7.0	Pandya et al.[71]
Total plate count	Banana purée	517 MPa	10 min	3.0	Palou et al.[64]
		689 MPa	10 min		
Yeast and molds	Banana purée	517 MPa	10 min	2.0	Palou et al.[64]
		689 MPa	10 min		
Total plate count	Avocado purée	345 MPa	10 min	3.0	López-Malo et al.[19]
			20 min		
			30 min		
Yeast and molds	Avocado purée	345 MPa	10 min	1.0	López-Malo et al.[19]
Total plate count	Guacamole	689 MPa	5 min	4.0	Palou et al.[20]
Yeast and molds	Guacamole	345 MPa	10 min	1.0	Palou et al.[20]
Yeast and molds	Freshly cut pineapple	200 MPa	60 min	0.8	Alemán et al.[72]
		340 MPa	15 min	3.3	
Total plate count	Freshly cut pineapple	200 MPa	5 min	0.6	Alemán et al.[72]
		270 MPa	5 min	1.8	
		(≅ 4°C)	15 min	1.6	
		340 MPa	5 min	1.9	
		(≅ 4°C)	15 min	3.0	
			40 min	2.9	
Listeria innocua	Minced beef muscle	330 MPa	10 min	≅ 2.0	Carlez et al.[42]
			20 min	≅ 3.0	
			30 min	≅ 5.0	
		360 MPa	5 min	≅ 1.0	
			10 min	≅ 2.0	
			20 min	≅ 4.0	
			30 min	≅ 6.0	
Byssochlamys nivea	Apple and cranberry juices	689 MPa (60°C)	1 s 3 cycles	4.0	Palou et al.[47]

sure resistance.[39] The baroprotective effect of reduced a_w reveals that inhibition of microorganisms by high pressure depends not only on pressure and extent of the treatment, but also on the interactions with other intrinsic and extrinsic variables that influence microbial response.[17]

Most minimally processed fruits and vegetables have $a_w > 0.98$. Therefore, a_w may not affect microbial baroresistance in these products.

In biologic systems, volume changes associated with ionization can be involved in the mechanism of microbial inactivation.[7,16] Enhanced ionization under high-pressure treatments is reported for water and acid molecules.[40] Palou et al.[18] mentioned that, during pressurization, a decrease are the pKa of the acids and pH reduction is expected, and a temporary reduction in pH and an increase in the dissociated form of the acid can be present during pressurization. The pH changes could enhance the effects of HHP treatments on microorganisms and favor inactivation. Kinetic studies at pressures over the pressure threshold are needed. With confident kinetic data, pressure sterilization or pasteurization of foods can be predicted and achieved.[4,7]

Temperature during pressurization can have an important effect on the inactivation of microbial cells. Several authors[4,41–43] observed that resistance to pressure of endogenous or inoculated flora is maximal at normal temperatures (15–30°C) and decreases significantly at higher or lower temperatures. Freezing temperatures (–20°C) in treatments ranging from 100 to 400 MPa for 20 minutes enhance microbial inactivation, when compared with HHP treatments at 20°C.[44] Hashizume et al.[37] reported that HHP treatments at elevated (40°C) or subzero (–10 and –20°C) temperatures more effectively inactivated *S. cerevisiae* cells. The decrease in pressure resistance of vegetative cells at low temperatures (< 5°C) may be due to changes in the membrane structure and fluidity, weakening of hydrophobic interactions, and crystallization of phospholipids.[4] On the other hand, moderate heating (40–60°C) can also enhance the pressure microbial inactivation, resulting, in some cases, in a lower minimal inactivation pressure. Ogawa et al.[36,45] reported an enhanced inactivation of natural and inoculated microorganisms in mandarin juice when treated at 40°C, in combination with pressures in the range of 400–450 MPa. Spores from yeast and molds are easily inactivated at pressures of 300 (*Aspergillus oryzae*) or 400 MPa (*Rhizopus javanicus*) at ambient temperatures.[4] However, Butz et al.[46] and Palou et al.[47] demonstrated that ascospores of heat-resistant molds, such as *Byssochlamys nivea*, are extremely pressure resistant. Temperature during pressurization has an important effect on the inactivation of heat-resistant fungi. For the inactivation of *B. nivea* ascospores, pressures above 600 MPa and temperatures above 60°C were needed. No effects on spore viability after treatment at 70°C and 500 MPa for 60 minutes were observed by Butz et al.,[46] a slight inactivation (around 3 log cycle reduction) at 70°C and 600 MPa after 60 minutes, approximately 5 log cycles were reduced at 70°C and 700 MPa, and total and rapid inactivation of *B. nivea* within a few minutes (< 10) was observed at 70°C and 800 MPa. Inactivation of heat-resistant molds in fruit products by thermal processing affects sensory quality of the products, due to severe time–temperature requirements for ascospore inactivation. Preservation by HHP, in combination with moderately elevated temperatures, may be an alternative for fruit juice processing.

Palou et al.[47] reported that *B. nivea* ascospores suspended in a_w 0.98 and 0.94 apple and cranberry juices survived continuous pressure treatments at 60°C and 689 MPa, even after 25 minutes. Similar observations were made by Maggi et al.[48] for *B. nivea* and *B. fulva* ascospores pressurized at 900 MPa for 20 minutes at 20°C. Maggi et al.[48] reported for *B. nivea* ascospores suspended in apricot nectar that there was no inactivation after a 7-minute HHP treatment at 600 MPa and 60°C; complete inactivation was obtained after 1 or 2 minutes if the pressure was raised to 700 MPa at the same temperature. Butz et al.[46] observed no effect on mold ascospore viability after a high-pressure treatment at 500 MPa and 70°C for 60 minutes. Applying another approach, Hayakawa et al.[49] reported successful inactivating HHP treatments for *B. stearothermophilus* spores when oscillatory pressure treatments were combined with 70°C. Four to six compression–decompression cycles at 70°C with 5 minutes of holding time at 600 MPa in each cycle reduced the initial inocula of *B. stea-*

rothermophilus by 4–6 log cycles. Using this approach, Palou et al.[47] reported successful inactivation treatments for *B. nivea* ascospores subjected to three or five cycles of pressurization at 689 MPa and 60°C. Oscillatory pressurization at 60°C was more effective on spore inactivation than was the continuous high-pressure treatment.[47] Oscillatory pressure treatments combined with heat are potentially a sound approach to inactivate heat-resistant molds in diluted fruit juices. However, for concentrated fruit juices, higher pressures, more cycles of compression–decompression, and/or higher temperatures need to be evaluated.

Microbial Inactivation Kinetics

For a broader use of HHP in food processing, it is of special interest to determine the process conditions for pressure pasteurization in view of industrial applications. To increase microbial safety and assure microbial stability of foods processed by HHP, the pressure treatment must ensure a satisfactory reduction in the initial microbial counts. Thus, kinetic analysis and the pressure dependence of microbial inactivation rates are needed. The patterns of HHP inactivation kinetics observed with different microorganisms are quite variable.[7] Some investigators indicate first-order kinetics in the case of several bacteria and yeasts.[18,37,41,42,50] Other authors observed a change in the slope and a two-phase inactivation phenomenon, the first fraction of the population being quickly inactivated, whereas the second fraction appears to be much more resistant.[4] Pressure, temperature, and composition of the medium[7] also influence the pattern of inactivation kinetics.

Determination of the microbial pressure inactivation kinetics depends on pressure increase and decrease rates, and includes a decompression step for sampling. Therefore, sampling, treatment evaluation, and survivor enumeration are not continuous.[7] The pressure increase and decrease are not always reported, and the come-up time to reach the pressure (or pressure buildup velocity) is not always taken into account in the logarithmic representation of the survivors. In many reported HHP survival curves, it is not clear whether time zero experiments are growth controls without pressure treatment or are pressure treatments that take into account only the come-up time to reach the working pressure.[47]

Cheftel[4] and Palou et al.[16,47] reported that the rate of pressure increase and decrease is often neglected as an experimental variable in HHP microbial inactivation studies, and the initial population (N_0) can be notably reduced during the come-up time. Table 12–2 presents the *S. cerevisiae* survival fraction after HHP treatments that account for the time needed to reach 175, 241, and 345 MPa.[50] Pressure come-up times exert an important effect on yeast survival fraction; counts decreased as pressure increased; and, at 345 MPa, a reduction of around 0.5 log cycle was observed. Palou et al.[16,18] reported an important effect of the come-up time at pressures of 345 and 517 MPa on *Z. bailii* log reductions in food model systems with a_w 0.98 or 0.95 and pH 3.5.

The logarithm of *S. cerevisiae* and *Z. bailii* survival fraction (N/N_0) decreased linearly with time, indicating apparent first-order kinetics (Figure 12–3). Yeast cells were inactivated more rapidly with increasing pressure treatments. The experimental points with pressure treatment duration of "0 minutes" express the effect of come-up time to reach the working pressure

Table 12–2 Pressure Come-Up Time and *Saccharomyces cerevisiae* Survival Fraction Suspended in Model Systems of a_w 0.98 and pH 3.5

Pressure (MPa)	*Come-Up Time (min)*	*Survival Fraction (N_0/N_i)**
175	1.9	0.85
241	2.4	0.77
345	3.1	0.39

*N_0 = yeast count (CFU/mL) after pressure come-up time, N_i = yeast initial population (CFU/mL).

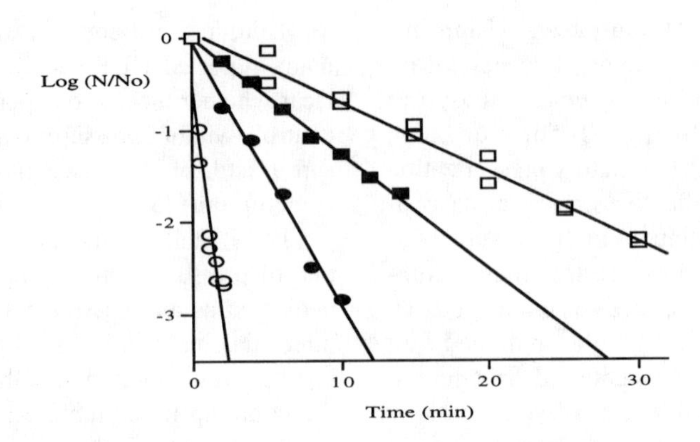

Figure 12–3 *Saccharomyces cerevisiae* (■, ●) and *Zygosaccharomyces bailii* (□, ○) first-order pressure (■, □, 241 MPa; ●, ○ 345 MPa) inactivation kinetics in media of a_w 0.98 and pH 3.5.

and correspond to the initial population (N_0) for the kinetic analysis.[18,50] Hashizume et al.[37] also observed first-order pressure inactivation kinetics for *S. cerevisiae* at 25°C and $a_w \sim 0.99$. To compare the effectiveness of pressure treatments and optimize process conditions, the calculation of D values can be used to compare the resistance of microorganisms. The calculated D values are presented in Table 12–3 and represent the holding time needed at each pressure to reduce 90% of the initial population, considering that the pressure is reached instantaneously. Decimal reduction time obtained by Pérez et al.[50] were greater than those reported by Hashizume et al.[37]; this can be attributed to the pH of model system (3.5), which could favor inactivation.

HIGH HYDROSTATIC PRESSURE AND FOOD-QUALITY-RELATED ENZYMES

Enzymatic reactions are a key problem area to address in high-pressure processing of fruits. For the development of high-pressure processes for fruit products, it is essential to understand the influence of pressure on deteriorative enzymes. Hendrickx et al.[51] stated that the advantages of HHP are that food quality characteristics, such as flavor and vitamins, are unaffected or only minimally altered by high-pressure processing while microorganisms, as well as enzymes re-

Table 12–3 *Zygosaccharomyces bailii* and *Saccharomyces cerevisiae* Pressure Decimal Reduction Times

a_w	Pressure (MPa)	D (min)
*Z. bailii**		
0.98	241	13.1
	276	4.8
	310	2.0
	345	0.8
0.95	414	2.6
	431	2.1
	517	0.9
S. cerevisiae		
≅ 0.99†	210	94.0
	240	34.6
	250	24.6
	270	12.3
0.98‡	175	31.8
	241	18.2
	345	8.0

* Palou et al.[18]
† Hashizume et al.[37]
‡ Pérez et al.[50]

lated to food safety and quality, can be deactivated. Vapor or hot-water blanching generally inactivates fruit and vegetable enzymes. Disadvantages of blanching include thermal damage, leaching of nutrients, and possible environmental pollution, due to the high biochemical oxygen demand effluent produced. HHP treatment can fulfill the requirements of hot-water blanching while avoiding mineral leaching and accumulation of waste water. HHP treatment produces less effluent because less water is required than in hot-water blanching.[52]

Exposure to HHP may activate or inactivate enzymes. However, enzyme activation due to pressure treatment has been reported in only a few cases.[53–55] Pressure inactivation of enzymes is influenced by the pH, substrate concentration, subunit structure of enzyme, and temperature during pressurization.[9,51,56] Pressure effects on enzyme activity are expected to occur at the substrate–enzyme interaction. If the substrate is a macromolecule, the effects may occur on the conformation of the macromolecule, which can make the enzyme action easier or more difficult.[31] Pressure enzyme inactivation can also be attributed to alteration of intermolecular structures or conformational changes at the active site. Enzymes are a special class of proteins in which biologic activity arises from an active site; even small changes in the active site can lead to a loss of enzyme activity.[51] Inactivation of some enzymes pressurized to 100–300 MPa is reversible. Reactivation after decompression depends on the degree of distortion of the molecule. The chances of reactivation decrease with an increase in pressure beyond 300 MPa.[57]

Quaglia et al.[58] reported that pressure treatment at 900 MPa for 10 minutes reduces by 88% the peroxidase activity in green peas, comparable with traditional water blanching. However, pressurization treatment resulted in greater ascorbic acid and firmness retention. Lower pressure levels decreased less than 50% the enzyme activity, even when pressure was combined with moderate temperatures (39–60°C). Anese et al.[55] also observed, for peroxidase from a carrot cell-free extract, that a complete loss of enzyme activity was achieved only when the pressure treatment was applied at 900 MPa for 1 minute. Enzyme activation was observed for treatments in the range of 300–500 MPa. For polyphenoloxidase (PPO) from apple cell-free extract, it was observed that, at pH 7.0, 5.4 and 4.5, a significant reduction in enzyme activity occurred in pressure treatments at 900 MPa for 1 minute. For both enzymes, a pH dependence on residual activity after the pressure treatment was observed. Eshtiaghi and Knorr[52] reported that addition of citric acid could lead to increased polyphenoloxidase inactivation because pH reduction enhances the pressure effects on enzyme inactivation.

Denaturation and inactivation of enzymes occur only when very high pressure treatments are applied; the activation effects that could be presented at relatively low pressures could be attributed to reversible configuration and/or conformation changes on enzyme and/or substrate molecules.[55] Enzyme inactivation by HHP requires a minimum pressure. Below this pressure, no or little enzyme inactivation occurs; when pressure exceeds this value, enzyme inactivation increases.[51] Seyderhelm et al.[14] evaluated the effects of HHP treatments on selected enzymes, including catalase, phosphatase, lipase, pectinesterase, lypoxygenase, peroxidase, polyphenoloxidase, and lactoperoxidase, and reported that peroxidase was the most barostable enzyme, with 90% of residual activity after 30 minutes of treatment at 60°C and 600 MPa. Therefore, peroxidase could be selected as an enzyme indicator for HHP treatments.

In most fruits and vegetables, polyphenoloxidase is responsible for many undesirable color changes during storage; this enzyme can be considered as an HHP target for many minimally processed fruit and vegetables. It is well established that polyphenoloxidases from different sources may have different molecular sizes and conformations. Thus, it is expected that the polyphenoloxidases may respond differently during and after high-pressure treatments.[19,51,56,59] It is also anticipated that important differences will occur when the enzyme activity is analyzed in whole foods, extracts, or commercial enzymes. In untreated onion cells, phenolic compounds are confined to vacuoles and spatially separated from the polyphenoloxidase by the tonoplast; after pressurization (> 100 MPa), the cell and the tonoplast are disrupted, phenolic oxidation products are formed, polyphenoloxidase is no longer separated from the substrate, and enzymatic browning begins.[60] Gomes and Ledward[61] reported a reduction in polyphenoloxidase activity from a crude potato extract with increasing pressure (400–800 MPa for 10 minutes). In contrast, when the crude extract of mushroom was treated at 400 MPa for 10 minutes, an enhancement in the activity was observed. Cano et al.[62] studied the combination of HHP and temperature on peroxidase and polyphenoloxidase activities of fruit-derived products. Optimal inactivation of peroxidase in strawberry purée was achieved using 230 MPa and 43°C. Pressurization–depressurization treatments caused a significant loss of strawberry polyphenoloxidase up to 230 MPa. Combinations of high pressure and 35°C effectively reduced peroxidase in orange juice.

Residual PPO activity in fruit purées after HHP treatments suggests that inhibition of undesirable enzymatic reactions, such as browning, requires the combination of pressurization with one or more additional factors, such as low pH, blanching, and refrigeration temperatures to inhibit (or at least significantly reduce) enzyme activity.[19,20,59] López-Malo et al.[19] reported that polyphenoloxidase in avocado purée can be partially inactivated to about 15–20% of its original activity when treated with HHP (689 MPa) at three pH levels (3.9, 4.1, or 4.3). Pressure and initial pH of the avocado purée significantly ($p < 0.05$) affect residual PPO activity. However, process holding time was not significant ($p < 0.05$) when 345 or 517 MPa was applied. Residual PPO activity in avocado purée (pH 4.1) was 24.7%, 21.8%, and 15.6%, when treated at 689 MPa for 10, 20, and 30 minutes, respectively. This substantial reduction in PPO activity was not enough to avoid avocado purée browning during further storage. However, a significant delay in brown color development was achieved when HHP-treated avocado purée was stored at 5°C.[19] Residual polyphenoloxidase activity of 86% and 63% in guava purée pressure treated at 400 and 600 MPa for 15 minutes and 25°C was reported by Yen and Lin.[15] During 4°C storage, the lightness and greenness of the pressure-treated guava purée decreased continuously. Yen and Lin[15] concluded that, during storage at 4°C, guava purée treated at 400 and 600 MPa at 25°C for 15 minutes maintained acceptable quality for 20 and 40 days, respectively.

Several reports indicate blanching to be a near-prerequisite for pressure treatment of fruit and vegetables to minimize enzymatic and oxidative reactions.[63] The effects of blanching and HHP treatments on polyphenoloxidase activity of banana purée adjusted to pH 3.4 and a_w 0.97 were evaluated by Palou et al.[64] PPO activity was reduced during steam blanching and further reduced after HHP treatments. Thus, browning of purée during storage is diminished, extending the acceptable shelf life. Longer browning induction times and slower browning rates were obtained as residual PPO activity in the banana purée was reduced. Although an important reduction of banana purée PPO activity was observed when a 7-minute steam blanch was combined with a 689 MPa treatment for 10 minutes, the residual PPO activity (< 5%) was sufficient to initiate browning, noticeable after 6 days of storage at 25°C. Pressure resistance of polyphenoloxidase at ambient temperatures was corroborated by Palou et al.,[64] suggesting that application of high pressures to control enzymatic browning in fruit products must be accompanied by other factors, such as blanching pretreatments, elevated temperatures during pressure treatment, and/or refrigerated storage.

It has been suggested that the efficiency of high-pressure enzyme inactivation be improved by applying pressure cycles. Successive applications of HHP treatments resulted in higher inactivation of many enzymes.[51] The enzyme activity retention after a multicycle process was lower than that of a single-cycle process with the same total duration.[65] Recently, Palou et al.[20]

evaluated the effects of continuous or oscillatory HHP treatments on polyphenoloxidase and lipoxygenase (LOX) activities in guacamole. Guacamole preparation included the homogenization of avocado pulp with 1% (w/w) dried onion, 1.5% (w/w) sodium chloride, and citric acid to attain pH 4.3. Two types of HHP treatments were considered: continuous (689 MPa with holding times of 5, 10, 15, or 20 minutes) and oscillatory (2, 3, or 4 cycles at 689 MPa with holding times of 5 or 10 minutes each). Significantly less ($p < 0.05$) residual PPO and LOX activities were obtained with increasing process time and number of pressurization-decompression cycles (Figure 12–4 and Table 12–4). LOX was inactivated with a 15-minute treatment of oscillatory HHP (Table 12–4). The lowest residual PPO activity value (15%) was obtained after 4 HHP cycles or 5 minutes each (Figure 12–4). However, residual PPO activity after HHP treatments suggests that, to avoid quality deteriorative enzymatic reactions during storage, a combination of HHP with additional factors, such as a chilling storage, is required.

HIGH HYDROSTATIC PRESSURE IN COMBINATION WITH OTHER TREATMENTS

The increasing demand for foods with reduced amounts of chemical additives and less physical damage is opening new opportunities for the hurdle technology concept of food preservation.[2,40] The commercial challenge of minimally processed foods provides a strong motivation to study food preservation systems that combine traditional microbial stress factors or hurdles while introducing "new" variables for microbial control, such as high pressure. High pressure presents unique advantages over conventional thermal treatments.[13] However, many reported data indicate that commercial pasteurization or sterilization of low-acid foods using high pressure is very difficult without using some additional factors to enhance the inactivation rate. Factors such as heat, antimicrobials, ultrasound, and ionizing radiation can be used in combination with high pressure. These approaches will not only help to accelerate the rate of inactivation, but can also be useful to reduce the pressure

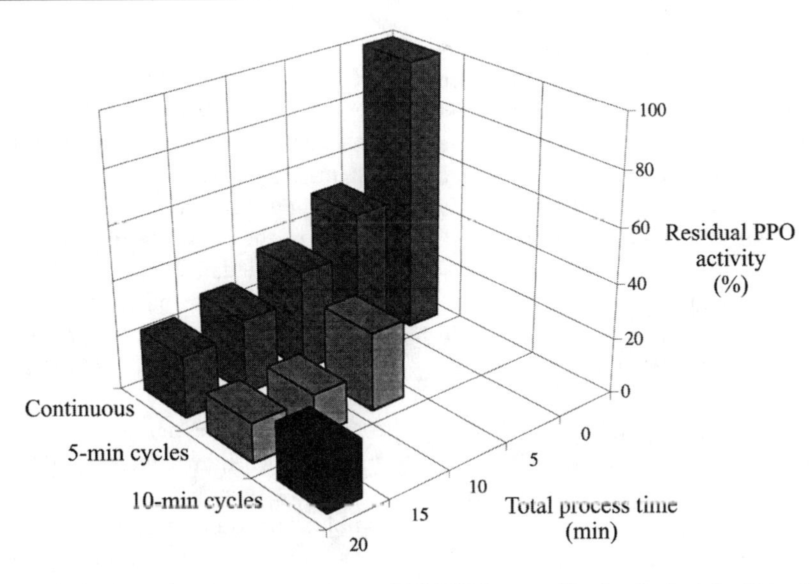

Figure 12–4 High hydrostatic pressure treatments (689 MPa) on residual polyphenoloxidase (PPO) activity of pH 4.3 guacamole.

Table 12–4 Residual Lipoxygenase Activity of High Hydrostatic Pressure-Treated Guacamole

Treatment	*Residual Activity (%)*
Control (without HHP)	100.0
5 min at 689 MPa	41.2
10 min at 689 MPa	4.9
15 min at 689 MPa	ND*
20 min at 689 MPa	ND
2 pressure cycles (5 min at 689 MPa each)	ND
3 pressure cycles (5 min at 689 MPa each)	ND
4 pressure cycles (5 min at 689 MPa each)	ND
2 pressure cycles (10 min at 689 MPa each)	ND

*ND = not detected.

level and, hence, the cost of the process, while eliminating the commercial problems associated with sublethal injury and survivor tails.

Crawford et al.[66] evaluated the combination of HHP, heat, and irradiation to eliminate *Clostridium sporogenes* spores in chicken breast. These authors reported no significant differences in the number of surviving spores between samples that were first irradiated and then pressurized, or vice versa. However, there was a significant difference between samples exposed to combined treatments and those that were only irradiated, showing the combined processes to be more effective. Total inactivation (no survivors) of the initial inoculated spores was accomplished with a 6-kGy irradiation dose, followed by pressurization at 690 MPa and 80°C for 20 minutes. Crawford et al.[66] concluded that a combination of lower doses of irradiation and high pressure is more useful in eliminating *C. sporogenes* spores than is the application of either process alone.

High pressure can be used to reduce the severity of the factors traditionally used to preserve foods. The use of high pressure, in combination with mild heating, has considerable potential.[7,59] Studies have demonstrated that the antimicrobial effect of high pressure can be increased with heat, low pH, carbon dioxide, organic acids, and bacteriocins, such as nisin.[16–18,67,68] Knorr,[12,13] Papineau et al.,[67] and Popper and Knorr[69] reported enhanced pressure inactivation of microorganisms when combining pressure treatments with additives such as acetic, benzoic, or sorbic acids, sulfites, some polyphenols, and chitosan. These combination treatments allow lower processing pressure, temperature, and/or time of exposure.

Tauscher[70] mentioned that the carboxylic acids commonly used as food preservatives show enhanced ionization when subjected to high pressure. However, Palou et al.[16] demonstrated that increased antimicrobial effects can be obtained when combining high pressure and potassium sorbate to inactivate *Z. bailii* in laboratory model systems with reduced a_w and pH. Citric and sorbic acids presented a temporary reduction in pH, and an increase in the dissociated form of the acids could be present, which will depend on the pressure level. This effect could decrease the antimicrobial effectiveness of potassium sorbate during pressurization, because the major antimicrobial action is attributed to the undissociated form of the acid. However, the result of HHP treatments would depend not only on the previously mentioned effects, but also on the consequences of high pressure in the biologic systems involved. For the same pressure level, the holding time required to inhibit *Z. bailii* is shorter in the presence of potassium sorbate. High-pressure damage to *Z. bailii* renders cells more susceptible to other antimicrobial agents (low pH, potassium sorbate), probably due to the exposure of critical cell surface targets. These effects can contribute in a cooperative manner to enhance the pressure effects on microorganisms, and the knowledge and understanding of their effects may aid in the design of effective combined pressure processes.

The hurdle concept could be applied to the optimization of HHP for low-acid foods. A combination of moderate treatments, including pressure, can lead to a food preservation method effective against bacterial spores. Acid foods could be protected from spore outgrowth with the combined treatment. For most of the possible combined processes, the primary goal consists of identifying the factors or treatments that could sensitize microorganisms to pressure[7] or recognizing the factors or treatments that could cause the microbial death in sublethal pressure-injured microbial cells. However, the protective effects that could be exerted on food components make necessary the assessment of each combination process for each particular food product.

CONCLUSION

High hydrostatic pressure has a promising future as a preservation method in the modern food industry. The unique physical and sensory properties of food processed by high pressure technology offer new challenges for food product development. Identification of commercially feasible applications is probably the most difficult of the challenges for high-pressure technology. High pressure, in combination with other preservation factors, can be used to inactivate microorganisms. Future work must be focused on the application of high pressure in the context of minimal processing, including studies on the interaction of pressure with other preservation factors, such as pH, ionic strength, and temperature. There is a need to study synergistic combinations of microbial stress factors, as well as the interactions of several preservation factors to improve food quality and stability of pressure-processed foods. Also, more stability studies are needed to identify the main degradation mode during storage of pressure-processed foods. Oscillatory pressure treatments increased the effectiveness of high-pressure processing. However, the characteristics of the oscillatory pressure treatments need to be further optimized, and it will be necessary to conduct in-depth studies to determine commercial feasibility.

For better and more efficient use of hurdle technology in fruit and vegetable minimal processing, further research is needed on the mechanism of action of traditional and emerging preservation factors on microorganisms, enzymes, and deteriorative reactions, which continues to be an important issue for the development of minimal processing systems. A better understanding regarding these areas will help to identify key factors and their combined effect on product safety, stability, and quality.

REFERENCES

1. Gould GW. The microbe as a high pressure target. In: Ledward DA, Johnston DE, Earnshaw RG, Hasting APM, eds. *High Pressure Processing of Foods*. Nottingham: Nottingham University Press; 1995:27–36.
2. Leistner L. Principles and applications of hurdle technology. In: Gould GW, ed. *New Methods of Food Preservation*. New York: Blackie Academic and Professional; 1995:1–21.
3. Hayashi R. Advances in high pressure processing technology in Japan. In: Gaonkar AG, ed. *Food Processing: Recent Developments*. London: Elsevier Science; 1995:185–195.
4. Cheftel JC. Review: High-pressure, microbial inactivation and food preservation. *Food Sci Technol Int*. 1995;1:75–90.
5. Pothakamury UR, Barbosa-Cánovas GV, Swanson BG, Meyer RS. The pressure builds for better food processing. *Chem Engr Prog*. 1995;March:45–53.
6. Barbosa-Cánovas GV, Pothakamury UR, Palou E, Swanson BG. *Nonthermal Preservation of Foods*. New York: Marcel Dekker; 1998.
7. Palou E, López-Malo A, Barbosa-Cánovas GV, Swanson BG. High pressure treatment in food preservation. In: Rahman MS, ed. *Handbook of Food Preservation*. New York: Marcel Dekker; 1999:533–576.
8. Anonymous. Flow International Co., Technical Data; 1998.
9. Hoover DG, Metrick C, Papineau AM, Farkas DF, Knorr D. Biological effects of high hydrostatic pressure on food microorganisms. *Food Technol*. 1989;43:99–107.

10. Farr D. High pressure technology in the food industry. *Trends Food Sci Technol.* 1990;1:14–16.

11. Ledward DA. High pressure processing—the potential. In: Ledward DA, Johnston DE, Earnshaw RG, Hasting APM, eds. *High Pressure Processing of Foods.* Nottingham: Nottingham University Press; 1995:1–6.

12. Knorr D. Hydrostatic pressure treatment of food: Microbiology. In: Gould GW, ed. *New Methods of Food Preservation.* New York: Blackie Academic and Professional; 1995:159–175.

13. Knorr D. High pressure effects on plant derived foods. In: Ledward DA, Johnston DE, Earnshaw RG, Hasting APM, eds. *High Pressure Processing of Foods.* Nottingham: Nottingham University Press; 1995:123–136.

14. Seyderhelm I, Bouguslawski S, Michaelis G, Knorr D. Pressure induced inactivation of selected enzymes. *J Food Sci.* 1996;61:308–310.

15. Yen GC, Lin HT. Comparison of high pressure treatment and thermal pasteurization effects on the quality and shelf-life of guava purée. *Int J Food Sci Technol.* 1996;31:205–213.

16. Palou E, López-Malo A, Barbosa-Cánovas GV, Welti-Chanes J, Swanson BG. High hydrostatic pressure as a hurdle for *Zygosaccharomyces bailii* inactivation. *J Food Sci.* 1997;62:855–857.

17. Palou E, López-Malo A, Barbosa-Cánovas GV, Welti-Chanes J, Swanson BG. Effect of water activity on high hydrostatic pressure inhibition of *Zygosaccharomyces bailii. Lett Appl Microbiol.* 1997;24:417–420.

18. Palou E, López-Malo A, Barbosa-Cánovas GV, Welti-Chanes J, Swanson BG. Kinetic analysis of *Zygosaccharomyces bailii* inactivation by high hydrostatic pressure. *Lebensm -Wiss u-Technol.* 1997;30:703–708.

19. López-Malo A, Palou E, Barbosa-Cánovas GV, Welti-Chanes J, Swanson BG. Polyphenoloxidase activity and color changes during storage of high hydrostatic pressure treated avocado purée. *Food Res Int.* 1998;31:549–556.

20. Palou E, Hernández- Salgado C, López-Malo A, Barbosa-Cánovas GV, Swanson BG, Welti J. High pressure-processed guacamole. *Innovative Food Sci Emerg Technol.* 2000; 1: in press.

21. Gould GW. Industry perspectives on the use of natural antimicrobials and inhibitors for food applications. *J Food Prot.* 1996;Suppl:82–86.

22. Parish ME. Public health and nonpasteurized fruit juices. *Crit Rev Microbiol.* 1997;23:109–119.

23. Reyes VG. Improved preservation systems for minimally processed vegetables. *Food Austr.* 1996;48:87–90.

24. Nguyen-the C, Carlin F. The microbiology of minimally processed fresh fruits and vegetables. *Crit Rev Food Sci Nutr.* 1994;34:371–401.

25. Patterson MF, Quinn M, Simpson R, Gilmour A. Sensitivity of vegetative pathogens to high hydrostatic pressure treatment in phosphate-buffered saline and foods. *J Food Prot.* 1995;58:524–529.

26. Isaacs NS, Chilton P, Mackey B. Studies on the inactivation by high pressure of micro-organisms. In: Ledward DA, Johnston DE, Earnshaw RG, Hasting APM, eds. *High Pressure Processing of Foods.* Nottingham: Nottingham University Press; 1995:65–79.

27. Shimada S, Andou M, Naito N, Yamada N, Osumi M, Hayashi R. Effects of hydrostatic pressure on the ultrastructure and leakage of internal substances in the yeast *Saccharomyces cerevisiae. Appl Microbiol Biotechnol.* 1993;40:123–131.

28. Mackey BM, Forestiere K, Isaacs NS, Stenning R, Brooker B. The effect of high hydrostatic pressure on *Salmonella thompson* and *Listeria monocytogenes* examined by electron microscopy. *Lett Appl Microbiol.* 1994;19:429–432.

29. Sato M, Kobori H, Ishijima SA, Feng ZH, Hamada K, Shimada S, et al. *Schizosaccharomyces pombe* is more sensitive to pressure stress than *Saccharomyces cerevisiae. Cell Structure Function.* 1996;21:167–174.

30. Perrier-Cornet JP, Marénchal PA, Gervais P. A new design intended to relate high pressure treatment to yeast cell mass transfer. *J Biotechnol.* 1995;41:49–58.

31. Heremans K. High pressure effects on biomolecules. In: Ledward DA, Johnston DE, Earnshaw RG, Hasting APM, eds. *High Pressure Processing of Foods.* Nottingham: Nottingham University Press; 1995:81–98.

32. Smelt JP. Some mechanistic aspects of inactivation of bacteria by high pressure. *Proceedings of European Symposium—Effects of High Pressure on Foods.* University of Montpellier, February 16–17; 1995.

33. Earnshaw RG. Kinetics of high pressure inactivation of microorganisms. In: Ledward DA, Johnston DE, Earnshaw RG, Hasting APM, eds. *High Pressure Processing of Foods.* Nottingham: Nottingham University Press; 1995:37–46.

34. Palou E, López-Malo A, Barbosa-Cánovas GV, Welti-Chanes J, Davidson PM, Swanson BG. High hydrostatic pressure come-up time and yeast viability. *J Food Prot.* 1998;61:1657–1660.

35. Smelt JP, Wouters PC, Rijke AGF. Inactivation of microorganisms by high pressure. In: Reid DS, ed. *The Properties of Water in Foods, ISOPOW 6.* London: Blackie Academic and Professional; 1998:398–417.

36. Ogawa H, Fukuhisa K, Fukumoto H. Effect of hydrostatic pressure on sterilization and preservation of citrus juice. In: Balny C, Hayashi R, Heremans K, Masson P, eds. *High Pressure and Biotechnology.* France: INSERM & John Libbey; 1992:269–278.

37. Hashizume C, Kimura K, Hayashi R. Kinetic analysis of yeast inactivation by high pressure treatment at low temperatures. *Biosci Biotech Biochem.* 1995;59: 1455–1458.
38. Oxen P, Knorr D. Baroprotective effects of high solute concentrations against inactivation of *Rhodotorula rubra. Lebensm-Wiss u-Technol.* 1993;26:220–223.
39. Knorr D. Effects of high-hydrostatic pressure process on food safety and quality. *Food Technol.* 1993;47: 156–161.
40. Earnshaw RG, Appleyard J, Hurst RM. Understanding physical inactivation process: Combined preservation opportunities using heat, ultrasound and pressure. *Int J Food Microbiol.* 1995;28:197–219.
41. Smelt JP, Rijke G. High pressure treatment as a tool for pasteurization of foods. In: Balny C, Hayashi R, Heremans K, Masson P, eds. *High Pressure and Biotechnology.* France: INSERM & John Libbey; 1992:361–363.
42. Carlez A, Rosec JP, Richard N, Cheftel JC. High pressure inactivation of *Citrobacter freundii*, *Pseudomonas fluorescens* and *Listeria innocua* in inoculated minced beef muscle. *Lebensm-Wiss u-Technol.* 1993;26:357–363.
43. Carlez A, Rosec JP, Richard N, Cheftel JC. Bacterial growth during chilled storage of pressure-treated minced meat. *Lebensm-Wiss u-Technol.* 1994;27:48–54.
44. Takahashi Y, Ohta H, Yonei H, Ifuku Y. Microbicidal effect of hydrostatic pressure on satsuma mandarin juice. *Int J Food Sci Technol.* 1993;28:95–102.
45. Ogawa H, Fukuhisa K, Kubo Y, Fukumoto H. Inactivation effect of pressure does not depend on the pH of the juice. *Agric Biol Chem.* 1990;54:1219–1225.
46. Butz P, Funtenberger S, Haberdtzl T, Tauscher B. High pressure inactivation of *Byssochlamys nivea* ascospores and other heat resistant moulds. *Lebensm-Wiss u-Technol.* 1996;29:404–410.
47. Palou E, López-Malo A, Barbosa-Cánovas GV, Welti-Chanes J, Davidson, PM, Swanson BG. Effect of oscillatory high hydrostatic pressure treatments on *Byssochlamys nivea* ascospores suspended in fruit juice concentrates. *Lett Appl Microbiol.* 1998;27:375–378.
48. Maggi A, Gola S, Spotti E, Rovere P, Mutti P. Tratamenti ad alta pressione di ascospore di muffe termoresistenti e di patulina in nettare di albicocca e in acqua. *Ind Conserve.* 1994;69:26–29.
49. Hayakawa I, Kanno T, Yoshiyama K, Fuji Y. Oscillatory compared with continuous high pressure sterilization on *Bacillus stearothermophilus* spores. *J Food Sci.* 1994;59:164–167.
50. Pérez J, López-Malo A, Palou E, Barbosa-Cánovas GV, Swanson BG, Vélez J, et al. *IFT Annual Meeting.* Orlando, FL. June 14–18, 1997: No. 59E-19.
51. Hendrickx M, Ludikhuyze L, Van den Broeck I, Weemaes C. Effects of high pressure on enzymes related to food quality. *Trends Food Sci Technol.* 1998;9:197–203.
52. Eshtiaghi MN, Knorr D. Potato cubes response to water blanching and high hydrostatic pressure. *J Food Sci.* 1993;58:1371–1374.
53. Asaka M, Hayashi R. Activation of polyphenol oxidase in pear fruits by high pressure treatment. *Agric Biol Chem.* 1991;55:2439–2440 .
54. Aoyama Y, Asaka M, Nakunishi R. Effect of high pressure on activity of some oxidizing enzymes. In: Yano T, Matsuno R, Nakamura K, eds. *Developments in Food Engineering.* Proceedings of the 6th International Congress on Engineering and Food. London: Blackie Academic and Professional; 1994:861–863.
55. Anese M, Nicoli MC, Dall'Aglio G, Lerici CR. Effect of high pressure treatments on peroxidase and polyphenoloxidase activities. *J Food Biochem.* 1995;18:285–293.
56. Weemaes C, Ludikhuyze L, Van den Broeck I, Hendrickx M. High pressure inactivation of polyphenoloxidases. *J Food Sci.* 1998;63:873–877.
57. Jaenicke R. Enzymes under extreme conditions. *Ann Rev Biophys Bioeng.* 1981;10:1–67.
58. Quaglia GB, Gravina R, Paperi R, Paoletti F. Effect of high pressure treatments on peroxidase activity, ascorbic acid content and texture in green peas. *Lebens-Wiss u-Technol.* 1996;29:552–555.
59. López-Malo A, Palou E, Barbosa-Cánovas GV, Swanson BG, Welti-Chanes J. Minimally processed foods and high hydrostatic pressure. In: Lozano J, Añon C, Parada-Arias E, Barbosa-Cánovas GV, eds. *Current Trends in Food Engineering.* Gaithersburg, MD: Aspen Publishers; 2000:267–286.
60. Butz P, Koller WD, Tauscher B, Wolf S. Ultra-high pressure processing of onions: Chemical and sensory changes. *Lebensm -Wiss u- Technol.* 1994;27:463–467.
61. Gomes MRA, Ledward DA. Effect of high-pressure treatment on the activity of some polyphenoloxidases. *Food Chem.* 1996;56:1–5.
62. Cano MP, Hernández A, De Ancos B. High pressure and temperature effects on enzyme inactivation in strawberry and orange products. *J Food Sci.* 1997;62:85–88.
63. Hoover DG. Pressure effects on biological systems. *Food Technol.* 1993;47:150–155 .
64. Palou E, López-Malo A, Barbosa-Cánovas GV, Welti-Chanes J, Swanson BG. Polyphenoloxidase activity and color of blanched and high hydrostatic pressure treated banana purée. *J Food Sci.* 1999;64:42–45.
65. Ludikhuyze L, Van den Broeck I, Weemaes CA, Hendrickx ME. Kinetic parameters for pressure-temperature

inactivation of *Bacillus subtilis* α-amylase under dynamic conditions. *Biotechnol Prog*. 1997;13:617–623.

66. Crawford YJ, Murano EA, Olson DG, Shenoy K. Use of high hydrostatic pressure and irradiation to eliminate *Clostridium sporogenes* spores in chicken breast. *J Food Prot*. 1996;59:711–715.
67. Papineau AM, Hoover HG, Knorr D, Farkas DF. Antimicrobial effect of water-soluble chitosans with high hydrostatic pressure. *Food Biotechnol*. 1991;5:45–57.
68. Mallidis CG, Drizou D. Effect of simultaneous application of heat and pressure on the survival of bacterial spores. *J Appl Bacteriol*. 1991;71:285–288.
69. Popper L, Knorr D. Applications of high-pressure homogenization for food preservation. *Food Technol*. 1990;44:84–89.
70. Tauscher B. Pasteurization of food by hydrostatic pressure: Chemical aspects. *Z Lebensm Unters Forsch*. 1995;200:3–13.
71. Pandya Y, Jewett FF, Hoover DG. Concurrent effect of high hydrostatic pressure, acidity and heat on the destruction and injury of yeast. *J Food Prot*. 1995;58:301–304.
72. Alemán GD, Farkas DF, Torres JA, Wilhelmsen E, Mcintyre S. Ultra-high pressure pasteurization of fresh cut pineapple. *J Food Prot*. 1994;57:931–934.

CHAPTER 13

Processing Fruits and Vegetables by Pulsed Electric Field Technology

Gustavo V. Barbosa-Cánovas, M. Marcela Góngora-Nieto, and Barry G. Swanson

INTRODUCTION

Since food preservation became a concern to society, scientists, processors, and consumers have been searching for a panacea method to preserve the high-quality attributes of food products for comparably long times with the guarantee that no illnesses will be developed by their consumption. In addition, the perfect technology must comply with the demands of business minds looking for energy-saving processes. No food processing technology gathers all of these attributes for all possible products, although nonthermal methods have been proven to overcome most of the disadvantages produced by thermal processes.

Consumer demand for healthy, fresh, minimally processed, preservative-free, and long shelf-life products is challenging traditional heat pasteurization treatments, as well as becoming a driving force for the development of new processing technologies. Although natural fruit juices are an excellent source of freshlike taste, aromatic flavor, and healthy high-vitamin content, unfortunately, these attributes are quickly lost if the products are not refrigerated, frozen, or processed immediately. Heat pasteurization, followed by refrigerated storage, allows higher product shelf life, but also causes freshlike taste and vitamin content to deteriorate dramatically. Nonthermal pasteurization methods, such as pulsed electric fields (PEF), can overcome such heat treatment disadvantages because they permit a low pasteurization temperature that inactivates spoilage and pathogenic flora, while preserving the freshlike flavor and vitamin content of processed juice. Due to their electrical and chemical characteristics, fruit juices are one of the most promising products to be processed by PEF technology in the new millennium.

PEF processing originated in the early 1900s, when the use of alternating current was claimed to pasteurize milk.[1] This and other inactivation studies were conducted on liquid products placed in between a pair of electrodes. Since then, it has become well known that electric currents significantly influence the metabolisms and viability of microorganisms. Conversely to the first electric food treatments, PEF uses electricity in the form of high-voltage short pulses that minimize the ohmic heating effect of current flow through the product that took place in the early electric pasteurization of milk.

Microbial inactivation by PEF has been proven to be mainly a function of the particular electric field and number of pulses applied (or total treatment time). Other factors, such as pulse wave shape and frequency, product composition, product physical and chemical characteristics, processing temperature, and type of target microorganism, also have an influence on the effectiveness of the process. This chapter reviews the basics of PEF processing, proposed inactivation mechanisms and related evidence, inactivation kinetics of predominant flora and enzymes in juices, effect of PEF on the overall quality and shelf life of fruit juices, and the energy demand of the process to pasteurize fruit juices.

PEF SYSTEM

PEF is one of the soundest technologies racing to be implemented in the near future. Supported by laboratory and pilot plant research results, PEF has successfully preserved the freshlike characteristics of fruit juices while inactivating some enzymes and microbiologic flora responsible for spoilage and quality degradation. A high-intensity PEF can be easily generated with a simple electric system consisting of a high-voltage source, capacitor bank, high-voltage switch, treatment chamber, and pulse-shaping inductors and resistances. The lethality of a PEF treatment corresponds to the energy delivered by each pulse, which can be monitored and measured using voltage and current probes interfaced with an oscilloscope. As with most food processing technologies, it is advisable to use an aseptic packaging system immediately after PEF processing.

PEF processing of food involves the application of short pulses of high electric fields with duration of micro- to milliseconds and intensity on the order of 20–80 kV/cm. The processing time is calculated as the number of pulses times its duration. As in traditional heat pasteurization, the lethality of the process is proportional to the treatment or residence time in the processing equipment or treatment zone. In thermal processing, this is generally known as a *heat exchanger*, whereas, in PEF, it is a *treatment chamber*. The many existing treatment chamber designs are basically a pair of electrodes—one connected to a high-voltage source and the other to the ground. Parallel plates and wires, concentric (coaxial) cylinders, rod-plate, and co-field are some possible electrode configurations of such treatment chambers.[2] The high electric field generated within the gap of the electrodes conforming the chamber is the most relevant feature in PEF. The simplest definition of an electric field from an electrical engineering viewpoint is "the voltage drop across the electrode gap." High electric field pulses are generated when the high-voltage electrode is charged and discharged in fractions of a second, so that a food product will be exposed to such electric field pulses when it is placed or pumped through the electrode gap. The high-voltage pulses have different wave shapes; in descending order, from the most to the least studied, these are: exponential, square-wave, oscillatory, bipolar, and instant reverse charge (Figure 13–1). Oscillatory pulses seem to be the least efficient for microbial inactivation. Exponentially decaying pulses are produced by a very simple electric circuit that allows high versatility because a variety of foods, including dairy, fruit, and poultry products, can be treated using the same system configuration. Therefore, this wave shape is suitable for evaluating a wide range of experimental conditions. Although the energy efficiency distribution of exponentially decaying pulses can generate ohmic-heating problems in products with high electrical conductivity, in general, square-wave pulses are more energy-efficient and lethal than are exponentially decaying pulses. However, the PEF electric circuit that they require is more complex because it involves a series of capacitor–inductor units that emulate a transmission line. The stress caused by bipolar pulses yields higher microbial inactivation levels, compared with monopolar pulses, and the former also offer the advantages of minimum energy utilization, reduced deposition of solids on electrode surfaces, and reduced food electrolysis.[3] Instant reverse charge pulses are one of the latest-developed pulse wave shapes. As such, they are considered to be the most energy-efficient, based on their capability to generate membrane stresses followed by membrane breakdown, and it has been claimed that they can achieve more than 99.9% spore inactivation.[4]

As with any emerging technology, PEF has some drawbacks. One of the most significant is dielectric breakdown of foods, which is characterized by a spark. Dielectric breakdown is attributed to local electric field enhancement, due to differences in dielectric properties. This phenomenon can be minimized during pulsing by

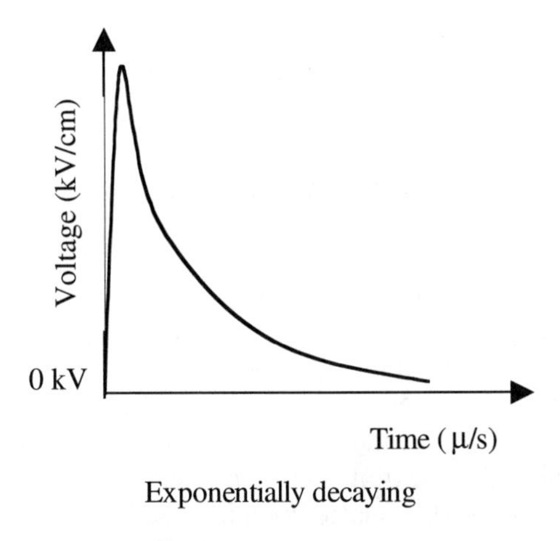

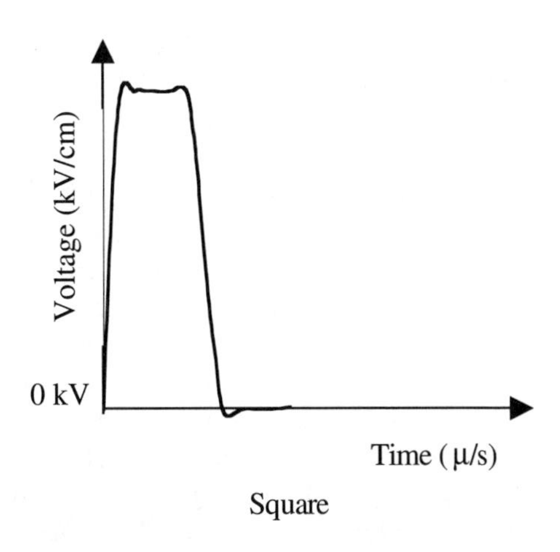

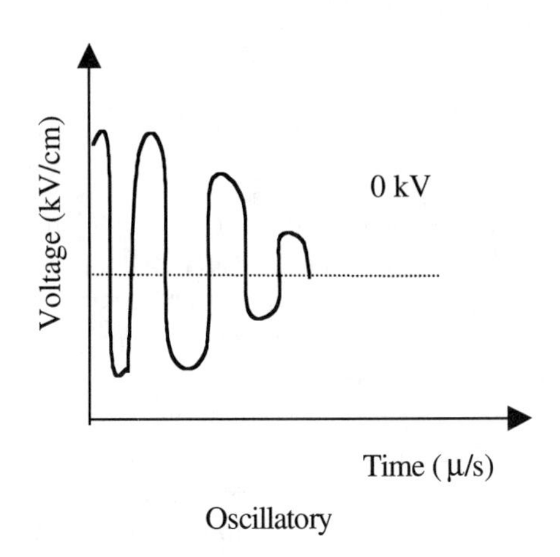

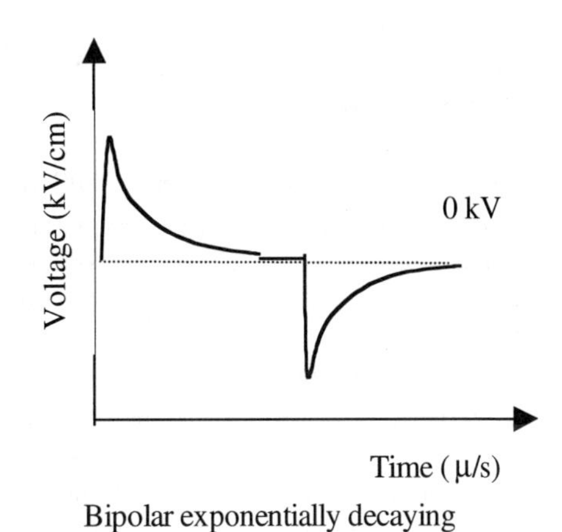

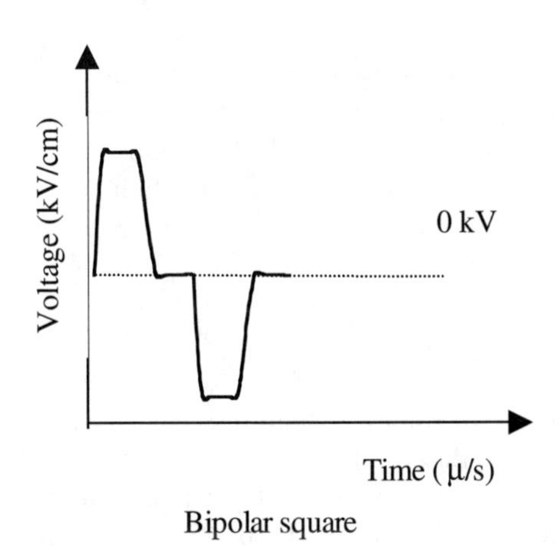

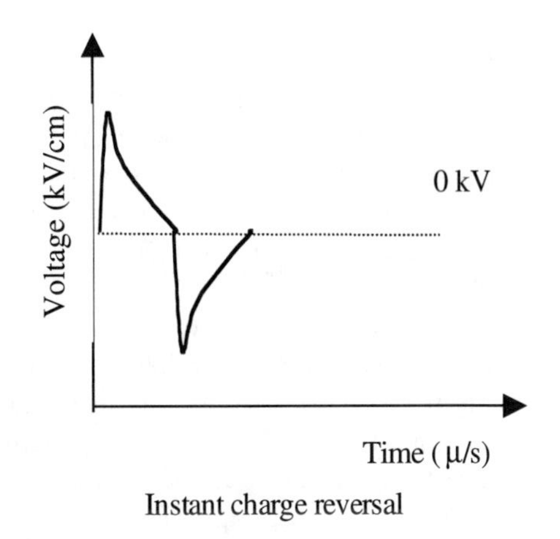

Figure 13–1 Possible wave forms in PEF technology.

pressuring the system, using smooth electrode surfaces, round electrode edges, and degassing the chamber to provide a uniform electric field.

INACTIVATION MECHANISMS

Based on the dielectric rupture theory, an external electric field can induce an electric potential difference across cell membranes. This is known as *transmembrane potential*.[5] When the transmembrane potential reaches a critical or threshold value higher than a cell's natural potential of ≈ 1 Volt,[6] electroporation or pore formation in the membrane occurs, and cell membrane permeability thus increases.[7] It is important to know, however, that the threshold transmembrane potential depends on the specific microorganism, as well as on the medium in which the microorganisms are present.

The increase in membrane permeability is reversible if the external electric field strength is equal to or slightly exceeds the critical value. Gásková et al.[8] report that a pulse-induced increase in membrane permeability for small species of inorganic ions is sufficient to cause cell death. Cell membrane breakdown or irreversible electroporation occurs under the application of high-intensity electric field pulses.[9] Microscopy observations by Barsamian and Barsamian[10] on fruit tissue samples and by Lubicki and Jayaram[11] on the bacterium *Yersinia enterocolitica* reveal irreversible membrane disruption to be an inactivation mechanism. Some of the ultrastructural changes observed in *Saccharomyces cerevisiae* cells after PEF treatment by scanning electron microscopy (SEM) include surface roughening and pinholes, craters or holes, elongation, wrinkling, and increased bud scar formation. Although transmission electron microcopy (TEM) observations of PEF-treated *S. cerevisiae* cells provide little evidence to support the electroporation inactivation theory as the major mode of yeast inactivation, this technique has made it possible to observe frequent disruption of cellular organelles and the absence of ribosome bodies. Such intracellular damage is evident in most of these cases, where less than 1% of the cells are disrupted (pored). These findings suggest cytologic disruption as an alternative inactivation mechanism to electroporation.[12]

MICROBIAL INACTIVATION BY PEF

When a suspension of microorganisms is treated at a constant electric field for several seconds, the decrease in number of viable organisms follows a first-order reaction. This same behavior is exhibited if the microbial suspension is treated for the same amount of time (or number of pulses) at different electric fields. Hülsheger et al.[13] related the microbial survival rate S (survivors after treatment/microbial population before treatment), with the electric field strength E (kV/cm), according to the following equation:

$$\text{Ln } S = -b_e(E - E_c) \qquad \textbf{(1)}$$

where b_e (cm/kV) is the electric field rate constant, E (kV/cm) the applied electric field, and E_c the extrapolated value of E for 100% survival, otherwise known as the *critical* or *threshold electric field* (Figure 13–2). In a similar way, Hülsheger et al.[13] related the microbial survival rate to the treatment time t (μs; $t = \tau^*n$) [defined as the product of the pulse width τ(μs) and number of pulses] with the following equation:

$$\text{Ln } S = -b_t(t - t_c) \qquad \textbf{(2)}$$

where b_t (1/s) is the time rate constant and t_c the extrapolated value of the treatment time for 100% survival (Figure 13–2). Equations 1 and 2 can be combined to express the relation among these variables:

$$S = k\left(\frac{n\tau}{t_c}\right)^{(E-E_c)} \qquad \textbf{(3)}$$

where k is an empirical constant.

In fruit juice pasteurization, the principal target is the inactivation of yeast and molds. The high sensitivity of these microorganisms and the low conductivity of fruit juices work

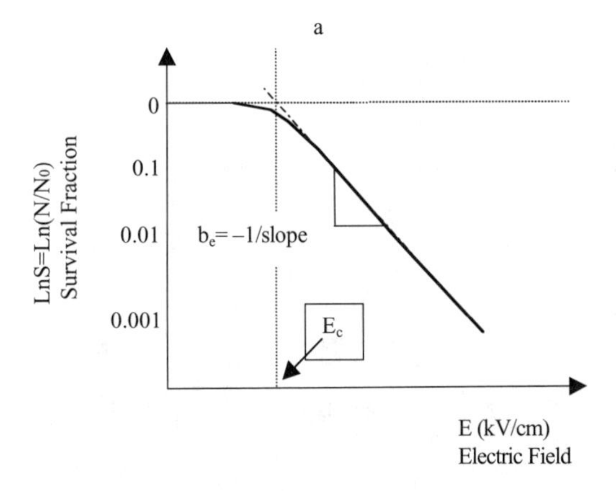

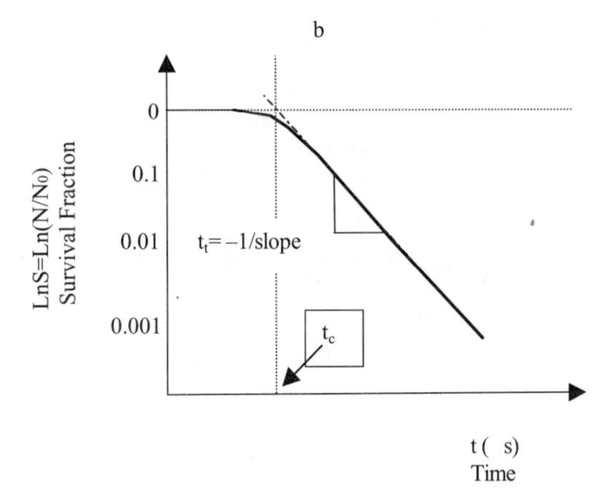

Figure 13–2 Dependence of microbial survival fraction on the (a) electric field and (b) treatment time.

together to enhance PEF inactivation efficiency. The majority of inoculation studies conducted to evaluate the feasibility of juice pasteurization by PEF have been performed using strains of *S. cerevisiae,* molds, and some *Lactobacillus* species. Most of the current studies deal with the inactivation effect of the main PEF factors (i.e., electric field, number of pulses, wave shape, processing temperature, and microbial cell growth stage). Although many of these studies were conducted in different PEF treatment chambers, which might have an affect on the final inactivation level, this has not yet been clearly quantified (Table 13–1).

The work of Gásková et al.[8] indicates that a substantial killing effect on *S. cerevisiae* suspended in distilled water can be obtained if rectangular electric pulses of 4–28 kV/cm with a pulse width of 10 μs are used. This study also evaluated the critical electric field, which was found to be about 2 kV/cm and the transmembrane breakdown voltage of 0.75 V. Because this critical potential for membrane breakdown decreases when temperature increases,[14] both moderate temperature and electric field treatments exhibit synergistic effects on the inactivation of microorganisms. Some examples of this can be found in the experiments conducted by Ohshima et al.,[15] where the minimum survival ratio of *S. cerevisiae* was determined to increase proportionally with the temperature from about 10^{-6} units when 300 J/mL was applied at 50°C to 10^{-1} units at 10°C. It was also verified that thermal effect without pulse treatment was not observed below 45°C. Although inactivation has been shown to increase with an increase in temperature, it should be noted that temperatures must be held far below those used in pasteurization to minimize heat effects on product quality and to claim a nonthermal pasteurization.

Zhang et al.[16] achieved a 4-log cycle reduction of *S. cerevisiae* suspended in apple juice after 20 pulses of 12 kV/cm. The final inactivation levels in this study were nearly the same for square and exponential decay pulse wave shapes, although the pulse energy efficiencies were 91% and 64%, respectively, considering that the effective energy came from electric fields higher than a critical electric field of 8 kV/cm.

A comparative study conducted by Raso et al.[17] revealed the effectiveness of PEF treatment in the inactivation of *Zygosaccharomyces bailii* vegetative cells and ascospores suspended in a variety of fruit juices. The study evaluated the effects of mild heat (50°C), high hydrostatic pressure (300 MPa at 25°C), and PEF treatments (two exponential decay pulses of 32–36 kV/cm with a pulse width of 2–3.3 μs). The results showed that ascospores were up to eight times more heat-resistant than were vegetative cells,

Table 13–1 Summary of *Saccharomyces cerevisiae* Inactivation with PEF

Source	*Suspension Media*	*Maximum Log Reduction*	*Treatment Vessel (B = batch, C = continuous)*	*Process Conditions (temperature, peak electric field, pulse width, number of pulses and shape, t = treatment time)*
Jacob et al.[33]	0.9% NaCl	1.3	B, 3 mL, d = 0.5 cm	35 kV/cm, 20 μs, 4 pulses
Dunn and Pearlman[29]	Yogurt	3	B	55°C, 18 kV/cm
Hülsheger et al.[13]	Phosphate buffer, pH 7.0	3 (stationary-phase cells) 4 (exponential-phase cells)	B, 4 mL, d = 0.5 cm	20 kV/cm, 36 μs, 30 pulses, t = 1,080 μs
Mizuno and Hori[33]	Deionized water	6	0.77 cal/cm³/pulse, B, parallel plate, 0.5 cm³, d = 0.8 cm	20 kV/cm, 160 μs 175 pulses, exponential decay
Matsumo et al.[35]	Phosphate buffer	5	B	30 kV/cm
Yonemoto et al.[36]	0.85% NaCl	2	B, parallel plate, 2 mL, d = 0.55 cm	5.4 kV/cm, 90 μs, 10 pulses
Zhang et al.[37]	Potato dextrose agar	5.5	62 J/mL, B, 14 mL	15 ± 1°C, 40 kV/cm, 3 μs, 16 pulses
Qin et al.[3]	Apple juice	4	270 J/pulse, B, parallel plate	< 30°C, 12 kV/cm, 20 pulses, exponential decay
Qin et al.[3]	Apple juice	4.2	270 J/pulse, B, parallel plate	< 30°C, 12 kV/cm, 20 pulses, square-wave
Zhang et al.[16]	Apple juice	4	260 J/pulse, B, parallel plate, 25 mL, d = 0.95 cm	4–10°C, 12 kV/cm, 90 μs, 6 pulses, exponential decay
Zhang et al.[16]	Apple juice	4	260 J/pulse, B, parallel plate, 25 mL, d = 0.95 cm	4–10°C, 12 kV/cm, 60 μs, 6 pulses, square-wave
Zhang et al.[16]	Apple juice	3–4	558 J/pulse, B, parallel plate, 25.7 mL, d = 0.95 cm	< 25°C, 25 kV/cm, 5 pulses
Qin et al.[38]	Apple juice	7	C, coaxial, 29 mL, d = 0.6 cm, 0.2 μF, 1 Hz	< 30°C, 25 kV/cm, 2–20 μs, ± 150 pulses, exponential decay
Qin et al.[38]	Apple juice	6	28 J/mL, C, coaxial, 30 mL, 2–10 1/min	22–29.6°C, 50 kV/cm, 2.5 μs, 2 pulses
Grahl et al.[39]; Grahl and Märkl[40]	Orange juice	5	B, 25 mL, d = 0.5 cm, Ec = 4.7	6.75 kV/cm, 5 pulses

and a similar behavior was observed in the HHP treatments where, after 5 minutes, the vegetative cell population decreased almost 5 log cycles, but the ascospore population decreased only 1 log cycle. Conversely, PEF was very effective in inactivating both vegetative cells and ascospores, because just two pulses decreased the vegetative cells up to 5 log cycles, and the ascospores count was reduced up to 4.2 log cycles (Table 13–2). Although no viable yeast was detected after the vegetative cells received three pulses (from an initial population of 10^6 CFU/mL), the higher resistance of the ascospores produced some tailing in the inactivation curves (from an initial population of 10^5 CFU/mL), which tended to disappear when the number of pulses was increased. It is also worth mentioning that higher heat and PEF resistances were found for both vegetative cells and ascospores when the microorganisms were suspended in pineapple juice. When the inactivation of *Neosartorya fischeri* ascospores and *Byssochlamys fulva* (*Paecilomyces fulvus*) conidiospores in fruit juices was evaluated by Raso et al.,[18] the results indicated that mold conidiospores are very sensitive to PEF and can be inactivated with a small number of pulses of moderate electric field intensity (Table 13–3). However, *N. fischeri* ascospores were found to be resistant to PEF, based on their structure and the presence of a very thick intermediate space between the cell wall and their cytoplasmatic membrane.

ENZYME INACTIVATION BY PEF

To prevent denaturation, an enzyme has to maintain its native structure. Such globular proteins (secondary, tertiary, and quaternary) are stabilized by hydrophobic interactions, hydrogen bonding, van der Waal interactions, ion pairing, electrostatic forces, and steric constraints. Because a change in the magnitude of any of these forces could cause denaturation, the application of PEF may affect them. Furthermore, external electric fields are also found to influence the conformational state of a protein through charged, dipole, or induced dipole chemical reactions.[19] Such charged groups and structures are highly susceptible to various types of electric field perturbations. An external electric field may also induce association and dissociation of ionizable groups, movements of charged side chains, changes in the packing and alignment of helices (helical or sheet content), and the overall shape of a protein.[20]

Proteases, pectin methylesterases, and oxidases are heat-stable enzymes present in fruit and vegetable juices that may cause browning, off flavors, loss of vitamins, cloud losses, and gelation during juice product storage. The effect of PEF over the conformational state of these proteins can be used to prevent such detrimental reactions. PEF research in the area of enzyme inactivation is still limited but has been successful in the inactivation of important enzymes

Table 13–2 Inactivation by PEF of *Zygosaccharomyces bailii* Vegetative Cells and Ascospores in Fruit Juices

Fruit Juice	*Electric Conductivity (μS/cm)*	*Peak Voltage (kV)*	*Electric Field Intensity (kV/cm)*	*Width Pulse (μs)*	*Inactivated Vegetative Cells (log_{10})**	*Inactivated Ascospores (log_{10})*
Grape	2.7×10^3	21	35	2.3	5	3.5
Apple	2.5×10^3	19.4	32.3	2.5	4.8	3.6
Orange	3.6×10^3	20.6	34.3	2.0	4.7	3.8
Cranberry	1.1×10^3	21.9	36.5	3.3	4.6	4.2
Pineapple	3.1×10^3	19.8	33	2.2	4.3	3.5

*Inactivation levels correspond to a treatment with two pulses and a maximum temperature in all cases of 22°C.

Table 13–3 Inactivation of Mold Spores in Fruit Juices by PEF

Fruit Juice	*pH*	*Electric Conductivity (µS/cm)*	*Peak Voltage (kV)*	*Electric Field Intensity (kV/cm)*	*Pulse Width (µs)*	*Inactivated* Byssochlamys fulva (Paecilomyces fulvus) *Conidiospores (log₁₀)**	*Inactivated* Neosartorya fischeri *Ascospores (log₁₀)**
Cranberry	3.0	1.1×10^3	21.9	36.5	3.3	5.9†	0.1
Grape	3.9	2.7×10^3	21	35	2.3	5.5	0.4
Pineapple	3.5	3.1×10^3	19.8	33	2.2	4.8	—
Orange	3.9	3.6×10^3	20.6	34.3	2.0	3.7	0.1
Apple	4.1	2.5×10^3	19.4	32.3	2.5	3.5	—
Tomato	4.1	4.8×10^3	18	30	2.0	3	—

*Inactivation levels correspond to a treatment with ten pulses and a maximum temperature in all cases of 23°C.
†Treatment of just two pulses.

present in milk, such as plasmin and alkaline phosphatase.[6,21] In the search for a better understanding of the effect of PEF on the enzymes of major importance in the food industry, Ho et al.[22] applied instant-charge-reversal pulse wave shapes to inactivate different enzymes suspended in buffer solutions, using a static parallel plate treatment chamber with a 0.3-cm gap. Thirty pulses of 2 ms width and 13–87 kV/cm yielded inactivations up to 70–85% for lipase, glucose oxidase, and α-amylase, whereas only 30–40% was obtained for peroxidase and phenol oxidase (Figure 13–3). In contrast, lysozyme and pepsin presented an increased activity in certain ranges of the applied voltage. Pepsin and lysozyme also exhibited an inhibitory effect, once a particular field was reached. Because the degree of denaturation varied from enzyme to enzyme, it can be concluded that more research is necessary to determine whether the inactivation level is due to the structure of the protein, creation of active sites, concentration

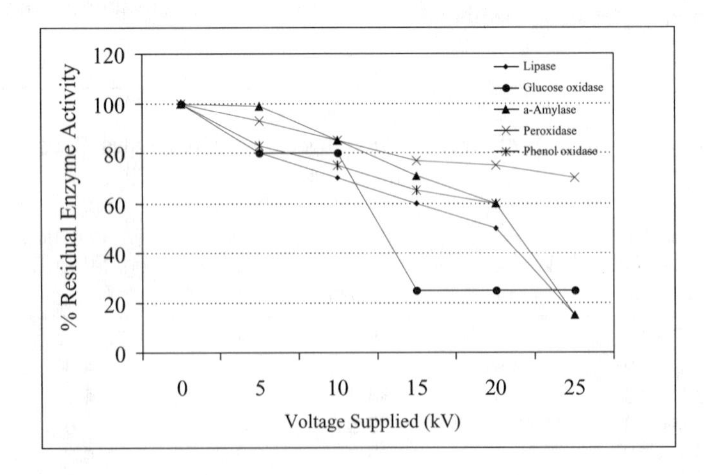

Figure 13–3 Effect of 30 instant charge reverse 2 µs pulses over selected enzymes suspended in buffer media.

of the treated enzyme, local heating effects, or some combination of each of these factors.

Yeom et al.[23] demonstrated the irreversible inactivation of papain (a cysteine protease in papaya) by a PEF treatment with 50 kV/cm and 500 pulses equivalent to a total treatment of 2 ms. After PEF treatment, the papain activity was nearly 10%, determined by calculating the initial rates of release of para-nitroalinile (product) as an increase in absorbance at 410 nm for 5 minutes, using N-benzoyl-D-Larginine-nitronilide as a substrate. (The final residual activity of the enzyme is defined as the percentage of the ratio between the specific activity of papain after PEF and the specific activity of control papain.) In contrast, to achieve similar inactivation levels of this enzyme by a short-time heat treatment, temperatures above 90°C are required because those near 80°C necessitate more than 48 hours to reduce enzyme activity below 60%.

Giner et al.[24] reported a PEF inactivation of pectin methylesterase (PE), which catalyzes pectin de-esterification to low-ester pectins or pectic acids and free methanol and, thus, produces loss of body in tomato juices or gelation in tomato paste. Samples were submitted to exponential decay pulses of 0.02 and 0.04 ms pulse width. The number of pulses applied to the samples in mono- and bipolar mode was up to 400, and the electric field intensities ranged from 5 to 24 kV/cm. As a result, a 93.8% maximum reduction of the initial enzymatic activity was achieved. The parameter values of critical electric field intensity, critical treatment time, and the independent constant factor from Hülsheger's equation were 0.7 kV/cm, 0.48 ms, and 39 kV/cm, respectively. These preliminary enzyme inactivation results expand the application spectrum of PEF by emphasizing its potential to process fruit juices under nonthermal conditions.

EVALUATION OF PEF AS A PASTEURIZATION TECHNIQUE

The feasibility of PEF to pasteurize fruit juices has been determined not only by its success with the inactivation of inoculated or naturally present yeasts, molds, and vegetative bacteria (pathogenic and spoilage), but the excellent physical, chemical, nutritional, and sensory characteristics of the products after PEF treatment. Such overall quality can be assessed by the evaluation of the microbiologic and physicochemical attributes of a product that are related to hygiene and safety or nutritional, sensory, and functional characteristics, respectively.

As mentioned before, electric field treatment is also associated with energy efficiency. In the treatment of apple juice, energy utilized with the PEF technology is 90% less than the amount needed for high-temperature short-time (HTST) processing.[25] Furthermore, the development and utilization of low-energy instant-charge-reversal pulses by one of the leading groups at Guelph University allow energy consumption to be less than 7 J/mL for the PEF processing of products such as waste brine, orange juice, and apple cider.[26]

PROCESSING OF APPLE JUICE

Juice commercialization is one of the most important areas in beverage production but, as previously indicated, one of the drawbacks has been the loss of freshlike flavor. The nonthermal attributes of PEF allow treatments at low temperatures that do not require the harmful effects of hot filling, as is the case in traditional processing. Qin et al.[27] reported some of the earliest studies of the effect of PEF pasteurization on the quality attributes and shelf life of fruit juices (Table 13–4). They processed two commercially available juices (reconstituted from concentrated apple juice and freshly prepared apple juice) that had both been stored at 4–6°C beforehand. The concentrate was reconstituted with one part concentrated juice and six parts water at room temperature (22–25°C) before processing, then subjected to 10 pulses with an electric field intensity of 50 kV/cm and pulse duration of 2 μs. The initial process temperature was 8.5 ± 1.5°C, with the maximum temperature during the process increasing to 45 ± 5°C.

Table 13–4 PEF Processing Conditions for Selected Fruit Juice Products

Treatment Conditions	*Apple Juice from Concentrate*	*Fresh Apple Juice*	*Orange Juice*	*Orange Juice*	*Orange Juice*
Peak electric field (kV/cm)	50	50	62.5	35	37
Pulse duration (μs)	2	2	0.57	0.93	0.96
Pulse number	10	16	—	—	—
Wave shape	exponential decay	exponential decay	exponential decay	square	charge reversal
Initial temperature (°C)	8.15 ± 1.5	8.15 ± 1.5	—	—	—
Treatment temperature (°C)	45 ± 5	45 ± 5	—	—	—
Storage temperature (°C)	22–25	4–6	22	22	22
Shelf life (days)	28	21	7	34	26
Source	Qin et al.[27]	Qin et al.[27]	Zhang et al.[30]	Zhang et al.[30]	Zhang et al.[30]

After treatment, the apple juice was aseptically filled into packages of 250 mL for shelf-life studies. The bags were opened and filled directly from the treatment chamber outlet while being flushed with purified nitrogen gas, due to product oxygen sensitivity. Qin et al.[27] then allowed the commercially available fresh apple juice to set for 24 hours at 4–6°C before processing to allow for sedimentation of particulates. Bulky dregs settled on the bottom of the container, then the upper layer of clear juice was collected for PEF treatment. The processing required three steps of PEF exposure to prevent an increase in temperature beyond 45°C during treatment. The first step consisted of six pulses, and the two additional steps were five pulses, with each utilizing an electric field intensity of 50 kV/cm and pulse duration of 2 μs. The processed juice was then filled into bags directly from the outlet tubing of the treatment chamber.

Simpson et al.[28] evaluated the physical and chemical attributes of PEF-treated apple juice from concentrate stored at 4°C and found no physical or chemical changes in ascorbic acid or sugars (glucose, fructose, sucrose). However, the pH of the treated and untreated control juices varied from 4.10 to 4.36, and the conductivity of untreated apple juice was slightly higher than that of the treated juice (1,097 μS/cm vs. 1,300 μS/cm). The difference in conductivities may be attributed to the greater mineral concentration (Ca, Mg, Na, and K) in the untreated juice. The shelf life of the treated apple juice from concentrate was as long as 4 weeks, and that for the fresh apple juice was extended by 3 weeks. A sensory panel found no significant differences between the untreated and PEF-treated juice from concentrate or freshly prepared juices.

Improvements to PEF techniques and expertise are even more evident when comparing these results with those obtained by Vega-Mercado et al.,[21] where apple juice from concentrate after PEF treatment was stored at room temperature (22–25°C) for more than 8 weeks, with no apparent change in its physicochemical and sensory properties. Likewise, fresh apple juice also remained unaffected after being processed with 16 pulses and stored for 32 days.

PROCESSING OF ORANGE JUICE

The main disadvantages of traditional orange juice heat treatments, by either HTST or ultrahigh temperatures (UHT), are the loss of vitamin

C, changes in color, and destruction of fresh flavor. Dunn and Pearlman[29] conducted a study of PEF effects on commercial, freshly squeezed, high-pulp orange juice with a limited shelf life by subjecting it to 35 exponentially decaying pulses of electric fields ranging between 33.6 and 35.7 kV. Its native microbiologic population consisted of a mixture of yeasts, molds, and bacteria, but after PEF treatment, a 5-log inactivation was obtained, and its shelf life was extended by more than a week. The PEF-treated orange juice was also acceptable in terms of taste and odor after 10 days, whereas the untreated juice was unacceptable after only 4 days.

Zhang et al.[30] evaluated the shelf life of orange juice treated with an integrated PEF pilot plant system. Single-strength (11.8°Bx), pulp-free orange juice reconstituted from frozen concentrate was processed in a PEF system with a series of co-field chambers at a flow rate of 75–85 L/hour; to maintain treatment temperatures near ambient (22 to 25°C), cooling devices were used in between the chambers. Three different wave shape pulses were used to compare the effectiveness of the processing conditions:

a. square waves with a peak electric field of 35 kV/cm, effective pulse width of 37.22 μs, and pulse rise time of 60 ns;
b. exponentially decaying waves with a peak electric field of 62.5 kV/cm, effective pulse width of 0.57 μs, and pulse rise time of 40 ns; and
c. charge-reversal waves with a peak electric field of 37 kV/cm, effective pulse width of 0.96 μs, and pulse rise time of 400 ns.

After treatment, the juice was aseptically packaged (200-mL plastic cups) and stored under refrigeration at 4, 22, and 37°C for microbial evaluations of total aerobic plate and fungi counts, as well as vitamin C and color retention levels. A PEF processing assessment was achieved by comparing these attributes to a control sample of orange juice pasteurized in a conventional HTST plate heat exchanger with a minimum shelf life of 5 months at a storage temperature of 4°C. The square-wave pulses were found to be most effective, yielding products with longer shelf life than those products treated with exponentially decaying and charge-reversal pulses (Table 13–4).

The follow-up analysis of vitamin C loss was higher in juices that were heat-treated, compared with those that were PEF-treated. The authors evaluated the kinetics of degradation and concluded that it followed a pseudo first-order reaction, as described in Equation (4). Furthermore, they evaluated the relation between the reaction rate constant k in Equation (4) and activation energy using Arrhenius' equation to come up with Equation (5):

$$C = C_0 \exp(-kt) \quad \textbf{(4)}$$

$$k = A \exp(-E_a / RT) \quad \textbf{(5)}$$

where C is the vitamin C concentration at time t, C_o is the initial vitamin C concentration at t zero, k is the reaction rate constant (1/day), A is a frequency factor, E_a is the activation energy (kJ/mol K), and T is the absolute temperature (K). The activation energy (E_a) that defines Equation (5) was found to be 28.6 kJ/mol with a Q_{10} equal to 1.3.

Another part of this comprehensive study by Zangh et al.[30] was a color evaluation of the orange juice that revealed a better preservation for the PEF-treated product, compared with the heat-treated samples, during the initial storage period. The authors declare the color change to be insignificant between the two processes during subsequent periods and explain the increase in the a* values during storage to be due to the degradation of ascorbic acid to furfural, which is a browning product. In another study conducted by Zhang,[31] a much better flavor was found in orange juice processed by PEF, compared with heat-treated juice, when the flavor evaluation was based on the relative content and loss percentage of key components such as d-limonene and ethyl butyrate (Table 13–5).

Table 13–5 Comparison of Orange Juice Pasteurized by PEF and Heat

	Relative % Content Lost After Treatment	
Processing Type	*d-Limonene*	*Ethyl Butyrate*
Control, freshly squeezed	0	0
PEF (35 kV/cm, 240 μs)	12	0
PEF (35 kV/cm, 480 μs)	16.7	1.2
Heat pasteurization (91°C, 30 sec)	44.9	21.5

CONCLUSION

The application of PEF for the inactivation of yeast, molds, their sporulated forms, and some highly acid-resistant vegetative bacteria has proved to be an attractive competitor to conventional heat processes for juice pasteurization. The success of laboratory and pilot plant sensory and shelf-life tests is highly encouraging of future investment in this technology. Furthermore, the appealing nonthermal characteristic of PEF can yield a product-saving pasteurization process that is sound enough to be implemented in the area of juice processing in the next millennium. However, regulation approval is still under way and must be completely addressed before the final implementation of PEF in the food industry. The U.S. Food and Drug Administration (FDA) in 1996 released a "letter of no objection" for liquid eggs treated with PEF. However, to meet the FDA requirements to introduce a new and novel process, it is necessary to follow and fulfill several evaluation procedures to determine whether the aseptically produced food product poses a potential public health hazard.[2,5]

REFERENCES

1. Getchell BE. Electric pasteurization of milk. *Agric Engr.* 1935;16:408–410.
2. Barbosa-Cánovas GV, Góngora-Nieto MM, Pothakamury UR, Swanson BG. *Preservation of Foods with Pulsed Electric Fields.* San Diego: Academic Press. 1999.
3. Qin BL, Zhang Q, Barbosa-Cánovas GV, Swanson BG, et al. Inactivation of microorganism by pulsed electric fields with different waveforms. *IEEE Trans Dielec Electr Insul.* 1994;1:1047–1057.
4. Marquez VO, Mital GS, Griffiths MW. Destruction and inhibition of bacterial spores by high voltage pulsed electric fields. *J Food Sci.* 1997;62:399–409.
5. Barbosa-Cánovas GV, Pothakamury UR, Palou E, Swanson BG. *Nonthermal Preservation of Foods.* New York: Marcel Dekker; 1997.
6. Castro-Castillo AJ. Pulsed electric field modification of activity and denaturation of alkaline phosphatase. Ph.D. Dissertation. Pullman, WA: Washington State University; 1994.
7. Glaser RW, Leikin SL, Chernomordik LV, Pastushenko VF, et al. Reversible electrical breakdown of lipid bilayers: Formation and evolution of pores. *Biochim Biophys Acta.* 1988;940:275–287.
8. Gásková D, Sigler K, Janderova B, Plasek J. Effect of high-voltage electric pulses on yeast cells: Factors influencing the killing efficiency. *Bioelectrochem Bioenerg.* 1996;39:195–202.
9. Ho SY, Mittal GS. Electroporation of cell membranes: A review. *Crit Rev Biotechnol.* 1996;16:349–362.
10. Barsamian ST, Barsamian TK. Dielectric phenomenon of living matter. *IEEE Trans Dielec Electric Insul.* 1997;4:629–643.
11. Lubicki, P, Jayaram S. High voltage pulse application for the destruction of the gram negative bacterium *Yersinia enterocolitica. Bioelectrochem Bioenerg.* 1997;43:135–141.
12. Harrison SL, Barbosa-Cánovas GV, Swanson BG. *Saccharomyces cerevisiae* structural changes induced by pulsed electric field treatment. *Lebens-Wiss u-Technol.* 1997;30:236–240.
13. Hülsheger H, Potel J, Niemann EG. Killing of bacteria with electric pulses of high field strength. *Rad Environ Biophys.* 1981;20:3–65.

14. Liu X, Yousef AE, Chism GW. Inactivation of *Escherichia coli* O157:H7 by the combination of organic acids and pulsed electric fields. *J Food Safety.* 1997;16:287–299.
15. Ohshima T, Sato K, Terauchi H, Sato M. Physical and chemical modifications of high-voltage pulse sterilization. *J. Electrostatics.* 1997;42:159–166.
16. Zhang Q, Monsalve-González A, Barbosa-Cánovas GV, Swanson BG. Inactivation of *E. coli* and *S. cerevisiae* by pulsed electric fields under controlled temperature conditions. *J Food Proc Engr.* 1994;17:469–478.
17. Raso J, Calderón ML, Góngora-Nieto MM, Barbosa-Cánovas GV, et al. Inactivation of *Zygosaccharomyces bailii* in fruit juices by heat, high hydrostatic pressure and pulsed electric fields. *J Food Sci.* 1998;63:1042–1044.
18. Raso J, Calderón ML, Góngora-Nieto MM, Barbosa-Cánovas GV, et al. Inactivation of mold ascospores and conidiospores suspended in fruit juices by pulsed electric fields. *Lebensm-Wiss u-Technol.* 1998;31:668–672.
19. Tsong TY, Astunian RD. Absorption and conversion of electric field energy by membrane bound ATPases. *Bioelectrical Bioenerg.* 1986;15:457–476.
20. Tsong TY. Electrical modulation of membrane proteins: Enforced conformational oscillations and biological energy and signal transductions. *Ann. Rev Biophys Chem.* 1990;19:83–106.
21. Vega-Mercado H, Powers JR, Martín-Belloso O, Luedecke OL, et al. Effect of pulsed electric fields on the susceptibility of proteins to proteolysis and inactivation of an extracellular protease from *P. fluorescens* M 3/6. *Seventh International Congress on Engineering and Food.* The Brighton Center, UK: 13–17 April, 1997:C73–C76.
22. Ho SY, Mittal GS, Cross JD. Effects of high field electric pulses on the activity of selected enzymes. *J Food Engr.* 1997;31:69–85.
23. Yeom HW, Zhang QH, Dunne CP. *Inactivation of Enzyme Papain by Pulsed Electric Field in a Continuous System.* Paper presented at the annual meeting of the Institute of Food Technologists. Atlanta, GA: No. 59C-14, 1998:150–151.
24. Giner J, Gimeno V, Élez P, Espachs A, et al. *Inhibition of Tomato* (Licopersicon esculentum Mill.) *Pectin Methylesterase by Pulsed Electric Fields.* Paper presented at the annual meeting of Food Technologists. Chicago: July, 1999:21–26.
25. Qin BL, Chang FJ, Barbosa-Cánovas GV, Swanson BG. Nonthermal inactivation of *Saccharomyces cerevisiae* in apple juice using pulsed electric fields. *Lebensm-Wiss u-Technol.* 1996;28:564–568.
26. EPRI. *Pulsed Electric Field Processing in the Food Industry*: A status report on PEF. Report CR-109742. Palo Alto, CA: Industrial and Agricultural Technologies and Services; 1998.
27. Qin BL, Pothakamury UR, Vega-Mercado H, Martín-Belloso OM, et al. Food pasteurization using high-intensity pulsed electric fields. *Food Technol.* 1995;12:55–60.
28. Simpson MV, Qin BL, Barbosa-Cánovas GV, Swanson BG. Pulsed electric field processing and the chemical composition of apple juice. *Internal Research Report.* Washington State University; 1996.
29. Dunn JE, Pearlman JS. *Methods and Apparatus for Extending the Shelf Life of Fluid Food Products.* Washington, DC: US Patent; 1987:4,695,472.
30. Zhang QH, Qiu X, Sharma SK. Recent developments in pulsed electric field processing. In: *New Technologies Yearbook.* Washington, DC: National Food Processors Association; 1997:31–42.
31. Zhang QH. Integrated pasteurization and aseptic packaging using high-voltage pulsed electric fields. *Seventh International Congress on Engineering and Food.* The Brighton Center, UK: 13–17 April, 1997:K3–K15.
32. Wouters PC, Smelt JPP. Inactivation of microorganisms with pulsed electric fields: Potential for food preservation. *Food Biotechnol.* 1997;11:193–229.
33. Jacob HE, Forster W, Berg H. Microbiological implications of electric field effects. II. Inactivation of yeast cells and repair of their cell envelope. *Zeit für Allge Microbiol.* 1981;21:225–233.
34. Mizuno A, Hori Y. Destruction of living cells by pulsed high-voltage application. *IEEE Trans Ind Appl.* 1988;24:387–394.
35. Matsumoto Y, Satake T, Shioji N, Sakuma A. Inactivation of microorganisms by pulsed high voltage applications. *IEEE Industrial Applications Society Annual Meeting.* 1991:652–659.
36. Yonemoto Y, Yamashita T, Muraji M, Tatebe W, et al. Resistance of yeast and bacterial spores to high voltage electric pulses. *J Ferment Bioeng.* 1993;75:99–102.
37. Zhang Q, Monsalve-González A, Barbosa-Cánovas GV, Swanson BG. Inactivation of *E. coli* and *S. cerevisiae* by pulsed electric fields under controlled temperature conditions. *Trans ASAE.* 1994;37:581–587.
38. Qin BL, Vega-Mercado H, Pothakamury UR, Barbosa-Cánovas GV, et al. Application of pulsed electric fields for inactivation of bacteria and enzymes. *J Franklin Inst.* 1995:332a:209–220.
39. Grahl T, Sitzmann W, Märkl H. Killing of microorganisms in fluid media by high voltage pulses. *10th DECHEMA Biotechnology Conference Series,* 5B. 1992:675–678.
40. Grahl T, Märkl H. Killing of microorganisms by pulsed electric fields. *Appl Microbiol Biotechnol.* 1996;45:148–157.

CHAPTER 14

Natural Antimicrobials from Plants

Aurelio López-Malo, Stella M. Alzamora, and Sandra Guerrero

INTRODUCTION

Food preservation is based primarily on inactivation, growth delay, or growth prevention of spoilage and pathogenic microorganisms and works through factors that influence microbial growth and survival. Major food preservation technologies can be classified as those that act mainly by preventing or slowing down microbial growth (low temperature, reduced water activity [a_w], less oxygen, acidification, fermentation, modified atmosphere packaging, addition of antimicrobials, compartmentalization in water-in-oil emulsions), those that act to inactivate microorganisms (heat pasteurization and sterilization, microwave heating, ionizing radiation, high hydrostatic pressure, pulsed electric fields), and those that prevent or minimize entry of microorganisms into food or remove them (aseptic handling or packaging, centrifugation, filtration).[1,2] In addition, techniques in combination, based on the hurdle technology concept,[3] may act to inhibit or inactivate microorganisms, depending on the combination of hurdles applied to achieve food preservation.

Among these techniques, the use of chemical agents exhibiting antimicrobial activity (by inhibiting and/or reducing microbial growth or even by inactivating undesirable microorganisms) is one of the oldest and most traditional. As early as 8000 B.C., people preserved excess meat and fish by smoking and dry salting, and during 6000–1000 B.C., sulfur dioxide by burning sulfur was employed to sanitize the equipment used in the fermentation and storage of wine in Egypt.[4] Antimicrobial agents may be either synthetic compounds intentionally added to foods or naturally occurring, biologically derived substances (the so-called naturally occurring antimicrobials), which may be used commercially as additives for food preservation, besides exhibiting antimicrobial properties in the biologic systems where they are originally found.[2]

Concerns about the use of antimicrobial agents in food products have been debated in the public domain for decades.[5] Both the increasing demand for reduced-additive (including antimicrobial preservatives) and more "natural" foods and the increasing demand for greater convenience have provoked in the food industry and researchers the search for alternative antimicrobial agents or combinations. In this search, a wide range of natural systems from animals, plants, and microorganisms is being studied.[6–8] However, mainly economic aspects originated in the strict requirements to obtain approval, and efforts to get the product onto the market

The authors acknowledge the financial support from CYTED Program, the European Union (STD-3 Program), Universidad de las Américas (Puebla) and CONACyT of México, and Universidad de Buenos Aires and CONICET of República Argentina.

restrict the specter of new chemical compounds that can help in the preservation of foods; furthermore, the approval process is very long (10–12 years). These obstacles have originated the search of emerging preservatives by examining compounds already used in the food industry, perhaps for other purposes, but with potential as antimicrobials, approved and not toxic in the used levels, many of them classified as generally recognized as safe (GRAS). Within these compounds are, for example, the so-called green chemicals present in plants that are utilized as flavor ingredients.[7]

Naturally occurring antimicrobial compounds are abundant in the environment. Pathogens are a ubiquitous fact of life, and the ability to resist their attentions constitutes one of the keys to biologic success. Plants, animals, and microorganisms have evolved a number of strategies for attack, defense, and counterattack. Moreover, several individual components of host–pathogen interactions are shared, either conceptually or mechanistically, in the two major branches of the eukaryotic lineage.[9] Rarely, the survival of a plant, animal, or microorganism depends on just a single defense but, commonly, a physical defense (skin, mucous membrane, integument, etc.) acting as a barrier to the ingress of aliens functions collaboratively with a cellular and/or chemical defense system by sequestering essential nutrients for the alien, inhibiting the physiology of the alien cell, increasing the permeability of the alien outer membrane, etc.[10] As Board and Gould[11] pointed out, this "cocktail" approach (that is, structural, constitutive and induced, or preinfectional and postinfectional defense mechanisms working together) developed by living organisms should be taken into account for minimizing the useful level of the food preservation factors, including natural and traditional antimicrobial compounds.

Major antimicrobial systems naturally present in plants and animals, produced by microorganisms, or considered within this classification are shown in Table 14–1. Some of these natural antimicrobial systems are already employed for food preservation, and many others are just being studied for use in foods.[12] Development, exploration, and use of naturally occurring antimicrobials in foods, as well as chemistry and food safety/toxicity aspects, antimicrobial activity, and mechanisms of action are covered in various reviews by, among others, Wilkins and Board,[10] Zaika,[13] Shelef,[14] Conner,[15] Nychas,[7] Beuchat and Golden,[16] Branen and Davidson,[17] Board,[6] and Smid and Gorris,[18] and, more recently, by the fine work of Sofos et al.,[2] who prepared their contribution following a recommendation by the Council for Agricultural Science and Technology of the United States.

Although many natural systems have potential to be used as antimicrobials, this chapter will focus mainly on natural antimicrobials from plants and their possible application in minimally processed fruit and vegetables.

SOURCES OF NATURAL ANTIMICROBIALS FROM PLANTS

Plants, herbs, and spices, as well as their derived essential oils and isolate compounds, contain a large number of substances that are known to inhibit various metabolic activities of bacteria, yeasts, and molds, although many of them are yet incompletely exploited. Wilkins and Board[10] reported that more than 1,340 plants are known to be potential sources of antimicrobial compounds; about 60 are mentioned by Nychas[7] and Beuchat.[19] The antimicrobial compounds in plant materials are commonly contained in the essential oil fraction of leaves (rosemary, sage), flowers and flower buds (clove), bulbs (garlic, onion), rhizomes (asafoetida), fruit (pepper, cardamom), or other parts of the plant.[7,14] Table 14–2 presents a list of selected plants, herbs, and spices that have been reported as sources of natural antimicrobials. This list does not pretend to include all possible sources of natural antimicrobials from plants but shows some of those more highly recognized. These compounds may be lethal to microbial cells or they may simply inhibit the production of a metabolite, e.g., my-

Table 14–1 Major Components and Strategies of Natural Defense Systems

Origin	*Example*	*Antimicrobial Agent*
Animals—constitutive	Milk	Lactoperoxidase, lactoferrin
	Eggs	Lysozyme, ovotransferrin, avidin
	Serum	Transferrins
	Phagosomes	Myeloperoxidase
Animals—inducible	Immune system	Antibodies
	Insects	Attacins, cecropins
Microorganisms	Lactic acid bacteria	Nisin, pediocin, other bacteriocins
	Other microorganisms	Pimaricin, subtilin, natamycin, diacetyl
		Bacteriofagos
		Yeast "killer toxins"
		Other low-MW metabolites (e.g., ethanol, H_2O_2)
Plants—preinfectional	Herbs, spices, and other plants	Organic acids (e.g., citric, succinic, tartaric, benzoic, malic)
		Phenolic compounds (e.g., catechol, DOPA, caffeic acid, eugenol, thymol)
		Methylated flavones (e.g., tangeretin, nobiletin)
		Flavonols (e.g., morin, hesperitin)
		Alkaloids (e.g., α-tomatine)
		Hydroxyphenyl—threne derivatives (e.g., hircinol, isobatasin)
		Lactones (e.g., borbonol)
		Proteinlike compounds (e.g., thaumathin-like proteins, zeamatin)
		Glucosides
		Glycosides
		Dienes
Plant—postinfectional	Infected or injured plants	Phenolic compounds (e.g., from hydrolyzable tannins, gallotannins, and ellagitannins)
		Sulfoxides (e.g., allicin)
		Isothiocyanates
		Phytoalexins (e.g., scoparone, resveratrol, pisatin)

cotoxins.[19] Major components with antimicrobial activity found in plants, herbs, and spices are phenolic compounds, terpenes, aliphatic alcohols, aldehydes, ketones, acids, and isoflavonoids. As a rule, it has been reported that the antimicrobial activity of essential oils depends on the chemical structure of their components and on their concentration. Shelef[14] mentioned that simple and complex derivatives of phenol are the main antimicrobial compounds in essential oils from spices. Katayama and Nagai[20] recognized eugenol, carvacrol, thymol, and vanillin as active antimicrobial compounds from plant essential oils. Aliphatic alcohols and phenolics were also reported as fungal growth inhibitors by Farag et al.[21] The chemical structures

Table 14–2 Components with Antimicrobial Activity Found in Plants, Herbs, and Spices

Plant, Herb, or Spice	*Major Component(s)*	*Other Component(s)*
Allspice (*Pimenta dioica*)	eugenol	methyl ether cineol
Basil (*Ocimum basilicum*)	d-linalool, methyl chavicol	eugenol, cineol, geraniol
Black pepper (*Pipper nigrum*)	monoterpenes, sesquiterpenes, oxygenated compounds	
Bay (*Laurus nobilis*)	cineol	l-linalool, eugenol, geraniol
Caraway seed (*Carum carvi*)	carvone	limonene
Celery seed (*Apium graveolens*)	d-limonene	
Cinnamon (*Cinnamomum zeylanicum*)	cinnamic aldehyde	l-linalool, p-cymene, eugenol
Clove (*Syzygium aromaticum*)	eugenol	cariofilene
Coriander (*Coriandum sativum*)	d-linalol	d-α-pinene, β-pinene
Cumin (*Cuminum cyminum*)	cuminaldehyde	p-cymene
Fennel (*Foeniculum vulgare*)	anethole	
Garlic (*Allium sativum*)	diallyl disulfide, diallyl trisulfide	diethyl sulfide, allicin
Lemongrass (*Cymbopogon citratus*)	citral	geraniol
Marjoram (*Origanum majorana*)	linalool, cineol, eugenol, terpinineol	methyl chavicol
Mustard (*Brassica hirta, B. juncea, B. nigra*)	allyl-isothiocyanate	
Onion (*Allium cepa*)	d-n-propyl disulfide, methyl-n-propyl disulfide	
Oregano (*Origanum vulgare*)	thymol, carvacrol	α-pinene, p-cymene
Parsley (*Petroselinum crispum*)	α-pinene, fenol-eter-apiol	
Rosemary (*Rasmarinus officinalis*)	borneol, cineol	camphor, α-pinene, bornyl acetate, terpinol
Sage (*Salvia officinalis*)	thujone, cineol, borneol	thymol, eugenol
Tarragon (*Artemisia dracunculus*)	methyl chavicol	anethole
Thyme (*Thymus vulgaris*)	thymol	carvacrol, l-linalool geraniol, p-cymene
Vanilla (*Vanilla planifolio, V. pompona, V. tahilensis*)	vanillin	vanillic, p-hidroxibenzoic and p-coumaric acids

of selected antimicrobial compounds from plant origin are presented in Figures 14–1 and 14–2.

Essential oils of a large number of plants possess useful biologic and therapeutic activities, and the oils are extensively utilized in the preparation of pharmacologic drugs. They are commercially recovered from plant materials primarily by steam distillation, and their use in the food industry is influenced by the nature of their constituents.

Antimicrobial activity of essential oils and extracts from plants, herbs, and spices depends not only on the extraction method, but also on the initial quantity of essential oil in the plant. Within the same spice or plant, the levels of constituents and, therefore, active antimicrobial groups, can substantially vary. Also, the geographic zone of the cultivation may influence the extract composition. Mishra and Dubey[22] reported that lemongrass essential oil varied in its effectiveness as an antimicrobial, depending on the harvesting time within the year. During May to December, it was more effective, inhibiting 100% of the evaluated microbial strains, but the essential oil

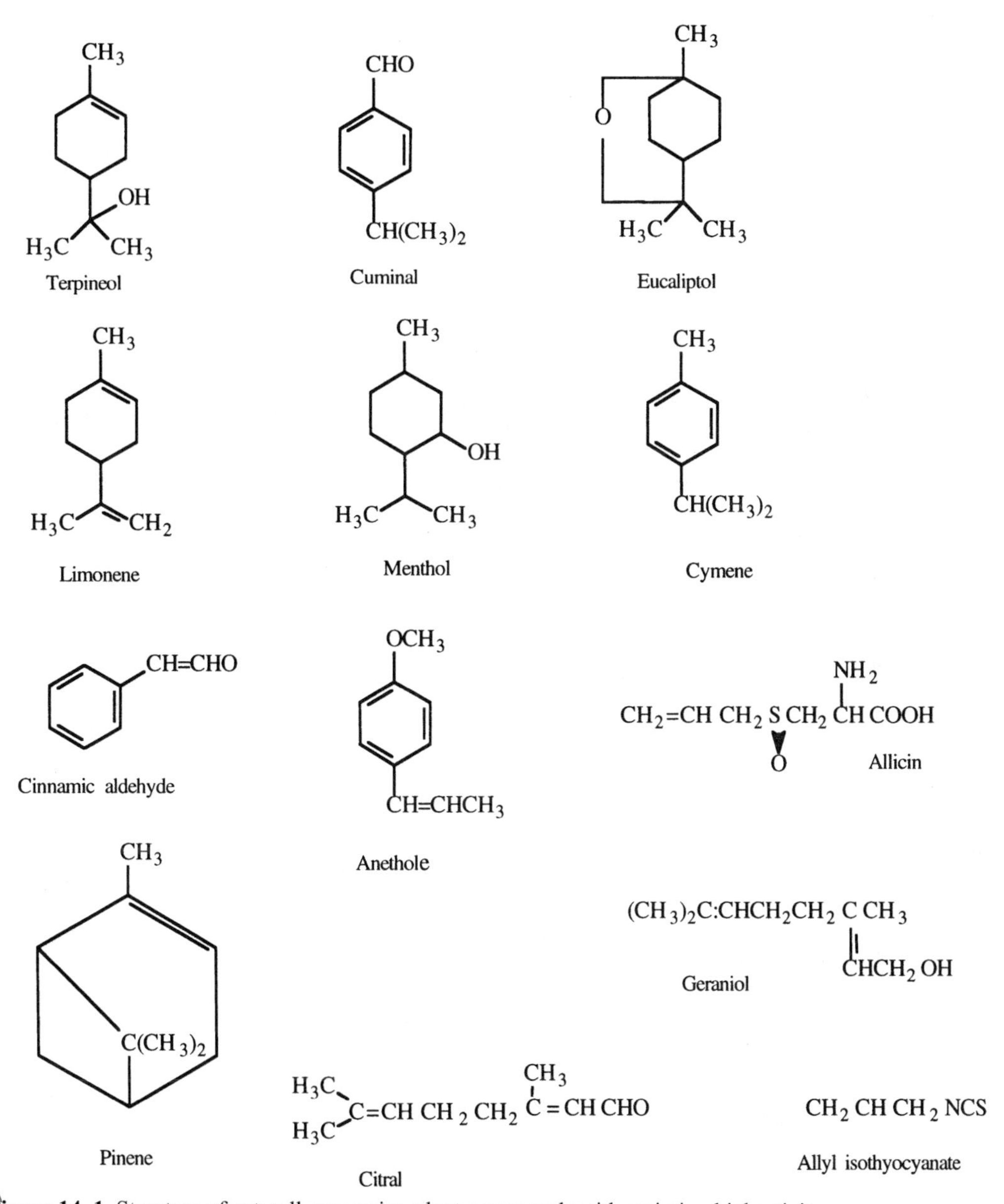

Figure 14–1 Structure of naturally occurring plant compounds with antimicrobial activity.

from plants collected during February to April was only 73–80% effective. Therefore, there is a necessity to establish methods to fix or standardize essential oil purity or concentration of active components. The knowledge of the qualitative and quantitative composition of essential oils will allow the collection of valuable data in systematic biologic studies of plants as antimicrobial agents. Differences in the antimicrobial effectiveness of plant extract compounds from different sources, extraction methods, and geographic zones can be diminished if the active antimicrobial com-

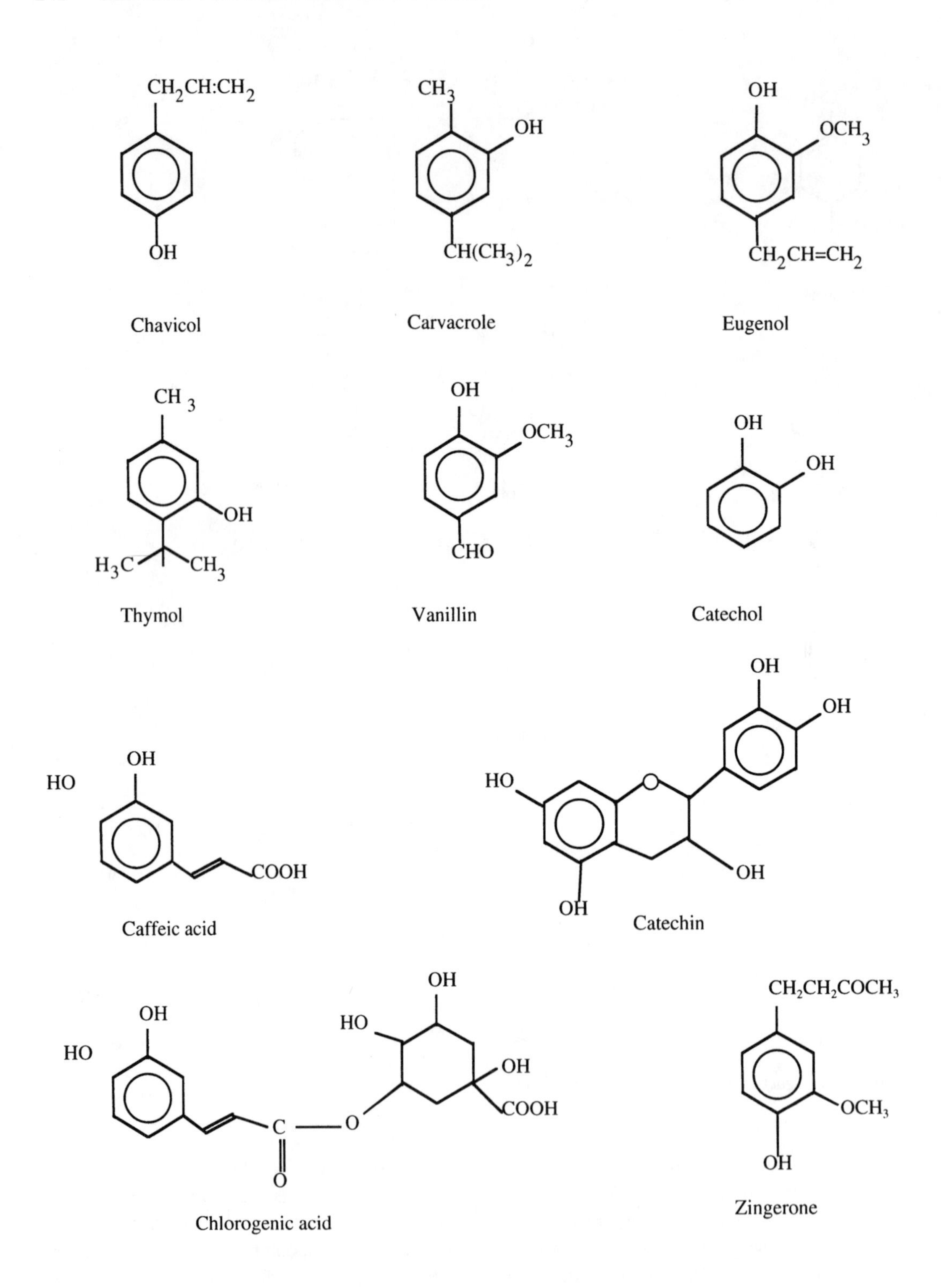

Figure 14–2 Structure of naturally occurring phenolic compounds with antimicrobial activity.

pounds from extracts or essential oils are identified. The adoption of plant or spice essential oils as alternatives to other preservatives will depend on the antimicrobial characteristic uniformity and/or on the analytical method availability to normalize their antimicrobial potency. In Table 14–2, major and minor antimicrobial constituents of some plant extracts are reported. The concentration varies, depending on the type of plant; among the richest are found clove, nutmeg, and laurel, and the major constituents in most cases are phenolic compounds.

METHODS FOR TESTING THE EFFICACY OF ANTIMICROBIALS

An important number of publications about the antimicrobial activity of extracts, oils, spices, and herbs have been reported. However, it is difficult to make quantitative comparisons of the effects, due at least partially to the great variety of methods employed to evaluate its efficiency.[13] The numerous tests that have been used, as well as the factors that are important to be considered in determining the effectiveness of natural antimicrobials in foods, have been critically discussed by Zaika,[13] Davidson and Parish,[23] and Parish and Davidson.[24] The classification of some of the methods most used to evaluate the antimicrobial efficiency in foods is shown in Exhibit 14–1. Screening or in vitro tests (generally made in laboratory media or in model systems) provide preliminary information about potential antimicrobial activity: the end-point tests, in which the microorganisms are challenged for an arbitrary period of incubation time, give qualitative information about approximate effective concentration; and the descriptive screening methods, in which the microorganism growth is analyzed over time, give quantitative information about the growth dynamics. Finally, in applied tests, the antimicrobial is applied to the actual food, and the results allow the evaluation of some factors that can affect the efficacy of the natural antimicrobial (e.g., binding to food components, inactivation by other additives, pH effects on antimicrobial stability and activity, uneven distribution in the food matrix, poor solubility, etc.). In vitro and in vivo tests must be used together for quantifying the antimicrobial effect. The measurements of the interactions—additive, synergistic, or antagonic—when combining antimicrobials are usually made by agar diffusion and dilution tests or by determining the inhibition curves, and results are visualized in isobolograms constructed using minimal inhibitory concentrations (MICs) or fractional inhibitory concentrations (FICs).

A combination of end-point and descriptive point methods has been used to determine the effects of natural antimicrobials against fungi. In these methods the substance to evaluate is mixed with the test medium, inoculated, and mold growth is measured. Thompson and Cannon[25] used this technique to evaluate the efficiency of 40 essential oils against 20 fungi strains. Thompson[26] described a procedure to evaluate essential oils' effect on mold spore germination, registering germination through microscopic observation of the spores.

Exhibit 14–1 Methods for Testing the Efficacy of Food Antimicrobials

In vitro or exploratory tests
- End-point screening methods
  - agar diffusion
  - agar and broth dilution
  - gradient plates
  - sanitizer and disinfectant tests
- Descriptive screening methods
  - turbidimetric assays
  - inhibition curves

Applied tests
- End-point methods
- Inhibition curves

Combined antimicrobials tests
- Agar diffusion
- Dilution assays
- Inhibition curves

Because many of the constituents of the extracts and evaluated oils are phenols, the phenolic coefficient technique is one of the end-point methods for testing disinfectants that has been used to quantify their effects. This term, proposed by Martindale,[27] is expressed as the ratio between minimal concentrations of phenol and the evaluated substance that inhibits a specific microorganism in a given incubation time. For this test, one of the microorganisms approved as a reference (*Salmonella typhi, Staphylococcus aureus,* or *Pseudomonas aeruginosa*) is exposed to several dilutions of the compound and phenol for 5, 10, and 15 minutes. The microorganism is plated in a free antimicrobial media, and the phenolic coefficient is calculated with the dilutions that inhibit microbial growth in a 10-minute treatment and do not inhibit it after 5 minutes.[23] Constituents of essential oils of oregano, thyme, cinnamon, and clove were found to be much more active than phenol, with phenolic coefficients greater than 1.[27] In addition, Müller[28] reported phenolic coefficients of 0.4, 10.0, 7.1, 5.4, 14.0, 0.4, and 13.4 for essential oils of anise, cardamom, cinnamon, coriander, fennel, lime, and thyme, respectively. Essential oils with high phenolic coefficients would have potential as antimicrobial agents, at least for the bacteria used as reference microorganisms.

Table 14–3 presents a summary of the most used methods for efficiency evaluation of phenolic antioxidants and essential oils of plants and spices. A great number of these studies has been accomplished in vitro and only a few of them in foods. Within the evaluation methods in model systems, descriptive (inhibition curves) methods have been commonly used with bacteria and yeasts and final point tests with bacteria, yeasts, and molds.

Davidson and Parish[23] mentioned that, for the application of these methods, other factors that can affect the microbial response (temperature, pH, a_w, and nutrients) must be controlled. The observed response will also depend on the type, genus, species, and strain of microorganism. The initial number of cells or spores used during the trial must be consistent to assure that results will be reproducible.[13]

Procedures in vitro give information about preservative performance, but these results do not necessarily duplicate all of the variability that can exist in a food. Therefore, it is necessary that its evaluation be done in real systems, assessing the effects of some components, such as protein, fat, fiber, pH, etc. In these kinds of tests, natural flora, isolated microorganisms, and/or some relevant spoilage or pathogenic microorganism should be inoculated, and environmental factors should reflect conditions of use and abuse.[23]

After an established incubation period, if no growth is observed, it should not be assumed that the spice, oil, or extract has a lethal effect on the microorganism. It might happen that an inhibitory effect was present. Therefore, microorganisms should be plated in a free "stress" media to prove its viability, or the test should be extended to confirm, at least during that lapse, no growth and to confirm whether the evaluated compound will lose its antimicrobial ability.

Zaika[13] concluded that many factors affect antimicrobial activity of spices, extracts, and essential oils, and mentioned that several aspects must be considered and reported. The observed inhibition depends on the evaluation method. Microorganisms differ in their resistance toward a spice or herb, and a given microorganism will differ in its resistance to different spices or herbs. Food components can affect, increasing (by the presence of acids, humectants, antimicrobials, etc.) or reducing (by partitioning of active components into the lipid phase, etc.) the antimicrobial capacity. The antimicrobial efficiency of a spice or herb depends on its origin, handling, processing, and storage.

PHENOLIC COMPOUNDS

Phenolic compounds have been used as antimicrobial agents since the early use of phenol as sanitizer in 1867.[29] As food antimicrobials, phenolic compounds can be classified, following Davidson,[29] as those currently approved (parabens), those approved for other uses (antioxi-

Table 14–3 Most Commonly Used Methods for Determining Antimicrobial Activity of Phenolic Compound

Methods	*Microorganism*	*Antimicrobial*	*Reference*
Exploratory tests			
Inhibition curves	Bacteria	BHA, TBHQ	Rico-Muñoz and Davidson[76]
		Mint essential oil	Tassou et al.[68]
		Selected spices	Beuchat[102]
		Phenolic compounds from spices	Al-Khayat and Blank[50]
		BHA	Robach et al.[75]
			Stern et al.[69]
			Chang and Branen[34]
			Shelef and Liang[78]
		Oleuropein	Tassou and Nychas[47]
		Sage	Shelef et al.[73]
		Essential oils from spices	Aureli et al.[77]
	Yeasts	BHA, TBHQ	Rico-Muñoz and Davidson[76]
		Vanillin	Cerrutti and Alzamora[41]
End point	Bacteria	Plant essential oils	Deans and Ritchie[103]
		Antioxidants	Sankaran[36]
		Thyme essential oil, thymol, and carvacrol	Juven et al.[59]
		Flavoring agents	Jay and Rivers[104]
		Oregano, thyme, and essential oils	Paster et al.[54]
		Fragrances	Morris et al.[105]
	Yeasts	Flavoring agents	Jay and Rivers[104]
		Oregano, thyme, and essential oils	Paster et al.[54]
		Essential oils	Conner and Beuchat[53]
	Molds	Flavoring agents	Jay and Rivers[104]
		BHA, BHT	Fung et al.[38]
		BHA	Thompson et al.[106]
			Thompson[37]
			Ahmad and Branen[35]
			Chang and Branen[34]
		BHA, TBHQ	Rico-Muñoz and Davidson[76]
		Garlic	Graham and Graham[107]
		Essential oils from citric	Karapinar[108]
		Cinnamon	Bullerman[109]
		Oregano, thyme, and essential oils	Paster et al.[54]
		Cinnamon, clove, and essential oils	Bullerman et al.[48]
		Thymol and carvacrol	Thompson[67]
		Essential oils	Thompson[26]
			Mahmoud[52]
		Cinnamon	Sebti and Tantaoui-Elaraki[71]
		Spices and herbs	Llewellyn et al.[110]
			Hitokoto et al.[111]
			Azzous and Bullerman[70]
			Akgül and Kivanc[112]

continues

Table 14–3 continued

Methods	*Microorganism*	*Antimicrobial*	*Reference*
		Thymol	Buchanan and Sheperd[51]
		Vanillin	López-Malo et al.[42–44]
Applied tests	Bacteria	Sage	Shelef et al.[73]
		BHA	Shelef and Liang[78]
			Robach et al.[75]
		Oleuropein	Tassou and Nychas[47]
		Several essential oils	Aureli et al.[77]
	Vanillin	Castañón et al.[90]	
	Molds	BHA and BHT	Fung et al.[38]
		BHA	Ahmad and Branen[35]
		Vanillin	López-Malo et al.[42]
		Vanillin	Cerrutti et al.[89]
		Oregano and thyme essential oils	Paster et al.[49]
	Yeast	Vanillin	Cerrutti and Alzamora[41]
		Vanillin	Cerrutti et al.[89]

dants), and those found in nature (polyphenolics, phenol). Parabens (methyl, propyl, and heptyl esters of p-hydroxybenzoic acid) are allowed in many countries as food antimicrobials for direct usage. Phenolic antioxidants, butylated hydroxytoluene (BHT), butylated hydroxyanisole (BHA), propyl gallate (PG), and tertiary butylhydroquinone (TBHQ), are approved as food antioxidants to prevent rancidity in fats, oils, and lipidic foods.[30] BHA, BHT, and TBHQ are also recognized as compounds that have antimicrobial activity against bacteria, fungi, virus, and protozoa.[17] Ward and Ward[31] were the first to report the antimicrobial activity of this kind of compound.[17,32] Naturally occurring phenolic compounds are widespread in plants and may be found in a great variety of food systems, and, as phenol derivatives, they may have antimicrobial activity. These naturally occurring phenols and phenolic compounds may be classified into the following groups:

- simple phenols and phenolic acids (e.g., p-cresol, 3-ethylphenol, vanillic, gallic, ellagic, hydroquinone);
- hydroxycinnamic acid derivatives (e.g., p-coumaric, caffeic, ferulic, sinapic);
- flavonoids (e.g., catechins, proanthocyanins, anthocyanidins and flavons, flavonols and their glycosides); and
- "tannins" (e.g., plant polymeric phenolics with the ability to precipitate protein from aqueous solutions).

The three groups of phenolic compounds share antimicrobial activity, and the mode of action against microorganisms may be deduced from the results found with one group. Therefore, although the main subject of this chapter is natural antimicrobials, some discussion will be based on findings using other synthetic phenolic compounds, especially with phenolic antioxidants.

Phenolic Antioxidants as Antimicrobials

The antimicrobial activity of phenolic antioxidants has been proven against several bacteria, recognizing that gram-positive bacteria are generally more sensitive to these compounds. MacNeil et al.[33] reported that chicken meat treated with 0.003% BHA reduced the standard count during a 3-week storage at 3°C. In a review made by Branen et al.,[17] research work was done on the inhibitory effect of phenolic antioxidants against the following bacteria: *S. senftenberg, S. typhimurium, Staphylococcus aureus, Escherichia coli, Vibrio parahaemolyticus, Clostridium perfringens, Pseudomonas fluorescens,* and *P. fragi.* Kabara[32] reported minimal inhibitory concentrations of BHA in the range of 125–250 ppm for *Streptococcus mutans, S. agalactiae,* and *S. aureus* in laboratory broth with pH 7.0. The antimicrobial effects of phenolic antioxidants are not limited to bacteria; they also have effects on molds and yeasts. Chang and Branen[34] reported that 250 ppm of BHA inhibited mold growth and aflatoxin production in a mycologic broth. Ahmad and Branen[35] confirmed these results, demonstrating that BHA is an effective mold growth inhibitor: *Penicillium expansum, P. notatum, P. claviforme, P. utricae, P. roqueforti, Aspergillus flavus, A. niger,* and *Geotrichum candidum* were inhibited with 150–200 ppm BHA in a glucose-salts broth. The addition of 750 ppm BHA in a solid medium delayed germination of *A. flavus* and *A. fumigatus* spores for 512 days.[36] Thompson[37] reported that 200 mg/mL BHA were effective to prevent for 48 hours the germination of *A. flavus* and *A. parasiticus,* and suggested that BHA can be considered as an effective fungistatic agent, inhibiting germination of mold spores. Fung et al.[38] found that BHA in concentrations of 0.01 and 0.02 g in approximately 20 mL of agar has significant effects on growth, sporulation, pigmentation, and toxin production of six toxigenic and six nontoxigenic strains of *A. flavus* during 4 days at 28°C. Ahmad and Branen[35] reported that 200 ppm BHA inhibited the growth of *P. expansum* and *A. flavus* in apple sauce when BHA was mixed into the product whereas only 150–200 ppm were required when BHA was applied to the surface of the product. Knox et al.[39] compared the efficacy of BHA and TBHQ against *S. cerevisiae* growth in apple or grape juice incubated for 24 hours at 32°C. MICs for BHA and TBHQ were 140 ppm and 600 ppm in apple juice and 150 ppm and > 950 ppm in grape juice, respectively. When compared in inhibitory action, BHA was less effective than sodium metabisulfite but more effective than potassium sorbate.

Raccah[40] reviewed the literature on the antimicrobial activity of phenolic antioxidants. Antimicrobial activity appeared to be strongly dependent on, among others, microbial species (strain and concentration); type and concentration of phenolic antioxidants; combination of phenolic antioxidants; combination with other antimicrobials; temperature; and food additives and components. The concentration of phenolic antioxidants with antimicrobial activity in food products was in the range of 30–10,000 ppm, whereas these compounds are permitted as antioxidants in concentrations generally up to 200 ppm, based on the fat or oil content of the food product. To comply with federal regulations, he suggested the use of phenolic antioxidants in combination with other antimicrobials.

Natural Phenolic Compounds from Spices, Extracts, and Essential Oils

It has been reported that some of these phenolic compounds have a wide antimicrobial spectrum, such as thymol extracted from thyme and oregano, cinnamic aldehyde extracted from cinnamon, and eugenol extracted from cloves.[10,16,29] Vanillin, a phenolic compound present in vanilla pods, is structurally similar to eugenol and has antifungal activity.[16,41–44] Another phenolic compound present in plants with antimicrobial activity is oleuropein, obtained from green olive extracts.[7,16,45–47]

Despite this, the antimicrobial activity of naturally occurring phenols and phenolic compounds has been examined to a limited extent, except for those coming from olives, tea, and coffee.[7] As reviewed by Nychas,[7] phenolic extracts of black and green tea and/or coffee have been found to be bactericidal or bacterostatic against *Campylobacter jejuni, Campylobacter coli, S. mutans, V. cholerae, S. aureus, S. epiderimidis, Plesiomonas shigelloides, S. typhi, S. typhimurium, S. enteriditis, Shigella flexneri,* and *S. dysenteriae. Pseudomonas fragi, Lactobacillus plantarum, S. aureus, S. carnosus, Enterococcus faecalis,* and *Bacillus cereus* were inhibited by ethyl acetate extracts from olives, and *B. subtilis, B. cereus,* and *S. aureus* were inhibited by oleuropein.

It is important to remark that the phenolic extracts from spices and plants may contain not only phenolics but also other compounds that may have antimicrobial activity. Bullerman et al.[48] reported that cinnamic aldehyde and eugenol, the major constituents of essential oils from cinnamon and clove, respectively, were the active antimicrobial compounds of the oils. However, they do not discard other minor constituents that may also have antimicrobial activity. Furthermore, Paster et al.[49] reported that carvacrol and thymol had lesser inhibitory effects on mold growth than did extracts from oregano and clove, suggesting that the antimicrobial activity may be due to several compounds from the extracts.

Eugenol has been reported as one of the most effective natural antimicrobials from plant origin, acting as a sporestatic agent: Significant reductions of viable *B. subtilis* spores were obtained when exposed to 0.1–1.0% eugenol for 8 days at 37°C.[50] Gingeron, zingerone, and capsaicin have been also found to be sporostatic for *B. subtilis*.[50] Buchanan and Shepherd[51] found that 100 ppm thymol inhibited *A. parasiticus* growth for 7 days at 28°C. Mahmoud[52] reported also that cinnamic aldehyde in lower concentrations than the minimal inhibitory delayed *A. flavus* growth for 8 days at 28°C. When studying the relationship between structure and inhibitory actions, Katayama and Nagai[20] assayed 32 pure phenol compounds. They found that 0.05% eugenol, carvacrol, isoborneol, thymol, vanillin, and salicyldehyde in agar were inhibitory against *B. subtilis, S. enteritidis, P. aeruginosa, P. morganii,* and *E. coli*, concluding that the presence of a hydroxyl group enhanced the antimicrobial activity. López-Malo et al.[42] demonstrated that vanillin concentration and the type of agar significantly ($p < 0.05$) affected the radial growth rate of aspergilli. Also, differences among mold responses were reported; the most resistant mold to the conditions assayed was *A. niger,* followed by *A. parasiticus, A. flavus,* and *A. ochraceus.* The presence of 1,000 ppm natural vanillin inhibited *A. ochraceus* growth for more than 2 months at 25°C in potato dextrose agar (PDA), whereas growth of *A. niger, A. flavus,* and *A. parasiticus* was inhibited by 1,500 ppm. Results obtained in vitro were corroborated for *A. flavus* and *A. parasiticus* in all fruit-based (apple, banana, mango, papaya, pineapple) agars and for *A. ochraceus* and *A. niger* in papaya-, pineapple-, and apple-containing agars. On the other hand, inhibitory concentrations of vanillin for *A. niger* and *A. ochraceus* in mango- and banana-based agars were greater than those found in PDA.

Information about antimicrobial effect of essential oils is more abundant. As we said before, although essential oils contain a variety of compounds of different chemical classes, natural phenolics have been reported as the main antimicrobial compounds. Conner and Beuchat[53] found, studying essential oils of plants and spices against several food spoilage yeasts, that each studied yeast responded in a different way, indicating that extracts can present various modes of action or that there are diverse yeast metabolism responses to the antimetabolic effect of essential oils. *Rhodotorula rubra* was highly sensitive to cinnamon, clove, garlic, onion, oregano, and thyme oils in concentrations between 25 and 200 ppm; however, *Klocckera apiculata* was moderately sensitive, and *Torulopsis glabrata* was resistant to most of the oils, excluding those of garlic and onion. Paster et al.,[54] evaluating the antimycotic capacity of essential oil of oregano and clove against three strains of *Aspergillus*, indicated that the studied molds differed in their sensibility to the extracts and found that *A. flavus* was sensitive to the essential oil of oregano. Zaika, Kissinger, and Wasserman[55] found that little, if any, inhibition of growth and acid production by *L. plantarum* and *Pediococcus acidilactiti* was noted in the presence of 40 ppm oregano oil, whereas levels > 200 ppm were bactericidal to both organisms. Also, they reported that bacteria exposed to sublethal concentrations of oregano oil were able to overcome the inhibition and to develop resistance to the toxic effect of oregano oil or oregano.

Essential oils of thyme and oregano were inhibitory to *V. parahaemolyticus,* 25 genera of bacteria, and *S. aureus;* essential oil of sage was inhibitory to *V. parahaemolyticus, B. cereus, S. aureus,* and *S. typhimurium;* and rosemary spice extract at 0.1% substantially inhibited growth of *S. typhimurium* and *S. aureus.*[2] Resnik et al.[56] analyzed the effect of the concentration of vanillin on the growth rate and aflatoxin accumulation of *A. parasiticus.* The growth rate decreased abruptly in the presence of 250 ppm vanillin, whereas 1,500-ppm vanillin inhibited mold growth during at least 37 days of storage at 28°C. However, 500 ppm vanillin enhanced the AFB1 and AFG1 accumulation, the toxin levels exceeding those of the control.

Mode of Action

The possible modes of action of phenolic compounds have been reported in different reviews.[7,15,29] However, these mechanisms have not been completely elucidated. Prindle and Wright[57] mentioned that the effect of phenolic compounds is concentration-dependent. At low concentration, phenols affected enzyme activity, especially of those enzymes associated with energy production, although at greater concentrations they caused protein denaturation. The effect of phenolic antioxidants on microbial growth and toxin production could be the result of the ability of phenolic compounds to alter microbial cell permeability, permitting the loss of macromolecules from the interior. They could also interact with membrane proteins, causing a deformation in its structure and functionality.[38]

Conner and Beuchat[53,58] suggested that antimicrobial activity of essential oils on yeasts could be the result of disturbance in several enzymatic systems involved in energy production and structural components synthesis. Once the phenolic compound crossed the cellular membrane, interactions with membrane enzymes and proteins would cause an opposite flow of protons, affecting cellular activity. Juven et al.[59] found that increasing thyme essential oil, thymol, or carvacrol concentration was not reflected in a direct relationship with antimicrobial effects. However, they reported that after exceeding a certain concentration (critical concentration), a rapid and drastic reduction in viable cells of *S. typhimurium* was observed. Phenolic compounds could sensitize cellular membranes, and when sites were saturated, a serious damage and rapid collapse of cytoplasmatic membrane integrity could be presented, with the consequent loss of cytoplasmatic constituents.[59] Ruiz-Barba et al.,[60] using scanning electron microscopy (SEM), showed that cells without treatment were smooth, compared with those treated with phenols for 24 hours, which appeared rugged and with irregular surfaces. Kabara and Eklund[61] mentioned that phenolic compound effects could be at two levels, on cellular wall and membrane integrity and on microbial physiologic responses. Phenolic compounds could also denaturalize enzymes responsible for spore germination or interfere with amino acids necessary in germination processes.[7] Rico Muñoz et al.,[62] after finding that phenolic compounds from olives did not have an effect at the membrane level on *S. aureus*, concluded that phenolic compounds would not share a common mechanism of action and that there might be various targets associated with their antimicrobial effect.

OTHER ANTIMICROBIAL COMPOUNDS FROM PLANTS

Among the plants widely consumed in the human diet, garlic, onion, and leek have been recognized and studied extensively for their antimicrobial properties.[16] Many food-borne pathogenic bacteria are sensitive to onion and garlic extracts. *Staphylococcus aureus, B. cereus, C. botulinum, S. typhimurium,* and *E. coli* have been adversely affected by garlic extracts.[16] The antimicrobial effects reported for garlic and onion were attributed to the allicin concentrations and other sulfur compounds present in their essential oils. Allicin inhibits several metabolic enzymes, inactivates proteins with -SH groups by oxidation, and inhibits in a competitive form the activity of selected compounds.[63,64]

Mahmoud[52] reported that 1,000-ppm geraniol, nerol, and citronelol inhibited *A. flavus* growth in nutritive broth (pH 5.5) during 15 days of incubation at 28°C, the MIC being 500 ppm for the three alcohols.

López-Malo and Argaiz[65] evaluated the effects of pH (6.5, 5.5, 4.5, or 3.5) and citral (3,7-dimethyl-2,6-octadienal) concentration (0; 500, 1,000, 1,500, or 2,000 ppm) on the growth of *A. flavus, A. parasiticus, P. digitatum,* and *P. italicum* in PDA adjusted with sucrose to a_w 0.97. Molds' radial growth rate increased as pH increased and citral concentration decreased. Conversely, lowering pH and increasing citral concentration increased mold germination time. Inhibitory citral concentration depended on pH and differed among molds. *P. digitatum* and *P. italicum* were inhibited with 500 ppm citral at pH 3.5 and 6.5, whereas 1,000 ppm was required to inhibit *A. flavus* and *A. parasiticus*. At pH 4.5 and 5.5, inhibitory concentrations varied from 1,000 ppm for *Penicillium* to 2,000 ppm for *Aspergillus*.

Phytoalexins, low-molecular-weight compounds produced by higher plants in response to microbial infections or treatment of plant tissues with biotic or abiotic elicitors, are also broad-spectrum antimicrobial agents that probably act by altering the microbial plasma membranes. Their antimicrobial activity is directed against fungi but they also have effect toward bacteria.[2,18] More than 200 compounds have been isolated from more than 20 families of plants, the most important chemical classes being isoflavonoids and, to a lesser extent, proteins (chitinases, thionins, zeamatins, thaumatins, etc.). However, the high concentrations needed to exert antimicrobial action in food matrices explain the few examples of the actual use in foods.[2]

INTERACTION WITH pH AND OTHER PRESERVATION FACTORS

The antimicrobial activities of extracts from several types of plants and plant parts used as flavoring agents in foods have been recognized for many years. However, not many data have been reported on the effect of extracts in combination with other factors on microbial growth. Its potential as a total or partial substitute for common preservatives to inhibit growth of spoilage and pathogenic microorganisms needs to be evaluated alone and in combination with traditional preservation factors or hurdles (mainly storage temperature, pH, water activity, other antimicrobials, modified atmospheres). These results could be very useful in terms of allowing research workers involved in product development under the concept of hurdle technology to assess quickly the impact of altering any combination of the studied variables.

pH Effect

The effect of pH on the antimicrobial activity of natural phenolic compounds is not clearly understood. However, some authors reported a greater effect as the pH was lowered. Sykes and Hooper[66] found greater effects of phenolics at acid pH values, and they attributed it to the increased solubility and stability of these compounds at low pH. Juven et al.[59] reported that phenolic active compounds sensitize the cell membrane of *S. typhimurium* and *S. aureus* and, when saturation of the sites of action occurs, there is gross damage and a sudden collapse of the integrity of the microbial cell. They mentioned that at low pH values, the thymol molecule is mostly undissociated and may bind better to hydrophobic regions of the membrane proteins and dissolve better in the lipid phase of the membrane.

Thompson[67] reported that 1.0 mM carvacrol inhibited at least 7 days at 27°C the growth of *Aspergillus* when the pH of the medium was 4.0 or 8.0; at pH 6.0, only mycelium production reductions (about 50–80%) were observed. Thymol

(1.0 mM) addition inhibited studied molds at the three evaluated pH levels. Al-Khayat and Blank[50] reported that 0.03% eugenol reduced *B. subtilis* spores capable of germination while lowering the pH of culture medium from 8.0 to 6.0 and mentioned that eugenol was effective at low pH. Juven et al.[59] found an increased antimicrobial activity of essential oils of thyme and thymol at pH 5.5, in comparison with the one at pH 6.5, and attributed these effects to changes in polar group distribution of phenolic constituent of oils between the cytoplasm membrane and the external medium. At low pH, the thymol molecule is largely undissociated and, therefore, more hydrophobic and can be joined better to hydrophobic regions of membrane proteins and dissolved more easily in the lipidic phase. Kabara[61] mentioned that undissociated phenolic groups are more active as antimicrobials than are dissociated forms, suggesting that phenols can act on a wide pH range (3.5–8.0). Tassou et al.[68] found synergistic effects between pH and mint essential oil in *S. enteritidis* and *L. monocytogenes* inhibition in foods and laboratory media.

Stern et al.[69] found that 50 ppm BHA proved inhibitory against *S. aureus,* at least by 48 hours, and reported that the effect was significantly greater when reducing the pH of the medium from 7.0 to 5.0 and increasing the concentration of sodium chloride.

López-Malo et al.[44] reported second-order polynomial equations for describing the lag time for mold (*A. ochraceus, A. niger, A. parasiticus,* and *A. flavus*) growth under the effects of pH and vanillin concentration in laboratory media. Contour plots for fixed lag times were obtained with predictive models (Figure 14–3). For the four molds assayed, the lag time increased as vanillin concentration increased and pH decreased. The effects of the variables were more important for *A. flavus* and *A. ochraceus*. For *A. niger* and *A. parasiticus,* the greatest level of vanillin concentration evaluated (1,000 ppm) was not enough to cause an effective increment in lag time, even at pH 3.0. Results demonstrated that a combination of vanillin with pH reduction had an additive or synergistic effect on mold growth, depending on the *Aspergillus* species. *A. ochraceus* was inhibited at pH 3.0 with 1,000 ppm vanillin. Higher concentrations could inhibit the other molds.

Other Preservation Factors

The use of BHA, in combination with common additives such as NaCl and in the presence of acid, helped in reducing the normally used concentration of the compounds added to inhibit microbial growth in foods.[69] Tassou et al.[68] found that, in a salty food, 1.0% mint essential oil drastically affected *S. enteritidis* growth and reported that salt acted synergistically with mint essential oil.

López-Malo et al.[43] reported mold (*Aspergillus*) inhibitory concentrations of vanillin at selected pH and incubation temperatures. Figure 14–4 shows the variable combination at lower levels that inhibited growth for each mold. The inhibitory conditions (no growth after 30 days) depended on the type of mold. The most resistant one, *A. niger*, was inhibited with 1,000 ppm vanillin at pH ≤ 3.0 and incubation temperature ≤ 15°C or with 500 ppm at pH ≤ 4.0 and temperature ≤ 10°C. For *A. ochraceus*, the most sensible, the inhibitory conditions in systems containing 500-ppm vanillin were pH 3.0 and temperature ≤ 25°C or pH 4.0 with temperature ≤ 15°C. If the pH was raised to 4.0, the vanillin concentration could be ≤ 1,000 ppm with temperatures lower than 15°C or 500 ppm if the incubation temperature was reduced to 10°C.

Fungal inhibition can be achieved by combining spices and traditional antimicrobials, reducing the concentrations needed to achieve the same effect. Azzouz and Bullerman[70] reported that clove was an efficient antimycotic agent against *A. flavus, A. parasiticus, A. ochraceus,* and four strains of *Penicillium*, delaying mold growth by more than 21 days. These authors also found additive and synergic effects in combining 0.1% clove with 0.1–0.3% potassium sorbate, delaying mold germination time. Sebti and Tantaoui-Elaraki[71] reported that the combination

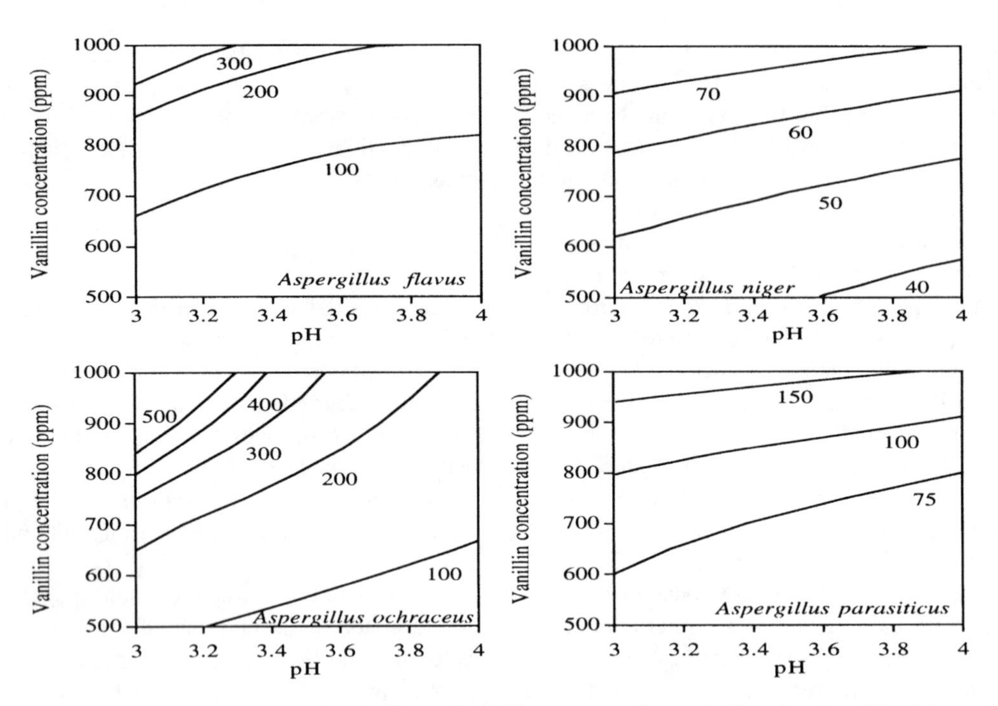

Figure 14–3 Contour plots for predicted effects of vanillin concentration and pH on lag time (h) of *Aspergillus flavus*, *A. niger*, *A. ochraceus*, and *A. parasiticus* in potato-dextrose agar with a_w 0.98.

of sorbic acid (0.75 g/kg) with an aqueous cinnamon extract (20 g/kg) was effective to inhibit growth of 151 mold and yeast strains isolated from a Moroccan bakery product when, to inhibit the studied microorganisms, 2,000 ppm sorbic acid were needed.

Matamoros-León et al.[72] evaluated the individual and combined effects of potassium sorbate and vanillin concentrations on the growth of *P. digitatum*, *P. glabrum,* and *P. italicum* in PDA adjusted to a_w 0.98 and pH 3.5. These authors found that 150 ppm potassium sorbate inhibited *P. digitatum,* whereas 700 ppm were needed to inhibit *P. glabrum*. Using vanillin, inhibitory concentration varied from 1,100 ppm for *P. digitatum* and *P. italicum* to 1,300 ppm for *P. glabrum*. When used in combination, MIC isobolograms show curves deviated to the left of the additive line (Figure 14–5). Also, calculated FIC index values varied from 0.60 to 0.84. FIC index and isobolograms show synergistic effects on mold inhibition when vanillin and potassium sorbate were applied in combination.

APPLICATION IN FOODS

Unfortunately, many of the published reports about the application of phenolic compounds (antioxidants and constituents of extracts and essential oils) as antimicrobials have been accomplished in model and laboratory systems, and there are few studies that have been carried out in real foods.[11] The essential oils of spices and plants, as well as their major components, are more effective in microbiologic media than when evaluated in real foods.[73] In most cases, the inhibitory concentrations found in model systems increase significantly when evaluated with the same microorganisms in actual foods and, in consequence, few of the applications of phenolic antioxidants as antimicrobials have been successful.[61] This reduction in the effectiveness observed *in vivo* represents an important limitation to the use of essential oils and phenolic antioxidants as antimicrobial agents in foods.[59] The interactions among phenolic groups and proteins, lipids, and aldehydes could ex-

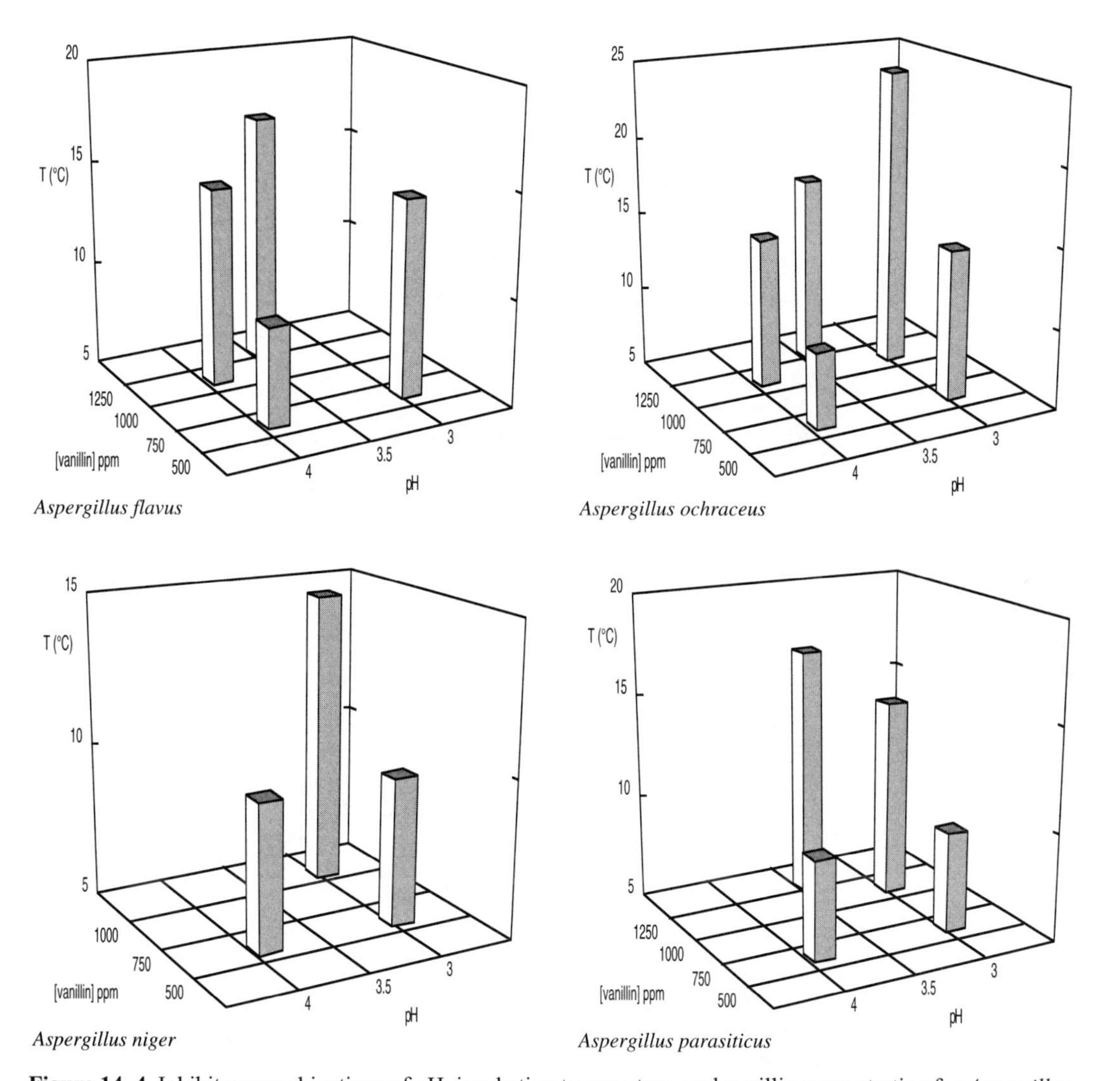

Figure 14–4 Inhibitory combinations of pH, incubation temperature, and vanillin concentration for *Aspergillus flavus*, *A. niger*, *A. ochraceus*, and *A. parasiticus* Growth in a_w 0.98 potato-dextrose agar.

plain, at least partially, the reduction of the antimicrobial effect of essential oils where the major constituents are phenols.

Juven et al.[74] showed that oleuropein antimicrobial activity could be reduced by the addition of triptone and/or yeast extract to the culture medium. Tassou and Nychas[47] demonstrated that inoculum size, oleuropein concentration, and pH significantly influenced *S. aureus* growth and lag time, and proved that the efficiency of phenolic compound antimicrobial action was reduced in foods with relatively low protein content. Robach et al.[75] reported that the reduction in BHA antimicrobial capacity in a crab homogenate was due to a partial inhibition of antioxidant properties by the presence of lipids and oxidation stage. Rico-Muñoz and Davidson,[76] studying casein and corn oil effects in phenolic antioxidant antimicrobial activity, reported that casein did not have effect on *Saccharomyces*

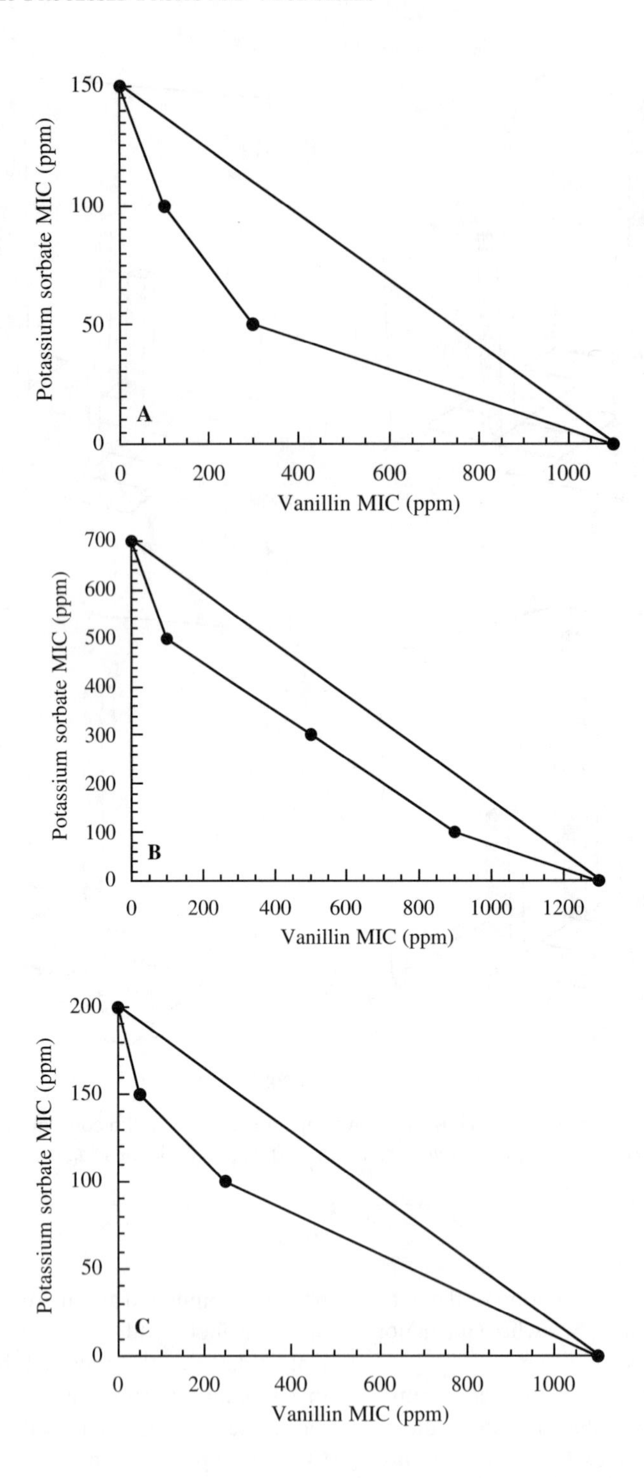

Figure 14–5 Minimal inhibitory concentration isobolograms for combination of potassium sorbate and vanillin in PDA (a_w 0.98, pH 3.5) against (A) *Penicillium digitatum*, (B) *P. glabrum,* and (C) *P. italicum*. Inhibition was defined as no growth after 30 days at 25°C.

cerevisiae or *S. aureus* growth and slightly reduced growth of *P. fluorescens*. However, casein presence dramatically reduced the antimicrobial activity of BHA. They found that casein addition stimulated the growth of *P. citrinum* and *A. niger,* and the activity of BHA in presence of protein depended on the species studied. For *P. citrinum,* the inhibitory activity of 200 ppm BHA was almost eliminated in the presence of 6% and 9% casein; on the other hand, for *A. niger,* the addition of 3% casein eliminated the antimycotic effect of the antioxidant. Aureli et al.,[77] evaluating the antimicrobial capacity of thyme against *L. monocytogenes* growth in model systems and in a real food, found that essential oil antilisteric efficiency decreased when used *in vivo* (ground pork meat), in comparison with the behavior in laboratory media (solid and liquid). Robach et al.[75] reported a decrease in antimicrobial activity of BHA in foods. They found that *V. parahaemolyticus* growth was inhibited with 50 ppm BHA in trypticase-soy broth, whereas 400 ppm were required to achieve the same effect in a crab homogenate. Shelef and Liang[78] reported an increase in the inhibitory concentration of BHA for several strains of *Bacillus*. More than 100-fold was required in chicken meat, in comparison with laboratory media. In studies with this antioxidant in milk defatted solids, Cornell et al.[79] found that BHA was bound to the casein through hydrophobic interactions and should probably decrease its antimicrobial activity.

Spencer et al.[80] reported that the interaction or complex between phenols and proteins depends partially on protein characteristics, on pH, and on phenolic group containing molecule. This interaction takes place by hydrogen bridges between phenolic groups and peptides, as well as by hydrophobic interactions.

The presence of corn oil or Tween 80 reduced or eliminated the antimicrobial activity of BHA and TBHQ. 200 ppm BHA inhibited the growth of the molds studied by Rico-Muñoz and Davidson.[76] However, inhibition of *P. citrinum* was reduced around 35%, 20%, and 19% when 1.5%, 3.0%, and 4.5% corn oil was added, and the antimicrobial activity was eliminated with 1.5% and 3.0% corn oil for *A. niger*.

The antimicrobial activity of BHA and TBHQ was affected by the presence of casein or corn oil in the laboratory media because TBHQ is less effective as an antimicrobial agent in the presence of fats or proteins, as compared with BHA.[76] The principal cause of antimicrobial activity loss could probably be the solubilization of these compounds in the lipidic phase of the medium, reducing its availability to act as antimicrobial. BHA is a lipophilic antioxidant with a low hydrophilic-lipophilic balance (HL), whereas TBHQ is more anphipatic with a greater HL balance. These differences in fat solubility could explain, at least partially, the greater decrease in antimicrobial activity of BHA in fat presence than that observed for TBHQ. Ahmad and Branen[35] demonstrated that small-quantity addition of lipids (0.25%) reduced antimycotic activity of BHA. They evaluated BHA activity in real foods and demonstrated that BHA inhibited *P. expansum* and *A. flavus* growth inoculated in spread cheese and apple pulp. Concentrations needed to inhibit mold growth were greater than those found in laboratory media. In apple pulp, 200 ppm were necessary, and in spread cheese, 400 ppm were required. These authors explained the activity reduction on solubility in the lipidic phase and/or on protein interactions.

The interactions of aldehydes with proteins have been extensively studied, because protein addition to aldehyde solutions can decrease the effective concentration of these groups.[81–83] Citral (lemon flavor component) concentration was reduced almost 100% when 5% of casein or soy protein isolate was added in aqueous solutions. Sixty-eight percent initial vanillin concentration, measured by HPLC, was lost after 26 hours in drinks containing aspartame.[84] Hussein et al.[85] reported that vanillin concentration in aqueous solutions was reduced 73% after 25 hours at 26°C when aspartame was added. The reduction in vanilla flavor by the reduction of vanillin concentration had been reported also when adding faba-bean proteins, sodium caseinates, or milk whey protein concentrate.[83,86–88]

Minimal Processing of Fruits and Vegetables

The levels of essential oils or plant extracts needed to cause a similar antimicrobial effect in food products, in comparison with laboratory media, are considerably higher. These greater concentrations of natural antimicrobials may modify the sensory characteristics of the product, making it unacceptable. The use of natural antimicrobials, in combination with other environmental stress factors, not only can enhance their antimicrobial properties, but makes possible the development of products that consumers are demanding that reduce the amounts of synthetic or natural antimicrobials needed to assure microbial stability.

It had been demonstrated that vanillin (4-hydroxy-3-methoxy-benzaldehyde) inhibited microbial growth in laboratory media and fruit purées stored at 25–27°C.[14,41,42] Promising results have been obtained by Cerrutti et al.[89] and Cerrutti and Alzamora[41] in strawberry (pH 3.4, a_w 0.95) and apple purées (pH 3.5, $a_w \cong$ 0.99) preserved by combined factors. Strawberry purée with 3,000 ppm vanillin and reduced a_w inhibited the native and inoculated flora (*S. cerevisiae*, *Zygosaccharomyces rouxii*, *Z. bailii*, *Schizosaccharomyces pombe*, *Pichia membranaefaciens*, *Botrytis* species, *Byssochlamys fulva*, *B. coagulans,* and *L. delbrueckii*) for at least 60 days of storage at 25°C. In apple purée with 2,000 ppm vanillin, a germicidal effect was observed on inoculated *S. cerevisiae*, *Z. rouxii,* and *D. hansenii*. Cerrutti and Alzamora[41] also reported that 3,000 ppm vanillin in a banana purée with pH 3.4 and a_w 0.98 were inhibitory for *S. cerevisiae*, *Z. rouxii,* and *D. hansenii.*

Castañón et al.[90] evaluated the effects of vanillin (1,000 or 3,000 ppm) or potassium sorbate (1,000 ppm) addition on the microbial stability during storage at 15, 25, or 35°C of banana purée preserved by combined methods (a_w 0.97, pH 3.4). Table 14–4 presents the native flora (standard plate, yeast, and mold) changes during storage at different temperatures in the control purées (without antimicrobial addition) and in those containing 1,000 ppm vanillin. In the purées used as control, there was an effect of the storage temperature on the microbial growth. However, a reduced storage temperature (15°C) was not enough to detain or delay the spoilage of the purée. After 6 days of storage at 15°C, native yeasts and molds reached counts of 10^4 CFU/g (about 3 log cycles more than the initial count), and the purées were at this time sensory-unacceptable (odor and textural changes accompanying the microbial spoilage). The addition of 1,000 ppm vanillin increased the lag phase up to $\cong$ 16 days at 15°C, and the time to detect the microbial spoilage was extended to around 21 days. In the presence of 3,000 ppm vanillin or 1,000 ppm potassium sorbate, no microbial growth (< 10 CFU/g) was detected after 6 days and up to 60 days of storage (15, 25, or 35°C). Furthermore, the initial counts of the native flora in these purées were lower, in comparison with control purées or those containing 1,000 ppm vanillin. Results obtained by Cerrutti and Alzamora,[41] Cerrutti et al.,[89] and Castañón et al.[90] demonstrated that addition of vanillin, in combination with a slight reduction of a_w and pH, may be a promising method for fruit purée preservation and confirmed antimicrobial properties of vanillin. They found that protein and fat contents of the fruit partially determined the vanillin concentration necessary to obtain a sound product. The relatively high protein (1.2%) and fat (0.3%) content in banana, compared with those of other fruits, explains the necessity for a higher vanillin concentration to obtain the same antimicrobial effects as in other fruits.

The results of sensory evaluation obtained by Castañón et al.[90] with banana purées containing 3,000 ppm vanillin or 1,000 ppm potassium sorbate are presented in Table 14–5. The mean scores corresponded to products with a good overall acceptability, with scores around 6 (like slightly). The purées were significantly different ($p < 0.05$) in odor (the one with vanillin was preferred) and flavor (the one with potassium sorbate was better), and there was no significant difference ($p < 0.05$) in color and overall acceptability. The flavoring characteristics of vanillin

Table 14–4 Native Flora Evolution (CFU/g) in Banana Purée (a_w 0.97, pH 3.4) without or with Vanillin (1,000 ppm) during Storage at 15, 25, or 35°C

	Storage Temperature					
	15°C		*25°C*		*35°C*	
Storage Time (day)	*Control*	*Vanillin*	*Control*	*Vanillin*	*Control*	*Vanillin*
Yeast and mold count						
0	3.5×10^1	1.0×10^1	3.5×10^1	1.0×10^1	3.5×10^1	1.0×10^1
3	NE	NE	1.3×10^5	3.7×10^2	4.6×10^6	1.6×10^3
4	NE	NE	NE	NE	a	2.1×10^6
5	NE	NE	8.5×10^6	9.1×10^4	NE	NE
6	7.3×10^4	1.7×10^3	a	6.3×10^5	a	a
16	a	4.6×10^4	a	a	a	a
21	a	7.6×10^5	a	a	a	a
32	a	a	a	a	a	a
Standard plate counts						
0	3.7×10^3	1.6×10^2	3.7×10^3	1.6×10^2	3.7×10^3	1.6×10^2
3	NE	NE	3.8×10^5	3.7×10^2	4.7×10^6	1.7×10^4
4	NE	NE	NE	NE	a	6.2×10^5
5	NE	NE	9.2×10^6	1.1×10^5	NE	NE
6	2.8×10^5	1.9×10^3	a	7.9×10^5	a	a
16	a	5.6×10^4	a	a	a	a
21	a	8.1×10^6	a	a	a	a
32	a	a	a	a	a	a

NE, Not evaluated; a, not determined, samples completely spoiled and discarded.

are well accepted and have demonstrated compatibility with many fruits in concentrations up to 3,000 ppm.[41]

The microbial stability of mango juice (pH 4.9), supplemented with extracts of ginger (antimicrobial compounds: zingerone, gingerol, and shogaol) and nutmeg (antimicrobial compounds: myristicin and sabinene), was investigated during 3 months at room temperature by Ejechi et al.[91] The combination of heating (15 minutes at 55°C) and 4% v/v of each spice inhibited microbial growth (yeasts and non-spore-forming bacteria) and produced a product with acceptable taste.

Surface disinfection of tomatoes using cinnamaldehyde was studied by Smid et al.[92] Whole tomatoes were dipped for 30 minutes in a solution containing 13 mM cinnamaldehyde, then stored at 18°C in sealed plastic bags. The combination of the treatment with the natural phenolic compound and packaging under modified atmosphere reduced spoilage-associated fungi and bacteria on tomato surfaces, increasing the shelf life up to 11 days.

TOXICOLOGIC ASPECTS

Phenolic compounds are naturally present in vegetable products, sometimes in considerable quantities. Few of the natural phenols are toxic, and only some have been reported as causes of allergic reactions in animals. In the few occasions in which phenolic compounds have been harmful to humans, an abnormal consumption of vegetable-origin phenols or consumption of abnormal phenols for the diet has been identi-

Table 14–5 Sensory Evaluation* of Banana Purées Preserved by Combined Methods (a_w 0.97, pH 3.4) with 3,000 ppm Vanillin or 1,000 ppm Potassium Sorbate

Attribute	*Vanillin*	*Potassium Sorbate*
Color	6.09	6.45
Odor	6.14	4.87
Flavor	5.75	6.78
Overall acceptability	6.07	6.42

*Represent mean values of 50 untrained judges.

Table 14–6 Allowed Daily Intake for Selected Antioxidants

Antioxidant	*Daily Allowed Dose (mg/kg Corporal Weight)*
Tocopherol	0.15–2.0
BHA	0.0–0.50
BHT	0.0–0.125
TBHQ	0.0–0.20
Propyl gallate	0.0–2.50

fied as possible causes.[93] Natural phenolic compounds can be classified as common, uncommon, and rare, depending on their occurrence in vegetables. Phenols that are highly toxic for animals fall in the third category, whereas those that are found in considerable quantities in nature ("common" phenols) have very low toxicity.[93]

It has been reported that phenolic compound consumption can affect nutritional aspects in human and animals. However, it has been found that many phenols exhibit beneficial effects in some disorders, such as cancer.[7] It is recognized that consumption of fruits and vegetables and other sources of phenolic compounds, such as tea, reduces propagation of various forms of human cancer and other diseases, such as cirrhosis, emphysema, and arteriosclerosis.[94,95]

The toxicologic and safety characteristics of phenolic antioxidants have been verified through studies with animals submitted to an excessive and extended consumption before their approval. An expert committee, created with members of the Food and Agriculture Organization, the World Health Organization, the European Commission, and the Scientific Committee for Foods, established the acceptable (and secure) daily dose for antioxidant consumption.[30] These doses are defined as 100 times less than the quantity, in a corporal weight base, that can be consumed daily without causing any damage and are shown in Table 14–6.

CONCLUSION

Plant-derived antimicrobials are not yet fully exploited. The use of spices, herbs, plants, essential oils, and related phenolic compounds is limited, due to the high MICs in actual foods with high protein and/or fat contents, which impart undesirable flavor or/and aroma. These undesirable effects can be minimized if the natural compound is used in combination with other environmental stress factors in the framework of hurdle technology. In this way, considering the consumer's interest in more "natural" foods, the potential for applications in minimally processed fruits and vegetables appears to be good.[96,97] However, for a wider and more rational use of these natural compounds, some points should be addressed:

1. the extraction methodology, the variation in composition within the plant or spice due to differences in geographical cultivation zones, climatic conditions, etc., as well as the chemical characterization of extracts or oils and the identification of

Exhibit 14–2 Factors that Influence Plant-Derived Antimicrobial Selection and Effectiveness

Composition of food
- moisture, fat, and protein content
- water activity and pH
- presence of other inhibitors (acids, salts, smoke, antimicrobials)
- interactions with food matrix and other food additives

Initial contamination level
- sanitary conditions of ingredients and raw materials
- sanitary conditions of equipment
- processing conditions
- type of potential growing microorganisms

Handling and distribution
- length of storage
- temperature of storage
- packaging

Possible synergistic, antagonic, or additive interaction effects with other antimicrobial factors

Toxicologic and legal aspects

Solubility

Sensory impact

Cost

the compounds with antimicrobial activity.

2. the response of key microorganisms to the multitarget preservation system *in vitro* and the evaluation of the efficacy *in vivo,* i.e., in the food product. The levels of the naturally occurring antimicrobials and of the other factors must be selected, taking into account the sensory impact on the minimally processed product and the desired shelf life.

This issue requires a better understanding of the modes of action of the natural antimicrobial system and/or their interactions with other preservation factors, as well as the knowledge of the interactions between the stress factors applied and the food matrix.[98]

Few studies in the literature have described the inhibitory activities of naturally derived antimicrobials used at low concentration (compatible with the food flavor), alone or in combination with other hurdles, nor have these studies been systematically designed to obtain a quantitative microbial response in terms of primary and secondary kinetics models that allow the design of the combined preservation systems.[99]

3. the incorporation of the active principles through the addition of the plant, the herb, or the spice, or, on the contrary, the corresponding "extract" in preference to the highly purified natural antimicrobial compound. In this last case, the compound not only must be approved as a food additive for use as food preservative, but it would have to be labeled as a chemical additive.[2,100]
4. the form in which the plant, herb, spice, essential oil, or component will be incorporated to the foods without adversely affecting sensory, nutritional, and safety characteristics, and without increasing significantly the formulation, processing, or marketing costs of the minimally processed product to which it is added.[16]

Exhibit 14–2 presents some general guidelines to be followed for selecting plant-derived antimicrobials.

REFERENCES

1. Gould GW. Overview. In: Gould GW, ed. *New Methods of Food Preservation.* Glasgow: Blackie Academic and Professional; 1995: xv–xix.
2. Sofos JN, Beuchat LR, Davidson PM, Johnson EA. *Naturally Occurring Antimicrobials in Food.* USA: Council for Agricultural Science and Technology. Task Force Report No 132; 1998.
3. Leistner L. Principles and applications of hurdle technology. In: Gould GW, ed. *New Methods of Food Preservation.* Glasgow: Blackie Academic and Professional; 1995:1–21.
4. Ray B, Daeschel M. *Food Biopreservatives of Microbial Origin.* Boca Raton, FL: CRC Press; 1992:2–23.
5. Parish ME, Carroll DE. Minimum inhibitory concentration studies of antimicrobial combinations against *Saccharomyces cerevisiae* in a model broth system. *J Food Sci.* 1988;53:237–239, 263.
6. Board RG. Natural antimicrobials from animals. In: Gould GW, ed. *New Methods of Food Preservation.* Glasgow: Blackie Academic and Professional; 1995: 40–57.
7. Nychas GJE. Natural antimicrobials from plants. In: Gould GW, ed. *New Methods of Food Preservation.* Glasgow: Blackie Academic and Professional; 1995: 58–89.
8. Hill C. Bacteriocins: Natural antimicrobials from microorganisms. In: Gould GW, ed. *New Methods of Food Preservation.* Glasgow: Blackie Academic and Professional; 1995:22–39.
9. Taylor CB. Defense responses in plants and animals—more of the same. *The Plant Cell.* 1998;10:873–881.
10. Wilkins KM, Board RG. Natural antimicrobial systems. In: Gould GW, ed. *Mechanisms of Action of Food Preservation Procedures.* New York: Elsevier Science; 1989:285–362.
11. Board RG, Gould GW. Future Prospects. In: Russel NJ, Gould GW, eds. *Food Preservatives.* Glasgow: Blackie and Son; 1991:267–284.
12. Gould GW. Industry perspectives on the use of natural antimicrobials and inhibitors for food applications. *J Food Prot.* 1996;Suppl:82–86.
13. Zaika LL. Spices and herbs: Their antimicrobial activity and its determination. *J Food Safety.* 1988;9:97–118.
14. Shelef LA. Antimicrobial effects of spices. *J Food Safety.* 1983;6:29–44.
15. Conner DE. Naturally occurring compounds. In: Davidson PM, Branen AL, eds. *Antimicrobials in Foods*, 2nd ed. New York: Marcel Dekker; 1993:441–467.
16. Beuchat LR, Golden DA. Antimicrobials occurring naturally in foods. *Food Technol.* 1989;43(1):134–142.
17. Branen AL, Davidson PM, Katz B. Antimicrobial properties of phenolic antioxidants and lipids. *Food Technol.* 1980;34(5):42–53.
18. Smid EJ, Gorris LGM. Natural antimicrobials for food preservation. In: Rahman MS, ed. *Handbook of Food Preservation.* New York: Marcel Dekker; 1999:285–308.
19. Beuchat LR. Antimicrobial properties of spices and their essential oils. In: Dillon VM, Board RG, eds. *Natural Antimicrobial Systems and Food Preservation.* Wallingford, England: CAB Intl; 1994:167–179.
20. Katayama T, Nagai I. Chemical significance of the volatile components of spices in the food preservative view point. VI. Structure and antibacterial activity of terpenes. *Bull Jpn Soc Sci Fisheries.* 1960;26:29–32.
21. Farag RS, Daw ZY, Hewedi FM, El-Baroty GSA. Antimicrobial activity of some Egyptian spice essential oils. *J Food Prot.* 1989;52:665–667.
22. Mishra AK, Dubey NK. Evaluation of some essential oils for their toxicity against fungi causing deterioration of stored food commodities. *Appl Environ Microbiol.* 1994;60:1101–1105.
23. Davidson PM, Parish ME. Methods for testing the efficacy of food antimicrobials. *Food Technol.* 1989;43(1): 148–155.
24. Parish ME, Davidson PM. Methods of evaluation. In: Davidson PM, Branen AL, eds. *Antimicrobials in Foods.* New York: Marcel Dekker; 1993:597–615.
25. Thompson DP, Cannon C. Toxicity of essential oils on toxigenic and nontoxigenic fungi. *Bull Environ Contam Toxicol.* 1986;36:527–532.
26. Thompson DP. Effect of essential oils on spore germination of *Rhizopus, Mucor* and *Aspergillus* species. *Mycologia.* 1986;78:482–485.
27. Martindale WH. Essential oils in relation to their antiseptic powers as determined by their phenolic coefficients. *Perfum Essent Oil Rec.* 1910;1:266–274.
28. Müller G. *Microbiología de los Alimentos Vegetales.* Zaragoza, España: Editorial Acribia; 1981.
29. Davidson PM. Parabens and phenolic compounds. In: Davidson PM, Branen AL, eds. *Antimicrobials in Foods*, 2nd ed. New York: Marcel Dekker; 1993:263–305.
30. Sahidi F, Janitha PK, Wanasundara PD. Phenolic antioxidants. *Crit Rev Food Sci Nutr.* 1992;32:67–103.
31. Ward MS, Ward BQ. Initial evaluation of the effects of butylated hydroxyanisole (BHA). *J Food Sci.* 1967;32:349–351.
32. Kabara JJ. Phenols and chelators. In: Russel NJ, Gould GW, eds. *Food Preservatives.* Glasgow: Blackie and Son; 1991:200–214.
33. MacNeil JH, Dimich PS, Mast MG. Use of chemical compounds and a rosemary spice extract in quality

maintenance of deboned poultry meat. *J Food Sci.* 1973;38:1080–1081.

34. Chang HC, Branen AL. Antimicrobial effects of butylated hydroxyanisole (BHA). *J Food Sci.* 1975;40: 349–351.
35. Ahmad S, Branen AL. Inhibition of mold growth by butylated hydroxyanisole. *J Food Sci.* 1981;46: 1059–1063.
36. Sankaran R. Comparative antimicrobial action of certain antioxidants and preservatives. *J Food Sci Technol.* 1976;13:203–204.
37. Thompson DP. Effect of butylated hydroxyanisole on conidial germination of toxigenic species of *Aspergillus flavus* and *Aspergillus parasiticus*. *J Food Prot.* 1991;54:375–377.
38. Fung DYC, Taylor S, Kahan J. Effects of butylated hydroxyanisole (BHA) and butylated hydroxitoluene (BHT) on growth and aflatoxin production of *Aspergillus flavus*. *J Food Safety.* 1977;1:39–51.
39. Knox TL, Davidson PM, Mount JR. Evaluation of selected antimicrobials in fruit juices as sodium metabisulfite replacements or adjuncts. *Annual Meeting Institute of Food Technology*. June 10–13, 1984. Anaheim, California.
40. Raccach M. The antimicrobial activity of phenolic antioxidants in foods: A review. *J Food Safety.* 1984;6:141–170.
41. Cerrutti P, Alzamora SM. Inhibitory effects of vanillin on some food spoilage yeasts in laboratory media and fruit purées. *Int J Food Microbiol.* 1996;29:379–386.
42. López-Malo A, Alzamora SM, Argaiz A. Effect of natural vanillin on germination time and radial growth of moulds in fruit-based agar systems. *Food Microbiol.* 1995;12:213–219.
43. López-Malo A, Alzamora SM, Argaiz A. Effect of vanillin concentration, pH and incubation temperature on *Aspergillus flavus, A. niger, A. ochraceus* and *A. parasiticus* growth. *Food Microbiol.* 1997;14:117–124.
44. López-Malo A, Alzamora SM, Argaiz A. Vanillin and pH synergistic effects on mold growth. *J Food Sci.* 1998;63:143–146.
45. Paster N, Juven BJ, Harshemesh H. Antimicrobial activity and inhibition of aflatoxin B1 formation by olive plant tissue constituents. *J Appl Bacteriol.* 1988;64:293–297.
46. Gourama H, Bullerman LB. Effects of oleuropein on growth and aflatoxin production by *Aspergillus parasiticus*. *Lebensm-Wiss u-Technol.* 1987;20:226–228.
47. Tassou CC, Nychas GJE. Inhibition of *Staphylococcus aureus* by olive phenolics in broth and in a model food system. *J Food Prot.* 1994;57:120–124.
48. Bullerman LB, Lieu FY, Seier SA. Inhibition of growth and aflatoxin production by cinnamon and clove oils. Cinnamic aldehyde and eugenol. *J Food Sci.* 1977;42:1107–1109, 1116.
49. Paster N, Menasherov M, Ravid U, Juven B. Antifungal activity of oregano and thyme essential oils applied as fumigants against fungi attacking stored grain. *J Food Prot.* 1995;58:81–85.
50. Al-Khayat MA, Blank G. Phenolic spice components sporostatic to *Bacillus subtilis*. *J Food Sci.* 1985;50: 971–974, 980.
51. Buchanan RL, Shepherd AJ. Inhibition of *Aspergillus parasiticus* by thymol. *J Food Sci.* 1981;46:976–977.
52. Mahmoud ALE. Antifungal action and antiaflatoxigenic properties of some essential oil constituents. *Lett Appl Microbiol.* 1994;19:110–113.
53. Conner DE, Beuchat LR. Effects of essential oils from plants on growth of food spoilage yeasts. *J Food Sci.* 1984;49:429–434.
54. Paster N, Juven BJ, Shaaya E, Menasherov M et al. Inhibitory effect of oregano and thyme essential oils on moulds and foodborne bacteria. *Lett Appl Microbiol.* 1990:11:33–37.
55. Zaika LL, Kissinger JC, Wasserman E. Inhibition of lactic bacteria by herbs. *J Food Sci.* 1983;48: 1455–1459.
56. Resnik SL, Pacin AM, Alvarez G, Alzamora SM. Effect of vanillin concentration on aflatoxins accumulation. *IX International Symposium on Mycotoxins and Phycotoxins*. May 27–31, 1996, Rome, Italy.
57. Prindle RF, Wright ES. Phenolic compounds. In: Block SS, ed. *Disinfection, Sterilization and Preservation.* Philadelphia: Lea & Febiger; 1977.
58. Conner DE, Beuchat LR. Sensitivity of heat-stressed yeasts to essential oils of plants. *Appl Environ Microbiol.*1984;47:229–233.
59. Juven BJ, Kanner J, Schved F, Weisslowicz H. Factors that interact with the antibacterial action of thyme essential oil and its active constituents. *J Appl Bacteriol.* 1994:76:626–631.
60. Ruiz-Barba JL, Rios-Sanchez RM, Fedriani-Triso C, Olias JM, et al. Bactericidal effect of phenolic compounds from green olives against *L. plantarum* system. *Appl Microbiol.* 1990;13:199–205.
61. Kabara JJ, Eklund T. Organic acids and esters. In: Russel NJ, Gould GW, eds. *Food Preservatives.* Glasgow: Blackie & Son Ltd.; 1991:44–71.
62. Rico-Muñoz E, Bargiota E, Davidson PM. Effect of selected phenolic compounds on the membrane-bound adenosine triphosphate of *Staphylococcus aureus. Food Microbiol.* 1987;4:239–249.
63. Barone FE, Tansey MR. Isolation, purification, identification, synthesis and kinetics of activity of the anticandidal component of *Allium sativum* and a hypothesis for its mode of action. *Mycologia* 1977;69:793–799.
64. Davidson PM. Chemical preservatives and natural antimicrobial compounds. In: Doyle MP, Beuchat

LR, Montville TJ, eds. *Food Microbiology—Fundamentals and Frontiers.* Washington, DC: ASM Press;1997:520–556.

65. López-Malo A, Argaiz A. *Citral and pH Synergistic Effects on Mold Growth.* Paper No. 37D-15, presented at IFT Annual Meeting. Chicago, July 24–28; 1999.
66. Sykes G, Hooper MC. Phenol as the preservative in insulin injections. *J Pharm Pharmacol.* 1954;6:552.
67. Thompson DP. Influence of pH on the fungitoxic activity of naturally occurring compounds. *J Food Prot.* 1990;53:482–429.
68. Tassou CC, Drosinos EH, Nychas GJE. Effects of essential oil from mint (*Mentha piperita*) on *Salmonella enteritidis* and *Listeria monocytogenes* in model food systems at 4° and 10°C. *J Appl Bacteriol.* 1995;78:593–600.
69. Stern NJ, Smoot LA, Pierson MD. Inhibition of *Staphylococcus aureus* growth by combinations of butylated hydroxyanisole, sodium chloride and pH. *J Food Sci.* 1979;44:710–712.
70. Azzous MA, Bullerman LB. Comparative antimycotic effects of selected herbs, spices, plant components and commercial antifungal agents. *J Food Prot.* 1982;45:1298–1301.
71. Sebti F, Tantaoui-Elaraki A. *In vitro* inhibition of fungi isolated from "pastilla" papers by organic acids and cinnamon. *Lebensm-Wiss u-Technol.* 1994;27:370–374.
72. Matamoros-León B, Argaiz A, López-Malo A. Individual and combined effects of vanillin and potassium sorbate on *Penicillium digitatum, P. glabrum* and *P. italicum* growth. *J Food Prot.* 1999: in press.
73. Shelef LA, Jyothi EK, Bulgarelli MA. Growth of enteropathogenic and spoilage bacteria in sage-containing broth and foods. *J Food Sci.* 1984;49:737–740, 809.
74. Juven B, Henis Y, Jacoby B. Studies on the mechanism of the antimicrobial action of oleuropein. *J Appl Bacteriol.* 1972;35:559–567.
75. Robach MC, Smoot LA, Pierson MD. Inhibition of *Vibrio parahaemolyticus* O4:K11 by butylated hydroxyanisole. *J Food Prot.* 1977;40:549–551.
76. Rico-Muñoz E, Davidson PM. Effect of corn oil and casein on the antimicrobial activity of phenolic antioxidants. *J Food Sci.* 1983;48:1284–1288.
77. Aureli P, Constantini A, Zolea S. Antimicrobial activity of some plant essential oils against *Listeria monocytogenes. J Food Prot.* 1992;55:344–348.
78. Shelef LA, Liang P. Antibacterial effects of butylated hydroxyanisole (BHA) against *Bacillus* species. *J Food Sci.* 1982;47:796–799.
79. Cornell DG, De Vilbiss ED, Pallansch MJ. Binding of antioxidants by milk proteins. *J Food Sci.* 1971;54:634–637.
80. Spencer CM, Cai Y, Martin R, Gaffney SH, et al. Polyphenol complexation: Some thoughts and observations. *Phytochemistry.* 1988;27:2397–2409.
81. Cha AS, Ho CT. Studies of the interaction between aspartame and flavor vanillin by high performance liquid chromatography. *J Food Sci.* 1988;53:562–564.
82. Montgomery MW, Day EA. Aldehyde-amine condensation reaction: a possible fate of carbonyls in foods. *J Food Sci.* 1965;31:829–832.
83. Hansen AP, Heinis JJ. Decrease of vanillin flavor perception in the presence of casein and whey proteins. *J Dairy Sci.* 1991;74:2936–2940.
84. Tateo F, Triangeli L, Panna E, Berte F, et al. Stability and reactivity of aspartame in cola-type drinks. In: Charambous G, ed. *Frontiers of Flavor.* Amsterdam: Elsevier Science; 1988.
85. Hussein MM, D'Amelia RP, Manz AL, Jacin H, et al. Determination of reactivity of aspartame with flavor aldehydes by gas chromatography, HPLC and GPC. *J Food Sci.* 1984;49:520–524.
86. Ng PKW, Hoehm E, Bushuk W. Binding of vanillin by fababeans proteins. *J Food Sci.* 1989;54:105–107.
87. Ng PKW, Hoehm E, Bushuk W. Sensory evaluation of binding of vanillin by fababeans proteins. *J Food Sci.* 1989:54:324–325, 346.
88. Barr A. Consumer motivational forces affecting the sale of light dairy products. *Food Technol.* 1990;44(10): 97–98.
89. Cerrutti P, Alzamora SM, Vidales SL. Vanillin as an antimicrobial for producing shelf-stable strawberry purée. *J Food Sci.* 1997;62:608–610.
90. Castañón X, Argaiz A, López-Malo A. Effect of storage temperature on the microbial and color stability of banana purées prepared with the addition of vanillin or potassium sorbate. *Food Sci Technol Int.* 1999; 5:53–60.
91. Ejechi BO, Sousey JA, Akpomedaye DE. Microbial stability of mango (*Mangifera indica* L.) juice preserved by combined applications of mild heat and extracts of two tropical spices. *J Food Prot.* 1998;61:725–727.
92. Smid EJ, Hendriks L, Boerrigter HAM, Gorris LGM. Surface disinfection of tomatoes using the natural plant compound *trans*-cinnamaldehyde. *Postharv Biol Technol.* 1996;8:343–348.
93. Singleton VL. Naturally occurring food toxicants: Phenolic substances of plant origin common in foods. In: Chichester CO, Mrak EM, Stewart GF, eds. *Advances in Food Research,* Vol. 27. New York: Academic Press; 1981.
94. Weisburger JH. Mutagenic, carcinogenic and chemopreventive effects of phenols and catechols. In: Huang M, Ho C, Lee CY, eds. *Phenolic Compounds in Food and Their Effect on Health II, Antioxidants and Cancer*

Prevention. Washington, DC: ACS Symposium Series; 1992.

95. Newmark HL. Plant phenolic compounds as inhibitors of mutagenesis and carcinogenesis. In: Huang M, Ho C, Lee CY, eds. *Phenolic Compounds in Food and Their Effect on Health II, Antioxidants and Cancer Prevention.* Washington, DC: ACS Symposium Series; 1992.
96. Alzamora SM, Cerrutti P, Guerrero S, López-Malo A. Minimally processed fruits by combined methods. In: Barbosa-Cánovas G, Welti J, eds. *Food Preservation by Moisture Control. Fundamentals and Applications.* Lancaster, PA: Technomic Publishing Co.; 1995:463–491.
97. Leistner L. Food preservation by combined methods. *Food Res Int.* 1992;25:151–158.
98. Gould GW, Jones MV. Combination and synergistic effects. In: Gould GW, ed. *Mechanisms of Action of Food Preservation Procedures.* New York: Elsevier Science ; 1989:401–422.
99. Gould GW, Russell NJ. Major food-poisoning and food-spoilage micro-organisms. In: Russel NJ, Gould GW, eds. *Food Preservatives.* Glasgow: Blackie & Son Ltd.; 1991:1–12.
100. Wagner MK, Moberg LJ. Present and future use of traditional antimicrobials. *Food Technol.* 1989;43(1): 143–147.
101. Farrell KT. *Spices, Condiments and Seasonings,* 2nd ed. New York: Van Nostrand Reinhold; 1990.
102. Beuchat LR. Sensitivity of *Vibrio parahaemolyticus* to spices and organic acids. *J Food Sci.* 1976;41: 899–902.
103. Deans SG, Ritchie G. Antibacterial properties of essential oils. *Int J Food Microbiol.* 1987;5:165–180.
104. Jay JM, Rivers GM. Antimicrobial activity of some food flavoring compounds. *J Food Safety.* 1984;6:129–139.
105. Morris JA, Khettry A, Seitz EW. Antimicrobial activity of aroma chemicals and essential oils. *J Am Oil Chem Soc.* 1979;56:595–603.
106. Thompson DP, Metevia L, Vessel T. Influence of pH alone and in combination with phenolic antioxidants on growth and germination of mycotoxigenic species of *Fusarium* and *Penicillium*. *J Food Prot.* 1993;56:134–138.
107. Graham HD, Graham EJF. Inhibition of *Aspergillus parasiticus* growth and toxin production by garlic. *J Food Safety.* 1987;8:101–108.
108. Karapinar M. The effects of citrus oils and some spices on growth and aflatoxin production by *Aspergillus parasiticus* NRRL 2999. *Int J Food Microbiol.* 1985;2:239–245.
109. Bullerman LB. Inhibition of aflatoxin production by cinnamon. *J Food Sci.* 1974;39:1163–1165.
110. Llewellyn GC, Burkett ML, Eadie T. Potential mold growth, aflatoxin production and antimycotic activity of selected natural spices and herbs. *J Assoc Off Anal Chem.* 1981;64:955–960.
111. Hitokoto H, Morozumi S, Wauke T, Sakai S, et al. Inhibitory effects of spices on growth and toxin production of toxigenic fungi. *Appl Environ Microbiol.* 1980;39:818–822.
112. Akgul A, Kivanc M. Sensitivity of four foodborne moulds to essential oils from Turkish spices, herbs and citrus peel. *J Sci Food Agric.* 1989;47:129–132.
113. Sauer F. Control of yeast and molds with preservatives. *Food Technol.* 1977;31(2):66–67.
114. Giese J. Antimicrobials: Assuring food safety. *Food Technol.* 1994;48(6):102–110.

CHAPTER 15

Use of Biopreservation in Minimal Processing of Vegetables

Marjon H.J. Bennik, E.J. Smid, and Leon Gorris

INTRODUCTION

Minimally processed vegetables include a range of perishable products for which mild preservation technologies are being employed to preserve the fresh characteristics of the products while assuring product safety. An extensive range of vegetables qualifies as minimally processed, including raw vegetables; washed, trimmed, and sliced vegetables (with or without dressings); *sous vide* cooked vegetables; and potato-based dishes. This chapter will mainly focus on fresh minimally processed produce with metabolic activity. Refrigeration is the main mild preservation technique that is used to ensure durability and safety of minimally processed vegetables. In addition, modified-atmosphere packaging is employed to prolong the shelf life of these products effectively. Favorable gas concentrations for modified atmosphere storage of vegetables are in the range of 1–5% O_2 and 5–10% CO_2. Such modified-atmosphere conditions, in combination with refrigeration, slow down physiologic processes that cause deterioration of plant tissues and can inhibit proliferation of the bacterial microflora on the produce.

The bacterial microflora of minimally processed vegetables consists of epiphytic microorganisms that can cause spoilage (e.g., *Enterobacteriaceae, Pseudomonas* species, lactic acid bacteria) but are considered harmless to the consumer (reviewed in Nguyen-the and Carlin[1]). Storage of such produce under modified atmospheric conditions has been reported to suppress the growth of epiphytic bacteria, several of which can cause soft rot.[2] Especially *Pseudomonas* species were found to have significantly reduced maximum specific growth rates at CO_2 concentrations that are suitable for modified-atmosphere packaged vegetables, whereas such gas conditions do not significantly reduce the growth of various *Enterobacteriaceae.*[3] In general, minimally processed vegetables have a fair safety record, despite the fact that they may incidentally harbor food-borne pathogens. Several studies have demonstrated the presence of *Listeria monocytogenes* on produce.[4–6] Other psychrotrophic (cold tolerant) pathogens, such as *Aeromonas hydrophila* and *Yersinia enterocolitica*, can be a concern as well.[7] Furthermore, mesophilic pathogens, such as *Salmonella typhimurium* and *Staphylococcus aureus*, have been detected on these products, and these microorganisms can proliferate at abuse temperatures (reviewed in Nguyen-the and Carlin[1]). Specific modified-atmosphere conditions, in combination with refrigeration, affect the growth of different pathogens to different extents. For instance, a psychrotrophic strain of *Bacillus cereus* was found to be more sensitive to the inhibitory effect of CO_2 than were strains of *A. hydrophila, Y. enterocolitica*, and *L. monocytogenes,* when tested under conditions that are suitable for minimally processed vegetables.[8]

Proliferation of the microflora on minimally processed vegetables should be of concern only toward the end of the shelf life of these products, when problem levels of bacteria may be reached with respect to quality or safety. To gain increased control of the outgrowth of spoilage and pathogenic bacteria in minimally processed vegetables that are stored under refrigeration, with or without modified-atmosphere packaging, biopreservation using bacteriocins produced by lactic acid bacteria (LAB) or using bacteriocin-producing LAB cultures may be suitable in specific cases, as discussed in the following paragraphs.

USE OF LACTIC ACID BACTERIA FOR BIOPRESERVATION

Microorganisms produce a wide range of compounds that can influence the growth of other microorganisms present in their environment. Often, such compounds render the producing organisms a competitive advantage over other species and are, therefore, important for the survival and proliferation of a microorganism. With regard to food preservation, LAB are the most important group of microorganisms to be considered for use in biopreservation. For centuries, LAB have been used in food fermentations to produce stable foods, e.g., dairy, yogurt, sausages, and sauerkraut. The extensive consumption of products containing LAB without negative health effects to consumers has given LAB the status of generally recognized as safe (GRAS).[9]

LAB can inhibit or eliminate the growth of many different microorganisms, including bacteria, yeasts, and fungi, through the production of organic acids, diacetyl, hydrogen peroxide, enzymes, defective phages, lytic agents, and antimicrobial peptides, or bacteriocins. There are considerable differences in the antimicrobial action spectra between the various compounds produced by LAB, and not all LAB produce the above-mentioned compounds to the same extent. For example, organic acids (i.e., lactic acid, acetic acid) affect the growth of many different microorganisms, whereas bacteriocins inhibit a relatively narrow range of microorganisms.[10,11]

Biopreservation using LAB might be achieved by applying an organism as a so-called protective culture to the food product, thereby relying on the proliferation and consequent competition of LAB with the microorganism(s) to be suppressed by means of the production of specific or a variety of antimicrobials. Alternatively, a purified preparation of an active antimicrobial compound can be used, with the advantage of an instant and more controlled effect. The procedures for approval of protective cultures in foods may be relatively short in most countries around the world, based on the fact that LAB are GRAS organisms. In contrast, the use of purified antimicrobial metabolites is commonly subject to specific rules and regulations in food legislation and, consequently, requires a lengthier approval procedure (e.g., bacteriocins) unless the compounds of interest have an approved or GRAS status already (e.g., lactic acid).

Obviously, the choice of a strategy for biopreservation using protective cultures or purified antimicrobials must be tailored to the target microorganism (type of pathogen or spoilage microorganism) and to the target minimally processed vegetable product. Furthermore, the extent to which a selected biopreservation agent might impact the organoleptic properties of the food product should be given consideration. Organic acids or LAB cultures producing these compounds, for instance, may have a marked effect on the taste and texture of minimally processed vegetables, whereas the tasteless and odorless bacteriocins may have no noticeable effect.

Despite a significant increase in the effort to develop protective cultures for biopreservation in recent years, until now, studies have mostly been performed at a laboratory scale and rarely relate to minimally processed vegetables. From studies of Cerny and Hennlich[12] on the use of LAB as protective cultures in potato salads to control food poisoning by salmonellae and toxin-producing staphylococci or clostridiae, several prospects became evident. In mayonnaise-based potato salads with pH values of 5.5–6.0 that were exposed to ambient temperatures for up to one week, the protective cultures greatly reduced the hygienic risks, although they did not increase the shelf life of those products. Hennlich[13] reported on the selection and evaluation of LAB isolated from potato salads as protective cultures for chilled delicatessen salads, assuming that they were ecologically well adapted. Important criteria for selection were minimum growth temperature, rate of acidification at refrigeration temperature, and rapid growth and acid formation at abuse temperature, mimicking interruption of the cold chain. These criteria were adequately met by *Lactobacillus casei* ILV 110 and *Lb. plantarum* ILV 3. When used as protective cultures (at 10^4 CFU/g as a minimum), these strains inhibited the normal spoilage flora of delicatessen salads and suppressed growth of *Escherichia coli* and *Cl. sporogenes* inoculated into meat salads during storage at chill temperature. One of the isolates, *Lb. plantarum* ILV 3, was also found to be suitable as a protective culture for weakly acidic delicatessen salads (pH 5.0–6.0).

Cerny[14] studied the inhibitory effect of a range of LAB (*Leuconostoc cremoris*, *Lactococcus lactis* var. *diacetylis*, *Lc. lactis* subsp. *lactis*, *Lc. lactis* subsp. *cremoris*, and *Lb. casei*) on the growth of several indicator microorganisms (*E. coli*, *Staphylococcus saprophyticus*, *Cl. sporogenes*) in mayonnaise-based meat and potato salads of pH 5.5–6.5 that were prepared using pasteurized ingredients to eliminate the endogenous LAB. It was found that addition of *Leuc. cremoris* as a protective culture to potato salads completely controlled the growth of *E. coli* and *Cl. sporogenes* at room temperature. *Lc. lactis* subsp. *lactis* (inoculation level 10^3–10^6 CFU/g) suppressed the growth of *E. coli* (10^2–10^4 CFU/g) in meat salad stored at room temperature. Importantly, it was concluded that the best protective effects were observed when the ratio of *Lc. lactis* subsp. *lactis* to *E. coli* was greater than 10:1. Vescovo et al.[15] studied the application of five psychrotrophic LAB isolated from commercial salads to inhibit a range of pathogens in ready-to-use vegetables. *Lb. casei*, *Lb. plantarum,* and *Pediococcus* species were found to inhibit *A. hydrophila*, *L. monocytogenes*, *Salmonella typhimurium,* and *S. aureus* on MRS agar, in salads, and in juice prepared from vegetable salads. One of the LAB strains, *Lb. casei* IMPCLC34, effectively reduced the total mesophilic bacteria count and the coliform count, whereas *A. hydrophila*, *S. typhimurium,* and *S. aureus* were no longer detectable after 6 days of storage. *L. monocytogenes* counts remained constant throughout the storage period. Various review papers on this topic are available with a focus on different product groups.[16–21]

BACTERIOCINS PRODUCED BY LACTIC ACID BACTERIA

The use of protective LAB cultures is, in most instances, aimed at controlling several undesired microorganisms simultaneously by various antimicrobials produced by LAB. A more specific control strategy is the rationale for using bacteriocins produced by LAB. Bacteriocins are antimicrobial peptides that are ribosomally synthesized by a range of bacteria.[22] In recent years, bacteriocins of LAB have provoked a great deal of interest for their potential as nontoxic preservatives, and a large number of chemically diverse bacteriocins of LAB have recently been identified and characterized.[23] Exhibit 15–1 shows how the various bacteriocins can be grouped into four classes (I to IV).[24]

Members of classes I and IIa have been isolated and characterized most frequently and have been intensively studied because they are prominent candidates for industrial application. These bacteriocins can permeate the cytoplasmic membranes of gram-positive bacteria by the formation of pores that may lead to cell death.[25–28] A prominent member of the class I bacteriocins, nisin, has a broad spectrum of activity toward gram-positive bacteria (including many spoilage bacteria and pathogens, such as *L. monocy-*

Exhibit 15–1 Classification of Bacteriocins Produced by Lactic Acid Bacteria

Class I: Lantibiotics. These are defined as small, heat-stable membrane-active peptides that contain posttranslationally modified amino acids.

Class II: Small (< 5 kDa), heat-stable membrane-active bacteriocins that are characterized by the absence of unusual amino acids and the presence of a double glycine processing site in the leader sequence of the precursor. Most of the bacteriocins of LAB that have been characterized to date belong to this class, in which three major subclasses are distinguished:

- IIa Bacteriocins with activity toward *L. monocytogenes*, containing a consensus sequence at the N-terminus: Tyr-Gly-Asn-Gly-Val-Xaa-Cys (Xaa is His, Thr, Ser, or Tyr).
- IIb Bacteriocins that form pores that require two peptides for activity.
- IIc Thiol-activated peptides that require reduced cysteines for activity.

Class III: Large (> 30 kDa) heat-labile protein bacteriocins.

Class IV: Bacteriocins composed of a protein plus one or more nonproteinaceous moieties (lipid, carbohydrates).

togenes, *S. aureus*, and *Clostridium botulinum*). Its antimicrobial spectrum, its heat stability at low pH, and its nontoxic nature have promoted the use of nisin in food processing and fermentation.[23,29–31] Nisin is produced by strains of *L. lactis* subsp. *lactis* and was introduced commercially as a food preservative in England some 30 years ago, where it was employed as a preservative in processed cheese products. Today, nisin is recognized as a safe food preservative in approximately 50 countries worldwide for several different applications in foods and beverages. Unfortunately, nisin is not suitable for use in all food systems because of its limited solubility above pH 5 and its instability in some foods.[32]

A number of bacteriocins of LAB other than nisin also exhibit fairly broad spectra of inhibition, which might make them suitable candidates for improvement of food safety as well. Of special interest are the heat-stable class IIa bacteriocins, which generally show antimicrobial activity toward *L. monocytogenes* but may also inhibit the growth of other gram-positive pathogens.[24] For industrial applications, it is noteworthy that these compounds are usually active over a wide pH range, whereas nisin has the highest antimicrobial activity at a low pH. Class IIa bacteriocins have been reported to be produced by a considerable range of LAB, including *Pediococcus,*[33] *Lactobacillus,*[34–37] *Carnobacterium,*[38–40] *Enterococcus,*[41,42] and *Leuconostoc.*[43,44]

The inhibitory spectrum of LAB bacteriocins other than members of classes I and IIa is usually limited to bacteria that are closely related to the producer. When the inhibitory spectrum includes specific food-borne pathogens (i.e., *L. monocytogenes*), the use of bacteriocins enables targeted intervention. In addition to deploying biopreservation to improve food safety, LAB strains that produce narrow spectrum bacteriocins are interesting from a food quality point of view: Such bacteriocinogenic strains can be used as (part of) starter cultures in fermentations, thereby minimizing interference of undesired spoilage LAB in the process. Many good recent reviews are available that give more detail on many relevant aspects of bacteriocins produced by LAB.[10,17,20,32,45–49]

STRATEGIES FOR THE APPLICATION OF BACTERIOCINS FROM LACTIC ACID BACTERIA FOR BIOPRESERVATION

The application of bacteriocins in food systems can be achieved by three basic methods:

1. Application of a bacteriocinogenic LAB that produces a desired bacteriocin—a pure culture of the viable bacteriocin-producing LAB can be applied in accordance with the protective culture concept. This offers an indirect way to incorporate bacteriocins in a food product. The success of this method depends on the ability of the LAB to grow and produce the bacteriocin to the desired extent in the food under the prevailing environmental conditions (e.g., composition of the food, temperature, pH, gas atmosphere). Conceivably, bacteriocinogenic LAB originating from food or environmental samples that, in ecologic terms, resemble the environmental conditions under which the application is pursued might be most suitable. Nevertheless, selection of suitable cultures should be followed by a thorough evaluation of their in situ performance.
2. Application of an unpurified culture broth preparation of a bacteriocinogenic LAB containing the desired bacteriocin—application of an unpurified bacteriocin preparation obtained by growing the bacteriocin-producing LAB on a complex, natural substrate (e.g., milk). This method is currently employed for the production of a nisin preparation on an industrial scale by growing nisin-producing LAB in milk whey at optimal temperature. During the course of incubation, nisin is secreted into the substrate. At sufficiently high nisin concentrations, the substrate is pasteurized, thereby killing the bacteria but not inactivating the heat-stable nisin. The advantage of this method is that the activity of the preparation is known and can be standardized; also, a shelf life of up to 1 year is possible. A drawback is that the preparation may contain compounds that do not contribute to the desired activity of the bacteriocin, e.g., salts and proteins. Also, because nisin and other bacteriocins of LAB are proteins, the concentration of active bacteriocin can gradually be reduced in time. The rate of this process is dependent on the presence and activity of proteolytic components in the food product and needs to be considered in selecting the relevant application design (food product, temperature, time, etc.).
3. Application of a purified preparation of the desired bacteriocin—application of a (semi)purified preparation of the bacteriocin. Here, the dosage of the bacteriocin is most accurate, and unnecessary accompanying components are absent. The effect of such preparations will, therefore, be the most predictable and devoid of undesired side effects. However, also in this case, the activity of the bacteriocin may decline in the food product in time. Furthermore, the production of the preparation involves higher costs than do either option 1 or 2, especially when extensive purification procedures are required to remove accompanying substances. There is currently no public information available on the stability of the bacteriocin activity in commercial, pure bacteriocin preparations to allow an estimate of possible shelf life. The application of this method is limited as a result of national regulations on food additives (see below).

To achieve a successful inhibition of the growth of pathogenic or spoilage bacteria in foods by the use of bacteriocins or bacteriocinogenic strains, multiple factors need to be taken into account. The inhibitory action of most bacteriocins of LAB is directed toward relatively closely related gram-positive bacteria and, in certain cases, affect pathogenic bacteria. Therefore, only those pathogenic bacteria that fall within the inhibitory spectrum of a certain bacteriocin can be controlled. Also, the inhibitory action of a bacteriocin can vary considerably between different genera, species of genera, identical species, and even for identical cultures under different environmental conditions.[24,50,51] Furthermore, intrinsic factors in foods (such as pH, presence of proteolytic components) and the storage conditions (temperature, gas atmosphere) can influence the bacteriocin production and/or the activity of bacteriocins.[52] Finally, bacteriocins secreted in the food matrix by a viable LAB culture or added in crude or purified form may be adsorbed to certain hydrophobic compounds present in the food, such as fats and phospholipids, thereby possibly diminishing their antimicrobial effect. From the above, it may be evident that the application of bacteriocins or bacteriocin-producing strains requires tailoring for different food products and should be extensively tested under specific practical conditions.

NATURAL OCCURRENCE OF BACTERIOCIN-PRODUCING LACTIC ACID BACTERIA

LAB naturally occur on many mildly or minimally processed food products. They can originate from raw materials and ingredients or are introduced during processing. The natural microflora of minimally processed vegetables generally consists of only about 1% LAB. In the last years, a great number of studies have been devoted to tracing bacteriocin-producing LAB in fresh and fermented food products. Although the detection methods were not always standardized, a general conception is that only a small percentage of LAB that are isolated from food products are able to produce bacteriocins and that the inhibitory spectra of different isolates are highly variable.

Vaughan et al.,[53] for instance, evaluated LAB isolates from cheese, milk, meat, fruits, and vegetables for bacteriocin production, using spoilage and pathogenic bacteria as target organisms. Approximately 1,000 isolates from each of the above-mentioned food categories were tested for bacteriocin activity against *S. aureus*, *Listeria innocua,* and *Pseudomonas fragi*. LAB isolated from cheese, milk, and meat samples inhibited *L. innocua* rather than the other target strains, whereas LAB isolates from vegetables generally inhibited *S. aureus*. Most of the bacteriocinogenic strains were active against only one indicator, but a small number had inhibitory activity toward two or three of the target microorganisms.

Coventry et al.[54] studied a total of 663,533 LAB isolates from 72 dairy and meat sources and found that only 0.2% of the isolates were positive for bacteriocin production, using direct plating techniques. Among another 83,000 colonies that were isolated from 40 fish and vegetable sources, 3.4% of the LAB isolates were able to produce bacteriocins according to selective enrichment procedures.

In a study conducted by Kelly et al.[55] LAB were isolated from 41 different foods sold in ready-to-eat form and screened for bacteriocin production. Here, a total of 22 bacteriocinogenic cultures were isolated from 14 of the foods. Bacteriocin-producing LAB that were typically associated with meat, fish, and dairy products were *Lactobacillus* and *Leuconostoc* species. Bacteriocinogenic LAB cultures isolated from fruit and vegetable products were, in all cases, identified as strains of *Lactococcus*, with several having activities similar to nisin. The authors noted that the relative ease in which bacteriocin-producing strains can be isolated from ready-to-eat food products implies that bacteriocins are already being safely consumed in foods. This highlights the potential for using bacteriocin-producing cultures for biopreservation, especially in association with minimally processed products. In another study by Uhlman et al.,[56] bacteriocinogenic LAB were isolated from mixed salads and fermented carrots: From a total of 123 LAB, two isolates were found to produce heat-stable bacteriocins, and these strains were identified as *Lc. lactis*. Their bacteriocins were active from pH 2 to 9 and inhibited species of *Listeria*, *Lactobacillus*, *Lactococcus*, *Pediococcus*, *Leuconostoc*, *Carnobacterium*, *Bacillus,* and *Staphylococcus*. Franz et al.[57] identified four more bacteriocin-producing LAB isolates from vegetables as strains of *Lc. lactis*. The heat-resistant bacteriocins produced had a wide spectrum of activity and affected growth of several LAB, *S. aureus,* and *L. monocytogenes*. The highest activity of the bacteriocins was observed at pHs below 5.0. They were inactivated by the proteolytic enzymes a-chymotrypsin and proteinase K, but not by lipase, a-amylase, catalase, or lysozyme. Each of the bacteriocinogenic *Lactococcus* strains was immune to its cognate bacteriocin and to commercial nisin.

Cai et al.[58] reported a study on the isolation of LAB from mung bean sprouts that produced bacteriocins with activity toward *L. monocytogenes.* Thirty-four of 72 isolates were found to inhibit the growth of *L. monocytogenes* Scott A. The isolate that showed the largest inhibition zone against *L. monocytogenes* Scott A in a well-diffusion assay was identified as *Lc. lactis* subsp. *lactis* producing nisin Z. In MRS broth, this isolate survived at 3–4.5°C for at least 20 days, grew at 4°C, and produced antilisterial compounds at 5°C. When cocultured with *L. monocytogenes* in MRS broth, the isolate inhibited the growth of the pathogen at 4°C after 14 days and at 10°C after 2 days. When coinoculated with 10^2 cells of *L. monocytogenes* per gram of freshly cut, ready-to-eat Caesar salad, *Lc. lactis* subsp. *lactis* (10^8 cells/g^{-1}) was able to reduce the number of the pathogen by 1 to 1.4 logs after storage for 10 days at 7 and 10°C.

Besides nisin-producing LAB, (minimally) processed vegetables have been shown to contain other bacteriocinogenic LAB. Olasupo[59] isolated a strain of *Lb. plantarum* that produced a bacteriocin with inhibitory activity against *L. monocytogenes*. This compound was resistant to heat, stable over a wide pH range (pH 2–10), and displayed a bactericidal mode of action. Growth inhibition of *L. monocytogenes* was dependent on the bacteriocin concentration and varied between different strains of *L. monocytogenes* but was not independent of the growth phase of this pathogen. Franz et al.[60] isolated a strain of *Lb. plantarum* from Waldorf salad that produced a heat-stable bacteriocin termed *plantaricin D*. This bacteriocin was active against several strains of *Lactobacillus sake* and *L. monocytogenes,* and stable between pH 2.0 and 10.0. Franz et al.[61] also reported the isolation of *Enterococcus faecium* that produces a bacteriocin termed *enterocin 900*. This heat-stable bacteriocin was active against *Lb. sake*, *Clostridium butyricum*, enterococci, and *L. monocytogenes,* and was active at pH values ranging from 2.0 to 10.0, with the highest activity at pH 6.0. Enterocin 900 was produced in the late exponential phase of growth when culture densities were ca. log 8.0 CFU/mL^{-1}. Enterocin 900 was produced in media with initial pHs ranging from 6.0 to 10.0 but not in media with a pH lower than 6.0. The composition of the growth medium (especially the concentrations of peptone and yeast extract) influenced bacteriocin production markedly: Bacteriocin activity was not detected when these compounds were absent. Whereas this trait may limit the applicability of the producing organism as a biopreservation agent, *E. faecium* is, furthermore, not a GRAS organism, and its application would require substantial legislatory scrutiny.

So far, the interaction between LAB and pathogens in foods has primarily been studied in a qualitative way, mainly to assess inhibition per se. Some recent studies, however, were aimed at obtaining a more quantitative insight in this interaction on the basis of mathematical modeling. Vescovo et al.,[62] for instance, studied the effects of carbon dioxide concentration, inoculum size of *Lb. casei,* and storage temperature on the growth of *A. hydrophila* in ready-to-use modified-atmosphere packaged mixed salad vegetables. Bacterial growth curve parameters, modeled according to the Gompertz equation, were analyzed to generate polynomial equations. The model obtained emphasized the role of the inoculum size of *Lb. casei* in controlling *A. hydrophila* and permitted identification of appropriate combinations of the selected variables to reduce survival of the pathogen. In another study, Breidt and Fleming[63] developed a mathematical model consisting of a system of nonlinear differential equations to describe the growth of competing cultures of *Lc. lactis* and *L. monocytogenes* grown in a vegetable broth medium. In the model, bacterial cell growth was essentially limited by the accumulation of protonated lactic acid and decreasing pH. Predictions of the model indicated that pH reduction caused by the nonbacteriogenic LAB was the primary factor that limited growth of *L. monocytogenes*. Further development of this model might incorporate the effects of additional inhibitors, such as bacteriocins, and might aid in the efficient selection of bacteriocinogenic lactic acid bacterium cultures for use in competitive inhibition of pathogens in minimally processed foods.

APPLICATION OF BACTERIOCINS OR BACTERIOCIN-PRODUCING LACTIC ACID BACTERIA TO MINIMALLY PROCESSED VEGETABLES

Bacteriocin-producing LAB have potential for the biopreservation of foods of plant origin, especially minimally processed foods, such as prepackaged mixed salads and fermented vegetables. Vescovo et al.[64] observed a reduction of the high initial bacterial loads of ready-to-use mixed salads when bacteriocin-producing LAB obtained from different sources were added to the refrigerated salad mixtures. It was noted that coliforms and enterococci were strongly reduced or eliminated from the products as a result of adding bacteriocinogenic LAB. In this respect, *Lb. casei* strains were more effective than pediococci. Furthermore, bacteriocin-producing starter cultures may be useful for the fermentation of sauerkraut[50,65] or olives to prevent the growth of spoilage organisms. In the fermentation of Spanish-style green olives, a bacteriocin-producing strain of *Lb. plantarum* dominated the endogenous LAB without adversely affecting the organoleptic properties of the product.[66] In contrast, a nonbacteriocin-producing variant of this strain was outnumbered by the "natural" *Lactobacillus* population.

As mentioned above, the primary requirement for using viable bacteriocinogenic LAB cultures is that optimal growth and bacteriocin production take place under the prevailing processing and/or storage conditions. A promising strategy would be to select for suitable protective cultures that are well adapted to the ecosystem in which they will be applied by isolating them from related food or environmental samples. Following this strategy, a study was performed in our laboratory to obtain potentially well-adapted bacteriocinogenic LAB that could be used for minimally processed vegetables that are packaged under modified atmospheres and stored under refrigeration.

Of a total of 890 LAB isolates from mung bean sprouts and chicory endive, only nine strains were found to produce bacteriocins.[42,67] Three of these strains exhibited antimicrobial activity toward a wide variety of gram-positive bacteria, including the food-borne pathogens *L. monocytogenes* and nonproteolytic *C. botulinum*. Two of these bacteriocinogenic isolates were identified as *Pediococcus parvulus*. The bacteriocin produced by both strains was identified and found to be identical to pediocin PA1.[67] The third broad-spectrum bacteriocin-producing strain was identified as *Enterococcus mundtii*, and its bacteriocin, mundticin, was identified as a novel class IIa bacteriocin.[42] In in vitro studies, both pediococci produced significant amounts of bacteriocin only at temperatures above 15°C, and, therefore, were not considered suitable for application at lower temperatures. *E. mundtii* produced significant amounts of bacteriocin, even at 4–10°C.[68] Although this bacterium does not have the GRAS status, it is a suitable candidate to test whether biopreservation can ascertain the required safety toward certain psychrotrophic pathogens. When applying the mundticin producer as a protective culture at 8°C on sterile vegetable agar medium, growth of *L. monocytogenes* was effectively suppressed. However, on fresh, nonsterile mung bean sprouts, no activity was found.[68] Most likely, either the production of mundticin on produce at low temperature is insufficient or the mundticin is inactivated after secretion by the producing strains, e.g., as a result of enzymatic inactivation or adsorption to produce. To inhibit the growth of *L. monocytogenes* on this product, more promising results were obtained when a partially purified bacteriocin solution was added to the product: A decline of 2 log units in the initial numbers was achieved when the produce was dipped in a solution of mundticin prior to contamination with the pathogen.[68] Identical results were obtained when the product was treated with a mundticin-containing alginate film. Noteworthy is that the counts of the pathogen did not exceed the initial inoculation level for approximately 8 days. Thus, use of food-approved bacteriocins as a dipping solution or as part of an edible coating may have good potential as a biopreservative treatment for minimally processed vegetables.

Leal et al.[69] recently discussed how vegetable-derived bacteriocinogenic LAB strains can be used effectively for food safety reasons and can also contribute to better control of traditional spontaneous fermentation systems, e.g., as are used for olives. Essentially, starter cultures with certain favorable fermentation traits and able to produce bacteriocins should have a good competitive advantage over spoilage microorganisms of the same or closely related species and, thus, could help to assure the desired product features to be developed. Using a culture medium referred to as *olive juice broth*, obtained from a Spanish-style green olive fermentation, they studied the interaction of a bacteriocin-producing strain of *Lb. plantarum* (LPCO10) with a nonproducing strain of the same species (strain 128/2). Strain LPCO10 was able to produce bacteriocin in olive juice broth throughout the incubation time (15 days) and, in mixed cultures, indeed dominated *L. plantarum* 128/2 strain. A strain of LPCO10 that had lost the capability to produce bacteriocins, *L. plantarum* 55–1 strain, did not show such capability. Other examples of controlled fermentation have been reported by other researchers.[50,65]

LEGISLATION ASPECTS OF BIOPRESERVATION

Existing food legislation in most countries will not favor the use of natural compounds when they are purified from their natural source unless these compounds have genuinely acquired the GRAS status. In most cases, the legislative point of view on such biopreservatives may be that they are new food additives or that they are applied for new purposes and, consequently, would require a nontoxicity record, despite a possible GRAS status. A more favorable way of application, therefore, would be the addition of the bacteriocin-producing strain to the food preparation, because this may still be regarded as the most natural source.

The regulatory considerations for use of bacteriocins in foods as they are concurrent in the United States were described by Fields.[70] Which authority is responsible for use of bacteriocins in foods is dependent on the foods in which the bacteriocin is used and the purpose for which it is used. Use of (a) purified bacteriocins, (b) bacteriocin-producing bacteria, or (c) genetic expression of bacteriocins in food-producing organisms to serve a preservative effect in processed foods is under the jurisdiction of the U.S. Food and Drug Administration (FDA) and is regulated as food ingredients under the Federal Food, Drug and Cosmetic Act (FFDCA). Under the FFDCA, those substances that are GRAS by qualified experts (either based on scientific principles or because they have been historically and safely present in food) are exempt from mandatory premarket approval. Substances used in processed food that are not GRAS are defined as "food additives" under the FFDCA and require premarket approval by the FDA. Bacteriocins used in meat products will require an additional suitability assessment by the U.S. Department of Agriculture (USDA) Food Safety and Inspection Service (FSIS). Bacteriocins that are used on whole fruits or vegetables (or genetically expressed in whole fruits and vegetables and intended to act in the whole food) fall within the definition of "pesticide" found in the Federal Insecticide, Fungicide, and Rodenticide Act (FIFRA) and are, therefore, regulated by the Environmental Protection Agency (EPA). Bacteriocins that are genetically expressed in food-producing domestic animals may be regulated as animal drugs if they are intended for use in preventing disease in animals.

The current regulatory status of bacteriocins and bacteriocin-producing organisms is a clear example of the current controversy between the use of the active compound or the natural source as a whole. In 1969, a joint Food and Agriculture Organization/World Health Organization expert committee accepted nisin as a legal food additive, although it was not until 1988 that it was approved in the United States by the FDA for use in certain pasteurized cheese spreads. Presently, nisin is permitted in at least 50 countries for the inhibition of clostridia in cheese and canned foods. None of the other bacteriocins known to date has a fully approved legal status as a food additive, although the application of a pediocin-producing strain of *Pediococcus acidilactici* has been approved by the USDA for use in reduced nitrite bacon to aid in the prevention of botulinum toxin production caused by the outgrowth of *C. botulinum* spores. The regulation of bacteriocin preparations from LAB stands in sharp contrast to the common use of these organisms as starter cultures. Moreover, LAB are commonly consumed in high numbers in fermented or cultured products and are often present as endogenous contaminants in many retail products. The general concept would be that the introduction of bacteriocins in foods at levels analogous to those capable of being produced by starter cultures should be as safe as the consumption of the cultured products themselves. In this respect, LAB and their antimicrobial compounds could play an important role as biopreservation agents to extend the storage life and safety of products, including modified atmosphere packaged vegetables.

REFERENCES

1. Nguyen-the C, Carlin F. The microbiology of minimally processed fresh fruits and vegetables. *Crit Rev Food Sci Nutr.* 1994;34:371–401.
2. Lund BM. Bacterial spoilage. In: Dennis C, ed. *Post-harvest Pathology of Fruits and Vegetables.* London: Academic Press; 1983:219–257.
3. Bennik MHJ, Vorstman W, Smid EJ, Gorris LGM. The influence of oxygen and carbon dioxide on the growth of prevalent *Enterobacteriaceae* and *Pseudomonas* species isolated from fresh and modified atmosphere stored vegetables. *Food Microbiol.* 1998;15:459–469.
4. Breer C, Baumgartner A. Occurrence and behaviour of *Listeria monocytogenes* on salads, vegetables, and in fresh vegetable juices. *Arch Lebensm.* 1992;43: 108–110.
5. Sizmur K, Walker CW. Listeria in prepacked salads. *Lancet.* 1988;1:1167.
6. Schlech WF, Lavigne PM, Bortolussi RA, Allen AC, et al. Epidemic listeriosis—Evidence for transmission by food. *New Engl J Med.* 1983;308:203–206.
7. Beuchat LR. Pathogenic microorganisms associated with fresh produce. *J Food Microbiol.* 1995;59:204–216.
8. Bennik MHJ, Smid EJ, Rombouts FM, Gorris LGM. Growth of psychrotrophic foodborne pathogens in a solid surface model system under the influence of carbon dioxide and oxygen. *Food Microbiol.* 1995;12:509–519.
9. Wood BJB, Holzapfel WH. *The Genera of Lactic Acid Bacteria.* Glasgow, UK: Blackie Academic and Professional, 1995.
10. Ray B, Daeschel M. *Food Biopreservatives of Microbial Origin.* Boca Raton, FL: CRC Press; 1992.
11. Lindgren SE, Dobrogosz WJ. Antagonistic activities of lactic acid bacteria in food and feed fermentations. *FEMS Microbiol Rev.* 1990;87:149–164.
12. Cerny G, Hennlich W. Minderung des Hygienerisikos bei Feinkostsalaten durch Schutzkulturen Teil II: Kartoffelsalat. ZFL *(Int Zeit Lebens Technol u Verfahr.)* 1991;42:1–2, 12.
13. Hennlich W. Sicherer Hygieneschutz. Leistungsanforderungen an Schutzkulturen in Feinkostsalaten. *Lebensm.* 1995;27:51–53.
14. Cerny G. Einsatz von Schutzkulturen zur Minderung des Hygienerisikos bei Lebensmitteln. *Lebensm.* 1991;23: 448, 450–451.
15. Vescovo M, Torriani S, Orsi C, Macchiarolo F, Scolari G. Application of antimicrobial-producing lactic acid

bacteria to control pathogens in ready-to-use vegetables. *J Appl Bact.* 1996;81:113–119.

16. Giraffa G. Lactic and non-lactic acid bacteria as a tool for improving the safety of dairy products. *Ind Aliment.* 1996;35:244–248, 252.
17. Holzapfel WH, Geisen R, Schillinger U. Biological preservation of foods with reference to protective cultures, bacteriocins and food-grade enzymes. *Int J Food Microbiol.* 1995;24:343–362.
18. Huss HH, Jeppesen VP, Johansen C, Gram L. Biopreservation of fish products—A review of recent approaches and results. *J Aq Food Prod Technol.* 1995;4:5–26, 37.
19. Rozbeh M, Kalchayanand N, Field RA, Johnson MC, et al. The influence of biopreservatives on the bacterial level of refrigerated vacuum packaged beef. *J Food Safety.* 1993;13:99–111.
20. Schillinger U, Geisen R, Holzapfel WH. Potential of antagonistic microorganisms and bacteriocins for the biological preservation of foods. *Trends Food Sci Technol.* 1996;7:158–164.
21. Weber H. Dry sausage manufacture. The importance of protective cultures and their metabolic products. *Fleischw.* 1994;74:278–282.
22. Tagg JR, Dajani AS, Wannamaker LW. Bacteriocins of gram-positive bacteria. *Bacteriol Rev.* 1976;40: 722–756.
23. Van den Bergh PA. Lactic acid bacteria, their metabolic products and interference with microbial growth. *FEMS Microbiol Rev.* 1993;12:221–238.
24. Klaenhammer TR. Genetics of bacteriocins produced by lactic acid bacteria. *FEMS Microbiol Rev.* 1993;12: 39–86.
25. Chikindas ML, Garcia Garcera MJ, Driessen AJM, Ledeboer AM, et al. Pediocin PA-1, a bacteriocin from *Pediococcus acidilactici* PAC1.0, forms hydrophilic pores in the cytoplasmic membrane of target cells. *Appl Environ Microbiol.* 1993;59:3577–3584.
26. Driessen AJM, Vandenhooven HW, Kuiper W, Vandekamp M, et al. Mechanistic studies of lantibiotic-induced permeabilization of phospholipid vesicles. *Biochem.* 1995;34:1606–1614.
27. Ruhr E, Sahl HG. Mode of action of the peptide antibiotic nisin and influence on the membrane potential of whole cells and on cytoplasmic and artificial membrane vesicles. *Antimicrob Ag Chemother.* 1985;27:841–845.
28. Venema K, Venema G, Kok J. Lactococcal bacteriocins: mode of action and immunity. *Trends Microbiol.* 1995;3:299–304.
29. Gorris LGM, Bennik MHJ. Bacteriocins for food preservation. *ZFL Int Zeit Lebensm Technol u Verfah.* 1994; 45:65–71.
30. Gould GW. Industry perspectives on the use of natural antimicrobials and inhibitors for food applications. *J Food Prot* 1996 (Suppl.); 82–86.
31. Harris LJ, Fleming HP, Klaenhammer TR. Developments in nisin research. *Food Res Int.* 1992;25:57–66.
32. Stiles ME. Biopreservation by lactic acid bacteria. *Antonie van Leeuwenhoek.* 1996;70:331–345.
33. Motlagh AM, Bhunia AK, Szostek F, Hansen TR, et al. Nucleotide and amino acid sequence of pap-gene (pediocin AcH production) in *Pediococcus acidilactici* H. *Lett Appl Microbiol.* 1992;15:45–48.
34. Ennahar S, Aoude-Werner D, Sorokine O, Van Dorsselaer A, et al. Production of pediocin AcH by *Lactobacillus plantarum* WHE 92 isolated from cheese. *Appl Environ Microbiol.* 1996;62:4381–4387.
35. Holck AL, Axelsson L, Huhne K, Krockel L. Purification and cloning of sakacin 674, a bacteriocin from *Lactobacillus sake* Lb674. *FEMS Microbiol Lett.* 1994;115:143–149.
36. Larsen AG, Vogensen FK, Josephsen J. Antimicrobial activity of lactic acid bacteria isolated from sour doughs—purification and characterization of bavaricin-A, a bacteriocin produced by *Lactobacillus bavaricus* MI401. *J Appl Bacteriol.* 1993;75:113–122.
37. Tichaczek PS, Vogel RF, Hammes WP. Cloning and sequencing of *sak p* encoding sakacin p, the bacteriocin produced by *Lactobacillus sake* LTH 673. *Microbiol.* 1994;140:361–367.
38. Bhugaloo-Vial P, Dousset X, Metivier A, Sorokine O, et al. Purification and amino acid sequences of *Piscicocins* V1a and V1b, two class IIa bacteriocins secreted by *Carnobacterium piscicola* V1 that display significantly different levels of specific inhibitory activity. *Appl Environ Microbiol.* 1996;62:4410–4416.
39. Jack RW, Wan J, Gordon J, Harmark K, et al. Characterization of the chemical and antimicrobial properties of piscicolin 126, a bacteriocin produced by *Carnobacterium piscicola* JG126. *Appl Environ Microbiol.* 1996;62:2897–2903.
40. Quadri LEN, Sailer M, Roy KL, Vederas JC, et al. Chemical and genetic characterization of bacteriocins produced by *Carnobacterium piscicola* LV17B. *J Biol Chem.* 1994; 269:12204–12211.
41. Aymerich T, Holo H, Havarstein LS, Hugas M, et al. Biochemical and genetic characterisation of enterocin A from *Enterococcus faecium*, a new antilisterial bacteriocin in the pediocin family of bacteriocins. *Appl Environ Microbiol.* 1996;62:1676–1682.
42. Bennik MHJ, Vanloo B, Brasseur R, Gorris LGM, et al. A novel bacteriocin with a YGNGV motif from vegetable-associated *Enterococcus mundtii*. full characterization and interaction with target organisms. *Biochim Biophys Acta.* 1998;1373:47–58.
43. Hastings JW, Sailer M, Johnson K, Roy KL, et al. Characterization of leucocin A-UAL187 and cloning of the

bacteriocin gene from *Leuconostoc gelidum*. *J Bacteriol.* 1991;173: 7491–7500.

44. Hechard Y, Derijard B, Letellier F, Cenatiempo Y. Characterization and purification of mesentericin y105, an anti-listeria bacteriocin from *Leuconostoc mesenteroides*. *J Gen Microbiol.* 1992;138:2725–2731.
45. Abee T, Krockel L, Hill C. Bacteriocins. Modes of action and potentials in food preservation and control of food poisoning. *Int J Food Microbiol.* 1995;28:169–185.
46. Carolissen-Mackay V, Arendse G, Hastings JW. Purification of bacteriocins of lactic acid bacteria: Problems and pointers. *Int J Food Microbiol.* 1997;34:1–16.
47. Delves-Broughton J, Gasson MJ. Nisin. In: Dillon VM, Board RG, eds. *Natural Antimicrobial Systems and Food Preservation.* Wallingford, UK: CAB International; 1994:99–132.
48. Ray B, Daeschel MA. Bacteriocins of starter culture bacteria. In: Dillon VM, Board RG, eds. *Natural Antimicrobial Systems and Food Preservation.* Wallingford, UK: CAB International, 1994; 99–132.
49. Schillinger U. Bacteriocins of lactic acid bacteria. In: Bills DD, Kung SD, eds. *Biotechnology and Food Safety.* Boston: Butterworth-Heinemann, 1990; 54–74.
50. Harris LJ, Fleming HP, Klaenhammer TR. Novel paired starter culture system for sauerkraut, consisting of a nisin-resistant *Leuconostoc mesenteroides* strain and a nisin-producing *Lactococcus lactis* strain. *Appl Environ Microbiol.* 1992;58:1484–1489.
51. Abee T, Rombouts FM, Hugenholtz J, Guihard G, et al. Mode of action of nisin Z against *Listeria monocytogenes* Scott A grown at high and low temperatures. *Appl Environ Microbiol.* 1994;60:1962–1968.
52. Kaiser AL, Montville TJ. The influence of pH and growth rate on production of the bacteriocin, bavaricin MN, in batch and continuous fermentations. *J Appl Bacteriol.* 1993;75: 536–540.
53. Vaughan EE, Caplice E, Looney R, O'Rourke N, et al. Isolation from food sources, of lactic acid bacteria that produced antimicrobials. *J Appl Bacteriol.* 1994;76:118–123.
54. Coventry MJ, Gordon JB, Wilcock A, Harmark K, et al. Detection of bacteriocins of lactic acid bacteria isolated from foods and comparison with pediocin and nisin. *J Appl Microbiol.* 1997;83:248–258.
55. Kelly WJ, Asmundson RV, Huang CM. Isolation and characterization of bacteriocin-producing lactic acid bacteria from ready-to-eat food products. *Int J Food Microbiol.* 1996;33:209–218.
56. Uhlman L, Schillinger U, Rupnow JR, Holzapfel WH. Identification and characterization of 2 bacteriocin-producing strains of *Lactococcus lactis* isolated from vegetables. *Int J Food Microbiol.* 1992;16:141–151.
57. Franz CMAP, Dutoit M, Vonholy A, Schillinger U, et al. Production of nisin-like bacteriocins by *Lactococcus lactis* strains isolated from vegetables. *J Bas Microbiol.* 1997;37:187–196.
58. Cai Y, Ng LK, Farber JM. Isolation and characterization of nisin-producing *Lactococcus lactis* subsp. *lactis* from bean-sprouts. *J Appl Microbiol.* 1997;83:499–507.
59. Olasupo NA. Inhibition of *Listeria monocytogenes* by plantaricin NA, an antibacterial substance from *Lactobacillus plantarum. Fol Microbiol.* 1998;43:151–155.
60. Franz CMAP, Dutoit M, Olasupo NA, Schillinger U, et al. Plantaricin D, a bacteriocin produced by *Lactobacillus plantarum* Bfe 905 from ready-to-eat salad. *Lett Appl Microbiol.* 1998; 26:231–235.
61. Franz CMAP, Schillinger U, Holzapfel WH. Production and characterization of enterocin 900, a bacteriocin produced by *Enterococcus faecium* BFE 900 from black olives. *Int J Food Microbiol.* 1996;29:255–270.
62. Vescovo M, Scolari G, Orsi C, Sinigaglia M, et al. Combined effects of *Lactobacillus casei* inoculum, modified atmosphere packaging and storage temperature in controlling *Aeromonas hydrophila* in ready-to-use vegetables. *Int J Food Sci Technol.* 1997; 32:411–419.
63. Breidt F, Fleming HP. Modeling of the competitive growth of *Listeria monocytogenes* and *Lactococcus lactis* in vegetable broth. *Appl Environ Microbiol.* 1998;64: 3159–3165.
64. Vescovo M, Orsi C, Scolari G, Torriani S. Inhibitory effect of selected lactic acid bacteria on microflora associated with ready-to-use vegetables. *Lett Appl Microbiol.* 1995;21:121–125.
65. Breidt F, Crowley KA, Fleming HP. Controlling cabbage fermentations with nisin and nisin-resistant *Leuconostoc mesenteroides*. *Food Microbiol.* 1995;12:109–116.
66. Ruiz-Barba JL, Cathcart DP, Warner PL, Jimenez-Diaz R. Use of *Lactobacillus plantarum* LPCO10, a bacteriocin producer, as a starter culture in Spanish-style green olive fermentations. *Appl Environ Microbiol.* 1994;60:2059–2064.
67. Bennik MHJ, Smid EJ, Gorris LGM. Vegetable-associated *Pediococcus parvulus* produces pediocin PA-1. *Appl Environ Microbiol.* 1997;63:2074–2076.
68. Bennik MHJ, Van Overbeek W, Smid EJ, Gorris LGM. Biopreservation in modified atmosphere stored mungbean sprouts: the use of vegetable-associated bacteriocinogenic lactic acid bacteria to control the growth of *Listeria monocytogenes*. *Lett Appl Microbiol.* 1999; 28;226–232.
69. Leal MV, Baras M, Ruizbarba JL, Floriano B, et al. Bacteriocin production and competitiveness of *Lactobacillus plantarum* LPCO10 in olive juice broth, a culture medium obtained from olives. *Int J Food Microbiol.* 1998;43:129–134.
70. Fields FO. Use of bacteriocins in food—regulatory considerations. *J Food Prot.* 1996;(Suppl. S):72–77.

Chapter 16

Minimal Processing of Fresh Produce

Raija Ahvenainen

INTRODUCTION

Minimal processing of raw fruits and vegetables has two purposes. First, it is important to keep the produce fresh but convenient without losing its nutritional quality. Second, the product should have a shelf life sufficient to make distribution feasible within a region of consumption.[1] The microbiologic, sensory, and nutritional shelf life of minimally processed vegetables or fruits should be at least 4–7 days, but preferably even longer, up to 21 days, depending on the market.[2,3] The aim of this contribution is to present an integrated approach to modern minimal processing of produce. The present know-how of all of the steps in the food chain, beginning with raw material and processing, and ending with packaging factors that affect the quality and shelf life of minimally processed fresh prepared fruits and vegetables, will be introduced.

THE KEY FACTORS IN THE MINIMAL PROCESSING OF FRESH PRODUCE

Minimal processing methods for fresh produce have been developed at different laboratories during the past few years. Recently, Ahvenainen[4] and Laurila et al.[5] have compiled extensive reviews on new approaches in improving the shelf life of minimally processed fruit and vegetables. This paper is based partly on those reviews and partly on literature that appeared after those articles.

Minimally processed vegetables can be manufactured on the basis of many different working principles (Table 16–1). If the principle is that the products are prepared today and they are consumed tomorrow, then very simple processing methods can be used. Most fruits and vegetables are suitable for this kind of preparation. These products are suitable for catering but not for retailing. The greatest advantage of this principle is the low need for investments. If the products need a shelf life of several days up to one week or even more, as is the case with the products intended for retailing, then more advanced processing methods and treatments using the hurdle concept[2,3,6] are needed, as well as correctly chosen raw material suitable for minimal processing. Not all produce is suitable for this kind of preparation.

All in all, a characteristic feature in minimal processing is an integrated approach, where raw material, handling, processing, packaging, and distribution must be properly considered to make shelf-life extension possible. Preservation is based on the synergies of all treatments. As examples, processing and packaging guidelines for prepeeled and sliced potato, grated carrot, and shredded Chinese and white cabbage are given in Exhibits 16–1, 16–2, and 16–3, respectively.

Key requirements in minimal processing of fresh produce are raw material of good quality, strict hygiene and good manufacturing practises, low temperatures during working, careful

Table 16–1 Requirements for the Commercial Manufacture of Prepeeled and/or Sliced, Grated, or Shredded Fruits and Vegetables

Working Principle	*Demands for Processing*	*Customers*	*Shelf Life (day) at 5°C*	*Examples of Suitable Fruits and Vegetables*
Preparation today, consumption tomorrow	• Standard kitchen hygiene and tools • No heavy washings for peeled and shredded produce; potato is an exception • Packages can be returnable containers	Catering industry Restaurants Schools Industry	1–2	Most fruits and vegetables
Preparation today, the customer uses the product within 3–4 days	• Disinfection • Washing of peeled and shredded produce, at least with water • Permeable packages; potato is an exception	Catering industry Restaurants Schools Industry	3–5	carrots cabbages iceberg lettuce potatoes beetroot acid fruits berries
Products are also intended for retailing	• Good infection • Chlorine or acid washing for peeled and shredded produce • Permeable packages; potato is an exception • Additives	In addition to the customers listed above, retail shops can also be customers	5–7*	carrots Chinese cabbage red cabbage potatoes beetroot acid fruits berries

*If longer shelf life up to 14 days is needed, the storage temperature must be 1–2°C.

cleaning and/or washing before and after peeling, good-quality water (sensory, microbiology, pH) used in washing, use of mild additives in washing for disinfection or browning prevention, gentle spin drying after washing, gentle peeling, gentle cutting/slicing/shredding, correct packaging materials and packaging methods, and correct temperature and humidity during distribution and retailing. Most of these factors are considered in detail in the following.

RAW MATERIAL

It is self-evident that vegetables or fruits intended for prepeeling and cutting must be easily washable and peelable, and they must be of top quality. The correct and proper storage of vegetables and careful trimming before processing are vital for the production of prepared vegetables of good quality.[2,3,7] The study of various cultivar varieties of eight different vegetables showed that not all varieties of the specified vegetable can be used for the manufacturing of prepared vegetables. The correct choice of variety is particularly important for carrot, potato, swede and onion. For example, with carrot and swede, the variety that gives the most juicy grated product cannot be used in the production of grated products that should have a shelf life of several days.[3] Another example is potato, with which poor color and flavor become problems if the variety is wrong.[8,9] Furthermore, climatic conditions, soil conditions, and agricultural practices such as fertilization and harvesting conditions can also significantly affect the behavior of vegetables, particularly that of potatoes, in minimal processing.[10] These aspects of minimal processing should be studied further.[4]

PEELING, CUTTING, AND SHREDDING

Some vegetables and fruits, such as potatoes, carrots, apples, and oranges, need peeling. There are several peeling methods available, but on an industrial scale, the peeling is normally accomplished mechanically (e.g., rotating carborundum drums), chemically, or in high-pres-

Exhibit 16–1 Processing Guidelines for Prepeeled and Sliced Potato

Processing temperature	4–5°C
Raw material	Suitable variety or raw material lot should be selected using a rapid storage test of prepared produce at room temperature. Attention must be focused on browning susceptibility.
Pretreatment	Careful washing with good-quality water before peeling. Damaged and contaminated parts, as well as spoiled potatoes, must be removed.
Peeling	1) One-stage peeling: knife machine. 2) Two-stage peeling: slight carborundum first, then knife peeling.
Washing	Washing immediately after peeling. The temperature and amount of washing water should be 4–5°C and 3 L/kg potato. Washing time 1 min. Observation: the microbiologic quality of washing water must be excellent. In washing water—in particular, for sliced potato—it is preferable to use citric acid with ascorbic acid (maximum concentration of both, 0.5%), possibly combined with calcium chloride, sodium bensoate, or 4-hexyl resorcinol to prevent browning.
Slicing	Slicing should be done immediately after washing with sharp knives.
Straining off	Loose water should be strained off in a colander.
Packaging	Packaging immediately after washing in vacuum or in a gas mixture of 20% CO_2 + 80% N_2. The headspace volume of a package 2 L/kg potato. Suitable oxygen permeability of packaging materials is 70 cm^3/m^2 24 hr 101.3 kPa, 23°C, RH 0% (80 μm nylon-polyethylene).
Storage	4–5°C, preferably in dark.
Other remarks	Good manufacturing practices must be followed (hygiene, low temperatures, and disinfection).
Shelf life	The shelf life of prepeeled whole potato is 7–8 days at 5°C. Due to browning, sliced potato has very poor stability; the shelf life is only 3–4 days at 5°C.

Exhibit 16–2 Processing Guidelines for Grated Carrot

Processing temperature	0–5°C
Raw material	Suitable variety or raw material lot should be selected using a rapid storage test of prepared produce at room temperature.
Pretreatment	Carrots must be washed carefully before peeling. Stems, damaged and contaminated parts, as well as spoiled carrots, must be removed.
Peeling	Peeling with knife or carborundum machine.
Washing	Immediately after peeling. The temperature and amount of washing water: 0–5°C and 3 L/kg carrot, respectively. The washing time 1 min. Observation: the microbiologic quality of washing water must be excellent.
	It is preferable to use 0.01% active chloride or 0.5% citric acid in washing water.
Grating	The shelf life of grated carrot: the shorter the shelf life, the finer the shredding grade. The optimum grate degree is 3–5 mm.
Centrifugation	Immediately after grating. Grate may be lightly sprayed with water before centrifugation. The centrifugation rate and time must be selected so that centrifugation removes only loose water but does not break vegetable cells.
Packaging	Immediately after centrifugation. Proper packaging gas is normal air, and the headspace volume of a package is 2 L/kg grated carrot.
	Suitable oxygen permeability of packaging materials is between 1,200 (e.g., oriented polypropylene) and 5,800, preferably 5,200–5,800 (e.g., polyethylene-ethylene vinyl acetate-oriented polypropylene) cm^3/m^2 24 hr 101.3 kPa, 23°C, RH 0%.
	Perforation (one microhole/150 cm^3) of packaging material is advantageous. The diameter of microhole is 0.4 mm.
Storage	0–5°C, preferably in dark.
Other remarks	Good manufacturing practices must be followed (hygiene, low temperatures, and disinfection).
Shelf life	7–8 days at 5°C.

sure steam peelers.[2] However, the ideal method would be hand peeling with a sharp knife because peeling should be as gentle as possible (Figure 16–1). Carborundum-peeled potatoes must be treated with a browning inhibitor, whereas water washing is enough for hand-peeled potatoes.[10] If mechanical peeling is used, it should resemble knife peeling. Carborundum, steam peeling, or caustic acid disturbs the cell walls of a vegetable, which enhances the possibilities of microbial growth and enzymatic changes. Many commercial companies selling vegetable processing equipment already have knife peelers available. In Finland, some potato processing companies peel potatoes successfully in two stages: rough peeling with a carborundum peeler and final peeling with knife peeler. Enzymatic peeling can also be used, e.g., for oranges. Pretel et al.[11] obtained shelf life at least as good at +4°C for enzymatically peeled whole oranges as for manually separated segments of oranges.

Many studies show that the cutting and shredding must be performed with knives or blades as sharp as possible, these being made from stainless steel. Carrots cut with a razor blade were more acceptable from a microbiologic and sensory point of view than were carrots cut with commercial slicing machines.[12] It is clear that slicing with dull knives impairs quality retention because of the breaking of cells and the release of tissue fluid to a great extent. Mats and blades used in slicing can be disinfected, for example, with a 1% hypochlorite solution. A slicing machine must be installed solidly, because vibrating equipment may impair the quality of sliced surfaces.[4]

CLEANING, WASHING, AND DRYING

Incoming vegetables or fruits, which are covered with soil, mud, and sand, should be carefully

Exhibit 16–3 Processing Guidelines for Shredded Chinese Cabbage and White Cabbage

Processing temperature	0–5°C
Raw material	Suitable variety or raw material lot should be selected using a rapid storage test of prepared produce at room temperature.
Pretreatment	Outer contaminated leaves and damaged parts, as well as stem and spoiled cabbages, must be removed.
Shredding	The shelf life of shredded cabbage: the shorter the shelf life, the finer the shredding grade. The optimum shredding degree is about 5 mm.
Washing of shredded cabbage	Immediately after shredding. The temperature and amount of washing water: 0–5°C and 3 L/kg cabbage, respectively. The washing time 1 min. Observation: the microbiologic quality of the washing water must be excellent.
	Washing should be done in two stages: 1) Washing with water containing 0.01% active chloride or 0.5% citric acid. 2) Washing with plain water (rinsing).
Centrifugation	Immediately after washing. The centrifugation rate and time must be selected so that centrifugation removes only loose water but does not break vegetable cells.
Packaging	Immediately after centrifugation. Proper packaging gas is normal air, and the headspace volume of a package is 2 L/kg cabbage.
	Suitable oxygen permeability of packaging material is between 1,200 (e.g., oriented polypropylene) and 5,800, preferably 5,200–5,800 (e.g., polyethylene-ethylene vinyl acetate-oriented polypropylene) cm^3/m^2 24 hr 101.3 kPa, 23°C, RH 0%.
	For white cabbage, perforation (one microhole/150 cm^3) can be used. The diameter of microhole is 0.4 mm.
Storage	0–5°C, preferably in dark.
Other remarks	Good manufacturing practices must be followed (hygiene, low temperatures, and disinfection).
Shelf life	7 days for Chinese cabbage and 3–4 days for white cabbage at 5°C.

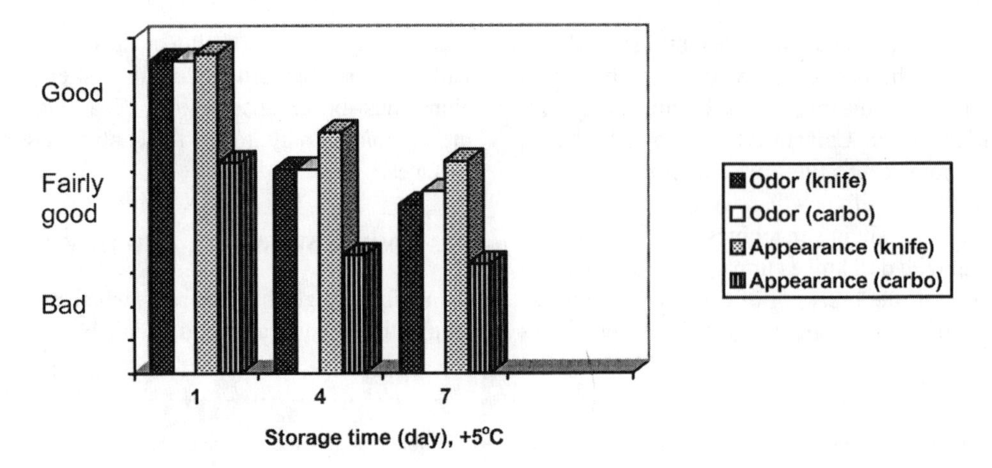

Figure 16–1 The effect of peeling method and storage time on the odor and appearance of potato packed in a gas mixture of 20% CO_2 + 80% N_2 and stored at 5°C.

cleaned before processing. The second washing must usually be done after peeling and/or cutting.[2,3] For example, Chinese and white cabbage must be washed after shredding, whereas carrot must be washed before grating.[13,14] Washing after peeling and cutting removes microbes and tissue fluid, thus reducing microbial growth and enzymatic oxidation during storage. Washing in flowing or air-bubbling waters is more preferable than merely dipping into water.[15] The microbiologic and sensory quality of the washing water used must be good and its temperature low, preferably below 5°C. The recommendable quantity of water to be used is 5–10 L/kg of product before peeling/cutting[1] and 3 L/kg after peeling/cutting.[13,14]

Preservatives can be used in the washing water for the reduction of microbial numbers and to retard enzymatic activity, thereby improving the shelf life of sensory quality. According to several researchers, 100–200 mg/L of chlorine or citric acid is effective in the washing water before or after peeling and/or cutting to extend the shelf life (Figure 16–2).[2,7,12–14] However, when chlorine is used, vegetable material should be rinsed. Rinsing reduces the chlorine concentration to the level of that in drinking water and improves the sensory shelf life.[13] The effectiveness of chlorine can be enhanced by using a low pH, high temperature, pure water, and correct contact time.[2,7] Chlorine is widely used in the industry of producing minimally processed vegetables and fruits. It seems that chlorine compounds reduce counts of aerobic microbes, at least in some leafy vegetables, such as lettuce,[2,16] but not necessarily in root vegetables or cabbages.[14,16] Its effectiveness in suppressing growth of *Listeria monocytogenes* in shredded lettuce[17] or Chinese cabbage[18] is limited. Francis and O'Beirne[19] have even observed that dipping lettuce in chlorine (100 ppm solution), followed by storage at 8°C, resulted in a significant increase ($p < 0.05$, by 2 log cycles) in *Listeria innocua* populations, compared with undipped samples. The result was similar with 1% citric acid dip and N_2 flushing. It is concluded that both N_2 flushing and use of antimicrobial dips, combined with storage at 8°C, enhanced the survival and growth of *Listeria* populations on shredded lettuce.

Another disadvantage with chlorine is that some food constituents may react with chlorine to form potentially toxic reaction products. Consequently, the safety of chlorine use for food or water treatment has been questioned, and future regulatory restrictions may require the development of alternatives. The alternatives for chlorine might be chlorine dioxide, ozone, trisodium phosphate, or hydrogen peroxide (H_2O_2).[20] The use of H_2O_2 as an alternative to chlorine for disinfecting freshly cut fruits and vegetables shows particular promise. H_2O_2 vapor treatment appears to reduce microbial populations on freshly cut products (cucumbers, green bell peppers, and zucchini) and to extend shelf life without leaving significant residues or causing loss of quality. However, according to Sapers and Simmons,[20] more definitive data are required to establish the technical and economic feasibility of the treatment. In particular, additional research is needed to optimize H_2O_2 treatments with respect to efficacy in delaying the growth of spoilage and pathogenic bacteria. The applicability of H_2O_2 treatments to a broad range of freshly cut commodities should also be determined.

It is recommended to gently remove the washing water from the product.[2] Centrifugation seems to be the best method. The centrifugation time and rate should be chosen carefully,[15,21,22] so that centrifugation only removes loose water and does not break vegetable cells.

BROWNING INHIBITION

With regard to fruits and vegetables such as prepeeled and sliced apple and potato, for which the main quality problem is browning—which causes particularly poor appearance—washing with water is not effective enough for the prevention of discoloration.[2,8] Traditionally, sulfites have been used for prevention of browning. However, the use of sulfites has some disadvantages. In particular, they can cause dangerous side effects for asthmatics. For this reason, the U.S. Food and Drug Administration (FDA) partly restricted the use of sulfites.[23] At the same time, interest in substitutes for sulfites is increasing.

Enzymatic browning requires four different components: oxygen, an enzyme, copper, and a substrate. To prevent browning, at least one component must be removed from the system. In theory, polyphenoloxidase (PPO)-catalyzed browning of vegetables and fruits can be prevented by heat inactivation of the enzyme, exclusion or removal of one or both of the substrates (O_2 and phenols), lowering the pH to 2 or more units below the optimum, reaction inactivation of the enzyme, or adding compounds that inhibit PPO or prevent melanin formation.[24] Many inhibitors of PPO are known, but only a few have been considered as potential alternatives to sulfites.[25] The most attractive way to inhibit browning would be "natural" methods, such as the combination of certain salad ingredients with each other. Pineapple juice appears to

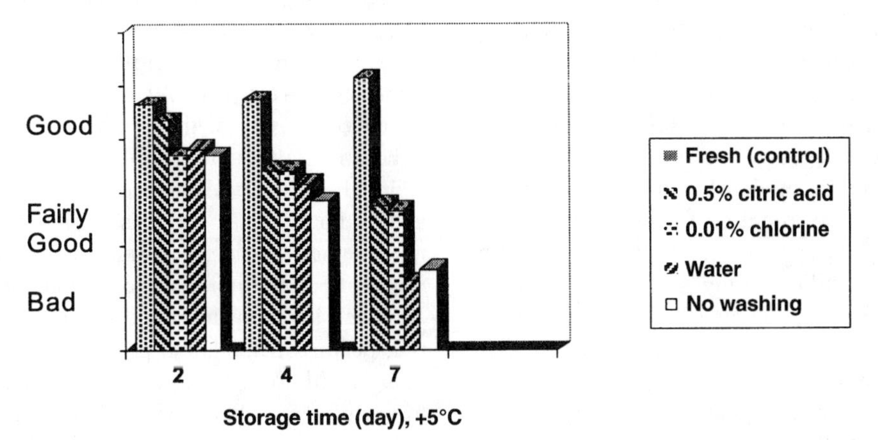

Figure 16–2 The effect of washing solution and storage time on the odor of grated carrots packed in air and stored at 5°C.

be a good potential alternative to sulfites for the prevention of browning in fresh apple rings.[26,27]

Probably the most often studied alternative to sulfite is ascorbic acid (AA). This compound is a highly effective inhibitor of enzymatic browning, primarily because of its ability to reduce quinones back to phenolic compounds before they can undergo further reaction to form pigments. Unfortunately, once added AA has been completely oxidized to dehydro-ascorbic acid (DHAA), quinones can accumulate and undergo browning. Digestion with hot AA/citric acid solutions improved the shelf life of prepeeled potatoes. A shelf life of about 2 weeks was obtained. However, high concentrations of AA (0.75%) imposed an unpleasant taste in the fruits.[28] Ascorbic acid derivatives, such as AA-2-phosphate (AAP) and AA-2-triphosphate (AATP), have been used as browning inhibitors alone or in combinations with other inhibitors for potatoes and apples.[29–32] Erythorbic acid, an isomer of AA, has been used as an inhibitor of enzymatic browning, in combination with AA or citric acid for potato slices[33] and for whole abrasion-peeled potatoes.[34] Citric acid acts as a chelating agent and acidulant, both functionalities inhibiting PPO. Reliable and promising results have been obtained using citric acid and the combinations of citric-AA and benzoic-sorbic acid as dipping treatments for minimally processed potatoes.[8]

4-Hexylresorcinol is a good inhibitor of enzymatic browning for apples, potatoes, iceberg lettuce, and cut pears.[24,28,32,35,36] It interacts with PPO and renders it incapable of catalyzing the enzymatic reaction. 4-Hexylresorcinol has several advantages over sulfites in foods, including its specific mode of inhibitory action, its lower use level required for effectiveness, its inability to bleach preformed pigments, and its chemical stability.[37] Ethylenediamine tetraacetic acid (EDTA), a complexing agent, has been used with potatoes[33,38] and iceberg lettuce,[35] in combinations with other browning inhibitors. Sporix, a chelating agent described by its supplier as an acidic polyphosphate, has been found to be an effective browning inhibitor in several fruits and vegetables.[29,39] Sulfhydryl-containing amino acids, such as cysteine, prevent brown pigment formation by reacting with quinone intermediates to form stable, colorless compounds.[40] Cysteine has been used as a browning inhibitor for potatoes, apples, and iceberg lettuce,[35,41] and it has also been used as an ingredient in a commercial browning inhibitor.[38]

Protease enzymes were found to be effective browning inhibitors for apples and potatoes.[42–44] It is believed that an effective protease acts to hydrolyze and, therefore, inactivate the enzyme or enzymes responsible for enzymatic browning. Of the proteolytic enzymes tested so far, mainly three plant proteases (ficin from figs, papain from papaya, and bromelain from pineapple) proved to be effective. All of the three proteases are sulfhydryl enzymes of broad specificity. According to Taoukis et al.,[42] ficin was as effective as sulfite for potatoes at 4°C, but slightly less effective than sulfite at 24°C. Papain was somewhat effective for potatoes at 4°C. Papain treatment can prevent enzymatic browning of apples about as well as sulfite treatment at both temperatures (4 and 24°C).

It seems that there is no one substance that, alone, can substitute for sulfites in browning inhibition, but combinations of various substances and proper packaging methods must be used. All commercially available sulfite-free antibrowning agents are combinations, and most of them are AA-based compositions.[37] A typical combination may include a chemical reductant (e.g., AA), an acidulant (e.g., citric acid), and a chelating agent (e.g., EDTA). In many cases, the enhanced activity of the combined ingredients is additive, although synergism has also been claimed for several blends of antibrowning agents. Laurila et al.[5] emphasize also that, when new methods replacing sulfites are looked for, it is particularly important to take an integrated approach by choosing proper potato raw material, peeling method, and processing and packaging conditions. Furthermore, it is necessary to examine the effects of these methods on all sensory and nutritional quality attributes, not only on color.

It is also possible to improve the retention of the sensory quality, particularly the appearance of prepacked shredded lettuce, by washing in glycine betaine (GB) solutions. The optimum concentration of GB is 0.2 mol/L.[45] Glycine betaine retards browning and wilting of lettuce.

PACKAGING

The final, but not the least important, operation in producing minimally processed fruit and vegetables is packaging. The most studied packaging method for prepared raw fruits and vegetables is modified atmosphere packaging (MAP). The basic principle in MAP is that a modified atmosphere can be created passively by using properly permeable packaging materials or actively by using a specified gas mixture, together with permeable packaging materials. The aim of both is to create an optimal gas balance inside the package, where the respiration activity of a product is as low as possible; on the other hand, the oxygen concentration and carbon dioxide levels are not detrimental to the product. In general, the aim is to have a gas composition where there is 2–5% CO_2, 2–5% O_2, and the rest nitrogen.[46–48]

Actually, to reach this aim is the most difficult task in manufacturing raw, ready-to-use, or ready-to-eat fruit and vegetable products of good quality and possessing a shelf life of several days. The main problem is that only a few packaging materials on the market are permeable enough[48] to match the respiration of fruit sand vegetables. Most films do not result in optimal O_2 and CO_2 atmospheres, especially when the produce has high respiration. However, one solution is to make microholes of defined sizes and a defined quantity in the material to avoid anaerobiosis.[49] For example, this procedure significantly improves the shelf life of grated carrots.[14] Other solutions are to combine ethylene vinyl acetate with orientated polypropylene and low-density polyethylene or to combine ceramic material with polyethylene. Both composite materials have significantly higher gas permeability than does polyethylene or the oriented polypropylene much used in the packaging of salads, even though gas permeability should be still higher. These materials have good heat-sealing properties, and they are commercially available.[3] The shelf life of shredded cabbage and grated carrots packed in these materials is 7–8 days at 5°C and, therefore, 2–3 days longer than in the oriented polypropylene that is generally used in the vegetable industry. The products can be packed in normal air.[13,14] Recently, a new breathable film has been patented that has a three-layer structure, consisting of a two-ply blown coextrusion about 25 mm thick, with an outer layer of K-Resin KR10 and an inner metallocene polyethylene layer. It is claimed that this film gives 16 days' shelf life at 1–2°C for fresh salads washed in chlorine solution.[50]

With fresh respiring products, it would be advantageous for the product shelf-life retention to have film permeability increased by temperature at least as much as the respiration rate increases to avoid anaerobic conditions. Unfortunately, the permeation rates of most packaging films are only modestly affected by temperature. However, Landec Company has developed films engineered with an adjustable "temperature-switch" point, at which the film's permeation changes rapidly and dramatically. Landec's technology uses long-chain fatty alcohol-based polymeric chains. Under the predetermined temperature-switch point, these chains are in a crystalline state, providing a gas barrier. At the specified temperature, the side chains melt to a gas-permeable amorphous state.[51] Some commercial packages where Landec's polymers have been utilized are already available.[52]

In another study, a temperature-sensitive film for respiring products has been studied at Brunel University in the United Kingdom. The film is made from two dissimilar layers or from two different thicknesses of the same material, both layers containing minute cuts. When the temperature rises or falls, the layers expand at different rates. As the temperature rises, the film at the cut edge retracts and curls upward to give enlarged holes, thus significantly increasing the film permeability.[53] Exama et al.[49] have also proposed the safety valve system, preventing excessive O_2 depletion and excessive CO_2 accumulation when a transient temperature increase occurs. This kind of system might be suitable, particularly in bulk and transport packages of fruit and vegetable products.

High O_2 MAP treatment has been found to be particularly effective at inhibiting enzymatic browning, preventing anaerobic fermentation reactions, and inhibiting aerobic and anaerobic microbial growth.[54] It is hypothesized that high O_2 levels may cause substrate inhibition of PPO or, alternatively, high levels of colorless quinones subsequently formed may cause feedback production of PPO. Carbon monoxide (CO) gas atmosphere was found to inhibit mushroom PPO reversibly. Use of this compound in an MAP system would require measures to ensure the safety of packing plant workers.

One interesting MAP method is moderate vacuum packaging (MVP).[55] In this system, respiring produce is packed in a rigid, airtight container under 40 kPa of atmospheric pressure and stored at refrigerated temperature (4–7°C). The initial gas composition is that of normal air (21% O_2, 0.04% CO_2, and 78% N_2) but at a reduced partial gas pressure. The lower O_2 content stabilizes the produce quality by slowing down the metabolism of the produce and the growth of spoilage microorganisms. Gorris et al.[55] have compared the storage of several whole and lightly processed fruits and vegetables under ambient conditions to MVP and found that MVP improved the microbial quality in red bell pepper, chicory endive, sliced apple, and sliced tomato; the sensory quality of apricot and cucumber; and the microbial and sensory quality of mung bean sprouts and a mixture of cut vegetables. Gorris et al.[55] also conducted pathogen challenge tests with *L. monocytogenes*, *Yersinia enterocolitica*, *Salmonella typhimurium,* and *Bacillus cereus* on mung bean sprouts at 7°C. All of the pathogens lost viability quickly during the course of storage.

Maybe the most challenging task is to design an MAP for a mixed vegetable salad, where respiration rates of all the components are different. Lee et al.[56] have been successful in optimizing packaging for the salad mixture containing cut carrot, cut cucumber, sliced garlic, and whole green pepper. A pouch form package made of 27 mm low-density polyethylene film was predicted to achieve a target atmosphere of O_2 3–4% and CO_2 3–5% in the design, and in actual storage test, the package attained an equilibrium gas composition of O_2 2.0–2.1% and CO_2 5.5.–5.7%—close to the target. The package showed some improved quality retention, such as sensory quality, compared with other designed packages.

When MAP or vacuum packaging is used, there is often a fear about toxin productions of nonproteolytic *Clostridium botulinum* or growth of other pathogens, e.g., *L. monocytogenes*, particularly when the temperature is over 3°C. This is the case very often in the distribution chain. If the shelf life of vacuum-packed or MAP products should be greater than 10 days and there is a risk that temperature is over 3°C, the products should meet one or more of the specific controlling factors detailed below:

- minimum heat treatment of 90°C for 10 minutes or equivalent
- pH of 5 or less throughout the food
- salt level of 3.5% (aqueous) throughout the food
- a_w of 0.97 or less throughout a food
- any combination of heat and preservative factors that has been shown to prevent growth of toxin production by *C. botulinum.*[57]

Practically, if the aim is to keep minimally processed produce in a freshlike state, the last factor mentioned above and various preservative factors are the only possibilities to increase shelf life and to assure microbiologic safety of MAP or vacuum-packed fresh produce.

THE FUTURE OF MINIMALLY PROCESSED FRESH PRODUCE

It is probable that, in the future, fruits and vegetables intended for minimal processing will be cultivated under specified controlled conditions, and, furthermore, plant geneticists will develop selected and created cultivars or hybrids adapted to the specific requirements of minimal processing.[58,59] The unit operations, such as peelers and shredders, need further development to make them more gentle. There is no sense in disturbing the quality of produce by rough treatment during processing and afterward to patch it up with preservatives.

Microbiologic safety of minimally processed produce is a unique challenge because microbial populations on (and in) fresh vegetables can range as low as 10^2 CFU/g to as high as 10^9. Currently, there is interest in possible use of lactic acid bacteria (LAB) as biocontrol agents to ensure safety of minimally processed fruits and vegetables. The objective in using biocontrol cultures is not to ferment foods, but to control the microbial ecology through competitive inhibition of pathogenic bacteria.[18,60] Lactic acid bacteria species can produce a variety of metabolites, including lactic and acetic acids, which lower pH, or bacteriocins. Although bacteriocins, e.g., nisin, cannot be used as the sole preservatives to enhance the safety of refrigerated MAP-stored minimally processed vegetables; however, they can contribute to tackling safety problems that may arise from certain cold-tolerant, gram-positive pathogens.[61]

Torriani et al.[62] have evaluated the effect of the addition of a *Lactobacillus casei* culture and its sterile permeate and of 0.5% or 1% lactic acid on the growth of microorganisms associated with ready-to-use salad vegetables containing carrots, endive, garden rocket, and green chicory at 8°C. The addition of 3% culture permeate to mixed salads reduced the total mesophilic bacteria counts from 6 to 1 log CFU/g and suppressed coliforms, enterococci, and *Aeromonas hydrophila* after 6 days. A similar effect was shown when *L. casei* culture was inoculated in the vegetables. One percent lactic acid had a bacteriostatic effect on the bacterial groups examined, except for total and fecal coliforms, which were reduced by about 2 and 1 log unit, respectively, while 0.5% lactic acid did not affect the indigenous microflora of the vegetables. This means that the inhibitory effect of *L. casei* on the microbial population of mixed salad is not due only to lactic acid.

Breidt and Fleming[60] have currently made an extensive review on the use of lactic acid bacteria to improve the safety of minimally processed fruits and vegetables. They concluded that biocontrol strategies for minimally processed fruit and vegetable products may include:

1. isolation of potential biocontrol LAB from the refrigerated product;
2. reduction of the total microflora in the vegetable product by one or more procedures, including heat, washing using chemical sanitizers, irradiation, or others;
3. addition of a bacteriocin-producing biocontrol culture to achieve an appropriate initial CFU/mL, as determined experimentally; and
4. storage of the product under refrigeration.

The product shelf life would then be dictated by growth of the biocontrol culture, and, under temperature abuse conditions, the biocontrol culture will grow rapidly and prevent growth of pathogens. Successful application of biocontrol cultures will require balancing quality (shelf life) and safety considerations. Measuring and understanding the factors affecting the competitive growth of bacteria will be required to make rational choices in selecting biocontrol cultures for food products.[60]

Nonthermal physical treatments, such as high-intensity pulsed electric fields, high-intensity pulsed light, and high hydrostatic pressure, are also promising methods, and, most probably in the future, they will be used to maintain freshness and to improve microbiologic safety of minimally processed fruits and vegetables.[63] Hagenmaier and Baker[64] have proposed low-dose irradiation of 0.5 kGy, combined with chlorine, as one alternative for sanitation of shredded carrots. Commercially prepared shredded carrots treated with irradiation and stored at 2°C had a mean microbial population of 1,300

CFU/g at the expiration date (9 days after irradiation), compared with 87,000 CFU/g for nonirradiated, chlorinated controls. Oxygen content of the head-space gas, ethanol content, or texture of the carrots were not significantly affected.

Active packaging, i.e., packaging with various gas absorbents and emitters, is an interesting packaging method for minimally processed fruits and vegetables.[48] Active packaging of these specified products is still in its infancy. However, it seems that it is possible to affect respiration activity, as well as microbial and plant hormone activity, by correct active packaging.

One possible "packaging" method for extending the postharvest storage life of lightly processed fruits and vegetables is the use of edible coatings, i.e., thin layers of material that can be eaten by the consumer as part of the whole food product. Antioxidants are added to edible coatings to protect against oxidative rancidity, degradation, and discoloration. At least theoretically, coatings have the potential to reduce moisture loss, restrict oxygen entrance, lower respiration, retard ethylene production, seal in flavor volatiles, and carry additives that retard discoloration and microbial growth.[65]

REFERENCES

1. Huxsoll CC, Bolin HR. Processing and distribution alternatives for minimally processed fruit and vegetables. *Food Technol.* 1989;43:124–128.
2. Wiley RC. *Minimally Processed Refrigerated Fruits and Vegetables.* New York: Chapman & Hall; 1994.
3. Ahvenainen R, Hurme E. Minimal processing of vegetables. In: Ahvenainen R, Mattila-Sandholm T, Ohlsson T, eds. *Minimal Processing of Foods.* Espoo, Finland: VTT Symposium 142, VTT, 1994;17–35.
4. Ahvenainen R. New approaches in improving the shelf life of minimally processed fruit and vegetables. *Trends Food Sci Technol.* 1996;7:179–187.
5. Laurila E, Kervinen R, Ahvenainen R. The inhibition of enzymatic browning in minimally processed vegetables and fruits. Review article. *Postharvest News Information.* 1998; 4:53N–66N.
6. Leistner L, Gorris LGM. Food preservation by hurdle technology. *Trends Food Sci Technol.* 1995;6:41–46.
7. Kabir H. Fresh-cut vegetables. In: Brody AL, ed. *Modified Atmosphere Food Packaging.* Herndon, VA: Institute of Packaging Professionals; 1994;155–160.
8. Mattila M, Ahvenainen R, Hurme E. Prevention of browning of pre-peeled potato. In: De Baerdemaeker J, McKenna B, Janssens M, et al, eds. COST 94 Post-harvest treatment of fruit and vegetables. *Proceedings of Workshop on Systems and Operations for Post-Harvest Quality.* Brussels, Belgium: Commission of the European Communities; 1995:225–234.
9. Laurila E, Hurme E, Ahvenainen R. Shelf-life of sliced raw potatoes of various cultivar varities—substitution of bisulphites. *J Food Prot.* 1998;61:1363–1371.
10. Ahvenainen R, Hurme EU, Hägg M, Skyttä E, et al. Shelf-life of pre-peeled potato cultivated, stored and processed by various methods. *J Food Prot.* 1998;61:591–600.
11. Pretel MT, Fernández PS, Romojaro F, Martínez A. The effect of modified atmosphere packaging on 'ready-to-eat' oranges. *Lebensm-Wiss u-Technol.* 1998;31:322–328.
12. O'Beirne D. Influence of raw material and processing on quality of minimally processed vegetables. *Progress Highlight* C/95 of EU Contract AIR1-CT92–0125: Improvement of the safety and quality of refrigerated ready-to-eat foods using novel mild preservation techniques. Brussels, Belgium: Commission of the European Communities; 1995.
13. Hurme E, Ahvenainen R, Kinnunen A, Skyttä E. Factors affecting the quality retention of minimally processed Chinese cabbage. COST 94: Post-harvest treatment of fruit and vegetables: Current status and future prospects. *Proceedings of the Sixth International Symposium of the European Concerted Action Program.* Oosterbeek, 19–22 October, 1994. Luxembourg: Office for Official Publications of the European Communities; 1998;85–90.
14. Ahvenainen R, Hurme E, Kinnunen A, Luoma T, et al. Factors affecting the quality retention of minimally processed carrot. COST 94 Post-harvest treatment of fruit and vegetables: Current status and future prospects. *Proceedings of the Sixth International Symposium of the European Concerted Action Program.* Oosterbek, 19–22 October 1994. Luxembourg: Office for Official Publications of the European Communities, 1998;41–47.
15. Ohta H, Sugawara W. Influence of processing and storage conditions on quality stability of shredded lettuce. *Nipp Shok Kog Gakkaishi.* 1987;34:432–438.
16. Garg N, Churey JJ, Splittstoesser DF. Effect of processing conditions on the microflora of fresh-cut vegetables. *J Food Protect.* 1990;53:701–703.
17. Beuchat LR, Brackett RE. Survival and growth of *Listeria monocytogenes* on lettuce as influenced by shred-

ding, chlorine treatment, modified atmosphere packaging and temperature. *J Food Sci.* 1990;55:755–758, 870.

18. Skyttä E, Koskenkorva A, Ahvenainen R, Heiniö RL, et al. Growth risk of *Listeria monocytogenes* in minimally processed vegetables. *Proceedings of Food 2000 Conference on Integrating Processing, Packaging, and Consumer Research.* Natick, MA, 19–21 October 1993. Hampton, VA: Science and Technology Corporation; 1996:785–790.
19. Francis GA, O'Beirne D. Effects of gas atmosphere, antimicrobial dip and temperature on the fate of *Listeria innocua* and *Listeria monocytogenes* on minimally processed lettuce. *Int J Food Sci Technol.* 1997;32:141–151.
20. Sapers GM, Simmons GF. Hydrogen peroxide disinfection of minimally processed fruit and vegetables. *Food Technol.* 1998;52(2):48–52.
21. Zomorodi B. The technology of processed/prepacked produce preparing the product for modified atmosphere packaging (MAP). *Proceedings of the 5th International Conference on Controlled/Modified Atmosphere/Vacuum Packaging.* CAP '90, San Jose, CA; 17–19 January 1990. Princeton: Schotland Business Research; 1990:301–330.
22. Bolin HR, Huxsoll CC. Effect of preparation procedures and storage parameters on quality retention of salad-cut lettuce. *J Food Sci.* 1991;56:60–67.
23. Anonymous. Sulphites banned. *Food Ingredients Process.* 1991;11:11.
24. Whitaker JR, Lee CY. Recent advances in chemistry of enzymatic browning. In: Lee CY, Whitaker JR, eds. *Enzymatic Browning and Its Prevention.* Washington, DC: ACS Symposium Series 600; 1995:2–7.
25. Vámos-Vigyázó L. Polyphenol oxidase and peroxidase in fruit and vegetables. *Crit Rev Food Sci Nutr.* 1981;15:49–127.
26. Lozano-de-González PG, Barrett DM, Wrolstad RE, Durst RW. Enzymatic browning inhibited in fresh and dried apple rings by pineapple juice. *J Food Sci.* 1993;58:399–404.
27. Meza J, Lozano-de-González P, Anzaldúa-Morales A, Torres JV, et al. Addition of pineapple juice for the prevention of discoloration and textural changes of apple slices. *IFT Annual Meeting 1995 Book of Abstracts.* 1995;68.
28. Luo Y, Barbosa-Cánovas GV. Inhibition of apple-slice browning by 4-hexylresorcinol. In: Lee CY, Whitaker JR, eds. *Enzymatic Browning and Its Prevention*, Washington, DC: American Chemical Society; 1995:240–250.
29. Sapers GM, Hicks KB, Phillips JG, Garzarella L, et al. Control of enzymatic browning in apples with ascorbic acid derivatives, polyphenol oxidase inhibitors, and complexing agents. *J Food Sci.* 1989;54:997–1002,1012.
30. Sapers GM, Miller RL. Enzymatic browning control in potato with ascorbic acid-2-phosphates. *J Food Sci.* 1992;57:1132–1135.
31. Sapers GM, Miller RL. Control of enzymatic browning in pre-peeled potatoes by surface digestion. *J Food Sci.* 1993;58:1076–1078.
32. Monsalve-Gonzalez A, Barbosa-Cánovas GV, Cavalieri RP, McEvily AJ et al. Control of browning during storage of apple slices preserved by combined methods. 4-Hexylresorcinol as anti-browning agent. *J Food Sci.* 1993;58:797–800, 826.
33. Dennis JAB. *The Effects of Selected Antibrowning Agents, Selected Packaging Methods, and Storage Times on Some Characteristics of Sliced Raw Potatoes.* Dissertation. Stillwater, OK: Oklahoma State University; 1993.
34. Santerre CR, Leach TF, Cash JN. Bisulfite alternatives in processing abrasion-peeled russet Burbank potatoes. *J Food Sci.* 1991;56:257–259.
35. Castañer M, Gil MI, Artes F, Tomas-Barberan FA. Inhibition of browning of harvested head lettuce. *J Food Sci.* 1996;61:314–316.
36. Sapers GM, Miller RL. Browning inhibition in fresh-cut pears. *J Food Sci.* 1998;63:342–346.
37. McEvily AJ, Iyengar R, Otwell WS. Inhibition of enzymatic browning in foods and beverages. *Crit Rev Food Sci Nutr.* 1992;32:253–273.
38. Cherry JH, Singh SS. *Discoloration Preventing Food Preservative and Method.* Patent No. 1990;4,937,085.
39. Gardner J, Manohar S, Borisenok WS. *Sulfite-Free Preservative for Fresh Peeled Fruit and Vegetables.* Patent No. 1991;4,988,523.
40. Dudley ED, Hotchkiss JH. Cysteine as an inhibitor of polyphenol oxidase. *J Food Biochem.* 1989;13:65–75.
41. Moler-Perl I, Friedman M. Inhibition of browning by sulfur amino acids. 3. Apples and potatoes. *J Agric Food Chem.* 1990;38:1652–1656.
42. Taoukis PS, Labuza TP, Lin SW, Lillemo JH. *Inhibition of Enzymatic Browning.* Patent WO 1989;89/11227.
43. Labuza TP, Lillemo JH, Taoukis PS. Inhibition of polyphenol oxidase by proteolytic enzymes. *Fruit Process.* 1992;2:9–13.
44. Luo Y. *Enhanced Control of Enzymatic Browning of Apple Slices by Papain.* Dissertation. Pullman, WA: Washington State University; 1992.
45. Hurme E, Kinnunen A, Heiniö RL, Ahvenainen R, et al. The sensory shelf-life of packed shredded iceberg lettuce dipped in glycine betaine solutions. *J Food Prot.* 1999, 4.363–367.
46. Kader AA, Zagory D, Kerbel EL. Modified atmosphere packaging of fruit and vegetables *Crit Rev Food Sci Nutr.* 1989;28:1–30.
47. Powrie WD, Skura BJ. Modified atmosphere pack-

aging of fruit and vegetables. In: Ooraikul B, Stiles ME, eds. *Modified Atmosphere Packaging of Food.* Chichester, West Succex, England: Ellis Horwood Lt; 1991:169–245.

48. Day BPF. Modified atmosphere packaging and active packaging of fruit and vegetables. In: Ahvenainen R, Mattila-Sandholm T, Ohlsson T, eds. *Minimal Processing of Foods*. Espoo, Finland: VTT Symposium 142, VTT, 1994;173–207.
49. Exama A, Arul J, Lencki RW, Lee LZ, et al. Suitability of plastic films for modified atmosphere packaging of fruit and vegetables. *J Food Sci.* 1993;58:1365–1370.
50. Anonymous. Bags extend salad shelf time. *Packaging Digest.* January 1996;66–70.
51. Anonymous. Temperature compensating films for produce. *Prepared Foods.* 1992;161:95.
52. Anonymous. "Membrane" controls veggie tray's MAP permeation. *Packaging Digest.* October 1998;3.
53. Anonymous. Permeable plastics film for respiring food produce. *Food, Cosmetics Drug Pack.* 1994;17:7.
54. Day BPF. High oxygen modified atmosphere packaging: A novel approach for fresh prepared produce packaging. In: Blakistone B, ed. *Packaging Yearbook 1996.* Washington, DC: NFPA National Food Processors Association, 1997;55–65.
55. Gorris LGM, de Witte Y, Bennik MJH. Refrigerated storage under moderate vacuum. *ZFL Focus Int.* 1994;45:63–66.
56. Lee KS, Park IS, Lee DS. Modified atmosphere packaging of a mixed prepared vegetable salad dish. *Int J Food Sci Technol.* 1996;31:7–13.
57. Betts G, ed. *Code of Practice for the Manufacture of Vacuum and Modified Atmosphere Packaged Chilled Foods with Particular Regard to the Risks of Botulism.* Guideline No. 11. Chipping Campden, England: Campden & Chorleywood Food Research Association; 1996:114.
58. Varoquaux P, Wiley R. Biological and biochemical changes in minimally processed refrigerated fruit and vegetables. In: Wiley RC, ed. *Minimally Processed Refrigerated Fruits and Vegetables.* New York: Chapman & Hall, 1994;226–268.
59. Martinez MV, Whitaker JR. The biochemistry and control of enzymatic browning. *Trends Food Sci Technol.* 1995;6:195–200.
60. Breidt F, Fleming HP. Using lactic acid bacteria to improve the safety of minimally processed fruit and vegetables. *Food Technol.* 1997;51:44–51.
61. Bennik MHJ. *Biopreservation in Modified Atmosphere Packaged Vegetables*. Thesis. Wageningen, The Netherlands: Agricultural University; 1997:96.
62. Torriani S, Orsi C, Vescovo M. Potential of *Lactobacillus casei*, culture permeate, and lactic acid to control microorganisms in ready-to-use vegetables. *J Food Prot.* 1997;60:1564–1567.
63. Hoover DG. Minimally processed fruit and vegetables: Reducing microbial load by nonthermal physical treatments. *Food Technol.* 1997;51:66–71.
64. Hagenmaier RD, Baker RA. Microbial population of shredded carrot in modified atmosphere packaging as related to irradiation treatment. *J Food Sci.* 1998;63:162–164.
65. Baldwin EA, Nisperos-Carriedo MO, Baker RA. Use of edible coatings to preserve quality of lightly (and slightly) processed products. *Crit Rev Food Sci Nutr.* 1995;35:509–524.
66. Ahvenainen R, Hurme E. Practical guidelines for minimal processing of vegetables. *Proceedings of International Symposium on Minimal Processing and Ready Made Foods*, SIK. Göteborg, Sweden: April 18–19;1996:31–44.

Part IV

Developments of Minimal Processing Technologies for Fruit Preservation in the Frame of Two Multinational Projects

CHAPTER 17

Minimally Processed Fruits Using Vacuum Impregnation, Natural Antimicrobial Addition, and/or High Pressure Techniques

Stella M. Alzamora, Pedro Fito, Aurelio López-Malo, María S. Tapia, and Efrén Parada Arias

INTRODUCTION

In general, any process that inhibits growth and/or inactivates microorganisms and prevents postcontamination can preserve foods. Traditional inactivating preservation processes, such as sterilization and pasteurization, inactivate microorganisms while altering nutrients and sensory attributes. Alternative approaches are under constant investigation because consumers increasingly demand foods with freshlike quality and natural ingredients. New processing methods are, therefore, being investigated and developed to meet these demands for maximum quality retention, freshlike characteristics, and added safety. This chapter includes a number of selected technologies for obtaining minimally processed fruits. These technologies are the result of two multinational projects, described next.

Project XI.3 of CYTED Program

CYTED is the Spanish acronym for "Science and Technology for Development," an Ibero-American Scientific and Technological Cooperation Program created in 1984 in Spain, when a general agreement between 19 Latin-American countries, Portugal, and Spain was signed. CYTED promotes among the 21 participant countries scientific and technological cooperation, applied research, technological development, and innovation to modernize the production and the life standards of the region. CYTED's subprograms include several areas of knowledge, and food treatment and preservation is one of them. Within this subprogram, CYTED has promoted and carried out nearly 15 precompetitive research projects. The first multinational project was called *Intermediate Moisture Foods Important to Ibero-America* and ran from 1986 to 1991. This project promoted research in the areas of intermediate moisture foods and combined methods technology.[1] The promising results of this first project gave rise to the need for research in combined methods or hurdle technology to develop high-moisture fruit preservation techniques. The second CYTED project was called *Bulk Fruit Preservation by the Combined Methods Technology*[2] and was carried out from 1991 to 1994. The main objective was to develop simple and inexpensive preservation processes for obtaining semielaborated or final products as a contribution to diversify the Ibero-American fruit industry and to reduce postharvest losses.

From 1995 to 1998, a third multinational collaborative project within the food preservation subprogram, *Development of Minimal Processing Technologies for Food Preservation*, was carried out. The main objectives of this concerted action project were:

a. Contribute basic knowledge about:

a.1. the effect of the different preservation factors and their interaction on microorganisms and sensory, physical, and chemical properties of selected food products, as well as on the ultrastructural alterations of their tissues.

a.2. the transport phenomena involved in the different stages of preservation procedures.

b. Develop minimal processing technologies to obtain:

b.1. high-moisture food products with physicochemical properties and sensory attributes similar to their fresh counterparts, stable at ambient temperature or under refrigeration, according to the proposed storage, distribution, and retail conditions.

b.2. traditionally preserved food products that maintain identity characteristics (flavor, color) of the "original" food or that exhibit optimal quality and sensory attributes.

Those technologies of minimal processing would use in the formulation of the conservation procedure the combination of different traditional and emerging preservation factors ("hurdle" approach).

Argentina, Chile, Mexico, Portugal, Spain, and Venezuela were the six countries involved in this project, which was coordinated by Stella M. Alzamora from the University of Buenos Aires of Argentina. Table 17–1 presents information about the participant research groups and their institutions.

The investigations performed dealt with a large variety of topics, such as:

- mass transport modeling in solid food-liquid operations (i.e., vacuum or atmospheric osmotic dehydration);
- physicochemical, biochemical, and structural characterization of minimally processed plant foods;
- response of pathogenic bacteria and spoilage microbial flora to emerging preservation factors and their interactions with other environmental stresses;
- ultrastructure-mechanical behavior relationships in fruit tissues, cheese, and egg gels, as affected by various processes;
- fruit and vegetable preservation using controlled/modified atmospheres;
- development and/or optimization of combined technologies for obtaining high-moisture minimally processed fruits using vacuum impregnation, edible films, natural antimicrobials and high hydrostatic pressures, and salting of fish and cheese by vacuum impregnation;
- application of high hydrostatic pressures to liquid whole egg, avocado, cheese, milk, recovery poultry meat, and yogurt, among others.

Project TS3-CT94–0333 of Life Sciences and Technologies for Developing Countries STD-3 Program of European Union

Development of Minimally Processed Products from Tropical Fruits, Using Vacuum Impregnation Techniques, coordinated by Dr. Pedro Fito of the Universidad Politécnica de Valencia, ran from January 1995 to December 1998 and involved institutions from Spain, Mexico, Venezuela, Argentina, and Portugal (Table 17–2). Most of the participants were also involved in the above-mentioned CYTED project. Most of the workshops, meetings, interactions between partners, and other activities, as well as dissemination of the scientific information that was generated, were organized and/or shared within the framework of both projects.

The main objective of the project was to develop minimally processed products from tropical fruits. The physical, chemical, and sensory characteristics of the final products would be as similar as possible to the raw material, and

Table 17–1 Summary of the Overall Participation in the Project "Development of Minimal Processing Technologies for Food Preservation"

Country	*Participant Institution*	*Institutional Coordinator*	*Number of Researchers*
Argentina	Universidad de Buenos Aires Facultad de Ciencias Exactas y Naturales Departamento de Industrias	Stella M. Alzamora– Lia N. Gerschenson	10
	Universidad Nacional de la Plata Centro de Investigación y Desarrollo en Criotecnología de Alimentos	Alicia Chavez	3
Chile	Universidad Católica de Valparaíso y Universidad de Santiago	Ismael Kasahara— Pedro Moyano	3
	Pontificia Universidad Católica de Chile Departamento de Ingeniería Química y Bioprocesos	José Manuel del Valle	2
	Universidad de Chile Facultad de Ciencias Agrarias Departamento de Agroindustrias y Tecnología de Alimentos	Marco Schwartz	1
Spain	Universidad Politécnica de Valencia Departamento de Tecnología de Alimentos	Pedro Fito- Amparo Chiralt	17
	Universidad de Córdoba Facultad de Veterinaria Tecnología de Alimentos	José Fernández Salguero	6
	Consejo Superior de Investigaciones Científicas. Instituto del Frío Departamento de Ciencia y Tecnología de Productos Vegetales	Pilar Cano	6
	Universidad Autónoma de Barcelona Facultad de Veterinaria Unidad de Tecnología de Alimentos	Buenaventura Guamis	8
Mexico	Instituto Politécnico Nacional Escuela Nacional de Ciencias Biológicas Departamento de Graduados e Investigación en Alimentos	Lidia Dorantes	3

continues

Table 17–1 continued

Country	*Participant Institution*	*Institutional Coordinator*	*Number of Researchers*
Mexico (cont'd)	Instituto Tecnológico de Tepic Departamento de Ingeniería Química y Bioquímica	Miguel Mata	4
	Universidad de las Américas-Puebla Departamento de Ingeniería Química y de Alimentos	Jorge Welti-Chanes	9
Portugal	Universidade Católica Portuguesa Escola Superior de Biotecnología	Alcina Morais	4
Venezuela	Universidad Central de Venezuela Instituto de Ciencia y Tecnología de Alimentos	María S. Tapia	10
	Universidad Simón Bolívar Departamento de Procesos Bioquímicos y Biológicos	Valentin Roa	10
	Instituto Universitario de Tecnología de Cumaná	Luis Elguezábal	6

they would be stable at room temperature or refrigeration, according to commercial practices in each country. Two types of products would be developed: high-moisture foods (HMF) similar to fresh foods, and intermediate-moisture foods (IMF), such as purées and jams, prepared with mild thermal treatments, showing optimum retention of fresh fruit color and flavor.

This general objective would be accomplished through the design and optimization of new processes that will render the products described above. The new processes would be based on vacuum impregnation operations (VI); vacuum osmotic dehydration (VOD), and pulsed vacuum osmotic dehydration (PVOD).[3,4]

The project comprised the following tasks:

a. determination of kinetics parameters during vacuum impregnation and osmotic dehydration processes,
b. basic microbiologic and physicochemical studies,
c. new products and process design, and
d. microbiologic, physicochemical, and sensory stability during processing and storage.

A major part of both projects focused on the optimization of combined preservation technologies for obtaining high-moisture fruit products (HMFP). Most of these technologies were developed as part of the previous multinational project of the CYTED Program: "Bulk preservation of fruits by combined methods technology."[2,5] Hurdles used for the stabilization of HMFP included reduction of water activity (a_w) and pH, mild heat treatment, and addition of preservatives. The preservation process combining these factors was very simple and consisted of fruit blanching, followed by an a_w depression

Table 17–2 Summary of the Overall Participation in the Project "Development of Minimally Processed Products from Tropical Fruits Using Vacuum Impregnation Techniques"

Country	*Participant Institution*	*Institutional Coordinator*	*Number of Researchers*
Spain	Universidad Politécnica de Valencia Departamento de Tecnología de Alimentos	Pedro Fito	17
Argentina	Universidad de Buenos Aires Facultad de Ciencias Exactas y Naturales Departamento de Industrias	Stella M. Alzamora	10
Mexico	Universidad de las Américas-Puebla Departamento de Ingeniería Química y de Alimentos	Jorge Welti-Chanes	9
Portugal	Universidade Católica Portuguesa Escola Superior de Biotecnología	Fernanda Oliveira	5
Venezuela	Universidad Central de Venezuela Instituto de Ciencia y Tecnología de Alimentos	María S. Tapia	10

step through osmotic dehydration with simultaneous incorporation of additives, achieving final values after equilibration of a_w 0.94–0.98, pH 3.0–4.1 (adjusted with citric or phosphoric acid), 400–1,000 ppm potassium sorbate or sodium benzoate, and generally 150 ppm sodium bisulfite (Flowchart 1 in Appendix 17–A).

The processing of some fruits (i.e., banana purée and pomalaca) included a slight thermal treatment after packing or a hot filling stage. The osmotic dehydration treatment was made at room temperature by placing fruit slices in concentrated sugar aqueous solutions (glucose, sucrose, maltodextrins, corn syrups, or their mixtures and additives).

In the cases of purées and some whole fruits (i.e., strawberries), the fruit, humectant(s), and additives were mixed in required proportions (blending).

In the optimization of these combined technologies, three aspects received consideration:

1. the equilibration stage, where vacuum techniques were used for the osmotic dehydration process and/or incorporation of additives,
2. the use of antimicrobials of natural origin as replacement (total or partial) of sorbates and other synthetic additives, and
3. the use of high pressure to process delicate fruit products and/or reduce or substitute synthetic additives.

This chapter includes a number of selected technologies for obtaining minimally processed fruits resulting from both projects, many of them being developed in collaboration between researchers of different institutions. The reported combined technologies can be classified as:

a. technologies for obtaining minimally processed fruits using vacuum impregnation techniques,
b. technologies for obtaining minimally processed fruits using natural antimicrobials, and
c. technologies for obtaining minimally processed fruits using high pressure.

In each case, a schematic diagram shows a possible process flow sheet, along with the operative conditions in each processing step. Comments about shelf-life evaluation, deterioration mode, and expected shelf life are included for each minimally processed fruit.

TECHNOLOGIES USING VACUUM IMPREGNATION TECHNIQUES

Food porous microstructure is an important property often neglected by food processors that usually are not aware that it can be useful in some food preservation operations and product development.[3] Foods exhibiting a porous microstructure can be impregnated, that is, their pores can be filled with a suitable solution, introducing solvents and solutes of choice into their porous spaces that are occupied by a certain amount of occluded gas.[6–9] The volume of this gas can be modified, substituting it with the impregnation solution as a result of capillary action or by the combined effect of capillary action and pressure gradients that are imposed on the system.[6] The impregnation produced by pressure gradients acting as driving forces can be controlled by the expansion or compression of the occluded gas.[6] A way to accomplish this is to apply vacuum to the product during short periods (i.e., 5–10 minutes) while immersed in the liquid, then to reestablish the atmospheric pressure.[4] These alternating pressures cause the gas to be expelled from the pores and be replaced by the entering liquid.[10]

A proper formulation of the impregnation solution allows expeditious compositional modifications of the solid matrix that may result in quality and stability enhancement of final products[11] without submitting the food structure to the eventual stress due to long exposure to gradient solute concentration.[9] During vacuum impregnation of porous fruits, important modifications in structure and composition occur as a consequence of external pressure changes. The final products may exhibit structural, physical, and chemical properties very different from those of atmospheric infused fruits.[9]

Fruits, in general, are good examples of foods having a microporous structure, which can be vacuum impregnated.[11] Vacuum impregnation shows faster water loss kinetics in short-time treatments, as compared with time-consuming atmospheric "pseudo-diffusional" processes, due to the occurrence of a specific mass transfer phenomenon, the hydrodynamic mechanism (HDM), and the result produced in the solid–liquid interface area.[12,13] This process could be appropriate in the development of new fruit products as minimally processed, as in our case, or in the development of improved pretreatments for such traditional preservation methods as canning, freezing, or drying[10,14] and also in high-quality jam process.[15]

High-Moisture Papaya

Flowchart 2 in Appendix 17–A presents the diagram to preserve shelf-stable high-moisture papaya using vacuum impregnation as a technique to incorporate the selected preservation factors in the formulation of this combined method. Tapia et al.[16] and López-Malo,[17] in their reports from which data have been taken, proposed a preservation process to achieve microbial and sensorial stability of papaya stored at 15°C for at least one month. The preservation process was based on the combined methods technology and includes the following factors or hurdles: a_w and pH reduction, the addition of potassium sorbate, and 15°C storage temperature. Pulsed vacuum osmotic dehydration was used to impregnate the fruit with sucrose, citric acid, and potassium sorbate. The obtained fruit stored at 15°C for one month was periodically evaluated for color and texture, aerobic plate, yeast and mold counts, and sensory quality changes. Moisture and soluble solids contents, pH, and a_w remained almost constant, and the product obtained was microbiologically sound during storage. Color, texture, and sensory evaluation revealed that there were not significant changes during storage. The papaya pieces showed a good overall acceptability (mean score

7.3) even after one month of storage at 15°C (mean score 6.7).

High-Moisture Melon

Flowchart 3 in Appendix 17–A presents a schematic diagram to prepare shelf-stable melon. The proposed preservation method was taken from data reported by Santacruz,[18] Vergara-Balderas et al.,[19] and Tapia et al.[20] Melon cylinders were submitted to pulsed vacuum osmotic dehydration in a 40°Brix sucrose syrup containing 0.6% w/w phosphoric acid, 1,000 ppm potassium sorbate, and 0.2% w/w calcium lactate to depress their a_w (0.98) and pH (4.3) and to incorporate potassium sorbate and calcium lactate. The product was placed either in glass jars covered with sucrose syrup (a_w 0.98, pH 4.3) or in polyethylene bags without syrup. Products were stored at 15 or 25°C for 45 days. Microbiologic analyses and a_w, pH, color (*L*, *a*, and *b*), and texture were periodically evaluated. Melon packed in plastic bags showed mold and yeast growth at 15 and 25°C presenting signs of fermentation after 10 days of storage. Product packed in jars with cover syrup did not show microbial growth. *L*, *a*, and *b* changes affected melon color acceptance after 30 days at 15 or 20 days at 25°C. Melon texture degradation was more noticeable when stored at 25°C. Melon packed in glass jars with cover syrup stored at 15 or 25°C was well accepted, according to the sensory studies during storage.

High-Moisture Orange Segments

A schematic diagram to preserve shelf-stable orange segments is presented in Appendix 17–A as Flowchart 4. Santacruz[21] and Welti-Chanes et al.[22] evaluated the effect of storage temperature and type of package on the stability of minimally processed orange segments. The peeled segments were vacuum impregnated in a 55°Brix sucrose syrup to depress a_w to 0.98, to maintain the pH in 3.6, and to incorporate potassium sorbate. The product was placed either in glass jars covered with sucrose syrup (a_w 0.98 and pH 3.6) or in polyethylene bags without syrup. The packaged products were stored at 5, 15, and 25°C for periods of up to 50 days. Microbiologic analysis, a_w, pH, color, and texture of the product were periodically determined. The orange segments packaged in polyethylene bags presented an important mold and yeast growth, provoking product fermentation after 3–15 days, depending on the storage temperature. The product packaged in glass jars did not present fermentation at any studied temperature, and after 50 days, the aerobic plate, yeast, and mold counts were lower than 100 CFU/g. Orange segments with a depressed a_w obtained by vacuum impregnation in a sucrose syrup were microbiologically stable and well accepted at up to 50 days if they were packaged in glass jars with cover syrup and stored at temperatures ≤ 25°C.

Minimally Processed Kiwifruit

Flowchart 5 in Appendix 17–A presents the proposed procedure for minimally processed kiwifruit, as reported by Leúnda et al.[23] In this study, a combined factor preservation technology involving blanching and vacuum solute (sucrose, potassium sorbate, ascorbic and citric acid, zinc chloride) impregnation was proposed to minimize color changes in minimally processed kiwifruit slices during one-month storage. Atmospheric impregnation was also studied to compare both impregnation techniques. A Box-Behnken design was adopted, and second-order polynomial models were computed for different storage times to relate some process variables (blanching time, zinc content, storage temperature) to a color function (Brown Index). As the storage time increased, the color response surface for vacuum-treated fruits was displaced to greater Brown Index values, whereas the response surface behavior for atmospheric impregnated fruits was less dependent on storage time. For vacuum-impregnated fruits, combinations of blanching and addition of zinc chloride improved the color of the finished product at all storage temperatures assayed, but these treatments were detrimental for atmospheric impregnated ones, increasing significantly the Brown Index values. The product was microbiologically sound, and the color changes determined its shelf life. After storage, total chlorophyll had been degraded between 70% and 90%, depending on the pretreatments. There did not appear to be any consistent relationship between the changes that occurred in the total chlorophyll content and color changes.

Minimally Processed Papaya Impregnated with Passion Fruit Juice

Passion fruit (*Passiflora edulis*) was selected as a model tropical fruit that seemed most apt and easy to be preserved by combined methods, due to its attractive exotic nature, wide acceptance, and very low natural pH. Thus, shelf-stable passion fruit ("parchita," as it is popularly known in Venezuela) proved to be microbiologically stable, with the following processing parameters: blanching (vapor), a_w 0.98, pH 2.7–3.0 (natural), and 100 ppm sodium bisulfite. The most important result of this investigation[24] was the following: No antimicrobial, such as potassium sorbate, was needed to obtain a microbiologically stable passion fruit preserved by combined methods. Thus, it appeared to be an interesting potential infusion system for vacuum impregnation techniques of tropical fruits of flavor comparable to that of passion fruit, as described next.

Flowchart 6 in Appendix 17–A presents the proposed methodology to preserve papaya pieces using passion fruit juice as impregnation solution.[25] Papaya cylinders impregnated in passion fruit proved to be an excellent example of a novel, minimally processed, refrigerated fruit product that exhibited attractive appearance and exotic flavor, because the high acidity of passion fruit not only improved stability by decreasing papaya's pH, but also improved flavor. Impregnation added a translucent appealing appearance to papaya pieces. They were packed in styrofoam trays and wrapped in commercial flexible autoadherent transparent film. Determinations of color, pH, a_w, ascorbic acid, and microbial counts were performed during refrigerated storage for 15 days and investigated for psychrotrophic and mesophilic aerobics, lactic acid bacteria, and yeast and molds. Results indicated that products kept well up to 8 days of storage at 8°C, when lactic acid bacteria proliferated and spoilage was evident.

TECHNOLOGIES USING NATURAL ANTIMICROBIALS

Governmental restrictions and consumers' demands for more natural foods have focused attention on naturally occurring antimicrobial systems derived from animals, plants, or microorganisms. Also, the use of antimicrobial agents to provide the desired safety in refrigerated foods is growing in importance, due to the potential microbial growth during temperature abuse situations. A growing number of such natural systems are being explored as potential substitutes (partial or total) for common synthetic preservatives to stop food poisoning.[26–28] Despite the increasing interest in the use of natural antimicrobials, they have not yet been exploited in a practical way of commercial interest.[26] The antimicrobial activities of extracts from several types of plant and plant parts used as flavoring agents in foods have been recognized for many years.[29] Among these, vanillin (4-hydroxy-3-methylbenzaldehyde), a major constituent of vanilla beans, was selected because of its several major advantages over other current spices. It is a generally recognized as safe (GRAS), effective flavoring crystalline agent widely used in ice creams, drinks, confectionery, cookies, etc., so its sensory characteristics are well accepted. Furthermore, it has been found to be compatible with the sensory characteristics of various fruits (apple, plum, pear, mango, papaya, pineapple, strawberry, banana) in concentrations up to 3,000 ppm.[30] Because it occurs as crystals, it cannot serve as substrate for microbial growth and toxin production, nor does its antimicrobial activity depend on their source and extraction process. The potential use of natural preservatives as one of the hurdles for fruit preservation (instead of potassium sorbate and sulfites) needs to be further investigated; two application examples are presented next.

High-Moisture Strawberry

Flowchart 7 in Appendix 17–A presents the schematic diagram proposed by Cerrutti et al.[31] to preserve high-moisture strawberry, using vanillin as antimicrobial. The combination of a mild heat treatment (blanching), addition of 3,000 ppm vanillin and 500 ppm ascorbic acid, and adjustment of a_w to 0.95 and pH to 3.0 prevented growth of both native (aerobic and anaerobic mesophilic bacteria, yeast, and molds) and inoculated flora (*Saccharomyces cerevisiae*, *Zygosaccharomyces rouxii*, *Z. bailii*, *Schizosaccharomyces pombe*, *Pichia membranaefaciens*, *Botrytis* species, *Byssochlamys fulva*, *Bacillus coagulans,* and *Lactobacillus delbrueckii*) for at least 60 days of storage at room temperature. Moreover, vanillin appeared to be very effective against *Z. bailii* and diminished greatly the rate of growth of *P. membranaefaciens*, both well-known preservative-resistant yeasts. For preventing the growth of *Z. rouxii*, the most common spoilage organism of osmotolerant yeasts, a combination of 100 ppm SO_2, 500 ppm sorbate, and pH $\leq$ 4.0 has been recommended in high-moisture fruit products because of its high sensitivity to SO_2. Vanillin may be a natural alternative for eliminating sulfite addition to control this yeast.

Color was better preserved at lower temperature. Thus, a maximum storage temperature of 10°C is recommended.

Minimally Processed Banana Purée

Flowchart 8 in Appendix 17–A presents the flow sheet to prepare minimally processed banana purée using vanillin as an antimicrobial agent. Castañón et al.,[32] in their report from which data have been taken, evaluated the microbial and color changes during storage at different temperatures (15, 25, and 35°C) of banana purée preserved by the combination of blanching, reduced pH (3.4) and a_w (0.97), and the addition of 3,000 ppm vanillin. The use of 3,000 ppm vanillin inhibited for at least 60 days of storage the native flora of the banana purée at any of the studied temperatures. The addition of 1,000 ppm vanillin was not enough to stop the microbial growth; however, at 15°C, the growth was delayed for 16 days. In the presence of 3,000 ppm vanillin, the banana purée stored at 15, 25, and 35°C presented several color changes that contributed to browning. Sensory evaluations performed with banana purées containing vanillin and with several browning degrees demonstrated that an acceptable color could be assured for 8 weeks when stored at 15°C. The results of sensory evaluation with banana purée containing 3,000 ppm vanillin show mean scores that correspond to products with a good overall acceptability, with scores around 6. The flavoring characteristics of vanillin are well accepted and have demonstrated compatibility with many fruits in concentrations up to 3,000 ppm, as mentioned before.

TECHNOLOGIES USING HIGH HYDROSTATIC PRESSURES

Two problems have been identified in high-moisture fruit products: the use of sulfites as antibrowning and antimicrobial agents and the possibility of growth of preservative-resistant-yeasts and heat-resistant molds. Thus, additional inactivation or stress factors should be used to control enzyme and microbial activity. From the emerging technologies, high hydrostatic pressure could be a good choice to circumvent the problems created by browning and specific microorganisms; however, the effect of this process needs to be investigated. High pressure treatments can be considered as a new hurdle that can be used in combination with other traditional microbial stress factors, such as pH, a_w, and preservatives.[33–36] The results of our research, focused on the use of high pressure as an additional hurdle for the manufacture of high-moisture fruit products, are described next.

Minimally Processed Avocado Purée

López-Malo et al.[37] evaluated the effects of high pressure treatments at 345, 517, or 689 MPa for 10, 20, or 30 minutes at initial pHs of 3.9, 4.1, or 4.3 on polyphenoloxidase (PPO) activity, color, and microbial inactivation in avocado purée during storage at 5, 15, or 25°C. The results were compared with untreated avocado purée. Standard plate, as well as yeast and mold counts of high-pressure-treated purées, were < 10 CFU/g during 100 days of storage at 5, 15, or 25°C. Significantly less ($p \leq 0.05$) residual PPO activity was obtained with increasing pressure and decreasing initial pH. Browning was related mainly with changes in the *a* color component. Flowchart 9 in Appendix 17–A presents as an example a proposed flow diagram to minimally process avocado purée using high pressure. Avocado purée with a residual PPO activity < 45% and stored at 5°C maintains an acceptable color for at least 60 days, and the product shelf life was assured for 35 days when stored at 15°C.

Minimally Processed Guacamole

The effects of continuous or oscillatory high-pressure treatments on PPO and lipoxygenase (LOX) activities; standard plate, yeast, and mold counts; sensory acceptability; and instrumental color in guacamole were evaluated by Palou et al.[33] Flowchart 10 in Appendix 17–A presents a schematic diagram summarizing the successful results to preserve minimally processed guacamole (a Mexican-style sauce) obtained by Palou et al.[33] Significantly less ($p < 0.05$) residual PPO and LOX activities were obtained by increasing process time and number of pressurization–decompression cycles. LOX was inactivated with a 15-minute continuous treatment or oscillatory high pressure. The lowest residual PPO activity value (15%) was obtained after four high-pressure cycles at 689 MPa with 5 minutes of holding time each. Standard plate, as well as yeast and mold counts, of high-pressure-treated guacamole were < 10 CFU/g. Sensory acceptability and color of high-pressure guacamole were not significantly different ($p > 0.05$) from that of guacamole control. Browning during storage was related mainly to changes in the hue attributed to a decrease in the green contribution to the color. An acceptable shelf life, based on color changes, can be obtained with the proposed methodology (20 days of storage at temperature < 15°C).

Minimally Processed Banana Purée

Enzymatic reactions are a key problem area to address in high-pressure processing of fruits. Studies indicate blanching to be a near prerequisite for pressure treatment of fruits and vegetables. The effects of blanching pretreatments and high pressure on PPO activity, color, and natural flora evolution of banana purée adjusted to pH 3.4 and a_w 0.97 were evaluated during storage at 25°C by Palou et al.[34] The proposed flowchart is presented in Flowchart 11 of Appendix 17–A. The edible portion of ripe bananas was sliced, blanched in saturated vapor for 0, 1, 3, 5, or 7 minutes, cooled, and homogenized with sucrose, ascorbic, and phosphoric acids. Each purée was poured into plastic bags, subjected to high-pressure treatments (517 or 689 MPa for 10 minutes) at 21°C, then stored at 25°C. PPO activity, color, and microbiologic analyses were periodically determined for two weeks. Standard plate, as well as yeast and mold counts, of high-pressure-treated purées were < 10 CFU/g during storage. PPO activity increased after high pressure in the purée prepared without blanching. Blanching time significantly ($p <$ 0.05) affected the color of the purées. PPO activity was reduced during blanching and further reduced after high-pressure treatments. Purée color was not significantly different ($p < 0.05$) between high-pressure treatments. Greater browning induction times and smaller browning rates were observed when a longer blanching time was combined with a 689-MPa pressure treatment. Reduced PPO activity in banana purée can be obtained when the effects of blanching and high-pressure treatments are combined. Thus, banana purée color changes during storage can be diminished, allowing it to maintain for a longer time an acceptable color during storage at 25°C.

CONCLUSION

Minimally processed foods are becoming an important component of the food supply. New food products and processes, including innovative methods and new packaging and distribution systems, are continuously devised. The effect of these minimal processing procedures on the growth and survival of both food-poisoning and food-spoilage microorganisms must be evaluated and, if risks are significant, further barriers must be incorporated into the preservation system design.[38] It is now well recognized that less processing requires more hurdles or barriers. Minimal processing technologies need to be established from this view to improve food safety and quality. The presented preservation technologies to minimally process fruits were designed to keep and retain freshlike characteristics, preserve sensory attributes, and assure safety and reasonable shelf life by the combination of several hurdles.

REFERENCES

1. Parada-Arias E. Food technology in CYTED, an Ibero-American R&D cooperative program. In: Barbosa-Cánovas GV, Welti-Chanes J, eds. *Food Preservation by Moisture Control. Fundamentals and Applications.* ISOPOW Practicum II. Lancaster, PA: Technomic Publishing Co.; 1995:449–462.
2. Welti-Chanes J, Vergara-Balderas F. Fruit preservation by combined methods: An Ibero-American research project. In: Barbosa-Cánovas GV, Welti-Chanes J, eds: *Food Preservation by Moisture Control. Fundamentals and Applications.* ISOPOW Practicum II. Lancaster, PA: Technomic Publishing Co.; 1995:449–462.
3. Fito P. Modeling of vacuum osmotic dehydration of food. *J Food Engr.* 1994;22:313–328.
4. Chiralt A, Fito P, Andrés A, Barat JM, et al. Vacuum impregnation: A tool in minimally processing of foods. In: Oliveira FAR, Oliveira JC, eds. *Processing of Foods: Quality Optimization and Process Assessment.* Boca Ratón, FL: CRC Press; 1999:341–356.
5. Alzamora SM, Cerrutti P, Guerrero S, López-Malo A. Minimally processed fruits by combined methods. In Barbosa-Cánovas GV, Welti-Chanes J, eds. *Food Preservation by Moisture Control. Fundamentals and Applications.* ISOPOW Practicum II. Lancaster, PA: Technomic Publishing Co.; 1995:463–492.
6. Fito P, Pastor R. On some nondiffusional mechanisms occurring during vacuum osmotic dehydration. *J Food Engr.* 1994;21:513–519.
7. Fito P, Andrés A, Chiralt A, Pardo P. Coupling of hydrodynamic mechanism and deformation-relaxation phenomena during vacuum treatments in solid porous food-liquid systems. *J Food Engr.* 1996;27:229–240.
8. Sousa R, Salvatori D, Andrés A, Fito P. Analysis of vacuum impregnation of banana (*Musa acuminata* cv. Giant Cavendish). *Food Sci Technol Int.* 1998;4:127–131.
9. Alzamora SM, Tapia MS, Leúnda A et al. Relevant results on minimal preservation of fruits in the context of the multinational project XI.3 of CYTED, an Ibero-American R&D cooperative program. In Lozano J, Añon C, Parada-Arias E, Barbosa-Cánovas GV, eds. *Current Trends in Food Engineering.* New York: Chapman & Hall; 1999: unpublished.
10. Fito P, Chiralt A, Serra J, Mata M, et al. *An Alternating Flow Procedure to Improve Liquid Exchanges in Food Products and Equipment for Carrying Out Said Procedure.* European Patent. 1994; 0 625 314 A2.
11. Salvatori D, Andrés A, Chiralt A, Fito P. The response of some properties of fruits to vacuum impregnation. *J Food Proc Engr.* 1998;21:59–73.
12. Fito P, Chiralt A. An update on vacuum osmotic dehydration. In Barbosa-Cánovas GV, Welti-Chanes J, eds. *Food Preservation by Moisture Control: Fundamentals and Applications.* ISOPOW Practicum II. Lancaster, PA: Technomics Publishing Co.; 1995:351–374.
13. Fito P, Chiralt A. Osmotic dehydration: An approach to the modeling of solid-liquid food operations. In: Fito P, Ortega-Rodríguez E, Barbosa-Cánovas GV, eds. *Food Engineering 2000.* New York: Chapman & Hall; 1997:231–252.
14. Martínez-Monzó J, Martínez-Navarrete N, Fito P, Chiralt A. Mechanical and structural changes in apple (var. Granny Smith) due to vacuum impregnation with cryoprotectants. *J Food Sci.* 1998;63:499–503.
15. Shi XQ, Chiralt A, Fito P, Serra J, Escoín C, Gasque L. Application of osmotic dehydration technology on jam processing. *Drying Technol.* 1996;14:841–847.

16. Tapia MS, López-Malo A, Consuegra R, Corte P, et al. Minimally processed papaya by vacuum osmotic dehydration (VOD) techniques. *Food Sci Technol Int.* 1999;5:43–52.

17. López-Malo A. *Personal communication*; 1999.

18. Santacruz V. *Diseño de un alimento mínimamente procesado a partir de melón.* Universidad de las Américas-Puebla, México. M.S. Thesis; 1998.

19. Vergara-Balderas F, Santacruz V, López-Malo A, Tapia MS, et al. *Stability of Minimally Processed Melon Obtained by Vacuum Dehydration (VOD) Techniques.* No. 20A-1 Presented at 1998 IFT Annual Meeting. Atlanta, GA: June 20–24, 1998.

20. Tapia MS, Ramírez MR, Castañón X, López-Malo A, et al. *Stability of Minimally Treated Melon (*Cucumis melon*, L.) During Storage and Effect of the Water Activity Depression Treatment.* No. 22D-13 Presented at 1999 IFT Annual Meeting. Chicago, IL: July 24–28, 1999.

21. Santacruz C. *Obtención de gajos de naranja mínimamente procesados.* Universidad de las Américas-Puebla. México. M.S. Thesis; 1998.

22. Welti-Chanes J, Santacruz C, López-Malo A, Wesche-Ebeling P. *Stability of Minimally Processed Orange Segments Obtained by Vacuum Dehydration Techniques.* No. 34B-8 Presented at 1998 IFT Annual Meeting. Atlanta, GA: June 20–24, 1998.

23. Leúnda MA, Guerrero SN, Alzamora SM. Color and chlorophyll content changes of minimally processed kiwifruit. *J Food Proc Pres.* 1999; accepted for publication.

24. Cardellichio, G. *Desarrollo de productos de parchita (*Passiflora edulis*) de alta humedad por métodos combinados.* Instituto de Ciencia y Tecnología de Alimentos, Universidad Central de Venezuela. B.S. Thesis; 1996.

25. Márquez Y. *Desarrollo de un producto de lechosa* (Carica papaya, L.) *utilizando jugo de parchita* (Passiflora edulis) *como solución osmótica y de impregnación.* Instituto de Ciencia y Tecnología de Alimentos, Universidad Central de Venezuela. B.S. Thesis; 1998.

26. Gould GW. Industry perspectives on the use of natural antimicrobials and inhibitors for food applications. *J Food Prot.* 1996;Suppl.:82–86.

27. Davidson PM. Chemical preservatives and natural antimicrobial compounds. In: Doyle MP, Beuchat LR, Montville TJ, eds. *Food Microbiology: Fundamentals and Frontiers.* Washington, DC: ASM Press; 1997:520–556.

28. Schillinger U, Geisen R, Holzapfel WH. Potential of antagonistic microorganisms and bacteriocins for the biological preservation of foods. *Trends Food Sci Technol.* 1996;7:158–164.

29. Wilkins KM, Board RG. Natural antimicrobial systems. In: Gould GW, ed. *Mechanisms of Action of Food Preservation Procedures.* New York: Elsevier Science; 1989:285–362.

30. Cerrutti P, Alzamora SM. Inhibitory effects of vanillin on some food spoilage yeasts in laboratory media and fruit purées. *Int J Food Microbiol.* 1996;29:379–386.

31. Cerrutti P, Alzamora SM, Vidales SL. Vanillin as antimicrobial for producing shelf-stable strawberry purée. *J Food Sci.* 1997;62:608–610.

32. Castañón X, Argaiz A, López-Malo A. Effect of storage temperature on the microbial and color stability of banana purée with addition of vanillin or potassium sorbate. *Food Sci Technol Int.* 1999;5:53–60.

33. Palou E, Hernández-Salgado C, López-Malo A, Barbosa-Cánovas GV, et al. High pressure treated guacamole. *J Innov Food Sci Emerg Technol.* 1999; in press.

34. Palou E, López-Malo A, Barbosa-Cánovas GV, Welti-Chanes J, et al. Polyphenoloxidase activity and color of blanched and high hydrostatic pressure treated banana purée. *J Food Sci.* 1999;64:42–45.

35. Palou E, López-Malo A, Barbosa-Canovas GV, Swanson BG. High pressure treatment in food preservation. In: Rahman MS, ed. *Handbook of Food Preservation.* New York: Marcel Dekker; 1999:533–576.

36. López-Malo A, Palou E. High hydrostatic pressure and fungal inactivation. In: Pandalai SG, ed. *Recent Research Developments in Agricultural and Food Chemistry.* India: Research Signpost;1999: 57–74.

37. López-Malo A, Palou E, Barbosa-Cánovas GV, Welti-Chanes J, et al. Polyphenoloxidase activity and color changes during storage of high hydrostatic pressure treated avocado purée. *Food Res Int.* 1999;31:549–556.

38. Alzamora SM, Tapia MS, Welti-Chanes J. New strategies for minimal processing of foods: The role of multitarget preservation. Food Sci Technol Int. 1998;4:353–361.

Appendix 17–A

Preservation Techniques for Minimally Processed Fruits

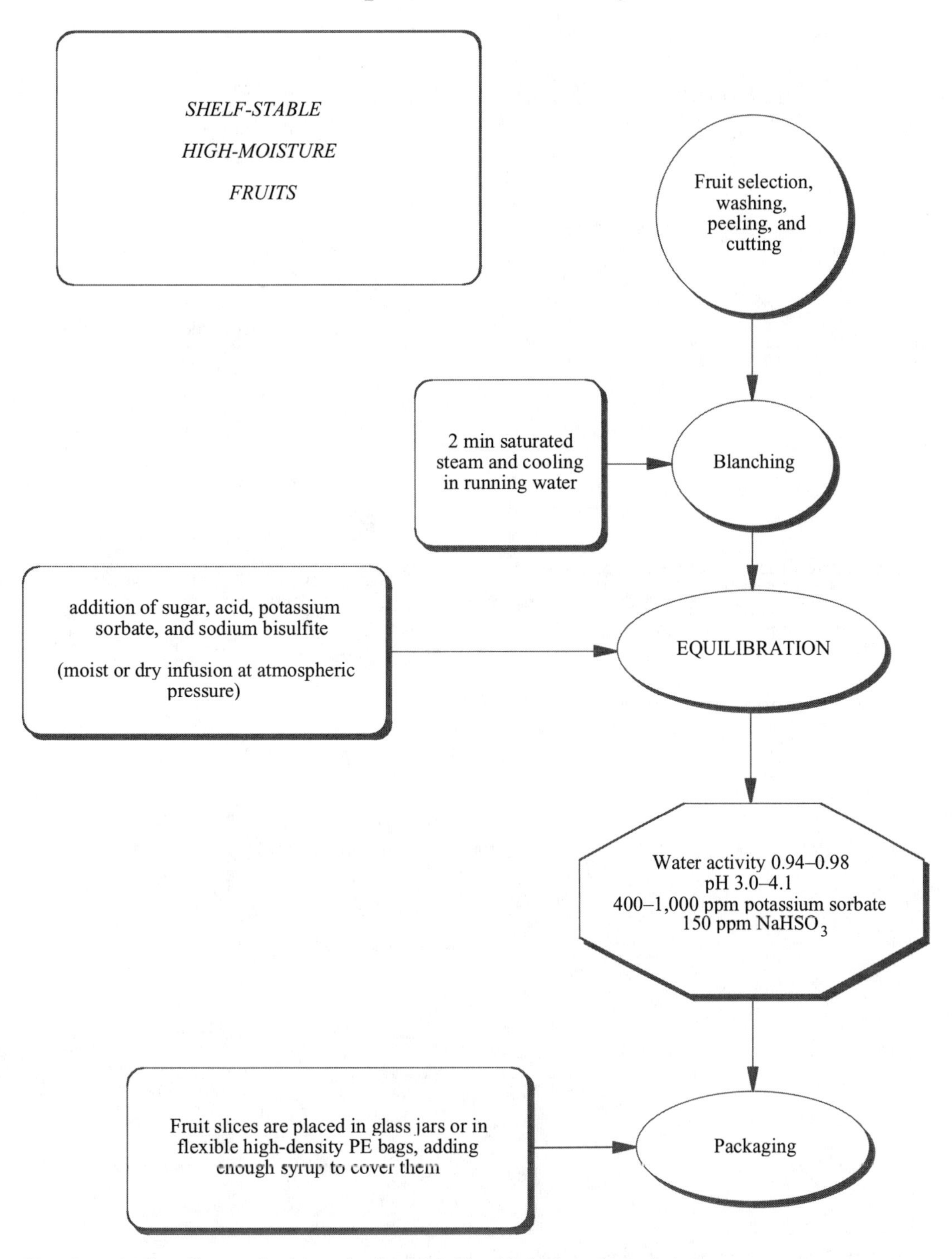

Flowchart 1 Flow diagram for the production of shelf-stable high-moisture fruits.

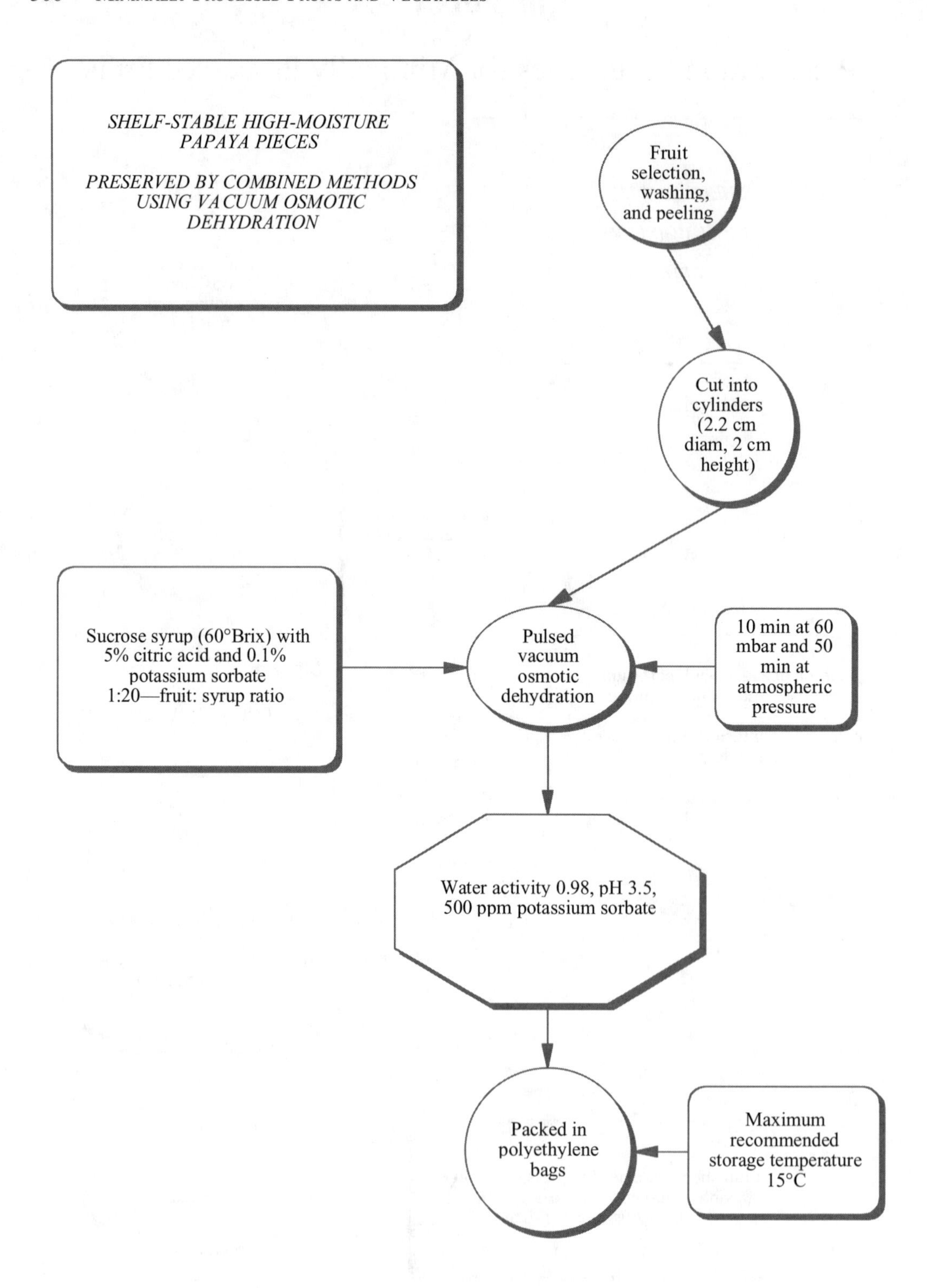

Flowchart 2 Schematic diagram to prepare high-moisture papaya.

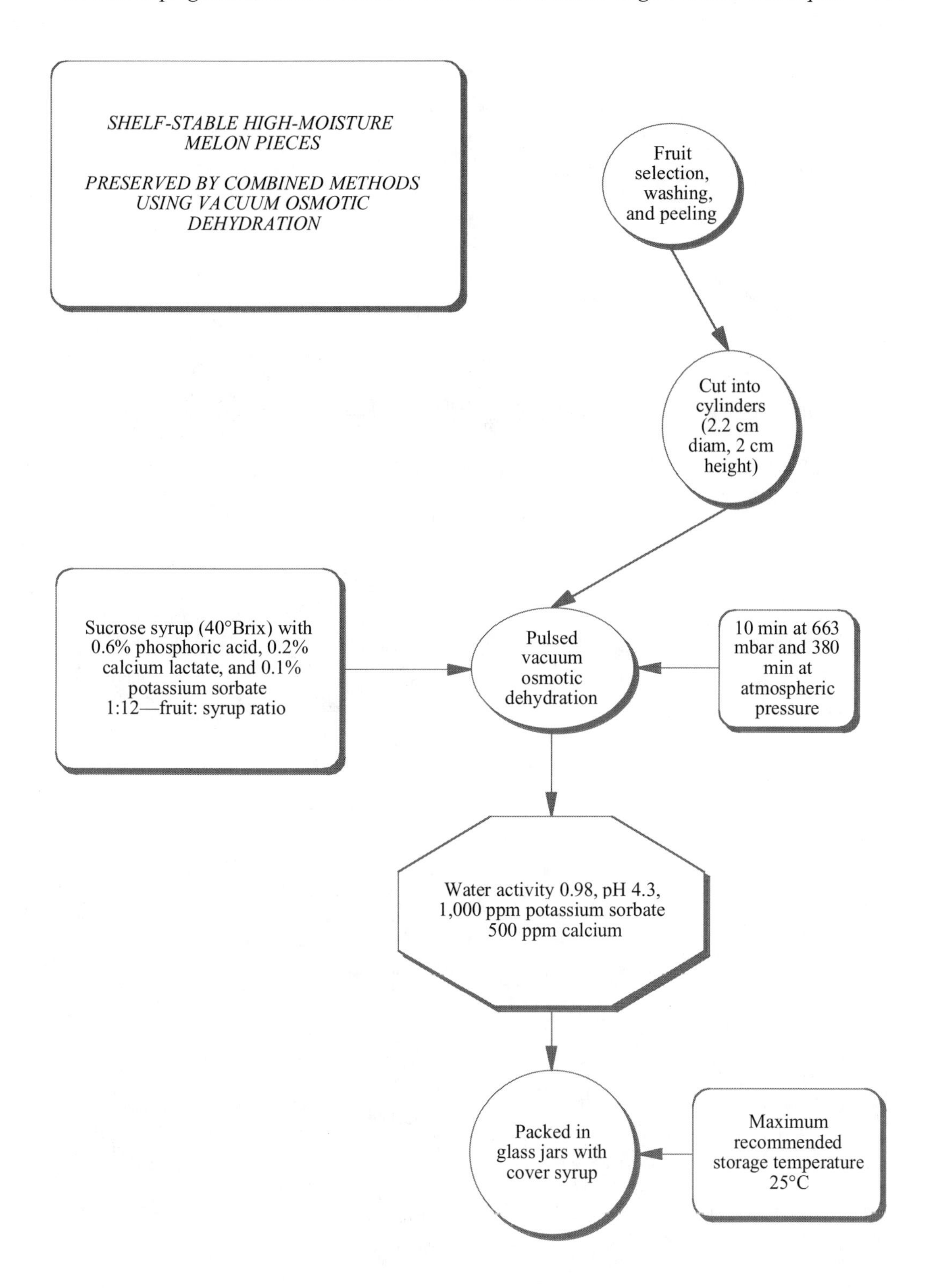

Flowchart 3 Schematic diagram to prepare high-moisture melon.

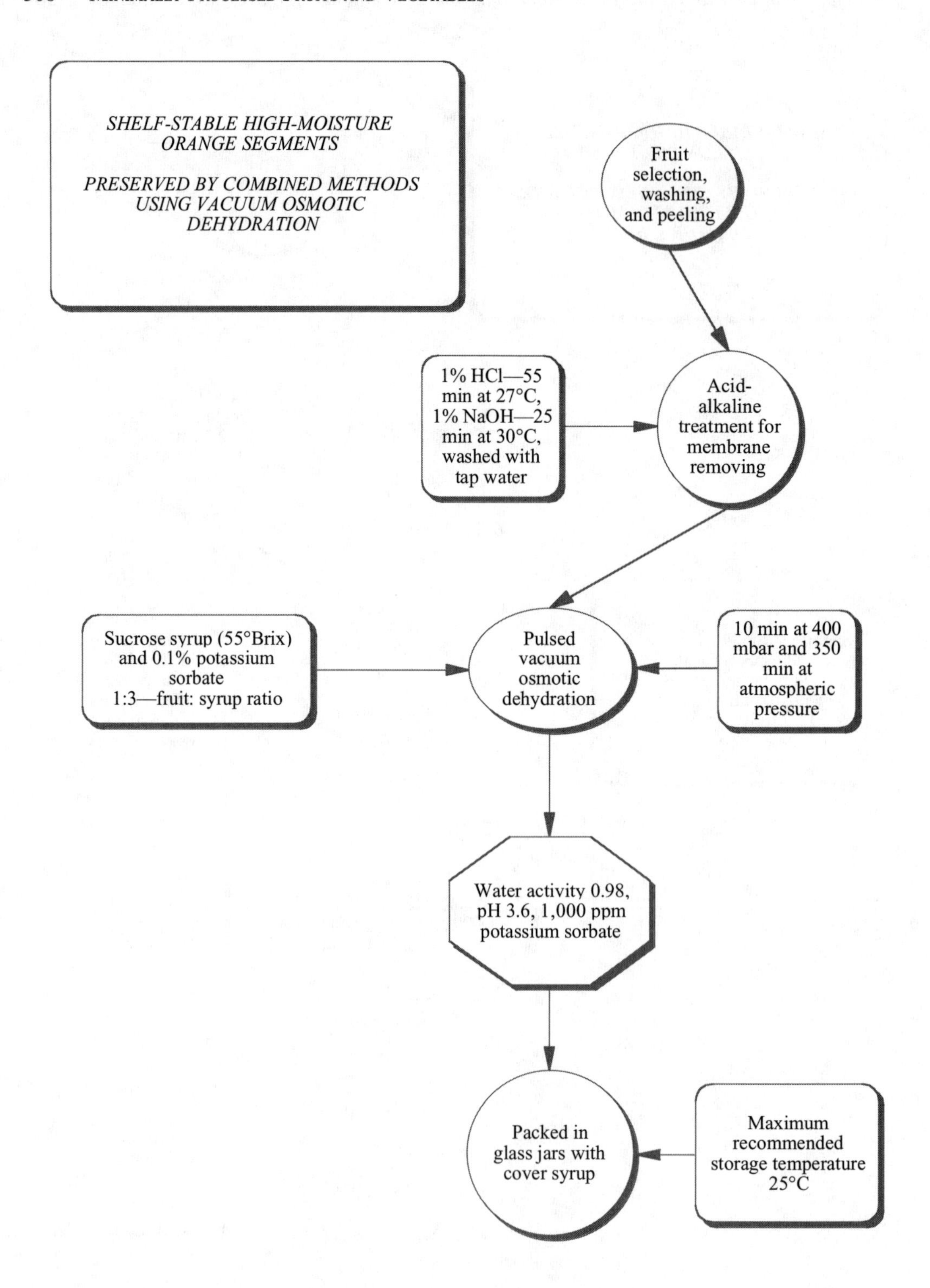

Flowchart 4 Schematic diagram to prepare high-moisture orange segments.

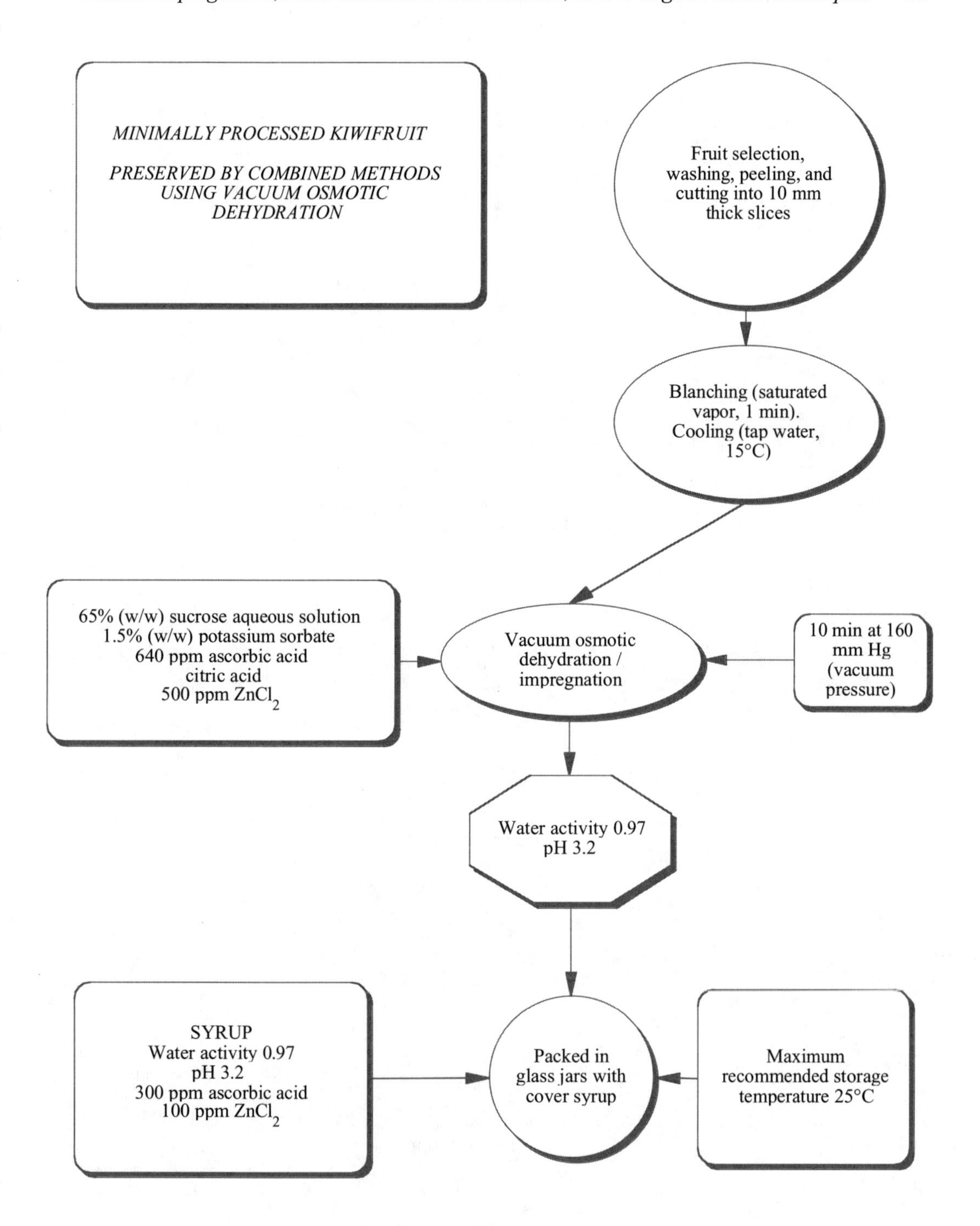

Flowchart 5 Schematic diagram to prepare minimally processed kiwifruit.

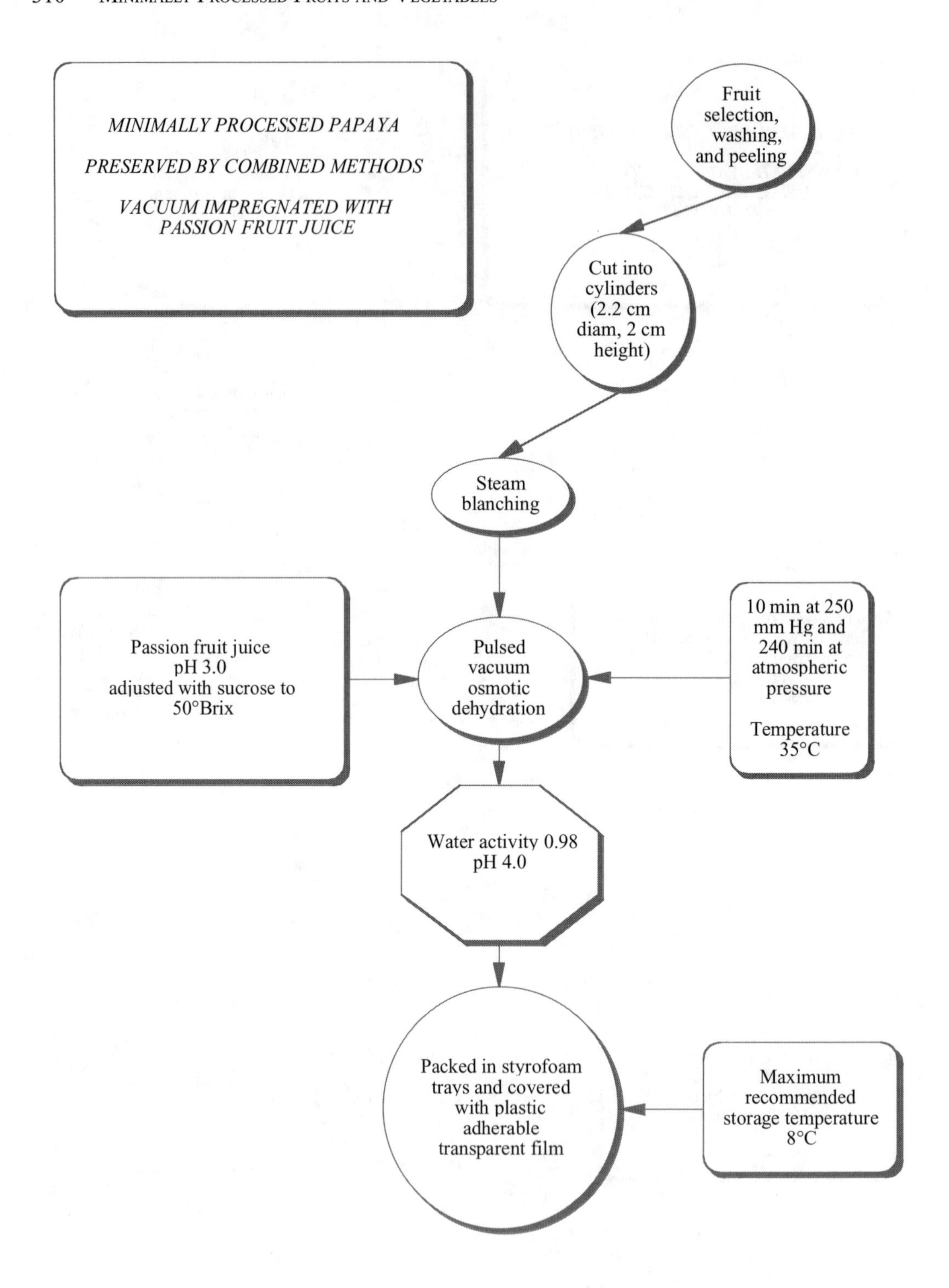

Flowchart 6 Schematic diagram to prepare minimally processed papaya impregnated with passion fruit juice.

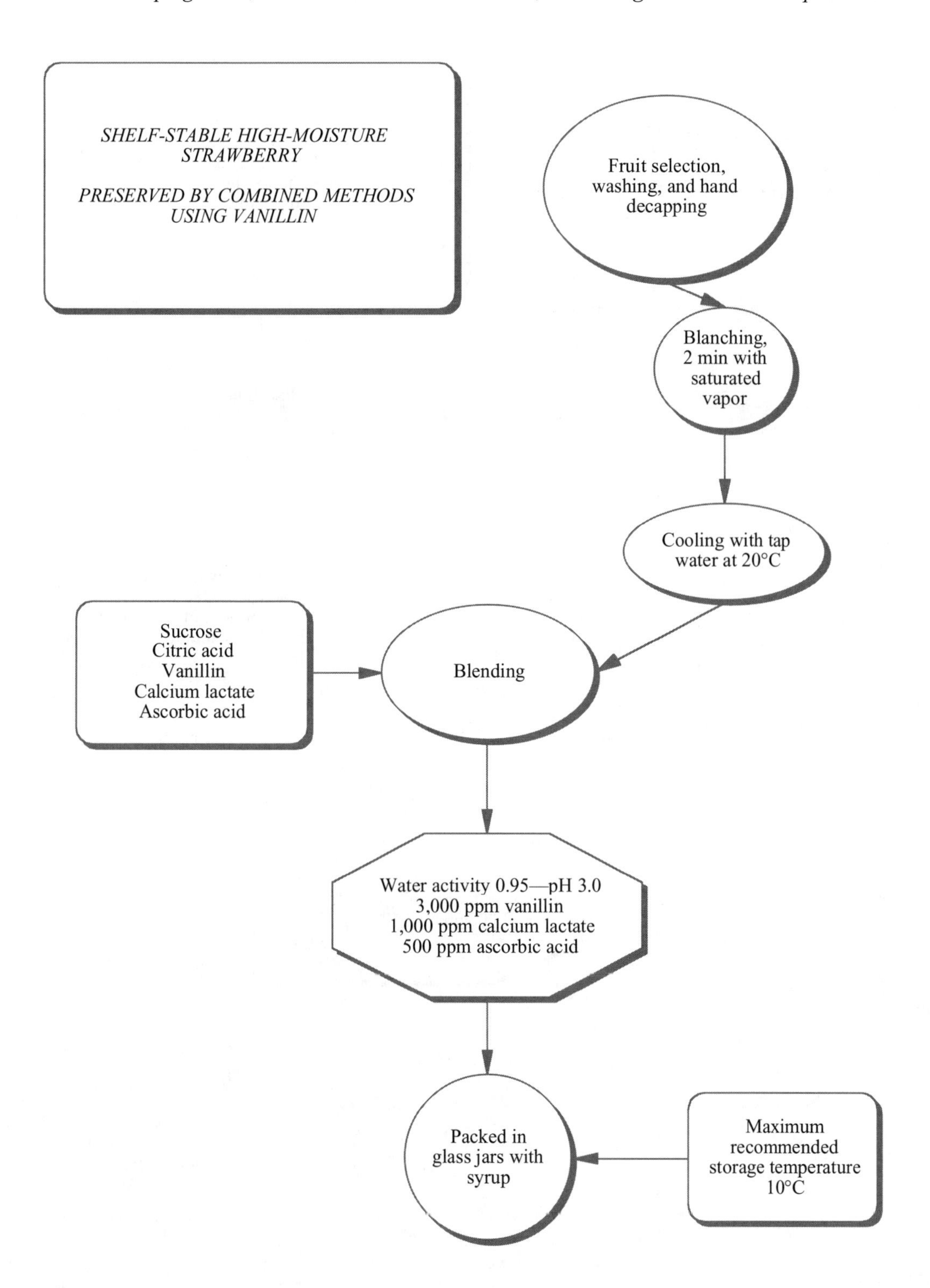

Flowchart 7 Schematic diagram to prepare high-moisture strawberry.

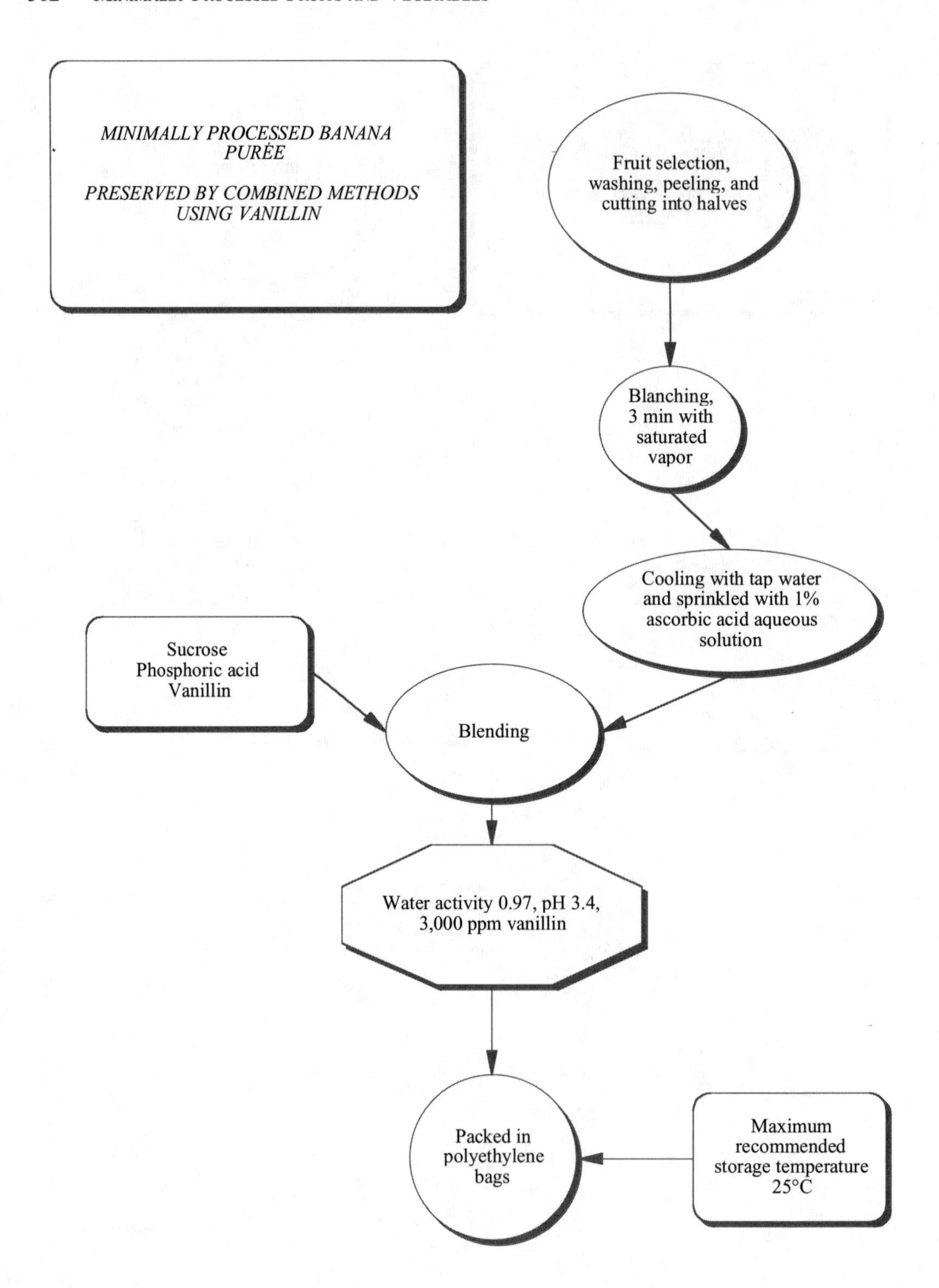

Flowchart 8 Schematic diagram to prepare banana purée with vanillin.

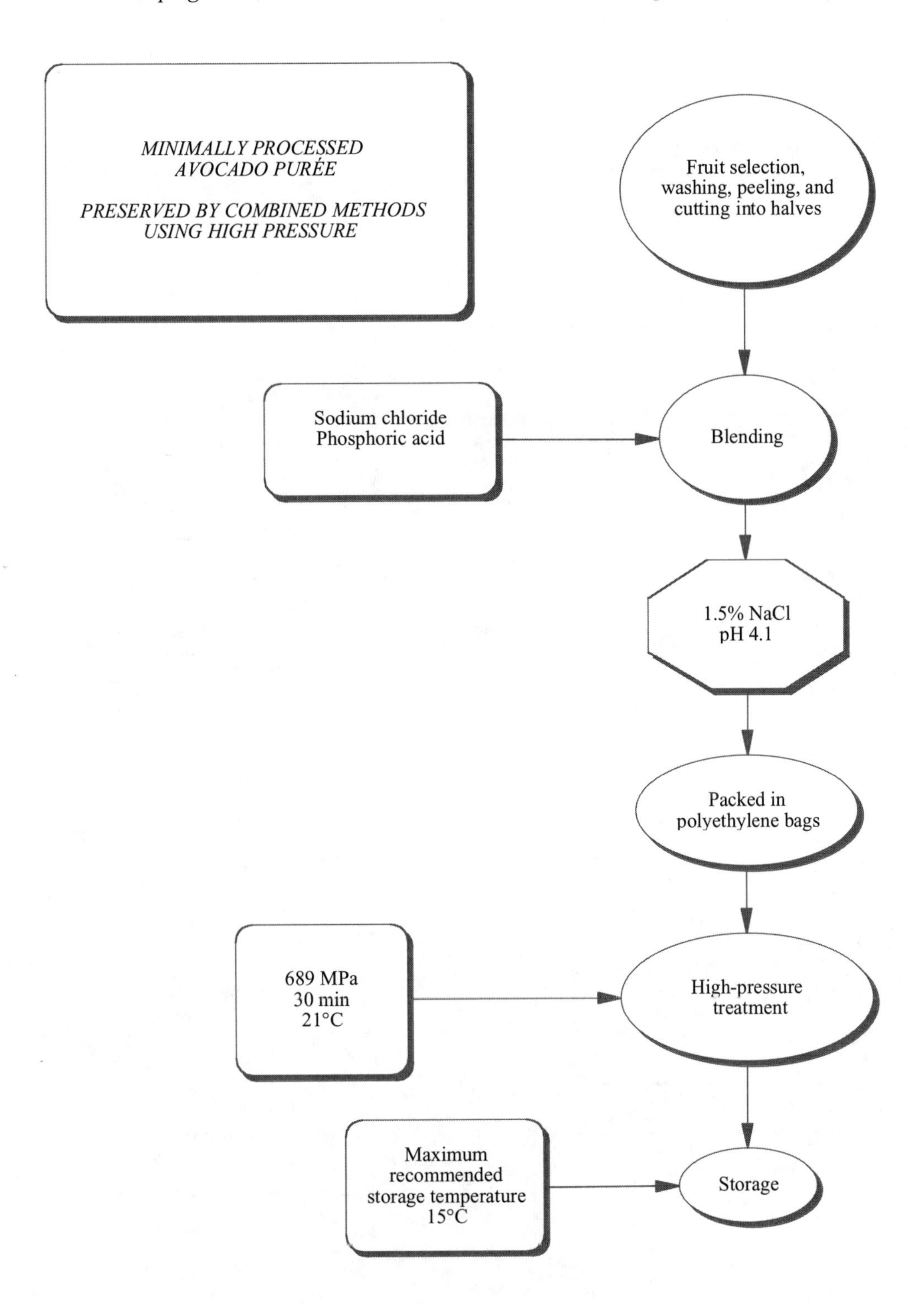

Flowchart 9 Schematic diagram to prepare high-pressure treated avocado purée.

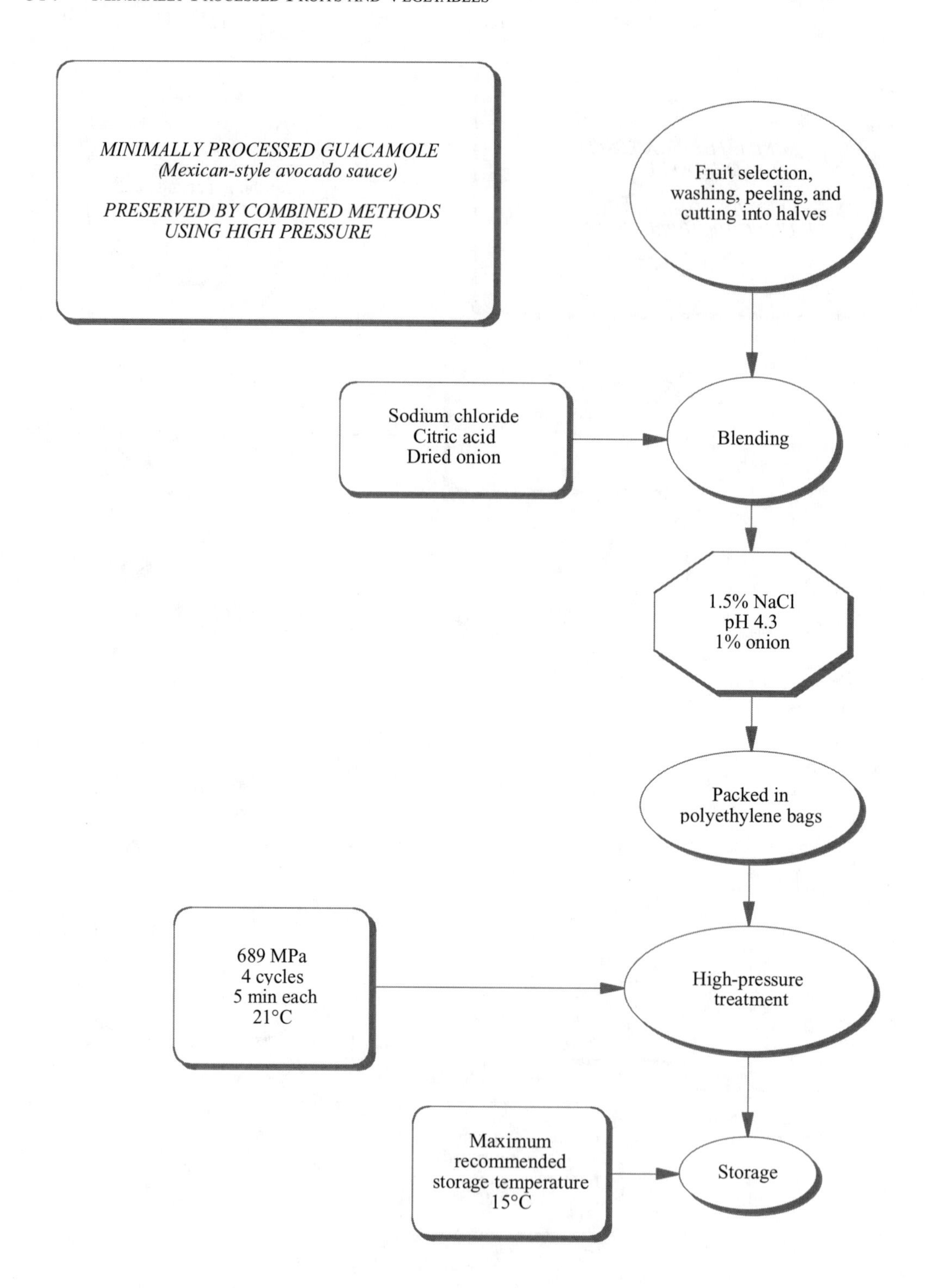

Flowchart 10 Schematic diagram to prepare high-pressure treated guacamole.

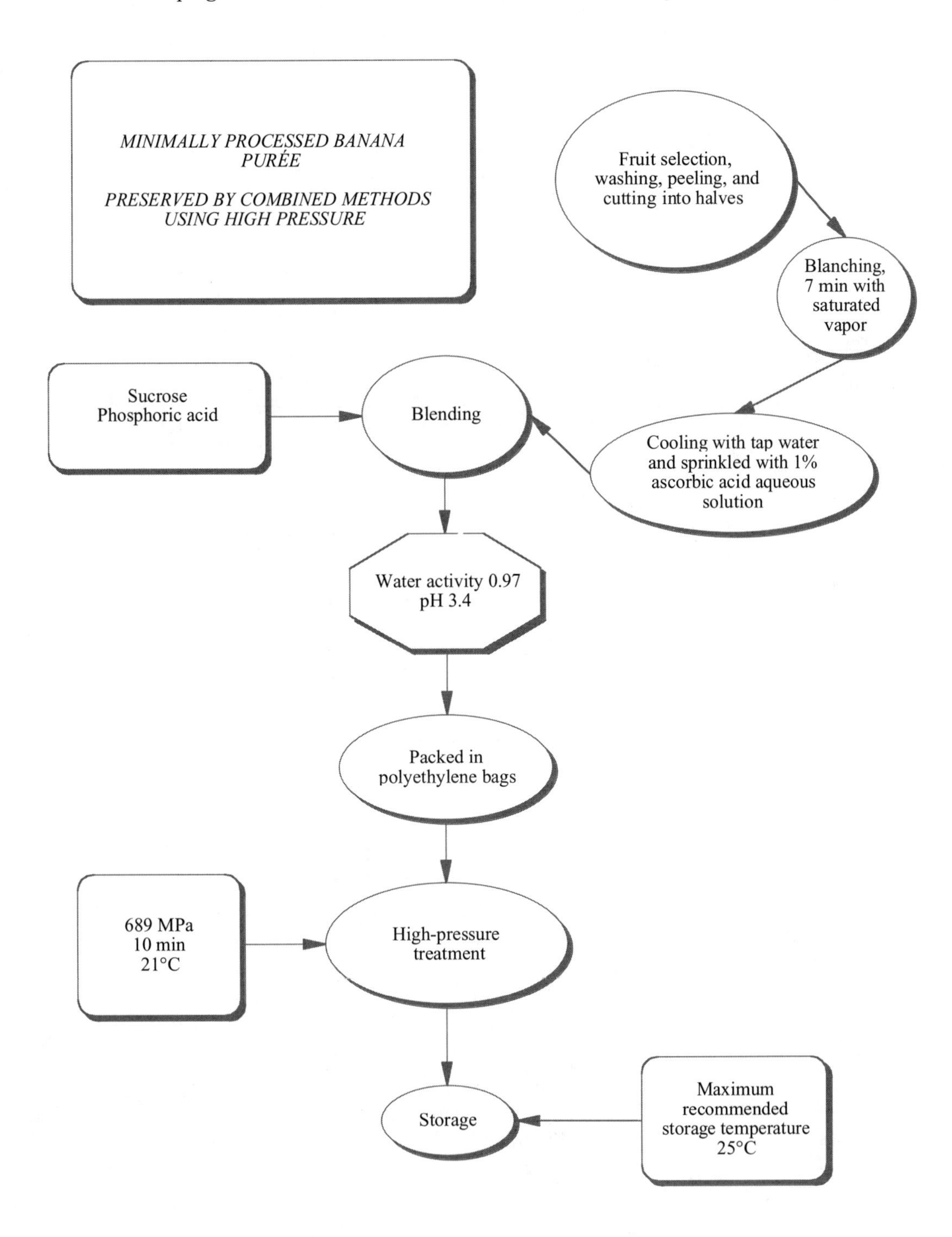

Flowchart 11 Schematic diagram to prepare high-pressure treated banana purée.

Part V

Legal Aspects of Minimally Processed Fruits and Vegetables

CHAPTER 18

Regulatory and Safety Aspects of Refrigerated Minimally Processed Fruits and Vegetables: A Review

J Andrés Vasconcellos

INTRODUCTION

Minimally processed foods have attracted interest in many areas of the food industry for their added value and convenience. The use of fresh or freshlike foods, particularly salad and fruit bars, has been increasing significantly in response to consumer health consciousness. Menu information is often geared to the degree of freshness of a vegetable salad and to enhance the concept of freshness; fresh flowers are often placed around food bars.

Some minimally processed food categories include vegetable salads, individual vegetables, pieced/sliced fruits, packaged sandwiches, chopped or shredded lettuce or cabbage, various forms of prepeeled fresh potatoes, fresh vegetable snacks (carrots, celery, etc.), fresh fish, meat or poultry, partially prepared or cooked entrees or complete meals, and milk and orange juice. Because of minimal processing, quality factors such as appearance, texture, and flavor are not stabilized, and product deterioration may proceed rapidly. Therefore, to extend shelf life, these products are usually stored and marketed under refrigeration, in combination with other treatments, such as modified-atmosphere or vacuum packaging.

The purpose of minimally processed foods is to deliver to the consumer a freshlike product with an extended shelf life while ensuring food safety and maintaining sound nutritional and sensory qualities. One of the fastest growing areas of minimally processing of foods is freshly cut produce, such as fruit and vegetable salads, projected to grow by 21% over the next five years.[1] Such trends and demands are the challenge of this type of technology.

Despite improved methods and efforts in maintaining quality and shelf life, the role of microorganisms in spoilage and the safety of minimally processed foods is still a limiting factor. Maintenance of quality requirements demands extreme care, so as to prevent contamination during harvesting, manufacturing, and processing, as well as the prevention of potential health risks due to improper handling during distribution or by the consumer.[2] Raw vegetables, particularly, are capable of supporting the growth of almost any type of microorganism, due to their high water activity (a_w), neutral pH, and nutrient content. Careful and minimal handling during harvest, proper culling of damaged produce, equipment cleanliness, facility sanitation, employee hygienic practices, postharvest handling, packaging, and storage are techniques that can be employed to minimize spoilage and maintain fruits and vegetables in optimal condition.[3–7]

Of all these areas, consumer handling constitutes the weakest point in the safe utilization of minimally processed foods. Five major conditions have been suggested as necessary when developing these products:[1]

1. they must be equal to restaurant quality;
2. they must be processed to survive distribution variability;
3. they must have increased shelf life;
4. they must show ease of preparation; and, most importantly,
5. they must be safe for consumers.

To ensure the safety of these products, it is necessary to develop manufacturing programs that allow the proper sanitary control from the point of collection in the field to the final destination on the consumer's table. The task is not an easy one because many factors play a role, mostly from a sanitary perspective. Regulatory programs must address agricultural practices, product handling on the way to the processing plant, integral good manufacturing practices at the plant, transportation to markets and distribution centers, product storage conditions, and, most importantly, consumer information and education. In the United States, efforts have been made toward this end; although no regulations presently exist addressing these issues, the publication for implementation of the so-called President's Produce Safety Initiative in 1998 is the beginning of what may soon be established regulations for the appropriate control of minimally processed food manufacture.

In September 1997, The FAIR Programme* of the European Commission issued its *Inventory Report on Harmonization of Safety Criteria for Minimally Processed Foods* (FAIR Concerted Action CT96–1020). The purpose was to investigate the current status of safety criteria for minimally processed foods as presented by the legislative branch and consumer organizations of the member states of the European Community and, based on the information gathered, to evaluate the scientific basis of good manufacturing practices and current legislation, so as to harmonize the criteria to an optimum level. At present, no known regulations pertaining to this area have been approved. It is expected that the effort will help to establish regulations throughout the European Community and that future interaction with American scientists will be of mutual benefit. Canada has followed with the issue (in February of 1999, by the Canadian Food Inspection Agency) of a guide for the safe manufacturing of minimally processed ready-to-eat vegetables.†

DEFINITION

Although the concept of minimally processed foods (MPFs) was originally linked to fresh meat and produce, during the last 10–15 years, the main developments in this type of technology have been with fruit and vegetable products; with these developments, the concept of minimal processing of foods has been better understood and, with that, its definition has been evolving, broadening its application. Besides physical methods to remove dirt and microorganisms, cold storage, and pH control, other more sophisticated technologies have been studied and developed as methods of preservation under the umbrella of the definition of minimally processed foods. Nonthermal processes have become an important area of research in this sense.[8] Barbosa-Cánovas, Qin, and Swanson studied the use of pulsed electric fields (PEF),[9] while other investigators actively studied other techniques, such as use of preservatives, oscillating magnetic fields (OMF),[10] high hydrostatic pressure (HHP),[11–12] light pulses,[13] and irradiation.[14] Tapia de Daza, Alzamora, and Welti-Chanes offered an example of the constantly evolving nature of minimally processed foods concept, as applied to high-moisture fruit products,[15] and Welti-Chanes, Vergara-Balderas, and López-Malo offered an excellent review of the historical evolution of the concept and definition of minimally processed foods as the technological reach of the definition presently applies to food processing.[16] In summary, although no formally accepted definition exists for minimally

*International initiative of European countries to disseminate R & D results to food factories, health professionals, and consumer groups.

†Code of Practice for Minimally Processed Ready-to-Eat Vegetables.

processed foods, perhaps the best and simplest one, in its broadest sense, can be expressed as follows: "products that have been processed by one or more techniques, causing minimum possible changes in the food characteristics in relation to their fresh counterparts, while improving their keeping characteristics under controlled conditions and preserving their nutritional and sensory qualities for a useful shelf-life."

The major differences between minimally processed fruits and vegetables and their raw counterparts are the processing and preservation steps taken. From a practical approach, minimally processed foods are prepared and handled in order to maintain their fresh nature while providing convenience to the consumer. Whole, intact apples stored in a controlled atmosphere would not be considered a minimally processed fruit, whereas refrigerated apple slices, treated with ascorbic acid and calcium salts, would easily be considered a minimally processed product.[17] More difficult to define involves washed, waxed cucumbers. This product might not be considered a minimally processed food because, although they are packaged (waxed), they are also intact and can be easily differentiated from precut, sliced cucumbers, which have a relatively short shelf life and may be considered a minimally processed food.

Some minimally processed foods are living tissues that have undergone normal postharvest handling treatments, such as cleaning, washing, and grading, and in which respiration is greatly increased by cutting, slicing, low-temperature heat treatments, and preservatives.[18] Whereas most food processing techniques stabilize the products and lengthen their storage and shelf life, minimal processing of fruits and vegetables actually increases their perishability by increasing the rate of the metabolic processes that cause deterioration of fresh products. The physical damage increases respiration and ethylene production with associated increases in the rate of other reactions responsible for biochemical changes in color, flavor, texture, and nutritional quality that can be greatly complicated by gas exchange during packaging. Control of the physical damage response is the key to providing a processed product of good quality. Refrigeration and packaging, including controlled-atmosphere or modified-atmosphere storage, may be optional for raw, fresh, intact fruits and vegetables but are mandatory for these minimally processed fruits and vegetables.[19] Temperature control after processing is critical in reducing damage-induced metabolic activity. Other techniques that can substantially reduce damage include the use of sharp knives, maintenance of stringent sanitary conditions in all steps of the agricultural and manufacturing operations, and the efficient removal of surface moisture from the cut product.

MINIMALLY PROCESSED FOODS IN THE MARKET, THE NEED FOR PRODUCT AND CONSUMER INFORMATION

Over the last several years, consumers in the United States and other countries have enjoyed a growing trend for fruit and vegetable salad bars.[20] Fresh produce offered in this form has been considered to be one of the safest food supplies and an important contributing segment to the health and well-being of the American consumer. The nature of the demand required these products to be visually appealing, have a fresh appearance, be of consistent quality, and be reasonably free of defects.[21]

Whereas the health benefits associated with the regular consumption of fresh fruits and vegetables have been clearly demonstrated, an increasing, although still small, proportion of reported outbreaks of food-borne illness has been associated with both domestic and imported fresh fruits and vegetables over the same period of time.[22–24] Recent outbreaks of food-borne illness associated with produce have raised concerns regarding the safety of fruits and vegetables that have not been processed to reduce or eliminate pathogens.[25] Several factors played a role in these incidents. Among them, perhaps the most important from the regulatory point of view, has been consumer abuse. For the most part, consumers are not aware that "fresh vegetables" may constitute a health risk. They consider that foods such as meat, fish, and most dairy products can be easily spoiled but not that this is also the case with fruits and vegetables. Consumers seldom read the storage or preparation instructions provided in the package, because these are often written in small print. Another limiting aspect is that, for the most part, consumers clean their household refrigerators too seldom. Frozen strawberries caused two widespread outbreaks of hepatitis A in the United States,[22–23] and an outbreak of hepatitis A in Scotland was linked to the consumption of mousse prepared from frozen raspberries.[26] These outbreaks were associated with the fact that freezing allows viruses to survive for a long time. Recently, Ponka et al.[27] reported the transmission of small round structured virus (SRSV) through water and through various foods—including salads, bakery products, freshly cut fruits, chicken, oysters, and mussels—attributed to infected food handlers. *Listeria monocytogenes* has long been known as a cause of infections in immunocompromised hosts. Research during the 1990s identified food as the most common source,[28] and this vehicle is now recognized as a major cause of the infection in humans.[29] Although *Listeria* will grow best in the pH range of 6–8, lower pH can allow survival and growth in products such as orange juice and salad dressings containing acids other than lactic acid.[30–31] *Listeria monocytogenes* is also capable of growing at 1°C aerobically[32] and under modified atmosphere.[33] Elimination of *L. monocytogenes* from the surface of vegetables by a chlorine wash is unpredictable and limited.[34–35] In the United States, 1,700 sporadic cases of listeriosis occur annually[36] with a mortality rate of 24–40%.[37–38]

In view of the increased consumption of minimally processed fruits and vegetables, it is necessary to provide the consumer with appropriate information regarding product handling and storage temperature. Because these products are relatively new, it is also important to provide the consumer with information about this new processing method and the length of time that an opened package can be kept under refrigeration. Nutritional information is equally important. A nutritional label must include a list of all raw materials, their percentages (or weight per serving), and a list of all additives used, because this type of food is becoming more valuable to institutional kitchens (hospitals, nursing homes, cafeterias, restaurants, etc.).

Limiting factors in dealing with minimally processed foods are also the manner in which these products have been named; descriptors such as "minimally processed," "as close to fresh as possible," "refrigerated foods with extended shelf life," *"sous vide"* ("vacuum packed"), etc., create confusion. To the consumer, these terms mean nothing. Some groups have already suggested classifying this new technology as an independent type, parallel to food drying, thermal processing, etc., using a specific technical descriptor to allow the consumer to identify it in the market. The name of the process must then be clearly exposed, along with consumer information and marketing and product labels, in terms that the consumer can fully understand without being misled.

REGULATORY ASPECTS

Procedures that should be followed to provide a safe product with extended shelf life include agricultural practices and sanitation, plant sanitation, equipment maintenance and sanitation, processing procedures, employee sanitation, product handling, waste disposal procedures, quality control, processing parameter control, and record keeping. At each of these points, it is necessary to determine appropriate control points so as to guarantee proper processing and the finished product's quality and safety. Because the processing of fresh produce is minimal, preservation mainly involves refrigeration without blanching. It must also be remembered that some fruits and vegetables may be easily chill-injured, which can also encourage microbial spoilage.[39]

Agricultural Considerations

When produce is destined to be eaten raw, sanitation and microbiologic considerations should begin with agricultural practices such as the use of raw animal manure as fertilizer and the treatment of land with sewage or land subjected to flooding with polluted water.[40] These are important concerns because these practices may result in serious microbial contamination of the produce (Exhibit 18–1). Environmental conditions are also important in regard to fruit and vegetable spoilage and the growth of pathogenic microorganisms.

The exteriors of fruits and vegetables are normally contaminated with bacteria and fungi that do not usually include types pathogenic to humans. Soil appears to be responsible for most microbial contamination, and vegetables can become contaminated from the soil or from manure used as fertilizer; *L. monocytogenes* has been isolated from soil, silage, food processing environments, and healthy humans and animals.[41–43] Poor management of human and other waste in the field significantly increases the risk of produce contamination. Intact fruits and vegetables are safe to eat partly because the surface peel is an effective physical and chemical barrier to most microorganisms. In addition, if the peel is damaged, the pulp's acidity prevents the growth of microorganisms other than the acid-tolerant fungi and bacteria associated with decay. On vegetables, the microflora is dominated by soil organisms. The normal spoilage flora, including the bacteria *Erwinia* and *Pseudomonas*, usually have a competitive advantage over other microorganisms that could be potentially harmful to humans.

Many organisms can survive for extended periods of time in the soil. Among the pathogenic organisms, *L. monocytogenes* can survive for years on decaying plant debris in the soil.[44] The potential for fruits and vegetables to become contaminated with pathogenic microorganisms is high because of their exposure to a wide variety of conditions during growth, harvesting, and distribution. Outbreaks of listeriosis were linked as early as 1982 to the consumption of raw cabbage contaminated with sheep feces containing the organism.[45–46] Lettuce, celery, and tomatoes were also implicated in listeriosis in the Boston area.[47] Sanitation and fresh produce quality can be affected by harvesting at the proper time, along with the maintenance and cleanliness of well-designed harvest equipment.

Exhibit 18–1 Sources of Field Microbial Contamination of Fruits and Vegetables

Air
Irrigation water
Environmental conditions (temperature, relative humidity)
Animals
Animal waste fertilizers
Harvesting equipment
Handling practices
Container contamination
Postharvest washing water
Processing equipment
Human sanitation practices

Preharvest and Harvest Practices

Microbial contamination or the cross-contamination of fresh produce during preharvest and harvest activities may result from contact with soils, fertilizers, water, workers, and harvesting equipment. Any of these may be a source of pathogenic microorganisms. To prevent contamination, common sense and the use of the guidelines listed in the President's Produce Safety Initiative[48] should be recommended. The guidelines, considered to be the basis for future regulations in this area of food processing, recommend the following precautions to be practiced in the field:

1. Harvest storage facilities should be cleaned prior to use. Facilities used to store fresh produce should be cleaned and inspected for evidence of pests such as rodents, birds, and insects, and, as necessary, disinfected prior to harvest.

2. Damaged containers, no longer cleanable, should be discarded to reduce the possible microbial contamination of fresh produce. Containers or bins, before used to transport fresh produce and ready-to-eat produce, should be routinely cleaned and sanitized.
3. Produce washed, cooled, or packaged in the field should not be contaminated in the process. Dirt and mud should be removed as much as practicable from the produce before it leaves the field.
4. Poor management of human and other wastes in the field can significantly increase the risk of contaminating produce. Proper measures to implement good management practices must be encouraged at all levels.

For a more in-depth review of the guidelines, the document is available on the Internet at: http:/www.foodsafety.gov/~dms/prodguid.html.

Irrigation Water. Water used in crop production involves numerous field operations, including irrigation, applications of pesticides and fertilizers, cooling, and frost control. Postharvest uses include produce rinsing, cooling, washing, and transport. Whenever water comes in contact with fresh produce, its quality dictates the potential for contamination. Typical sources of agricultural water include flowing-surface waters from rivers, streams, irrigation ditches, and open canals; impoundments, such as ponds, reservoirs, and lakes; groundwater from wells; and municipal supplies. Agricultural water quality varies, particularly surface water that may be subjected to intermittent contamination by waste-water discharge or polluted run-off from livestock operations. It is generally assumed that groundwater is less likely to be contaminated with high levels of pathogens than surface water. Potential sources of contamination should be assessed and controlled because water of inadequate quality can be a direct source of contamination and a vehicle for spreading contamination in the field, facility, or transportation environments. Water can be a carrier of pathogenic strains of *Escherichia coli*, *Salmonella* species, *Vibrio cholerae*, *Shigella* species, and hepatitis A viruses. Even small amounts of contamination with some of these organisms can result in food-borne illness. Practices to ensure adequate water quality should include properly constructed and protected wells, water treatment to reduce microbial loads, or the use of alternative methods to reduce or avoid water-to-produce contact.

Agricultural water can become contaminated, directly or indirectly, by improperly managed human or animal waste.[49] Human contamination may occur from improperly designed or malfunctioning septic systems and sewage treatment facility discharges. Contamination from animal waste occurs from animals pasturing in growing areas; manure storage adjacent to crop fields; leaking or overflowing manure lagoons; uncontrolled livestock access to surface waters, wells, or pump areas; and high concentrations of wildlife.

It is also necessary to consider that agricultural water is frequently a shared resource. Growers should evaluate their production areas in terms of their proximity to surrounding land uses that pose a potential for polluted run-off from heavy rainfall. Protection of surface waters, wells, and pump areas from uncontrolled livestock or wildlife access should be instituted as a good agricultural practice, so as to limit the extent of fecal contamination and minimize the potential for contaminated water to come in contact with the edible portion of the crop. Irrigation practices that expose the edible portion of plants to direct contact with contaminated water may increase microbial food safety risks. In some regions, agricultural water comes from surface waters that travel some distance before reaching the produce-growing area. Growers may not have control over factors that affect the watershed. Under such conditions, a proper program should be instituted to determine appropriate control options. Growers should also institute a program to test their water supplies for microbial contamination, on a periodic basis, using standard indicators of fecal pollution, such as *E. coli* tests.

However, bacterial safety of water does not necessarily indicate the absence of protozoa and viruses. Water quality, especially surface water quality, can vary with time (e.g., seasonally, or even hourly), and a single test may not indicate the potential for water to be contaminated. Furthermore, water testing may not reveal specific pathogens if they are present in low numbers. Growers should maintain constant and close contact with local water-quality experts, such as state or local environmental protection or public health agencies, extension agents, or land grant universities, for appropriate advice on individual operations.

Animal Waste Fertilization. Growers using manure or biosolids need to follow good agricultural practices to minimize microbial hazards. Properly treated manure or biosolids can be an effective and safe fertilizer; untreated, improperly treated, or recontaminated manure or biosolids contain pathogens that can contaminate produce. Crops in or near the soil are the most vulnerable to pathogens that may survive in the soil. Animal manure and human fecal matter represent significant sources of human pathogens. A particularly dangerous pathogen, *E. coli* O157:H7, originates primarily from ruminants. In addition, animal and human fecal matter are known to harbor *Salmonella, Cryptosporidium,* and other pathogens. Therefore, the use of biosolids and manure must be managed to limit the potential for pathogen contamination.

Growers must be alert to the presence of human or animal fecal matter that may be introduced into the produce by nearby municipal waste water storage, treatment, or disposal areas or by wildlife (nesting birds, heavy concentrations of migratory birds, bats, or deer). Possible contamination may also originate from manure in open fields, compost piles, or storage areas. Rainfall onto a manure pile can result in leachate that may pose a microbial pathogenic hazard similar to the manure from which it originates. A variety of treatments may be used to reduce pathogens in manure and other organic materials. Passive treatments rely on the passage of time in conjunction with environmental factors, such as natural temperature, moisture fluctuations, and ultraviolet (UV) irradiation to reduce pathogens. Active treatments include pasteurization, heat drying, anaerobic digestion, alkali stabilization, aerobic digestion, or combinations of these.

Composting is an active treatment used to reduce the microbial hazards of raw manure. It is a controlled and managed process in which organic materials are digested, aerobically or anaerobically, by microbial action. Treated manure can be recontaminated by birds and rodents. Equipment, such as tractors, that comes into contact with untreated or partially treated manure and is used in produce fields can also be a source of contamination.

It is recommended that manure be incorporated into the soil prior to planting to reduce microbial hazards and to maximize the time between the application of manure to production areas and harvest. In general, the shorter the time between the application of raw manure to a production area and the crop harvest, the greater will be the risk of pathogens being present in manure or soil and contaminating the crop. Growers purchasing manure should obtain a specification sheet from the manure supplier that contains information about the method of treatment.

Domestic animals should be excluded from fresh produce fields, vineyards, and orchards during the growing season. Livestock should be confined (in pens or yards) or prevented from entering the fields by using physical barriers, such as fences. Measures should be taken (ditches, mounds, grass/sod waterways, diversion berms, and vegetative buffer areas) to ensure that animal waste from adjacent fields or waste storage facilities does not contaminate the production areas during heavy rains. Federal, state, or local animal protection requirements must be considered in controlling wild animal populations from entering the field. Growers should redirect wildlife to areas with crops that are not destined for the fresh produce market.

Equipment Maintenance and Sanitation. Field equipment, such as harvesting machinery, knives, containers, tables, baskets, packaging materials, brushes, buckets, etc., can easily spread microorganisms to fresh produce. Thus, the following guidelines[39] are recommended:

1. Harvesting and packing equipment should be used appropriately and kept as clean as practicable.
2. Equipment used to haul garbage, manure, or other debris should not be used to haul fresh produce or come in contact with the containers or pallets used to haul fresh produce without first being carefully cleaned and sanitized.
3. Harvest containers must be kept clean to prevent cross-contamination of fresh produce. Harvest containers used repeatedly during a harvest should be cleaned prior to and after each use.
4. Responsibility for the equipment should be assigned to the person in charge, who needs to know how the equipment is used during the day and that it functions properly, and who ensures proper cleaning and sanitizing when it is needed.

Postharvest Practices

Postharvest Washing. Postharvest handling of fruits and vegetables often involves a high degree of water-to-produce contact. Although water is a useful tool for reducing potential contamination, it may also serve as a source of contamination or cross-contamination. Reusing processing water may result in the buildup of microbial loads, including undesirable pathogens from the crop.

Washing water can involve only water or may include a sanitizer, such as chlorine, which is ineffective when organic materials build up in recycled wash water. Chlorine levels of 200–300 mg/mL are necessary to meet the chlorine demand of large wash-water systems, which are often recirculated, but these levels can cause adverse discoloration and leave off-flavors in freshly processed produce.[7] Chlorine is used in varying forms to both flume freshly processed produce and sanitize the equipment. Bartz and Eckert[50] suggested that 50–200 ppm free available chlorine is necessary to destroy vegetative bacterial and fungal cells in commercial vegetable packinghouses. Guidelines for chlorine use are needed in the produce processing industry.

Water quality needs may vary, depending on where the water use falls within the series of processes. It may be greater for water used for a final rinse before packaging, compared with water in a dump tank, where field soil from arriving produce quickly mixes with the water. In general, it is recommended that, whenever possible, water flow counter the movement of produce through the different unit operations.

Antimicrobial chemicals in processing water are useful in reducing microbial buildup and may reduce the microbial load on the surface of produce, providing some assurance in minimizing the potential for microbial contamination. However, the effectiveness of an antimicrobial agent depends on its chemical and physical state, treatment conditions (water temperature, pH, contact time), resistance of pathogens, and the nature of the fruit or vegetable surface. Chlorine is commonly added to water at 50–200 ppm total chlorine, at a pH of 6.0–7.5, for postharvest treatments of fresh produce, with a contact time of 1–2 minutes. Ozone has also been used to sanitize wash and flume water in packinghouse operation.[51–52] Ultraviolet radiation may also be used to disinfect processing water. Chlorine dioxide, trisodium phosphate, and organic acids (lactic and acetic acids) have also been studied for use as antimicrobial agents in produce wash water, although more research needs to be done.

Washing fresh produce, even with antimicrobial chemicals, likely reduces but may not eliminate pathogens and, thus, the overall potential for microbial food safety hazards. Spray wash treatments may spread pathogens by splashing, by aerosol, or through food contact surfaces, such as brushes and utensils. Vigorous washing of produce not subject to bruising or injury may increase the likelihood of pathogen removal. Brush washing, although more effective than washing without brushes, requires the brushes to be cleaned frequently.

Good manufacturing practices (GMPs) for water used for food and food contact surfaces in processing facilities are found in Title 21 of the *Code of Federal Regulations* (CFR), sections 110.37(a) and 110.80(a)(1). The President's guidelines recommend the following to ensure and maintain water quality:

1. Periodic water sampling and microbial testing
2. Changing of water, as necessary, to maintain sanitary conditions
3. Development of standard operating procedures for all processes that use water
4. Cleaning and sanitizing of water contact surfaces as often as necessary to ensure the safety of produce
5. Installing backflow devices and legal air gaps, as needed, to prevent contamination of clean water with potentially contaminated water
6. Routinely inspecting and maintaining equipment designed to assist in maintaining water quality (chlorine injectors, filtration systems, backflow devices) to ensure efficient operation

Handling Practices. All operations in the receiving, inspecting, transporting, packaging, segregating, preparing, processing, and storing of food shall be conducted in accordance with adequate sanitation principles, and all reasonable precautions shall be taken to ensure that handling procedures during production do not contribute contamination with filth, harmful chemicals, undesirable microorganisms, or any other objectionable material. Such precautionary measures should include:

- Raw materials inspection and segregation as necessary to ensure that they are clean, wholesome, and fit as human food. Raw materials shall be washed or cleaned, as required, to remove soil or other contamination. Containers and carriers of raw materials should be inspected upon receipt to ensure that their condition has not contributed to the contamination or deterioration of the products.
- When ice is used in contact with food products, it shall be made from potable water and shall be used only if it has been manufactured in accordance with adequate standards and stored, transported, and handled in a sanitary manner.
- Food processing areas and equipment used for processing should not be used to process nonhuman food-grade animal feed or inedible products, unless there is no reasonable possibility for the contamination of the human food. Processing equipment shall be maintained in a sanitary condition through frequent cleaning, including sanitization where indicated.
- All food processing operations should be conducted under the conditions and controls necessary to minimize bacterial or other microbiologic growth, toxin formation, and deterioration or contamination of the processed product. Chemical, microbiologic, or extraneous-material testing procedures shall be utilized where necessary to identify sanitation failures or food contamination. Contaminated food material shall be rejected, treated, or processed to eliminate the contamination, when this may be properly accomplished.
- Packaging processes and materials shall not transmit contaminants or objectionable substances to the products. They shall conform to any applicable food additive regulation and should provide adequate protection from contamination.
- Meaningful coding of products shall be utilized to enable lot identification to facilitate, when necessary, the segregation of specific food lots that may have become contaminated or otherwise unfit for their intended use. Records should be retained

for a period of time that exceeds the shelf life of the product.

Equipment and Utensil Sanitation. Current GMP regulations recommend that all utensils and product-contact surfaces of equipment shall be cleaned as frequently as necessary to prevent contamination. Non-product-contact surfaces of equipment should be cleaned as frequently as necessary to minimize the accumulation of dust, dirt, food particles, and other debris. Other recommendations should be followed as established by GMPs, including:

- All plant equipment and utensils should be suitable for their intended use, designed and of material and workmanship as to be adequately cleanable and properly maintained. The design, construction, and use of such equipment and utensils shall preclude the adulteration of food with lubricants, fuel, metal fragments, contaminated water, or any other contaminants. All equipment should be installed and maintained so as to facilitate the cleaning of the equipment and of all adjacent spaces.
- Single-service articles (utensils intended for one-time use, paper cups, paper towels) should be stored in appropriate containers and handled, dispensed, used, and disposed of in a manner that prevents the contamination of food or food-contact surfaces.
- To prevent the introduction of undesirable microbiologic organisms into food products, all utensils and product-contact surfaces of equipment should be cleaned and sanitized prior to use. Where equipment and utensils are used in a continuous production operation, the contact surfaces of said equipment and utensils should be cleaned and sanitized on a predetermined schedule, using adequate methods.
- Cleaned and sanitized portable equipment and utensils with product-contact surfaces should be stored in a location and manner so that product-contact surfaces are protected from splashes, dust, and other contamination.
- The equipment in use in the food plant should be constructed not only for its functional use, but, most importantly, with due regard to its cleanability and protection from contamination. Construction materials should be smooth, hard, nonporous, and preferably of stainless steel. All pipe lines, fittings, etc., used in food handling and processing should be of the sanitary type. Sharp corners in tanks, flumes, and other equipment should be eliminated to facilitate cleaning and prevent spoilage organisms from building up. All equipment should be directly accessible for cleaning or proper provisions should be provided for cleaning in place (CIP).
- Open equipment, such as tanks, hoppers, buckets, elevators, etc., should be covered. Containers used to transport food materials should be kept clean and not used for other purposes. Nesting of pails, trays, etc., should not be allowed until they have been cleaned. Excess lubricant should be cleaned off after lubricating. Hoses and clean-up equipment should be properly put away after each use, and all unused equipment should be removed from the processing area.

Produce Storage Conditions. Storage of finished products should be under conditions so as to prevent contamination, including development of pathogenic or toxicogenic microorganisms, and should protect against undesirable deterioration of the product and the container. Storage conditions will also affect both the final populations and types of microorganisms that will grow on fresh produce.[5]

Temperature and relative humidity within the package are the two most important factors affecting the microflora in storage.[6] Controlled- or modified-atmosphere storage usually consists of a lower O_2 concentration and higher CO_2 concentration than is normally found in air. Under such conditions, respiration is lower, and, with all other physiologic functions altered, the result is a reduction in the ripening process and a retention of a higher quality for a longer period of time. However, the limitation of these storage processes is that, if maintained for relatively long periods of time at refrigeration temperatures, they can favor the growth of psychrotrophic microorganisms that could affect the safety of the products. Refrigeration storage of fresh produce is usually unfavorable for the growth of most pathogenic bacteria but favorable to spoilage microorganisms. Refrigeration alone is not sufficient to prevent pathogenic bacteria, such as *Salmonella, Shigella,* and *E. coli* O157:H7, from surviving and psychrotrophs such as *L. monocytogenes*, *Yersinia enterocolitica,* and *Aeromonas hydrophila* from growing on fresh produce. Thus, a very important aspect of a HACCP program is the control of proper temperatures and conditions for the storage of minimally processed foods.

Manufacturing Considerations

Plant Sanitation. In addition to employee training, minimally processed food producers need to develop more awareness of total plant sanitation. In many instances, floors, walls, ceilings, and drains may not be sanitarily designed. Areas such as rough welds need to be smoothed out to prevent bacterial growth sites. Particular emphasis must be placed on the sanitary design of operating plants and equipment.

In the plant, sanitation is every person's job and should be a part of the everyday policy of a company. When properly conducted, sanitation removes the worry about spreading communicable diseases or the potential of food poisoning. A company should organize its sanitation program within company guidelines and standards. The accomplishments should be regularly measured by a plant sanitation committee. A regular schedule of meetings will help to reinforce the importance of a sanitation program and of performance of all operations in clean surroundings with due regard to the basic principles of sanitation. The plant manager has the obligation to uphold the sanitary standards considered to be common practice for food-handling establishments.

The authority to uphold sanitary standards falls to the plant sanitarian, who is directly responsible to management. He or she is the key to the success of cleanliness and good operations in the food plant. This person's education requires a considerable amount of science and knowledge of the food industry, with an understanding of microbiology, chemistry, entomology, parasitology, and sanitary engineering. The responsibilities of the plant sanitarian include:

1. Supervision of matters of personal hygiene.
2. Maintenance of adequate plant cleanup.
3. Elimination of rodents and insects.
4. Supervision of water supply, sewage, and waste disposal.
5. Maintenance of the sanitation of rest rooms and toilets.
6. Supervision of sanitary storage of raw and finished products.
7. Corrective action to prevent any contamination.
8. Organization of training programs for plant personnel.
9. Individual inspections of the plant and reports to management.
10. Participation in general inspections with management and supervisory personnel.
11. Cooperation with local, state, and federal inspectors.

The success of a sanitation program depends in great part on training (courses, seminars, etc.), which should point out methods to be used and the responsibility of each individual to practice proper housekeeping in the food plant. Periodic refresher or follow-up sanitation training programs for the employees are recommended in any food-processing facility, including the requirements under the Occupational Safety and Health Act (OSHA) applicable to worker health and training.[53–54] Training and retraining of the sanitation personnel should be undertaken continuously. A manual should be developed that contains the minimum standards for each of the plant areas.

The final aspect of the sanitation program should be the "sanitation audit" or "sanitary evaluation" of the plant. The inspection, conducted by company personnel from either the home office or the local plant, or by outside agencies or groups, should include the outside, as well as the inside of the facility, and a final report should be written on all observed conditions, listed as critical, major, and minor observations, depending on the degree they pose for food adulteration and/or contamination.

A "critical" observation is defined as any condition of actual product or container contamination/adulteration or condition in which contamination/adulteration is imminent or inevitable. A "major" observation is any condition or practice that is a source of potential product or container contamination/adulteration. A "minor" observation is any other deviation from the current GMPs.

The most important part of the sanitation inspection or "sanitation audit" is the final written report, filed with plant management at the local level and the home office, conveying the pertinent concepts and knowledge about what is going on. The report should be acted upon accordingly by management, because it will be worthless unless the information is used advantageously.

The evaluation survey of a food plant will depend, in great part, on the standards of cleanliness that are established. These standards may be categorized as:

a. physical cleanliness, defined as the absence of visual product waste, foreign matter, slime, etc.;
b. chemical cleanliness, defined as the freedom from undesirable chemicals (contamination could occur from cleaning compounds, germicides, pesticides, etc., left near the product and near the processing equipment); and
c. microbiologic cleanliness (probably the most dominant factor in food plant sanitation today), which is defined by the amount of microorganisms that may be present on the product, the equipment, the building, and people.

The value of a planned sanitation program can be demonstrated by the manufacture of a better quality of product and greater employee productivity. Several studies have demonstrated that fewer accidents occur in food plants with top sanitary conditions (reason enough to have a planned sanitation program).

Packaging Technology. Changes in the environmental conditions surrounding a product may result in significant changes in microflora. The risk of pathogenic bacteria may increase with film packaging (high-humidity and low-oxygen conditions), the packaging of products with low salt content and high cellular pH, and the storage of packaged products at excessively high temperatures (> 5°C/41°F). Under such conditions, food pathogens, such as *Clostridium botulinum*, *Y. enterocolitica,* and *L. monocytogenes,* can potentially develop on minimally processed fruits and vegetables.[55–56]

The Federal Food, Drug and Cosmetic Act places the responsibility for assessing the safety of food packaging materials on the FDA. Unintentional food components, as a result of their use or presence in food packaging, are considered to be indirect food additives and legally require premarket safety approval and safety evaluation prior to their actual use. In general, increased migration of the components of food packaging materials results from their use at increased temperatures or with foods of increased fat content.

Products are often packaged under partial vacuum or after flushing with different mixtures of gases (oxygen, carbon dioxide, carbon monoxide, and/or nitrogen). Vacuum packaging and gas flushing establish the modified atmosphere quickly and increase the shelf life and quality of the processed products. Polyvinylchloride, used primarily for overwrapping, and polypropylene and polyethylene, used for bags, are the films most widely used for packaging minimally processed foods. Multilayered films, often with ethylene vinyl acetate, are manufactured with differing gas transmission rates. It has not yet been determined what are the ideal films and atmospheres for minimally processed products, because, obviously, there are different atmosphere requirements for different products; the specifics of the handling chains must be taken into account, especially their time delays and temperature fluctuations.[19]

Product Distribution. The proper transport of minimally processed foods from the processing plant to markets or to distribution centers helps to reduce the potential for microbial contamination. An active and ongoing discussion with the personnel responsible for transportation is essential for ensuring the success of any program designed to deliver safe foods to the consumer. The sanitation conditions during transport should be evaluated as microbial cross-contamination from other foods and nonfood sources, as well as from contaminated surfaces, may occur during loading, unloading, storage, and transportation. Workers involved in the loading and unloading of minimally processed foods during transport should practice good hygiene and sanitation practices. Transportation vehicles must be kept clean to reduce the risk of microbial contamination. Trucks and transport cartons must be inspected for cleanliness, dirt, or debris before loading. Trucks that have been recently used to transport animal products may increase the risk of contaminating minimally processed foods if the trucks have not been properly cleaned.

All minimally processed foods should be carefully loaded in trucks or transport cartons in a manner designed to minimize physical damage to the produce and to reduce the potential for contamination during transport. Operators should work with transporters to ensure adequate control of transport temperatures, which should be between 1 and 5°C (34–41°F). Loading the cargo should be done so as to allow proper refrigerated air circulation, because temperature maintenance is recognized as the most deficient factor. Transporters should be aware of temperature requirements and avoid delivery of mixed loads with incompatible refrigeration requirements. Product temperature should be checked by placing the thermometer probe between two packages and pressing these together until obtaining a temperature reading.

Personnel Sanitation and Health Considerations

Processors should operate their facilities or farms in accordance with the laws and regulations for field and facility sanitation practices. The field sanitation laws, prescribed under OSHA,[32] specify the appropriate number of toilets for a given number of workers, proper hand-washing facilities, maximum worker-to-restroom distance, and how often such facilities should be cleaned. OSHA standards also provide regulations relative to toilet facilities and other sanitation issues.[54]

Toilet facilities should be accessible. The more accessible the facilities, the greater the likelihood that they will be used when they are needed, reducing the incidence of workers relieving themselves in the field or outside of packing areas. However, it is necessary to pay attention that facilities in the field are not located near a water source used in irrigation or in a location that would subject such facilities to potential run-off, in the event of heavy rains. Toilet facilities should be well supplied with toilet paper; hand-washing stations should be equipped with a basin, water, liquid soap, sanitary hand-drying devices (such as disposable paper towels), and a waste container, and should be cleaned on a regular basis.

Management should have a plan for the containment and treatment of any effluent, in the event of leakage or a spill, and must guarantee that systems and practices are in place to ensure the safe disposal of waste water from permanently installed or portable toilets, preventing drainage into a field and fresh produce contamination. Operators should be made aware and be prepared in the event of any leakage incidence or spillage of effluent in a field and should follow EPA regulations for the use or disposal of sewage sludge. 40 CFR Part 503 (standards for the use or disposal of sewage sludge) should be enforced.

Disease-infected employees who work in the field with fresh produce during harvesting increase the risk of transmitting food-borne illnesses. Operators should place a high priority on ensuring the use of agricultural and management practices that minimize the potential for direct or indirect contact between fecal material and fresh fruits and vegetables. They should be aware of and follow the applicable standards for protecting worker health established under OSHA. In addition, CFR Title 21, Section 110.10 (21 CFR 110.10) prescribes worker health and hygienic practices within the context of GMPs in the manufacturing, packing, or holding of human food. Workers can unintentionally contaminate fresh produce, water supplies, and other workers, and can transmit food-borne illness if they do not understand and follow basic hygienic principles. It is important to ensure that all personnel, including those indirectly involved in fresh produce operations (such as equipment operators, potential buyers, and pest control operators), comply with established hygienic practices. All employees, including supervisors and full-time, part-time, and seasonal personnel, should have a good working knowledge of basic sanitation and hygiene principles. In most circumstances, single-service disposable gloves can be an important and effective hygienic practice, in combination with hand-washing. If gloves are used, they should be used properly. Their use, however, in no way lessens the need or importance of hand-washing and proper hygienic practices. Good hygienic practices must also be followed by visitors to the farm or to processing or packing facilities. Product inspectors, buyers, and other visitors must comply with the established hygienic practices of the plant.

A wide range of communicable diseases and infections may be transmitted by infected employees to consumers through food or food utensils. An important part of an ongoing program to ensure the safety of fresh produce is to institute a system of identifying any worker showing symptoms of an active case of illness and to exclude him or her from work assignments that involve direct or indirect contact with fresh produce or in the sorting and packing of products. The following is a partial list of infectious and communicable diseases transmitted through food and their symptoms: hepatitis A virus (fever, jaundice); *S. typhi* (fever); *Shigella* species (diarrhea, fever, vomiting); Norwalk and Norwalk-like viruses (diarrhea, fever, vomiting); *S. aureus* (diarrhea, vomiting); *S. pyogenes* (fever, sore throat).[57]

Hazard Analysis and Critical Control Points Program

The Hazard Analysis and Critical Control Points (HACCP) program systematically evaluates the potential for microbial, physical, and chemical hazards during food production and distribution, and identifies, to the best possible extent, the associated risk; critical control points are subsequently established; and the necessary monitoring programs are developed for hazard prevention. Sanitation, both at the individual and the physical plant levels, also plays a key integral role in an HACCP program. All aspects of the facility's construction must be considered from a microbial and pest-control point of view, identifying all potential routes of microbial contamination and pest entry to effectively integrate sanitary design. Because all three categories of hazards (microbiologic, physical, and chemical) are intimately associated with fresh produce, HACCP evaluations should be practiced for each specific minimally processed food product.

The industry has been conducting significant efforts in the areas of HACCP to ensure safe and wholesome minimally processed fruits and vegetables of high quality.[58] In addition to participating in working groups of the National Advisory Committee on Microbiological Criteria for Foods, the industry has been conducting other significant efforts. The work of the National Food Processors Association (NFPA) in this area has also been significant. NFPA recognizes the concerns and issues about minimally processed foods whose shelf lives are extended by refrigeration, in combination with other preservation procedures or techniques, and suggested certain factors to consider in establishing GMP.[59] The committee recommended the use of 4.4°C in place of 7.2°C as the upper limit for refrigerated products. Although it is recognized that 4.4°C may be unrealistic for practical applications, the committee endorsed the concept as a desirable goal. The *Manufacturing Guidelines for the Production, Distribution and Handling of Refrigerated Foods*[60] represents a significant industry effort. Addressing product and process development, predistribution, retailing, and food service with HACCP principles clearly emphasized, these guidelines represent a responsible approach to food safety. In addition to HACCP, the industry is initiating other steps to ensure the safe and economic production, distribution, and storage of minimally processed foods,[61–62] including consumer awareness programs.

In its publication, *Food Safety Guidelines for the Fresh-Cut Produce Industry*,[63] the International Fresh-Cut Produce Association suggested HACCP guidelines for fresh produce, as shown in Exhibit 18–2.

Consumer Practices

Refrigeration is one of the foremost tools for inhibiting or slowing the growth of pathogenic microorganisms. Unfortunately, refrigeration temperatures used in practice often exceed the recommended values of 7.2°C.[52] In a survey of the temperatures of home refrigerators, over 20% exceeded 10°C.[65] Temperatures in retail establishments vary and may depend on position within the case or proximity to fluorescent lights.[64]

Many refrigerated foods are marked with "use by" or "sell by" dates. Although these give the consumer guidelines for the use of a safe product, they do not give the temperature history. Spoilage and pathogen growth increase when temperature increases. Because temperature abuse is perhaps the most critical factor in the growth of pathogenic microorganisms, an ideal food packaging should contain some indication of past temperature history. Time–temperature indicators could be useful in this way, to reveal a product's temperature history. Unfortunately, not enough research information has been developed for minimally processed fruits and vegetables.[55]

Recognizing that consumers must bear some responsibility for protecting their own health, the USDA and the FDA are participating in a joint effort to study consumer attitudes and knowledge of the hazards posed by the improper handling of possibly hazardous foods, including in-home refrigeration temperatures, in-home food preparation, and handling and holding practices. It is expected that these studies can become the basis for the development of consumer education programs on how to store, handle, and prepare minimally processed fruits and vegetables.

Regulatory agencies will probably be expected to encourage and perhaps require greater reliance on "use by" or "sell by" dating, or some other similar mechanism, to ensure that extended shelf-life products remain safe and wholesome. Manufacturers will need a complete and thorough understanding of their products' shelf life to implement a useful and meaningful system. This will require an examination of distribution systems and monitoring of the effectiveness of in-store personnel, education of supermarket personnel, and the selection of dates, based not only on the product's microbiologic and quality attributes, but also on storage and handling practices. This is particularly important because consumers may perceive products as shelf-stable instead of shelf-life-extended.

Exhibit 18–2 Suggested Guidelines for an HACCP Program for Fresh Produce

1. Designate a person responsible for the HACCP plan and members of an HACCP "team" for the food facility and target products.
2. List target food products, describe each product, list raw materials and ingredients, and prepare a preliminary flow diagram.
3. Document the hazard analysis associated with the target products, their ingredients, and the hazards of the entire product manufacturing chain (Principle 1).
4. Develop individual flow diagrams for each product that document the location and type of CCPs for identified hazards (Principle 2).
5. Document descriptions of each CCP, including the type of hazard, procedures or processes to control the hazard, and definition of the critical limits or tolerances that apply to each CCP (Principles 2 and 3).
6. Document monitoring procedures for the CCPs and critical limits, monitoring frequency and the person(s) responsible for specific monitoring activities (Principle 4).
7. Document deviation procedures for each CCP that specify the actions to be taken if monitoring determines that a CCP is out of control. Actions must include the safe disposition of affected food product and correction procedures for the conditions that caused the situation (Principle 5).
8. Develop and document record-keeping systems for the HACCP program using Principle 6. Designate trained and responsible company personnel for management and sign-off of records.
9. Develop and document verification procedures based on Principle 7. Designate responsible company personnel to conduct verification of compliance with the HACCP program on a scheduled basis. Designate responsible persons to conduct verification who are not generally involved in the HACCP functions (such as corporate or division quality assurance personnel).
10. Document procedures for the revision and updating of the HACCP program any time there is a change of ingredients, products, manufacturing conditions, evidence of new potential or actual hazard risks, or any other reason that may influence the safety of the product(s). Otherwise, specify scheduled revision and updating.
11. Consult with the appropriate regulatory agency or agencies regarding company intention to develop an HACCP program and involve the agency or agencies in the development and approval of the HACCP program.

RESPONSIBILITIES

The Role of the Manufacturer

Persons working in the food industry and companies engaged primarily in the production of food for human consumption have a moral and legal obligation to perform all operations in clean surroundings and with due regard for the basic principles of sanitation. Each has the obligation to uphold sanitary standards in common practices for food handling establishments.

The Role of Government

As consumers move toward an increased consumption of partially and minimally processed food and meals, the risk of food-borne illnesses is increased due to potentially improper processing, packaging, and handling. This trend has created concern in the scientific community about the need for better manufacturing control of this new segment of the food industry. Study groups, conferences, and technical reviews of industry practices have started to appear and have created similar concerns at the highest levels of government. The present regulations do not specifically consider the various aspects of the minimally processed foods industry. Meat and poultry products are subject to USDA continuous inspection, and this offers a degree of protection to the manufacturer, as well as to the consumer. The applications of GMPs and HACCP concepts to minimally processed foods are based on the existing regulations for other areas of the food industry; however, although these are of use and can be applied to various aspects of minimal processing of foods, it is obvious that an exhaustive review of the regulations must be made and adaptations created for specific application to this relatively new and growing industry.

Dane Bernard, the vice president for food safety programs of the NFPA, believes that U.S. food regulators should adopt 100 colony-forming units of *L. monocytogenes* per gram as a tolerance for the food-borne pathogen in ready-to-eat foods. Bernard believes that processors are hesitant to monitor their products and plants for *L. monocytogenes* under the "zero tolerance" approach—absence of the organism in 25-g samples[30]—"because the common bacteria are bound to show up sooner or later." The FDA does not consider foods with a pH of 4.6 or less to be potentially hazardous, even though the food may contain infectious or toxicogenic organisms at a level sufficient to cause illness. In their study, Cook et al. challenged this supposition of the FDA.[66] Food safety recommendations related to temperature and acidity do not eliminate risks for certain pathogens,[67] whereas other studies have found that *Salmonella* species and *L. monocytogenes* are more tolerant to acid than previously believed.[68,69]

As a result of the work and concern of several consumer and scientific groups, President Clinton announced the Food Safety Initiative to improve the safety of the nation's food supply.[70] Because of the President's action, the subject of minimally processed food safety has received some attention at the federal level as congressional funding permits. Good manufacturing practices[71] can definitely help to minimize the risk. HACCP programs could be more effective than GMPs, but the reality is that, at present, there are no general federal requirements or regulations for HACCP specifically designed for the minimally processed food industry. Responsible processors must recognize and assume the responsibility of minimizing the risks to consumers. They are called upon to develop programs suitable and appropriate to their own manufacturing/processing operations. This, in itself, is evidence of the need for an urgent concerted action between government, industry, and academia to improve the safety situation in a segment of the food manufacturing industry that is growing at a fast pace.

In view of the situation, the FDA and the USDA address specific issues for minimally processed fruits and vegetables as they arise. They have already issued positions on certain minimally processed foods. For instance, the FDA established a regulation, effective November 5, 1998, requiring all packaged juice products not pasteurized or otherwise processed to prevent, reduce, or eliminate pathogens, to carry a warning statement informing consumers of the risk of food-borne illnesses, particularly to children, the elderly, and persons with weakened immune systems. Juices processed in a manner that achieves a 5-log reduction in pathogenic microorganisms are not required to bear the statement (63 *Federal Register* 37030–37042).[72] Citrus fruit juice manufacturers were given an additional eight months to comply with the new requirement for packaged fresh citrus fruit juices.

The FDA is prepared to charge that any product relying entirely on refrigeration for safety during storage, distribution, and retailing is adulterated under Section 402(a)(4) of the Food, Drug and Cosmetic Act if it has been prepared, packed, or held under unsanitary conditions, and it will be no longer considered a safe product, even if labeled "Keep Refrigerated," if the product does not contain an additional barrier to microbial growth.[64]

The FDA periodically dispatches unannounced inspectors with a written notice of inspection to inspect food production, storage, and distribution companies and report evidence of violation of the various acts enforced by the FDA. The agency maintains three current publications (*Inspection Operations Manual, Inspection Training Manual,* and *Inspector's Technical Guide*) that provide guidelines for an FDA inspector or consumer safety officer. On the basis of the guidelines and of enforcement programs for the specific inspection assignment, the officer examines the record(s) from which he/she should be able to understand the nature of the product, methods of operation, and last performance. In addition to plant sanitation, the officer is authorized to review the firm's analytical data; perform tests, whenever feasible, to determine product contamination and/or adulteration and ingredient stability; check the use and misuse of food ingredients, the accuracy of product labels, adequate packaging of the product; and document all evidence of violations. A common "food plant inspection checklist" includes inspection of:

1. raw materials receiving, handling, and storage;
2. plant construction and design;
3. plant sanitation practices;
4. personnel sanitary practices;
5. processing equipment and utensils maintenance, storage, and handling;
6. sewage and garbage disposal systems;
7. insects, rodents, vermin, and animal control;
8. pesticide usage and control; and
9. processing of food and control (including packaging, coding, and record keeping).

Section 704(b) of the Federal Food, Drug and Cosmetic Act (21 U.S.C. 374) further states that, upon completion of any inspection of a factory, warehouse, etc., and prior to leaving the premises, the officer making the inspection shall give the legal representative of the factory a written report, including all details of findings and observations, field test results, samples collected, and any photographic evidence of violative conditions which, in his/her judgment, indicate whether any food in the establishment consists, in whole or in part, of any filthy, putrid, or decomposed substance, or if it has been prepared, packed, or held under unsanitary conditions whereby it may have become contaminated with filth or have been rendered injurious to health. These regulatory measures, general for all food manufacturing facilities, are an excellent safeguard for the quality of minimally processed fruits and vegetables.

PRESENT STATUS ON REGULATORY ISSUES FOR MINIMALLY PROCESSED FOODS

The United States

Today's regulations regarding fruits and vegetables were influenced by President Clinton's health and safety initiative.

The President's Produce Safety Initiative

In May of 1997, as part of this initiative, the Department of Health and Human Services (DHHS), the U.S. Department of Agriculture (USDA), and the Environmental Protection Agency (EPA) sent the President a report identifying produce as an area of concern.[73] On October 2, 1997, President Clinton announced a plan entitled *Initiative to Ensure the Safety of Imported and Domestic Fruits and Vegetables* ("Produce Safety Initiative") to provide further assurance that fruits and vegetables consumed by U.S. citizens, whether grown domestically or imported, meet the highest health and safety standards.[74] In response to this, the FDA and the USDA issued *Guidance for Industry—Guide to Minimize Microbial Food Safety Hazards for Fresh Fruits and Vegetables*,[48] addressing microbial food safety hazards, good agricultural practices (GAPs), and GMPs common to the growing, harvesting, washing, sorting, packing, and transporting of most fruits and vegetables sold to consumers in an unprocessed or minimally processed (raw) form. Among the most important aspects of agricultural practices, the document covers control of water quality, manure and biosolids, workers' health and sanitation, field sanitation, and transportation.

The guide's purpose is to assist the United States and foreign produce industry in enhancing the safety of domestic and imported produce by addressing common areas of concern in the growing, harvesting, sorting, packing, and distribution of fresh produce. It identifies the broad microbial hazards associated with each area of concern, the scientific basis of that concern, and good agricultural and management practices for reducing the risk of contamination in fresh produce. It constitutes a voluntary, science-based set of concepts to be used by both domestic and foreign fresh fruit and vegetable producers to help ensure the safety of their produce, consistent with U.S. trade rights and obligations, and will not impose unnecessary or unequal restrictions or barriers on either domestic or foreign producers. Because it is a guidance and not a regulation, the guide does not have the force and effect of law and, thus, is not subject to enforcement. In no case do the recommendations in the guide supersede applicable federal, state, or local laws or regulations for U.S. manufacturers. Manufacturers outside the United States should follow corresponding or similar standards, laws, or regulations and should use the guide's general recommendations to tailor food safety practices appropriate to their particular operations.

The guide considers the fact that the scientific basis for reducing or eliminating pathogens in an agricultural setting is evolving and not yet complete. The examples of good agricultural practices and good management practices presented may not apply to all types of fresh and minimally processed produce. They need to be implemented when and where appropriate and are intended to build broad industry understanding and awareness of practices that individual growers, packers, and shippers may consider and incorporate in their own operations. In summary, the guide represents generally accepted, broad-based agricultural concepts developed from current knowledge of food safety practices of FDA and USDA, in cooperation with experts from other federal and state government agencies and the fresh produce industry. It does not address all microbiologic hazards potentially associated with fresh produce, but it provides the framework for identifying and implementing appropriate measures most likely to minimize risk on the farm, in the packinghouse, and during transport. Some important facts considered in the guide include:

- Focus on microbial hazards for fresh produce. It does not specifically address other areas of concern to the food supply or the environment (such as pesticide residues or chemical contaminants).
- Focus on risk reduction, not risk elimination, because current technologies cannot eliminate all potential food safety hazards associated with fresh produce that will be eaten raw.
- Provides broad, scientifically based principles. Manufacturers should assess microbiologic hazards within the context of the specific conditions (climatic, geographic, cultural, economic) that apply to their own operation and implement appropriate and cost-effective risk reduction strategies.

The guide is one of the first steps under the President's produce safety initiative to improve the safety of fresh produce as it moves from the farm to the table. It focuses on all stages of the farm-to-table food chain: grocery stores, institutions, restaurants, and other retail establishments.[75] The FDA is also actively seeking assistance from the Conference for Food Protection (a consortium of state, local, and federal agencies, academia, consumer and industry representatives) in identifying practical areas that may assist in reducing or eliminating microbial contamination of fresh produce at the retail level. In addition, as part of the President's food safety initiative, educational programs will promote consumer education to help improve safe food handling. Identifying and supporting research priorities designed to help fill gaps in food safety knowledge are another focus of the food safety initiative.[76] Research on and risk assessment of fresh produce will be incorporated in the multiyear food safety initiative research planning process. The overall goal is the development of cost-effective intervention and prevention strategies to reduce the incidence of food-borne illness. Research will also support development of improved detection methods targeted to sources of contamination.

The public health concerns stemming from minimally processed refrigerated fruits and vegetables are essentially microbiologic and chemical in nature. Safety issues and regulatory concerns have not been clearly identified or fully clarified so as to satisfy concerns of industry and public health officials about consumer protection. The safeguard against economic deception in food has always been truthful labeling. The microbiologic hazards associated with the production and distribution of minimally processed foods touch on the question of food additives, particularly indirect additives that might migrate from recent advances in packaging technology and help to make minimally processed refrigerated foods attractive choices. Minimally processed foods are not "commercially sterile." Because refrigeration is usually part of the preservation process, the foods are sensitive to temperature and consumer abuse and are limited in shelf life. The potential for consumers to mishandle this type of product is considerable because, in many instances, the products are perceived as shelf-stable. Optional storage temperatures for minimally processed foods range from 0°C to 4°C, with an upper maximal limit of around 7°C. Refrigeration, which is relied upon to ensure that minimally processed foods safely achieve their intended shelf life, is unfortunately the component, barrier, or part of the preservation feature that is most difficult to control. Exhibit 18–3 outlines the contents of the *Guide to Minimize Microbial Food Safety Hazards for Fresh Fruits and Vegetables*.

The Current Good Manufacturing Practices

The Current Good Manufacturing Practice Regulations[71] list what types of buildings (including building design and construction, lighting, ventilation), facilities (toilet and washing facilities), equipment cleaning and maintenance, handling of materials, and vermin control are needed and the errors to avoid to ensure good food plant sanitation. Further processing and manufacturing into finished foods in no way relieve raw materials from the requirements of cleanliness and freedom from deleterious impurities.

Foods that are free from contamination when they are shipped sometimes become contaminated en route and must be detained or seized. This emphasizes the importance of insisting on proper storage conditions in vessels, railroad cars, or other conveyances. Although the shipper may be blameless, the law requires action against illegal merchandise, no matter where it may have become illegal. All shippers should pack their products to protect them against spoilage or contamination en route and should urge carriers to protect the merchandise by maintaining sanitary conditions, segregating food from other cargo that might contaminate it. For example, vessels transporting foods may also carry other materials that can be contaminants or poisonous. Where import shipments become contaminated after customs entry, landing legal actions are not taken under the import provisions of the law but by seizure proceedings in a federal district court, as with domestic interstate shipments. The regulations governing current GMP are located in CFR Title 21, Part 110, as shown in Exhibit 18–4. This information has been expanded with the purpose of clarifying the provisions for the benefit of the reader.

The European Community

Many countries in Western Europe are also making significant strides toward solving the myriad problems associated with the manufacturing, distribution, and marketing of minimally processed fruits and vegetables. This appears to be due to strong governmental and food industry support for minimally processed food applications and a conducive regulatory climate in preparation for the expanding food markets that will be available within the European Community. A unified approach to classifying minimally processed foods as a specific food preservation industry/method is evident in many

Exhibit 18–3 Guide to Minimize Microbial Food Safety Hazards for Fresh Fruits and Vegetables

Preface

Introduction

Use of This Guide

Basic Principles

I. Definitions

II. Water
- A. Microbial Hazard
- B. Control of Potential Hazards
  - 1.0 Agricultural Water
  - 1.1 General Considerations
  - 1.2 Microbial Testing of Agricultural Water
  - 2.0 Processing Water
  - 2.1 General Considerations
  - 2.2 Antimicrobial Chemicals
  - 2.3 Wash Water
  - 2.4 Cooling Operations

III. Manure and Municipal Biosolids
- A. Microbial Hazard
- B. Control of Potential Hazards
  - 1.0 Municipal Biosolids
  - 2.0 Good Agricultural Practices for Manure Management
  - 2.1 Treatments to Reduce Pathogen Levels
    - 2.1.1 Passive Treatments
    - 2.1.2 Active Treatments
  - 2.2 Handling and Application
    - 2.2.1 Untreated Manure
    - 2.2.2 Treated Manure
  - 3.0 Animal Feces

IV. Worker Health and Hygiene
- A. Microbial Hazards
- B. Control of Potential Hazards
  - 1.0 Personal Health and Hygiene
  - 2.0 Training
  - 3.0 Customer-Pick Operations and Roadside Produce Stands

V. Sanitary Facilities
- A. Microbial Hazards
- B. Control of Potential Hazards
  - 1.0 Toilet Facilities and Hand-Washing Stations
  - 2.0 Sewage Disposal

VI. Field Sanitation
- A. Microbial Hazards
- B. Control of Potential Hazards
  - 1.0 General Harvest Considerations
  - 2.0 Equipment Maintenance

VII. Packing Facility Sanitation
- A. Microbial Hazard
- B. Control of Potential Hazards
  - 1.0 General Packing Considerations
  - 2.0 General Considerations for Facility Maintenance
  - 3.0 Pest Control

VIII. Transportation
- A. Microbial Hazard
- B. Control of Potential Hazards
  - 1.0 General Considerations
  - 2.0 General Transport Considerations

IX. Traceback

X. Conclusion

References

scientific communities; this would greatly assist in its development.

CONCLUSION

Protecting the safety of the U.S. food supply requires a comprehensive and coordinated effort throughout the food production and transportation system. The responsibility to safeguard our food supply is shared by everyone involved, from the grower to the consumer. This includes growers, farm workers, packers, shippers, transporters, importers, wholesalers, retailers, government agencies, and consumers.

Exhibit 18–4 Current Good Manufacturing Practice Regulations

Subpart A. General Provisions
110.3 Definitions
110.5 Good Manufacturing Practices
110.10 Personnel
- a) Disease Control
- b) Cleanliness
- c) Education and Training
- d) Supervision

110.19 Exclusions

Subpart B. Buildings and Facilities
110.20 Plant and Grounds
- a) Grounds
- b) Plan Construction and Design

110.35 Sanitary Operations
- a) General Maintenance
- b) Substances Used in Cleaning and Sanitizing. Storage of Toxic Materials
- c) Pest Control
- d) Sanitation of Food Contact Surfaces
- e) Storage and Handling of Cleaned Portable Equipment and Utensils

110.36 Sanitary Facilities and Controls
- a) Water Supply
- b) Plumbing
- c) Sewage Disposal
- d) Toilet Facilities
- e) Hand-Washing Facilities
- f) Rubbish and Offal Disposal

Subpart C. Equipment
110.40 Equipment and Utensils

Subpart E. Production and Process Controls
110.80 Processes and Controls
- a) Raw Materials and Other Ingredients
- b) Manufacturing Operations

110.93 Warehousing and Distribution

Subpart G. Defect Action Levels
110.110 Natural or Unavoidable Defects in Food for Human Use that Present No Health Hazard

The FDA/USDA guidance document, issued under the provisions of the President's produce safety initiative, provides some basic principles and recommended practices for operators to consider that will help to minimize microbial food safety hazards in the production, packing, and transport of fresh fruits and vegetables. Although research is ongoing and will continue to provide new information and improved technologies, the industry is urged to take a proactive role to minimize those microbial hazards over which they have control. Operators are encouraged to utilize this guide to evaluate their own operations and to assess site-specific hazards, so they can develop and implement reasonable and cost-effective agricultural and management practices to minimize microbial food safety hazards.

As outlined in the guide, analyzing the risk of microbial contamination includes a review of five major areas of concern. These involve:

1. water quality;
2. manure/municipal biosolids;
3. worker hygiene;
4. field, facility, and transport sanitation; and
5. traceback.

Growers, packers, and shippers should consider the variety of physical characteristics of produce and practices that affect the potential sources of microbial contamination associated with their operation and decide which combination of good agricultural and management practices is most cost-effective for them.

Once good agricultural and manufacturing practices are in place, it is important that the operator ensure that the process is working correctly. Operators should follow up with supervisors, or the person in charge, to be sure that regular monitoring takes place, equipment is working, and good agricultural and management practices are being followed. Without accountability to ensure that the process is working, the best attempts to minimize microbial food safety hazards in fresh fruits and vegetables are subject to failure.

REFERENCES

1. Mermelstein N. Minimal processing of produce. *Food Technol.* 1998;52(12):84–86.
2. Vetter J. *Personal communication.* 1999
3. García-Villanova B, Gálvez-Vargas R, García-Villanova R. Contamination of fresh vegetables during cultivation and marketing. *Int J Food Microbiol.* 1987;4:285–291.
4. Adams MR, Hartley AD, Cox LJ. Factors affecting the efficacy of washing procedures used in the production of prepared salads. *Food Microbiol.* 1989;6:69–77.
5. Wells JH, Singh RP. Quality management during storage and distribution. In: Taub IA, Singh RP, ed. *Food Storage Stability.* Boca Raton, FL: CRC Press; 1998:369–386.
6. Brackett RE. Shelf stability and safety of fresh produce as influenced by sanitation and disinfection. *J Food Prot.* 1992;55:808–814.
7. Hurst WC, Schuler GA. Fresh produce processing. An industry perspective. *J Food Prot.* 1992;55:824–827.
8. Barbosa-Cánovas GV, Pothakamury UR, Swanson BG. State of the art technologies for the stabilization of foods by non-thermal processes: Physical methods. In: Barbosa-Cánovas GV, Welti-Chanes J, ed. *Food Preservation by Moisture Control: Fundamentals and Applications: ISOPOW Practicum II.* Lancaster, PA: Technomic Publishing Co.; 1995:493–532.
9. Barbosa-Cánovas GV, Qin B, Swanson BG. The study of critical variables in the treatment of foods by pulsed electric fields. In: Fito P, Ortega-Rodríguez E, Barbosa-Cánovas GV, eds. *Food Engineering 2000.* New York: Chapman & Hall; 1997:141–159.
10. Pothakamury UR, Barbosa-Cánovas GV, Swanson BG. Magnetic field inactivation of microorganisms and generation of biological changes. *Food Technol.* 1993;47(12):85–92.
11. Esthiagi M, Knorr D. Potato cubes response to water blanching and high-hydrostatic pressure. *J Food Sci.* 1993;58:1371–1374.
12. Knorr D. Effects of high-hydrostatic pressure processes on food safety and quality. *Food Technol.* 1993;47(6): 156–161.
13. Dunn JE, Ott T, Clark W. Pulsed-light treatment of food and packaging. *Food Technol.* 1995;49(9):95–98.
14. Hayes D, Murano E, Murano P, Olson D, et al. Quality of irradiated foods. In: Murano E, ed. *Food Irradiation.* Iowa: Iowa State University Press; 1995:63–87.
15. Tapia de Daza MS, Alzamora SM, Welti-Chanes J. Minimally processed high-moisture fruit products by combined methods: Results of a multinational project. In: Fito P, Ortega-Rodríguez E, Barbosa-Cánovas GV, eds. *Food Engineering 2000.* New York: Chapman & Hall; 1997:161–180.
16. Welti-Chanes J, Vergara-Balderas F, Lopez-Malo A. Minimally processed foods. State of the art and future. In: Fito P, Ortega-Rodríguez E, Barbosa-Canovas GV, eds. *Food Engineering 2000.* New York: Chapman & Hall; 1997:181–212.
17. Ponting JD, Jackson R, Watters G. Refrigerated apple slices. Preservative effects of ascorbic acid, calcium and sulfites. *J Food Sci.* 1972;37:434–436.
18. Rolle RS, Chism GW III. Physiological consequences of minimally processed fruits and vegetables. *J Food Qual.* 1987;10:157–177.
19. Wiley RC. Introduction to minimally processed refrigerated fruits and vegetables. In: Wiley RC, ed. *Minimally Processed Refrigerated Fruits and Vegetables.* New York: Chapman & Hall; 1994:1–14.
20. Nutter J, Kotch A. Nutrition education strategies. A trip to the salad bar. *J Health Educ.* 1995;26:311–313.
21. Bruhn CM. Consumer perceptions of quality. In: Singh RP, Oliveira FAR, eds. *Minimal Processing of Foods and Process Optimization. An Interface.* Boca Raton, FL: CRC Press; 1994:493–504.
22. Hutin YJ, Paul V, Cramer EH, Nainan OV, et al. A multistate foodborne outbreak of hepatitis A. *N Engl J Med.* 1999;340:595–602.
23. CDC. Hepatitis A associated with consumption of frozen strawberries—Michigan, March 1997. Atlanta, GA: Centers for Disease Control and Prevention. *MMWR Morb Mortal Wkly Rep.* 1997;46(13):288,295.
24. Zepp G, Kuchler F, Lucier G. Food safety and fresh fruits and vegetables: Is there a difference between imported and domestically produced products? *Vegetables and Specialties, Situation and Outlook Report.* ERS/USDA, 1998; VGS-274:23–28, April.
25. Fain AR. A review of the microbiological safety of fresh salads. *Dairy Food Environ Sanit.* 1996;16:146–149.
26. Reid TM, Robinson HG. Frozen raspberries and hepatitis A. *Epidemiol Infect.* 1987;98:109–112.
27. Ponka A, Maunula L, von Bonsdorff CH, Lyytikainen O. Outbreak of calicivirus gastroenteritis associated with eating frozen raspberries. *Eurosurveillance.* 1999;4: 66–69.
28. Rocourt J, Bille, J. Foodborne listeriosis. *World Health Stat Q.* 1997;50:67–73.

29. Tappero JW, Schuchat A, Deaver KA, Mascola L, et al. Reduction in the incidence of human listeriosis in the United States. Effectiveness of prevention effort. The Listeriosis Study Group. *JAMA.* 1995;273:1118–1122.
30. Jay J. Radiation preservation of foods and nature of microbial radiation resistance. In: *Foodborne Listeriosis.* New York: Chapman & Hall; 1996.
31. Gahan C, O'Driscoll B, Hill C. Acid adaptation of *Listeria monocytogenes* can enhance survival in acidic foods and during milk fermentation. *Appl Environ Microbiol.* 1996;52:3129–3132.
32. Ryser E, Marth, E. *Listeria, Listeriosis and Food Safety.* New York: Marcel Dekker; 1991.
33. Hart CD, Mead GC, Morris AP. Effects of gaseous environment and temperature on the storage behaviour of *Listeria monocytogenes* on chicken breast meat. *J Appl Bacteriol.* 1991;70:40–46.
34. Zhang S, Farber JM. The effects of various disinfectants against *Listeria monocytogenes* on fresh-cut vegetables. *Food Microbiol.* 1996;13:311–321.
35. Beuchat L, Ryu J. Produce handling and processing practices. *Emerg Infect Dis.* 1997;3:1–10.
36. Bibb W, Gellin BG, Weaver R, Schwartz B, et al. Analysis of clinical and foodborne isolates of *Listeria monocytogenes* in the United States by multilocus enzyme electrophoresis and application of the method to epidemiologic investigations. *Appl Environ Microbiol.* 1990;56:2133–2141.
37. Farber J, Peterkin P. *Listeria monocytogenes*, a foodborne pathogen. *Microbiol Rev.* 1991;55:476–511.
38. Altekruse S, Cohen M, Swerdlow D. Emerging foodborne diseases. Centers for Disease Control and Prevention. *Emerg Infect Dis.* 1997;3(3):1–12.
39. Cheng TS, Shewfelt RL. Effect of chilling exposure of tomatoes during subsequent ripening. *J Food Sci.* 1988;53:1160–1162.
40. Nguyen-the C, Carlin F. The microbiology of minimally processed fresh fruits and vegetables. *Crit Rev Food Sci Nutr.* 1994;34:371–401.
41. Beuchat L. *Listeria monocytogenes*. Incidence in vegetables. *Food Control.* 1996;7:223–228.
42. Doyle MP. Fruit and vegetable safety. Microbiological considerations. *Hortsci.* 1990;25:1478–1482.
43. Geldreich EE, Bordner RH. Fecal contamination of fruits and vegetables during cultivation and processing for market. A review. *J Milk Food Technol.* 1971;34: 184–185.
44. Welshimer HJ, Donker-Voet J. *Listeria monocytogenes* in nature. *Appl Microbiol.* 1971;21:516–519.
45. Schlech WF 3d, Lavigne PM, Bortolussi RA, Allen AC, et al. Epidemic listeriosis. Evidence for transmission by food *Listeria monocytogenes*. *N Engl J Med.* 1983;308:203–206.
46. Wang W, Zhao, Doyle MP. Fate of enterohemorrhagic *Escherichia coli* O157:H7 in bovine feces. *Appl Environ Microbiol.* 1996;62:2567–2570.
47. Ho JL, Shands KN, Friedland G, Eckind P, et al. An outbreak of type 4b *Listeria monocytogenes* infection involving patients from eight Boston hospitals. *Arch Intern Med.* 1986;146:520–524.
48. FDA/USDA. Guidance for industry. Guide to Minimize Microbial Food Safety Hazards for Fresh Fruits and Vegetables. *Federal Register* 63 FR 58055. Oct. 29, 1998. <http://vm.cfscan.fda.gov/~lrd/fr981029.html>
49. Norman NN, Kabler PW. Bacteriological study of irrigated vegetables. *Sewage Ind Waste.* 1953;25:605–609.
50. Bartz JA, Eckert JW. Bacterial diseases of vegetable crops after harvest. In: Weichmann J, ed. *Postharvest Physiology of Vegetables.* New York: Marcel Dekker; 1987:351–376.
51. Hurst WC. Disinfection methods. A comparison of chlorine dioxide, ozone and ultraviolet light alternatives. *Cutting Edge Prod Assoc.* 1995; 9:4–5.
52. Graham DM. Use of ozone for food processing. *Food Technol.* 1997;51(6):72–75.
53. OSHA. 29 CFR 1928.110, Subpart I. *Field Sanitation. General Environmental Controls*. Washington, DC: Occupational Safety and Health Administration. U.S. Department of Labor; as amended, May 1987.
54. OSHA. 29 CFR 1910.141, Subpart J. *Sanitation. General Environmental Controls*. Washington, DC: Occupational Safety and Health Administration. U.S. Department of Labor; as amended, June 1998.
55. Prince TA. Modified atmosphere packaging of horticultural commodities. In: Brody AL, ed. *Controlled/Modified Atmosphere/Vacuum Packaging of Foods.* Trumbull: Food & Nutrition Press; 1989:67–100.
56. NACMCF. *Microbiological Safety Evaluations and Recommendations on Fresh Produce*. Report by the National Advisory Committee on Microbiological Criteria for Foods; 1998, Mar 5.
57. *USDA Agricultural Marketing Service Program*. Qualified through Verification for Fresh Cut Produce. Processed Products Branch, Fruit and Vegetable Programs. Washington, DC: Agricultural Marketing Service, USDA; 1997.
58. Corlett DA. Refrigerated foods and use of hazard analysis and critical control point principles. *Food Technol.* 1989;43(2):91–94.
59. Refrigerated Foods and Microbiological Criteria Committee of the National Food Processors Association. Safety considerations for new generation refrigerated foods. *Dairy Food Environ Sanit.* 1988;8:5–7.
60. National Food Processors Microbiology and Food Safety Committee. *Guidelines for the Development, Production and Handling of Refrigerated Foods.* Washington, DC; 1989.

61. Kalish F. Extending the HACCP concept to product distribution. *Food Technol.* 1991;45(6):119–120.

62. Keller S. *What Is R&D's Future Role in the Quality and Safety of Refrigerated Foods*. Presented at the Refrigerated Food Symposium. Rosemont, IL: 1989; April 5–6.

63. IFPA. *Food Safety Guidelines for the Fresh-Cut Produce Industry*, 3rd ed. Alexandria: International Fresh-Cut Produce Association (IFPA); 1996.

64. Van Garde SJ, Woodburn MJ. Food discard practices of householders. *J Am Diet Assoc.* 1987;87:322–329.

65. Dignan DM. Regulatory issues associated with minimally processed refrigerated foods. In: Wiley RC, ed. *Minimally Processed Refrigerated Fruits and Vegetables.* New York: Chapman & Hall; 1994:327–353.

66. Cook K, Dobbs TE, Hlady WG, Well JG, et al. Outbreak of *Salmonella* serotype *Hartford* infection associated with unpasteurized orange juice. *JAMA.* 1998;280:1504–1505.

67. IFT. Scientific Status Summary. Foodborne illnesses. Role of home food handling practices. Institute of Food Technologists' expert panel on food safety and nutrition. *Food Technol.* 1995;49(4):119–131.

68. Goverd KA, Beech FW, Hobbs RP, Shannon R. The occurrence and survival of coliforms and salmonella in apple juice and cider. *J Appl Bacteriol.* 1979;4:521–530.

69. Ita PS, Hutkins RW. Intracellular pH and survival of *Listeria monocytogenes* Scott A in tryptic soy broth containing acetic, lactic, citric and hydrochloric acids. *J Food Prot.* 1991;54:15–19.

70. The White House. Office of the Press Secretary. *Radio Address of the President to the Nation.* 1997; Jan. 25.

71. FDA. Current good manufacturing practices in manufacturing, packing or holding of human foods. Title 21. *Code of Federal Regulations.* Washington, DC: Food and Drug Administration, Department of Health and Human Services; as amended, April 1999.

72. FDA/CFSAN. *FDA Technical Scientific Workshop on How Citrus Juice Firms Can Achieve 5-Log Pathogen Reduction.* Washington, DC: U.S. Food and Drug Administration, Center for Food Safety and Applied Nutrition, Office of Food Labeling; 1998. <http//vm.cfsan.fda.gov/~dms/citrus.html>

73. EPA. *Food Safety from Farm to Table: A National Food-Safety Initiative. A Report to the President.* Washington, DC: U.S. Environmental Protection Agency, Department of Health and Human Services, and U.S. Department of Agriculture; May 1997.

74. The White House. *Memorandum for the Secretary of Health and Human Services*. Washington, DC: The Secretary of Agriculture, 1997; Oct. 2.

75. FDA. *1997 Food Code*. Washington, DC: Department of Health and Human Services, U.S. Public Health Service, Food and Drug Administration; 1997.

76. FDA/USDA. *Initiative to Ensure the Safety of Imported and Domestic Fruits and Vegetables: Status Report.* Washington, DC: 1998; Feb. 24.

List of Sources

OVERVIEW

Exhibit O–1 Adapted from J. Welti-Chanes, F. Vergara-Balderas, and A. López-Malo, Minimally Processed Foods: State of the Art and Future, in *Food Engineering 2000*, P. Fito-Maupoey, E. Ortega-Rodriguez, and G.V. Barbosa-Cánovas, eds., p. 183, © 1997, Aspen Publishers, Inc.

Figure O–1 Data from G.W. Gould, Overview, in *New Methods of Food Preservation*, G.W. Gould, eds., pp. XV-XIX, © 1995, Blackie Academic Professional; G.W. Gould, The Microbe as a High Pressure Target, in *High Pressure Processing of Foods*, D.A. Ledward et al., eds., © 1995, Nottingham University Press; and G.W. Gould, Industry Perspectives on the Use of Natural Antimicrobials and Inhibitors for Food Applications, *Journal of Food Protection*, Supplement, pp. 82–86, © 1996, International Association of Food Protection.

Table O–1 Adapted from J. Welti-Chanes, et al., Role of Water in the Stability of Minimally or Partially Processed Foods, in *Water Management in the Design and Distribution of Quality Foods*, Y.H. Roos, R.B. Leslie, and P.J. Lilford, eds., pp. 503–532, with permission from Technomic Publishing Co., Inc., Copyright 1999.

Table O–2 Adapted from S.M. Alzamora, M.S. Tapia, and J. Welti-Chanes, New Strategies for Minimal Processing of Foods: The Role of Multitarget Preservation, *Food Science Technology International*, Vol. 4, pp. 353–361, © 1998, Aspen Publishers, Inc.

CHAPTER 2

Figure 2–1 Adapted from I.R. Booth, Bacterial Responses to Osmotic Stress: Diverse Mechanisms to Achieve a Common Goal, in *Properties of Water in Foods ISOPOW 6*, D.S. Reid, eds., © 1998 Aspen Publishers, Inc.

Figure 2–2 Data from B.M. Mackey and C.M. Derrick, Elevation of the Heat Resistance *of Salmonella Typhimurium* by Sublethal Heat Shock, *Journal of Applied Bacteriology*, Vol. 61, pp. 389–393, © 1986; G.W. Gould, Heat-Induced Injury and Inactivation, in *Mechanisms of Action of Food Preservation Procedures*, G.W. Gould, eds., pp. 11–42, © 1989, Elsevier Applied Science; and W.S. Thompson, F.F. Busta, D.R. Thompson, and C.E. Allen, Inactivation of Salmonellae in Autoclaved Ground Beef Exposed to Constantly Rising Temperature, *Journal of Food Protection*, Vol. 42, pp. 410–415, © 1979, International Association of Food Protection.

Table 2–1 Adapted from *International Journal of Food Microbiology*, Vol. 33, G.W. Gould, Methods for Preservation and Extension of Shelf Life, p. 62, Copyright 1996, with permission from Elsevier Science.

CHAPTER 4

Table 4–2 Adapted with permission from R.-Y. Zhuang, L.R. Beuchat, and F.J. Angulo, Fate of *Salmonella Montevideo* on and in Raw Tomatoes as Affected by Temperature and Treatment with Chlorine, *Applied Environmental Microbiology*, Vol. 61, p. 2129, © 1995, American Society for Microbiology.

Table 4–3 Adapted with permission from S. Zhang and J.M. Farber, The Effects of Various Disinfectants Against Listeria Monocytogenes on Fresh-Cut Vegetables, *Food Microbiology*, Vol. 13, pp. 311–321, © 1996, Cornell University.

Table 4–4 Data from References 51–52 and 72–81.

Table 4–5 Adapted with permission from T. Deak and L.R. Beuchat, *Handbook of Food Spoilage Yeasts*, p. 49, Copyright 1996, CRC Press, Boca Raton, Florida.

CHAPTER 5

Figure 5–1 Adapted with permission from R. Stier, Practical Application of HACCP, M,C,P,S: Critical Control Points: Microbiological, Chemical, Physical, Sanitation, in *HACCP Principles and Applications*, M.D. Pierson and D.A. Corlett, eds., pp. 127–167, © 1992, AVI Publishing, Co., Inc.

Table 5–1 Data from References 17, and 28–36.

Table 5–2 Courtesy of Dr. van Gerwen, personal communication, 1999.

CHAPTER 6

Table 6–1 Reprinted from *Food Research International*, Vol. 29, C.A. Campos, A. M. Rojas, and L. N. Gerschenson, Studies of the Effect of Ethylene Diamine Tetraacetic Acid (EDTA) on Sorbic Acid Degradation, pp. 259–264, Copyright 1996 with permission from Elsevier Science.

Table 6–2 Reprinted from *Journal of the Science of Food and Agriculture*, Vol. 74, A.M. Rojas and L.N. Gerschenson, pp. 369–378, Copyright 1997 Society of Chemical Industry. Reproduced with permission. Permission granted by John Wiley & Sons, Ltd. on behalf of the SCI.

Table 6–3 Reprinted with permission from L.N. Gerschenson, S.M. Alzamora, and J. Chirife, Stability of Sorbic Acid in Model Food Systems of Reduced Water Activity: Sugar Solutions, *Journal of Food Science*, Vol. 51, No. 4, pp. 1028–1031, © 1986, Institute of Food Technologists.

CHAPTER 8

Figure 8–1 (A and B) Adapted from *Journal of Food Engineering*, Vol. 35, No. 1, L.M.M. Tijskens, et al., Kinetics of Polygalacturonase Activity and Firmness of Peaches During Storage, pp. 111–126, Copyright 1998 with permission from Elsevier Science.

Figure 8–2 Adapted from *Journal of Food Engineering*, Vol. 35, No. 1, L.M.M. Tijskens, et al., Kinetics of Polygalacturonase Activity and Firmness of Peaches During Storage, pp. 111–126, Copyright 1998 with permission from Elsevier Science.

Figure 8–3 (A and B) Adapted from *Journal of Food Engineering*, Vol. 34, No. 4, L.M.M. Tijskens, et al., The Kinetics of Pectin Methyl Esterase in Potatoes and Carrots During Blanching, pp. 371–381, Copyright 1997 with permission from Elsevier Science. **(C)** Adapted from *Journal of Food Engineering*, Vol. 39, No. 2, L.M.M. Tijskens, et al., Activity of Pectin Methyl Esterase During Blanching of Peaches, pp. 167–177, Copyright 1999 with permission from Elsevier Science.

Figure 8–4 (A) Adapted from *Journal of Food Engineering*, Vol. 34, No. 4, L.M.M. Tijskens, et al., Activity of Peroxidas During Blanching of Peaches, Carrots, and Potatoes, pp. 355–370, Copyright 1997 with permission from Elsevier Science. **(B)** Adapted from *Journal of Food Engineering*, Vol. 39, No. 2, L.M.M. Tijskens, et al., Activity of Pectin Methyl Esterase During Blanching of Peaches, pp. 167–177, Copyright 1999 with permission from Elsevier Science.

Figure 8–5 (A and B) Adapted from *Journal of Food Engineering*, Vol. 34, No. 4, L.M.M. Tijskens, et al., Activity of Peroxidas During Blanching of Peaches, Carrots, and Potatoes, pp. 355–370, Copyright 1997 with permission from Elsevier Science.

Figure 8–6 (A and B) Adapted from *Journal of Food Engineering*, Vol. 34, No. 4, L.M.M. Tijskens, et al., Activity of Peroxidas During Blanching of Peaches, Carrots, and Potatoes, pp. 355–370, Copyright 1997 with permission from Elsevier Science.

Figure 8–7 Adapted from *Journal of Food Engineering*, Vol. 34, No. 4, LM.M. Tijskens, et al., The Kinetics of Pectin Methyl Esterase on Potatoes and Carrots During Blanching, pp. 371–385, Copyright 1997 with permission from Elsevier Science.

Table 8–1 Adapted from *Journal of Food Engineering*, Vol. 35, No. 1, L.M.M. Tijskens, et al., Kinetics of Polygalacturonase Activity and Firmness of Peaches During Storage, pp. 111–126, Copyright 1998 with permission from Elsevier Science.

Table 8–2 Adapted from *Journal of Food Engineering*, Vol. 35, No. 1, L.M.M. Tijskens, et al., Kinetics of Polygalacturonase Activity and Firmness of Peaches During Storage, pp. 111–126, Copyright 1998 with permission from Elsevier Science.

Table 8–3 Adapted from *Journal of Food Engineering*, Vol. 34, No. 4, L.M.M. Tijskens, et al., The

Kinetics of Pectin Methyl Esterase in Potatoes and Carrots During Blanching, pp. 371–385, Copyright 1997 with permission from Elsevier Science; *Journal of Food Engineering*, Vol. 39, No. 2, L.M.M. Tijskens, et al., Activity of Pectin Methyl Esterase During Blanching, pp. 167–177, Copyright 1999 with permission from Elsevier Science.

Table 8–4 Adapted from *Journal of Food Engineering*, Vol. 34, No. 4, L.M.M. Tijskens, et al., Activity of Peroxidas During Blanching of Peaches, Carrots, and Potatoes, pp. 355–370, Copyright 1997 with permission from Elsevier Science.

Table 8–5 Adapted from *Journal of Food Engineering*, Vol. 34, No. 4, L.M.M. Tijskens, et al., Activity of Peroxidas During Blanching of Peaches, Carrots, and Potatoes, pp. 355–370, Copyright 1997 with permission from Elsevier Science.

CHAPTER 9

Figure 9–1 Courtesy of Dr. Elena Ancibor.

Figure 9–3 Adapted from S.L. Vidales, M.A. Castro, and S.M. Alzamora, The Structure-Texture Relationship of Blanched and Glucose Impregnated Strawberries, *Food Science Technology International*, Vol. 4, p. 173–176, © 1998, Aspen Publishers, Inc.

Figure 9–6 Adapted with permission from S.M. Alzamora, et al., Structural Changes in Minimal Processing of Fruits: Some Effects of Blanching and Sugar Impregnation, in *Food Engineering 2000*, P.Fito, E. Ortego-Rodríguez, and G.V. Barbosa-Cánovas, eds., p. 134, © 1997, Aspen Publishers, Inc.; and A. Nieto, et al., Structural Effects of Vacuum Solutes Infusion in Mango and Apple Tissues, in *Drying '98*, Vol. C, C.B. Akritidis, D. Marinos-Kouris, and G.D. Saravacos, eds., and A.S. Mujumdar, series editor, pp. 2134–2141, © 1998, Marcel Dekker, Inc., N.Y.

Figure 9–8 Reprinted from A.M. Rojas, et al., Firmness and Structural Characteristics of Glucose Impregnated Melon, Proceedings of the Poster Session, ISOPOW 7, Helsinki, Finland, June 30-July 4, 1998.

CHAPTER 10

Figure 10–1 Data from B.E. Proctor and S.A. Goldblith, Radar Energy for Rapid Cooking and Blanching and its Effect on Vitamin Content, *Food Technology*, Vol. 2, No. 3, pp. 95–104, © 1948; M.S. Eheart, Effect of Microwave vs. Water-Blanching on Nutrients in Broccoli, *Journal of American Dietetic Association*, Vol. 50, pp. 207–211, © 1967; and M.S. Eheart and C. Gott, Chlorophyll, Ascorbic Acid and pH Changes in Green Vegetables Cooked by Stir-Fry, Microwave and Conventional Methods and a Comparison of Chlorophyll Methods, *Food Technology*, Vol. 25, No. 1, pp. 185–188, © 1965.

CHAPTER 12

Figure 12–1 Adapted with permission from Anonymous, Flow International Co., Technical Data, 1998.

Figure 12–2 (A and B) Data from E. Palou, et al., High Hydrostatic Pressure as a Hurdle *for Zygosaccharomyces Bailii* Inactivation, *Journal of Food Science*, Vol. 62, pp. 855–857, © 1997, Institute of Food Technologists.

Figure 12–3 Data from E. Palou, et al., Kinetic Analysis of *Zygosaccharomyces Bailii* Inactivation by High Hydrostatic Pressure, *Lebensmittel-Wissenschaft und-Technologie*, Vol. 30, pp. 703–708, © 1997, Academic Press and J. Perez, et al., Paper No. 59E-19, IFT Annual Meeting, June 14–18, 1997, Orlando, Florida.

Figure 12–4 Adapted from *Innovative Food Science and Emerging Technologies*, E. Palou, et al., High Hydrostatic Pressure Treated Guacamole, Copyright 1999 with permission from Elsevier Science.

Table 12–1 Data from References 19–20, 36, 42, 44, 47, 64, and 71–72.

Table 12–2 Data from J. Pérez, et al., Paper No. 59E-19, IFT Annual Meeting, June 14–18, 1997, Orlando, Florida.

Table 12–3 Data from E. Palou, et al., Kinetic Analysis of *Zygosaccharomyces Bailii* Inactivation by High Hydrostatic Pressure, *Lebensmittel-Wissenschaft und-Technologie*, Vol. 30, pp. 703–708, © 1997, Academic Press; C. Hashizume, K. Kimura, and R. Hayashi, Kinetic Analysis of Yeast Inactivation by High Pressure Treatment at Low Temperatures, *Bioscience, Biotechnology, and Biochemistry*, Vol. 59, pp. 1455–1458, © 1995; and J. Pérez, et al., Paper No. 59E-19, IFT Annual Meeting, June 14–18, 1997, Orlando, Florida.

Table 12–4 Data from E. Palou, et al., High Hydrostatic Pressure Treated Guacamole, *Innovative Food Science and Emerging Technologies*, Vol. 1, © 1999, Elsevier Science.

CHAPTER 13

Table 13–1 Adapted from P.C. Wouters and J.P.P.M. Smelt, Inactivation of Microorganisms with Pulsed Electric Fields: Potential for Food Preservation, *Food Biotechnology*, Vol. 11, No. 3, pp. 193–229, © 1997, Marcel Dekker, Inc., N.Y.

Figure 13–3 Adapted from *Journal of Food Engineering*, Vol. 31, S.Y. Hoy, G.S. Mittal, and J.D. Cross, Effects of High Field Electric Pulses on the Activity of Selected Enzymes, pp. 69–85, Copyright 1997 with permission from Elsevier Science.

Table 13–2 Adapted with permission from J. Raso, et al., Inactivation of *Zygosaccharomyces Bailii* in Fruit Juices by Heat, High Hydrostatic Pressure and Pulsed Electric Fields, *Journal of Food Science*, Vol. 63, No. 1, pp. 1042–1044, © 1998, Institute of Food Technologists.

Table 13–3 Adapted with permission from J. Raso, et al., Inactivation of Mold Ascospores and Conidiospores Suspended in Fruit Juices by Pulsed Electric Fields, *Lebensmittel-Wissenschaft und-Technologie*, Vol. 31, No. 7/8, pp. 668–672, © 1998, Academic Press, Ltd.

Table 13–4 Data from B.L. Quin, et al., Food Pasteurization Using High-Intensity Pulsed Electric Fields, *Food Technology*, Vol. 12, pp. 55–60, © 1995 and Q.H. Zhang, X. Qui, and S.K. Sharma, Recent Developments in Pulsed Electric Field Processing, in *New Technologies Yearbook*, pp. 31–42, © 1997, National Food Processors Association.

CHAPTER 14

Exhibit 14–1 Data from F. Sauer, Control of Yeast and Molds with Preservatives, *Food Technology*, Vol. 31, No. 2, pp. 66–67, © 1977; and J. Giese, Antimicrobials: Assuring Food Safety, *Food Technology*, Vol. 48, No. 6, pp. 102–110, © 1994.

Figure 14–3 Adapted with permission from A. López-Malo, S.M. Alzamora, and A. Argaiz, Vanillin and pH Synergistic Effects on Mold Growth, *Journal of Food Science*, Vol. 63, pp. 143–146, © 1998, Institute of Food Technologists.

Figure 14–4 Adapted with permission from A. López-Malo, S.M. Alzamora, and A. Argaiz, Effect of Vanillin Concentration, pH and Incubation Temperature on Aspergillus Flavus, A. Niger, A. Ochraceus and A. Parasiticus growth, Food Microbiology, Vol. 14, pp. 117–124, © 1997, Cornell University.

Figure 14–5 Adapted with permission from B. Matamoros-Leon, A. Argaiz, and A. López-Malo, Individual and Combined Effects of Vanillin and Potassium Sorbate on *Penicillium Digitatum*, *P. Glabrum* and *P. Italicum* growth, *Journal of Food Protection*, Vol. 62, No. 5. Copyright held by the International Association of Food Protection, Des Moines, Iowa, U.S.A.

Table 14–1 Data from References 7–8, 10, 12, and 16–18.

Table 14–2 Data from L.A. Shelef, Antimicrobial Effects of Spices, *Journal of Food Safety*, Vol. 6, pp. 29–44, © 1983; and K.T. Farrell, *Spices, Condiments and Seasonings*, 2nd Edition, © 1990, Van Nostrand Reinhold.

Table 14–3 Data from P.M. Davidson and M.E. Parish, Methods for Testing the Efficacy of Food Antimicrobials, *Food Technology*, Vol. 43, No. 1, pp. 148–155, © 1989; and M.E. Parish and P.M. Davidson, Methods of Evaluation, in *Antimicrobials in Foods*, P.M. Davidson and A.L. Branen, eds., pp. 597–615, © 1993, Marcel Dekker, Inc.

Table 14–4 Data from References 26, 34–38, 41–44, 47–49, 50–54, 67–71, 73, 75, 77–78, 89–90, and 108–112.

Table 14–5 Adapted from X. Castañón, A. Argaiz, and A. López-Malo, Effect of Storage Temperature on the Microbial and Color Stability of Banana Purees Prepared with the Addition of Vanillin or Potassium Sorbate, *Food Science Technology International*, Vol. 5, No. 1, pp. 53–54, © 1999, Aspen Publishers, Inc.

Table 14–6 Adapted from X. Castañón, A. Agraiz, and A. López-Malo, Effect of Storage Temperature on the Microbial and Color Stability of Banana Purees Prepared with the Addition of Vanillin or Potassium Sorbate, *Food Science Technology International*, Vol. 5, No. 1, p. 58, © 1999, Aspen Publishers, Inc.

Table 14–7 Data from F. Sahidi, P.K. Janitha, and P.D. Wanasundara, Phenolic Antioxidants, *Critical Reviews in Food Science and Nutrition*, Vol. 32, pp. 67–103, © 1992, CRC Press, Inc.

CHAPTER 15

Exhibit 15–1 Data from T.R. Klaenhammer, Genetics of Bacteriocins Produced by Lactic Acid Bacteria, *FEMS Microbiology Review*, Vol. 12, pp. 39–86, © 1993.

CHAPTER 16

Figure 16–1 Adapted with permission from R. Ahvenainen, et al., Shelf-Life of Pre-Peeled Potato Cultivated, Stored, and Processes by Various Methods, *Journal of Food Protection*, Vol. 61, No. 5, pp. 591–600, © 1998. Copyright held by the International Association of Milk, Food and Environmental Sanitarians, Inc., Des Moines, IA, USA.

Figure 16–2 Adapted with permission from R. Ahvenainen, et. al., Factors Affecting the Quality Retention of Minimally Processed Carrot, in Proceedings of the Sixth International Symposium of the European Concerted Action Program COST 94 'Post-Harvest Treatment of Fruit and Vegetables,' *Current Status and Future Prospects*, Oosterbek, 19–22 October 1994, pp. 41–47, © 1998, Luxembourg Office for Official Publications of the European Communities.

Table 16–1 Reprinted from *Trends in Food Science & Technology*, Vol. 7, R. Ahvenainen, New Approaches in Improving the Shelf Life of Minimally Processed Fruit and Vegetables, pp. 179–187, Copyright 1996 with permission from Elsevier Science.

Table 16–2 Reprinted with permission from R. Ahvenainen, and E. Hurme, Practical Guidelines for Minimal Processing of Vegetables, *Proceedings of International Symposium on Minimal Processing and Ready Made Foods*, pp. 31–41, SIK, Goteborg, April 18–19, 1996.

Table 16–3 Reprinted with permission from R. Ahvenainen, E. Hurme, Practical Guidelines for Minimal Processing of Vegetables, *Proceedings of International Symposium on Minimal Processing and Ready Made Foods*, SIK, Goteborg, pp. 31–41, April 18–19, 1996.

Table 16–4 Reprinted with permission from R. Ahvenainen, E. Hurme, Practical Guidelines for Minimal Processing of Vegetables, *Proceedings of International Symposium on Minimal Processing and Ready Made Foods*, pp. 31–41, SIK, Goteborg, April 18–19, 1996.

CHAPTER 17

Appendix–A

Flowchart 1—Adapted from S.M. Alzamora, et al., Minimally Processed Fruits by Combined Methods, in *Food Preservation by Moisture Control: Fundamentals and Applications*, G.V. Barbosa-Cánovas, J. Welti-Chanes, eds., pp. 463–492, ISOPOW Practicum II, with permission from Technomic Publishing Co. Inc., Copyright 1995.

Flowchart 2—Data from M.S. Tapia, et al., Minimally Processed Papaya by Vacuum Osmotic Dehydration (VOD) Techniques, *Food Science Technology International*, Vol. 5, pp. 43–52, © 1999, Aspen Publishers, Inc. and A. López-Malo, personal communication, 1999.

Flowchart 3—Data from V. Santacruz, Diseno de un Alimento Minimamente Procesado a Partir de Melon, Universidad de las Americas-Puebla, Mexico, 1998 Thesis; F. Vergara-Balderas, et al., Stability of Minimally Processed Melon Obtained by Vacuum Dehydration (VOD) Techniques, Paper No. 20A-1, presented at the 1998 IFT Annual Meeting, June 20–24, 1998; and M.S. Tapia, et al., Stability of Minimally Treated Melon (*Cucumis Melon, L.*) During Storage and Effect of the Water Activity Depression Treatment, Paper No. 22D-13, presented at the 1999 IFT Annual Meeting, July 24–29, 1999.

Flowchart 4—Data from C. Santacruz, Obtención de Gajos de Naranja Minimamente Procesados, Universidad de las Américas-Puebla, Mexico, 1998, Thesis; and J. Welti-Chanes, et al., Stability of Minimally Processed Orange Segments Obtained by Vacuum Dehydration Techniques, Paper No. 34B-8, presented at the 1998 IFT Annual Meeting, June 20–24, 1998.

Flowchart 5—Adapted with permission from M.A. Leunda, S.N. Guerrero, and S.M. Alzamora, Color and Cholorphyll Content Changes of Minimally Processed Kiwifruit, *Journal of Food Processes Preservation*, © 1999, Food and Nutrition Press, Inc.

Flowchart 6—Data from Y. Márquez, Desarrollo de un Producto de Lechosa (*Carica Papaya L*) Utilizando Jugo de Parchita (*Passiflora Edulis*) Como Solución Osmótica y de Impregnación, Instituto de Ciencia y Tecnologia de Alimentos, Universidad Central de Venezuela, 1998, Thesis.

Flowchart 7—Adapted with permission from P. Cerrutti, S.M. Alzamora, and S.L. Vidales, Vanillin as Antimicrobial for Producing Shelf-Stable Strawberry Purée, *Journal of Food Science*, Vol. 62, pp. 608–610, © 1997, Institute for Food Technologists.

Flowchart 8—Adapted from X. Castañón, A. Argaiz, and A. López-Malo, Effect of Storage Temperature on the Microbial and Color Stability of Banana Purée with Addition of Vanillin or Potassium Sorbate, *Food Science Technology International*, Vol. 5, No. 1, pp. 51–58, © 1999, Aspen Publishers, Inc.

Flowchart 9—Adapted from *Food Research International*, Vol. 31, A. López-Malo, et al., Polyphenoloxidase Activity and Color Changes During Storage of High Hydrostatic Pressure Treated Avocado Purée, pp. 549–556, Copyright 1999 with permission from Elsevier Science.

Flowchart 10—Reprinted from *Journal of Food Science and Emerging Technologies*, E. Palou, et al., High Pressure Treated Guacamole, © 1999, with permission from Elsevier Science.

Flowchart 11—Adapted with permission from E. Palou, et al., Polyphenoloxidase Activity and Color of Blanched and High Hydrostatic Pressure Treated Banana Purée, *Journal of Food Science*, Vol. 64, pp. 42–45, © 1999, Institute of Food Technologists.

CHAPTER 18

Exhibit 18–2 Reprinted with permission from *Food Safety Guidelines for the Fresh-Cut Produce Industry, 3rd Edition*, © 1996, International Fresh-Cut Produce Association.

Exhibit 18–3 Reprinted from Guidance for Industry, Guide to Minimize Microbial Food Safety Hazards for Fresh Fruit and Vegetables, Federal Register 63, FR 58055, October 29, 1998, Food and Drug Administration, U.S. Department of Agriculture, http://vm.cfsan.fda.gov/~lrd/fr981029.html.

Exhibit 18–4 Reprinted from Current Good Manufacturing Practices in Manufacturing, Packing, or Holding of Human Foods, Title 21, Code of Federal Regulations, as amended, April 1999, Food and Drug Administration, Department of Health and Human Services.

Index

N

O

P

Q

R

W

X

Y

Z